Geometry of Lie Groups

Mathematics and Its Applications

Volume 393

Geometry of Lie Groups

by

Boris Rosenfeld

Pennsylvania State University, Department of Mathematics, University Park, PA, U.S.A.

SPRINGER-SCIENCE+BUSINESS MEDIA, B.V.

A C.I.P. Catalogue record for this book is available from the Library of Congress.

Rosenfeld, Boris.
Geometry of Lie groups / by Boris Rosenfeld.
p. cm. -- (Mathematics and its applications ; v. 393)
Includes index.
ISBN 978-1-4419-4769-7 ISBN 978-1-4757-5325-7 (eBook)
DOI 10.1007/978-1-4757-5325-7
1. Lie groups. 2. Geometry. I. Title. II. Series: Mathematics and its applications (Kluwer Academic Publishers) ; v. 393.
QA387.R67 1997
512'.55--dc21 96-40243

ISBN 978-1-4419-4769-7

Printed on acid-free paper

To my wife Lucy and my daughters Svetlana and Julia

CONTENTS

PREFACE

This book is the result of many years of research in Non-Euclidean Geometries and Geometry of Lie groups, as well as teaching at Moscow State University (1947-1949), Azerbaijan State University (Baku) (1950-1955), Kolomna Pedagogical College (1955-1970), Moscow Pedagogical University (1971-1990), and Pennsylvania State University (1990-1995).

My first books on Non-Euclidean Geometries and Geometry of Lie groups were written in Russian and published in Moscow: *Non-Euclidean Geometries* (1955) [Ro1], *Multidimensional Spaces* (1966) [Ro2], and *Non-Euclidean Spaces* (1969) [Ro3].

In [Ro1] I considered non-Euclidean geometries in the broad sense, as geometry of simple Lie groups, since classical non-Euclidean geometries, hyperbolic and elliptic, are geometries of simple Lie groups of classes B_n and D_n, and geometries of complex and quaternionic Hermitian elliptic and hyperbolic spaces are geometries of simple Lie groups of classes A_n and C_n.

[Ro1] contains an exposition of the geometry of classical real non-Euclidean spaces and their interpretations as hyperspheres with identified antipodal points in Euclidean or pseudo-Euclidean spaces, and in projective and conformal spaces. Numerous interpretations of various spaces different from our usual space allow us, like stereoscopic vision, to see many traits of these spaces absent in the usual space. After complex and quaternionic spaces I considered split complex and quaternionic spaces by means of which the interpretations of classical projective and symplectic spaces (whose groups of transformations also are simple Lie groups of classes A_n and C_n) as non-Euclidean spaces can be obtained. In [Ro1] the geometric interpretations of all simple Lie groups of classes A_n, B_n, C_n, and D_n (except one group of the class D_n) and of some exceptional groups of classes F_4 and E_6 are expounded.

[Ro2] is devoted to a detailed exposition of geometry of multidimensional real affine, Euclidean, pseudo-Euclidean, projective, conformal, and pseudoconformal geometries. [Ro2] differs from other books on these spaces by systematic usage of the linear operators and matrix coordinates.

[Ro3] is a detailed exposition of geometry of real non-Euclidean spaces. Next to classical elliptic and hyperbolic spaces, I also considered pseudoelliptic, pseudohyperbolic, and symplectic spaces with simple fundamental groups, and quasielliptic, quasihyperbolic, quasisymplectic, etc. spaces, whose fundamental groups are not simple but can be obtained from simple groups by limiting process (particular cases of these spaces are Euclidean and pseudo-Euclidean spaces), and r-quasielliptic, r-quasihyperbolic, r-quasisymplectic, etc. spaces, whose fundamental groups can be obtained from simple groups by r-fold limiting process; I call such groups quasisimple and r-quasisimple, respectively.

Later I wrote the third book of the series containing [Ro2] and [Ro3], *Geometry of Lie Groups*, on non-Euclidean spaces over fields and algebras with simple, quasisimple, and r-quasisimple fundamental groups, but this book was not published in the USSR; only the review of this book was published in the preface to [Ro3].

The present book with the same title as my unpublished book in the USSR is based on all these four books and on my lectures in the USA in 1991–1995. This book does not contain material of textbooks on non-Euclidean geometries, but contains all of the novelties of my former books: many geometric interpretations, systematic usage of linear operators and matrix coordinates, geometries over fields and algebras, geometry of quasisimple and r-quasisimple Lie groups.

This book contains 7 chapters (plus one introductory chapter). Unlike [Rol] there is no separation between real spaces and spaces over fields or algebras and there is a lot of new material. Chapter 0 contains necessary information from algebra, topology, and geometry of real spaces. Chapter 1 contains more specialized information on associative and nonassociative algebras and on Lie groups and algebras. In chapters 2, 3, 4 and 6, geometric interpretations of all simple Lie groups of classes A_n, B_n, C_n and D_n, as well as of finite groups of Lie type, are given. In Chapters 5 and 6 geometric interpretations of quasisimple and r-quasisimple Lie groups of the same classes are given. In Chapter 7, for the first time ever, geometric interpretations of *all* simple and quasisimple Lie groups of exceptional classes G_2, F_4, E_6, E_7, and E_8 are given.

This book contains many of my results published in various papers, as well as many findings of my students and colleagues.

Proofs of theorems which are already proved in the books by S. Helgason [He], J. A. Wolf [Wol] and in other books and papers are replaced by references to them. For historical-mathematical information about non-Euclidean geometries and geometries of Lie groups, see my book [Ro4].

In this book the role of exercises is played by the assertions and theorems given without a full proof, but with the indication that they can be proved analogously to already proved theorems.

The figures in this book were created by Boris Katok and Alexander Morgoulis. I greatly appreciate their work.

I owe thanks and gratitude to the editor of the Publishing House, Dr. Robert Hoksbergen, for his immense work on the manuscript of this book.

I thank Alexander Morgoulis of Pennsylvania State University who helped me to prepare the camera ready text of my book. I acknowledge the help of Moisey Guysinsky, Sergey Yaskolko, Anatoly Kouchnirenko, and Patricia Snare of the same university in the preparation of the AMS-TEX version of this book.

My special thanks go to my daughter Svetlana R. Katok, since this book would have been impossible without her constant support and help.

Last, but not least, I am very grateful to my wife Lucy Rosenfeld for her tremendous help in the preparation of this book for print. I thank her very much for her great work, encouragement, and inspiration.

CHAPTER 0

STRUCTURES OF GEOMETRY

§0.1. Algebraic Structures

0.1.1. Mathematical Structures. The objects of modern Geometry, *spaces*, like all objects of modern Mathematics, are *sets of elements* of arbitrary nature endowed with some *mathematical structure*. Such is the formulation in the best Encyclopaedia of Mathematics, the *Elements of Mathematics* by N. Bourbaki [Bou1-9]. Many books in this encyclopaedic series, in particular *Topological Vector Spaces* [Bou5], and *Lie Groups and Algebras* [Bou8], are closely connected with modern Geometry, and the title *Elements* of Bourbaki emphasizes that this mathematical Encyclopaedia is a successor of the classical mathematical Encyclopaedia of Antiquity, the *Elements* of Euclid [Euc], the main part of which is devoted to geometry.

0.1.2. Groups [Bou2]. The most important mathematical structures are *algebraic* and *topological structures*. One of the simplest algebraic structures is the *structure of a group*. A set G of elements of any kind is said to be a *group* if a group operation $a \circ b$ is defined in it satisfying the following axioms:

$G.1°$ For any two elements a and b there exists an element

$$c = a \circ b. \tag{0.1}$$

$G.2°$ This operation is associative, that is, for any three elements a, b, c,

$$(a \circ b) \circ c = a \circ (b \circ c). \tag{0.2}$$

$G.3°$ There exists a *neutral element* e, i. e. an element such that for every element a,

$$a \circ e = e \circ a = a. \tag{0.3}$$

$G.4°$ For each element a there exists a *symmetric element* $\bar{a}$ such that

$$a \circ \bar{a} = \bar{a} \circ a = e. \tag{0.4}$$

If the group operation $a \circ b$ is called *addition*, we write $c = a + b$ and the element c is called the *sum*, the neutral element is called *zero* and is written as 0, the symmetric element is called the *opposite* and is written as $-a$, and the group is called *additive*.

If the group operation $a \circ b$ is called *multiplication*, we write $c = a \cdot b$ or $c = ab$, the element c is called the *product*, the neutral element is called *unit* and is written as 1, the symmetric element is called the *inverse* and is written as a^{-1}, and the group is called *multiplicative*.

If the group satisfies in addition the axiom

$G.5°$. For any two elements a and b

$$a \circ b = b \circ a, \tag{0.5}$$

then the group is called *commutative* or *Abelian.*

A set of elements endowed with an operation $a \circ b$ without the properties $G.2°$, $G.3°$ and $G.4°$ is called a *magma*. A magma with the property $G.3°$ is called a *unital magma*, a magma with the property $G.2°$ is called a *semigroup*. A magma in which the equations $a \circ x = b$ and $x \circ a = b$ are solvable for all a and b is called a *quasigroup*. A unital semigroup is called a *monoid*, a unital quasigroup is called a *loop*. All these structures (as also the ones to be introduced yet) are termed infinite, respectively finite, if the underlying set is infinite, respectively finite.

For the history of group theory see [Ro4, pp. 327–333].

0.1.3. Rings and Fields. If, in a set of elements of any kind two operations $a + b$ and ab are defined such that

$R.1°$ The set is a commutative group with respect to the operation $a + b$;

$R.2°$ The set is a semigroup with respect to the operation ab;

$R.3°$ The operation ab is *distributive* with respect to the operation $a + b$:

$$a(b + c) = ab + ac, \quad (a + b)c = ac + bc, \tag{0.6}$$

the set is called a *ring.*

A ring in which the set of elements without 0 is a commutative group with respect to the operation ab is called a *field.*

A ring in which the set of elements without 0 is a noncommutative group with respect to the operation ab is called a *skew field.*

The ring $\mathbb{Z}$ of integers is a ring with 1, the rings $2\mathbb{Z}$ of even integers, $3\mathbb{Z}$ of integers of the form $3n$, n from $\mathbb{Z}$, and so on, are rings without 1.

The ring $\mathbb{Q}$ of rational numbers, the ring $\mathbb{R}$ of real numbers, the ring $\mathbb{C}$ of complex numbers, the ring $\mathbb{F}_p$ of residues modulo a prime integer p, and *the Galois field* $\mathbb{F}_q$ $(q = p^k)$ obtained from $\mathbb{F}_p$ by adjoining the roots of an irreducible algebraic equation of degree k over $\mathbb{F}_p$, are fields. The ring $\mathbb{H}$ of *quaternions* is a skew field. The fields $\mathbb{F}_p$ and $\mathbb{F}_q$ are finite and contain p and $q = p^k$ elements, respectively. The rings $\mathbb{Z}$, $\mathbb{Q}$, $\mathbb{R}$, $\mathbb{C}$, and $\mathbb{H}$ are infinite, and the three last fields are continuous (topological). A field $\mathbb{Q}(\alpha, \beta, \dots)$ of algebraic numbers, obtained from $\mathbb{Q}$ by adjoining the roots $\alpha, \beta, \dots$ of an irreducible algebraic equation over $\mathbb{Q}$, and the field $\mathbb{Q}_p$ of *p-adic numbers*, which is also an extension of $\mathbb{Q}$, are also infinite, and the last field is continuous.

For the history of ring theory see [Ro4, pp. 400–403].

0.1.4. Subgroups, Subrings, and Subfields. Subsets of groups, rings, and fields that are also groups, rings and fields, respectively, are called *subgroups, subrings, and subfields*, respectively.

A bijection between two sets endowed with algebraic structures which preserves the operations of these structures is called an *isomorphism*, sets between which there is an isomorphism are called *isomorphic.*

A surjection or injection between two sets endowed with algebraic structures which preserves the operations is called a *homomorphism*, sets between which there is a homomorphism are called *homomorphic*.

If two groups G and H are homomorphic, the elements of G corresponding to the neutral element of H form a subgroup N of G, such a subgroup is called an *invariant subgroup* or *normal subgroup*. In this case the group isomorphic to H is called the *quotient group* and is written as G/N.

If two rings $\mathcal{R}$ and $\mathcal{S}$ are homomorphic, the elements of $\mathcal{R}$ corresponding to the zero in $\mathcal{S}$ form a subring $\mathcal{J}$ of $\mathcal{R}$ called an *ideal*. In this case the ring isomorphic to $\mathcal{S}$ is called the *quotient ring* and is written as $\mathcal{R}/\mathcal{J}$.

The rings $2\mathbb{Z}$, $3\mathbb{Z}$ etc. are subrings and ideals of $\mathbb{Z}$. The ring $\mathbb{Z}$ is a subring of the field $\mathbb{Q}$, $\mathbb{Q}$ is a subfield of $\mathbb{R}$, $\mathbb{R}$ is a subfield of $\mathbb{C}$, $\mathbb{C}$ is a subring of $\mathbb{H}$.

The quotient ring $\mathbb{Z}/p\mathbb{Z}$ is isomorphic to the field $\mathbb{F}_p$.

If R and Z are the additive groups of the rings $\mathbb{R}$ and $\mathbb{Z}$, the quotient group R/Z is isomorphic to the multiplicative group T of complex numbers of the form e^{it} ($t \in \mathbb{R}$).

Isomorphisms of a group or a ring onto itself are called *group or ring automorphisms*, respectively; homomorphisms of a group or a ring into itself are called *group or ring endomorphisms*, respectively.

0.1.5. Direct Sums and Products. If G and H are two groups with operations $g_1 \circ g_2$ and $h_1 \circ h_2$, the pairs (g, h) with operation $(g_1, h_1) \circ (g_2, h_2) = (g_1 \circ g_2,\, h_1 \circ h_2)$ form a group. If G and H are both additive, respectively multiplicative, groups, this group is called a *direct sum*, respectively *direct product*, and is written as $G \oplus H$, respectively $G \otimes H$.

The direct sum $\mathcal{R} \oplus \mathcal{S}$ of rings $\mathcal{R}$ and $\mathcal{S}$ is similarly defined.

0.1.6. Linear Spaces. A set L^n of elements of any kind, called *vectors*, is said to be an *n-dimensional linear* (or *vector*) *space* if in this set the operations of *addition* and of *multiplication by scalars*, that is real numbers, are defined, satisfying:

$VI.1° - 5°$ Addition of vectors satisfies axioms $G.1 - 5°$ for a commutative group;

$VII.1°$ For any vector a and any scalar λ there exists a vector

$$b = a \cdot \lambda = a\lambda \tag{0.7}$$

called the *product of a by λ*;

$VII.2°$ Multiplication by 1 does not change a vector:

$$a \cdot 1 = a; \tag{0.8}$$

$VII.3°$ Multiplication of vectors by scalars is distributive with respect to addition of scalars:

$$a(\lambda + \mu) = a\lambda + a\mu; \tag{0.9}$$

$VII.4°$ Multiplication of vectors by scalars is distributive with respect to addition of vectors:

$$(a + b)\lambda = a\lambda + b\lambda; \tag{0.10}$$

$VII.5°$ Multiplication of vectors by scalars is associative:

$$(a\lambda)\mu = a(\lambda\mu); \tag{0.11}$$

and *axioms* $VIII.1^\circ - 2^\circ$ *of dimension*, which are based on the notions of linear independence and dependence of vectors. Vectors $a_1, a_2, ..., a_m$ are said to be *linearly independent* if a linear combination $a_1\lambda_1 + a_2\lambda_2 + ... + a_m\lambda_m$ is equal to zero only if all coefficients $\lambda_i = 0$, and *linearly dependent* if there are nonzero coefficients λ_i such that this linear combination is equal to zero.

$VIII.1^\circ$ There exist n linearly independent vectors;

$VIII.2^\circ$ Any $n+1$ vectors are linearly dependent.

If we have chosen n linearly independent vectors $e_1, e_2, ..., e_n$ in L^n, then each vector can be written as

$$x = \sum_i e_i x^i = e_i x^i. \tag{0.12}$$

The numbers x^i are called the *coordinates* of the vector x, the vectors e_i are called *basis vectors*. Later we will write the sums (0.12) only in the last form and when in our formulas the same upper and lower indices appear we will always mean summation with respect to these indices. This condition was proposed by A. Einstein [Ei2] and is called the *Einstein rule*.

For the history of the theory of linear spaces see [Ro4, pp. 251–253, 261–264].

Two linear spaces are called *isomorphic* if there is a bijection between them that preserves addition of vectors and multiplication of vectors by scalars, and *homomorphic* if there is a surjection or an injection between them that preserves these operations. Isomorphisms of a linear space onto itself and homomorphisms of a linear space into itself are called *automorphisms* and *endomorphisms* of this linear space respectively. Automorphisms and endomorphisms of a linear space are called *linear transformations* in it.

0.1.7. Algebras. If the set $\mathbb{A}$ is simultaneously a ring $\mathcal{R}$ and a linear space L^n and if the following axiom holds:

$A.1^\circ$ For any two elements a and b and any two scalars λ and μ we have

$$(a\lambda)(b\mu) = (ab)(\lambda\mu), \tag{0.13}$$

then the set $\mathbb{A}$ is called an n*-algebra*.

The product of two basis elements of an algebra can be written as

$$e_i e_j = C_{ij}^k e_k. \tag{0.14}$$

The numbers C_{ij}^k are called the *structure constants* of the algebra. In this book we will consider both associative and nonassociative algebras. The field $\mathbb{C}$ of complex numbers is a commutative associative algebra with basis $1, i \quad (i^2 = -1)$, the skew field $\mathbb{H}$ of quaternions is a noncommutative associative algebra with basis $1, i, j, k$ $(i^2 = j^2 = k^2 = -1,\ ij = -ji = k,\ jk = -kj = i,\ ki = -ik = j)$.

An algebra which is a field or a skew field is called a *division algebra*.

For the history of the theory of associative algebras see [Ro4, pp. 382–386, 388–397, 398–399].

Two algebras are called *isomorphic* if there is a bijection between them that preserves addition and multiplication of vectors, and multiplication of vectors by scalars, and *homomorphic* if there is a surjection or an injection between them that preserves these operations. Isomorphisms of an algebra onto itself and homomorphisms of an algebra into itself are *algebra automorphisms* and *algebra endomorphisms* of this algebra, respectively.

§0.2. Topological Structures

0.2.1. Topological Spaces [Bou3]. A set T of elements is said to be a *topological space* if in it subsets called *closed subsets* are singled out and the following axioms are fulfilled:

$T.1°$ The union of a finite number of closed subsets is closed;

$T.2°$ The intersection of arbitrary many closed subsets is closed;

$T.3°$ The whole space T is a closed set;

$T.4°$ The empty set $\emptyset$ is a closed set.

A topological structure also can be defined by means of *open sets*, which are complements of closed sets, by means of *closures* of sets (the closure $\overline{M}$ of a set M is the intersection of all closed sets containing M), by means of *interiors* of sets (the interior of a set M is the union of all open sets contained in M), and by means of *neighborhoods* (i.e. open sets such that any open set can be represented as a union of these sets).

The elements in topological spaces are called *points*, and a neighborhood U containing a point x is called neighborhood $U(x)$ of this point. A point x all neighborhoods $U(x)$ of which contain points of a set M different from x is called a *limit point* of M.

If the set of neighborhoods in a space T is countable, the space is called a *topological space with countable base.*

A topological space in which the only closed sets are the whole space and the empty set is called a *trivial space.* A topological space in which all subsets are closed is called a *discrete space.*

The most important topological spaces are the *Hausdorff spaces* and *regular spaces.* Hausdorff spaces satisfy the axioms:

$T.5°$ All points are closed subsets;

$T.6°$ Any two points in the space have disjoint neighborhoods.

Regular spaces satisfy the axiom $T.5°$ and

$T.6'$ Any point in the space and any closed set in this space which does not contain this point have disjoint open sets containing this point and this closed set, respectively.

The natural topology in the field $\mathbb{R}$, with closed and open sets defined as in usual real Calculus, can be specified by the countable system of neighborhoods consisting of the intervals with rational ends.

0.2.2. Subspaces of Topological Spaces. A subset of a topological space where for closed subsets we take the intersections with closed subsets of this space is called a *subspace.*

A topological space which cannot be divided into two closed non-empty subsets with empty intersection is called *continuous* or *connected.* A topological space admitting such a division is called *nonconnected* and consists of *connected components.*

A topological space, or a subset of it, is called *compact* if each infinite subset of it has a limit point. In every covering of a compact topological space we can choose a finite covering.

If $T_1, T_2, \ldots, T_n$ are topological spaces, the n-tuples $\{x_1, x_2, \ldots, x_n\}$ consisting of points x_i in T_i form a new topological space, whose closed sets are sets of such n-tuples for which every point x_i runs through a closed subset in T_i and arbitrary intersections of these sets. This new space is called the *topological product* of the spaces T_i.

0.2.3. Continuous Mappings and Homeomorphisms. A mapping from a topological space T onto a topological space T' is called *continuous* if for each neighborhood $V(x')$ of a point x' in T' there is a neighborhood $U(x)$ of the corresponding point x in T such that the images of all points in $U(x)$ belong to $V(x')$.

A bijection from T onto T' which is continuous together with its inverse bijection is called a *homeomorphism*; in this case the spaces T and T' are called *homeomorphic* or *topologically equivalent.* The sets of closed subsets of two homeomorphic spaces are mapped onto each other.

If f is a continuous mapping from T to T' and all preimages of points of T' in T are subspaces homeomorphic to a space S, the space T' is called the *quotient space* of T by S and is written as T/S.

For the history of the theory of topological spaces see [Ro4, pp. 304–309].

§0.3. Order Structures

0.3.1. Ordered Sets [Boul]. A set M is called *ordered* if a relation $x \leq y$ is defined for any two elements x and y in it, such that

$O.1°$ $x \leq x$ (*reflexivity axiom*);

$O.2°$ $x \leq y$ and $y \leq x$ imply $x = y$ (*antisymmetry axiom*);

$O.3°$ $x \leq y$ and $y \leq z$ imply $x \leq z$ (*transitivity axiom*).

These axioms are valid for the relation "less or equal" in the rings $\mathbb{Z}$, $\mathbb{Q}$, and $\mathbb{R}$.

0.3.2. Lattices. If the relation $x \leq y$ is defined not for all pairs of elements in a set, this set is called a *partially ordered set.* If in a partially ordered set for any two elements x and y there are a minimal element z, such that $z \leq x$, $z \leq y$, and a maximal element w, such that $x \leq w$, $y \leq w$, there are defined we write $z = x \cup y$ and $w = x \cap y$, and if the operations $\cup$ and $\cap$ satisfy the axioms:

$L.1°$ $x \cup y = y \cup x$, $x \cap y = y \cap x$ (*commutative law*);

$L.2°$ $x \cup (y \cup z) = (x \cup y) \cup z$, $x \cap (y \cap z) = (x \cap y)z$ (*associative law*);

$L.3°$ $x \cup (y \cap x) = (x \cap y) \cup x = x$ (*absorption law*),

then the partially ordered set is called a *lattice.*

If the operations $\cup$ and $\cap$ are distributive with respect to another, the lattice is called *distributive.* If in a distributive lattice there is a minimal element O which is $\leq$ than all elements and a maximal element I such that all elements are $\leq$ than it, and if for each element x there is a complementary element x' for which $x \cup x' = I$ and $x \cap x' = O$, the lattice is called a *Boolean lattice* or a *Boolean algebra.*

The totalities of all subsets of any set, of all subgroups or all invariant subgroups of any group, of all subrings or all ideals of any ring, and of all closed or open subsets of any topological space are lattices.

Boolean algebras are formed by the measurable sets in *Function theory*, by propositions in *Mathematical logic* and by events in *Probability theory*. In the first case I is the whole space and O is the empty set, in the second case I is the true proposition and O is the false proposition, in the third case I is a sure event and O is an impossible event.

§0.4. Incidence Structures

0.4.1. Blocks and Incidences. If subsets of a geometric space are determined by the same finite number of points, these subsets are called ***blocks incident with these points***. The structure defined by these blocks in the space is called an *incidence structure*.

0.4.2. Affine Space. The simplest incidence structure is the structure of the *affine space* E^n. The space E^n is the set of elements, called *points*, connected with vectors in a linear space L^n, satisfying the following axioms:

$A.1°$ For any two points X and Y there is a vector $a = \overrightarrow{XY}$ in L^n;

$A.2°$ For any point X and any vector a there is a point Y such that $a = \overrightarrow{XY}$;

$A.3°$ *Triangle axiom*: For any three points X, Y, Z:

$$\overrightarrow{XY} + \overrightarrow{YZ} = \overrightarrow{XZ} .$$

If we choose a point O in E^n and call it *origin*, then any point X in E^n determines a vector $x = \overrightarrow{OX}$, called *position vector* or *radius-vector* of this point. The coordinates x^i of this vector are called the *affine coordinates* of X.

If $a_1, a_2, \dots, a_m$ are m linearly independent vectors in L^n, then the points in E^n with position vectors

$$x = x_0 + a_1 t^1 + a_2 t^2 + \cdots + a_m t^m = x_0 + a_a t^a \tag{0.15}$$

form an *m-plane.* For $m = 1$ it is a *straight line*, for $m = 2$ it is a *plane*, for $m = n-1$ it is a *hyperplane*. Lines and m-planes are blocks in E^n. The m-planes in E^n are spaces E^m; they form a lattice. If A and B are an l-plane and an m-plane, $A \cup B$ is the *sum* of A and B, that is, the s-plane of minimal dimension containing A and B, and $A \cap B$ is the *intersection* of A and B.

Two m-planes $x = x_0 + a_a t^a$ and $x = x_1 + b_a u^a$ for which the vectors b_a are linear combinations of the vectors a_a, while x_1 cannot be written as $x_1 = x_0 + a_a t^a$ for some t^a, do not have points in common and are called *parallel.*

For the history of affine geometry see [Ro4, pp. 112–113, 130–133, 143–146, 149–150, 263–264].

0.4.3. Projective Space. If we regard the space E^n as a hyperplane in E^{n+1} and project this hyperplane from a point S in E^{n+1} onto another hyperplane, project this second hyperplane from a second point S' onto a third hyperplane, and after several such projections return to the first hyperplane, we obtain a transformation in E^n. This transformation is not a bijection since when the projection line is parallel to the hyperplane onto which we project, the points to be projected have no image. Therefore, for this transformation to become a bijection, all hyperplanes must be supplemented by new points so that the supplemented hyperplane will be in bijection with a bundle of lines through a point in E^{n+1}. This supplemented E^n is called *projective n-space* and is denoted by P^n. The new points are called the *points at infinity* and form the *hyperplane at infinity.*

The points in P^n are represented by nonzero vectors $x = \overrightarrow{SX}$ in E^{n+1} directed along the corresponding lines of the bundle. Therefore the space P^n is often defined as the set of 1-dimensional subspaces of the linear space L^{n+1}. The vectors in L^{n+1} representing points in P^n ("geometric points") are often called *analytic points* or *position vectors* of the points in P^n. The coordinates x^i of these vectors are called the *projective coordinates* of points X in P^n. Like the vectors x representing these points, these coordinates are determined up to a nonzero multiplier.

If $a_0, a_1, \ldots, a_m$ are $m+1$ linearly independent vectors in L^{n+1}, then the points in P^n represented by the vectors

$$x = a_0t^0 + a_1t^1 + \cdots + a_mt^m = a_at^a \tag{0.16}$$

form an *m-plane.* For $m = 1$ it is a *straight line*, for $m = 2$ a *plane*, for $m = n-1$ a *hyperplane.* Lines and m-planes are blocks in the space P^n. The m-planes in P^n are spaces P^m; they form a lattice. If A and B are an ℓ-plane and an m-plane, $A \cup B$ is the *sum* of A and B, defined as in E^n, and $A \cap B$ is their *intersection.*

The dimensions ℓ and m of A and B and the dimensions s and d of their sum and intersection, respectively, are related by

$$\ell + m = s + d. \tag{0.17}$$

We will encounter another kind of incidence structure, conformal and pseudoconformal spaces, in **0.5.7** and **0.5.8**.

For the history of projective geometry see [Ro4, pp. 116–121, 133–142, 143–146, 148–150].

§0.5. Metric Structures

0.5.1. Metric Spaces. A topological or geometric space is called a *metric space* if for any two points x and y in it there is a real nonnegative number $\rho(x, y)$, called *distance* between these points, satisfying the following axioms:

$M.1°$ $\rho(x, y) = 0$ if and only if $x = y$ (*reflexivity axiom*);

$M.2°$ $\rho(x, y) = \rho(y, x)$ (*symmetry axiom*);

$M.3°$ $\rho(x, z) \leq \rho(x, y) + \rho(y, z)$ (*triangle axiom* or *triangle inequality*).

0.5.2. Pseudometric Spaces. Suppose for any two points x and y in a topological or geometric space a number $\rho(x, y)$ is defined which is not necessarily real and satisfies instead of $M.1°$ only the axiom $\rho(x, x) = 0$, satisfies the axiom $M.2°$ and does not satisfy the axiom $M.3°$. This space is then called a *pseudometric space*, and this number $\rho(x, y)$ is also called the *distance* between x and y.

0.5.3. Euclidean Space. Suppose we can define in the affine space E^n a distance $\rho(x, y)$ between points by means of an *inner* (*scalar*) *product*, denoted by xy or (x, y), of vectors, and satisfying the following axioms:

$E.1°$ For any two vectors x and y there is a real number $s = xy$, called their *inner product*; we write x^2 for xx;

$E.2°$ $xy = yx$ (*commutative law*);

$E.3°$ $x(y+z) = xy + xz$ (*distributive law*);
$E.4°$ $(kx)y = k(xy)$, k a scalar (*associative law*);
$E.5°$ $x^2 > 0$ if $x \neq 0$, $x^2 = 0$ if $x = 0$ (*positivity axiom*).

The positive root of the inner square x^2 of a vector x is called the *modulus* $|x|$ of x. The *distance* $\rho(X,Y) = XY$ between two points X and Y with position vectors x and y is defined as the modulus $|y-x|$ of the vector $\overrightarrow{XY} = y - x$. The space E^n with such a distance between points is called *Euclidean space* and is denoted by R^n.

The space R^n is a metric space. The fulfillment of axiom $M.1°$ follows from axiom $E.5°$. The fulfillment of axiom $M.2°$ follows from the definition of the distance between points in R^n. In Theorems 0.1 and 0.2 we prove that axiom $M.3°$ holds.

Theorem 0.1. *In the space R^n we have the Cauchy inequality*

$$|x|\,|y| \leq |xy|. \tag{0.18}$$

PROOF. Since the inner square $(x-ty)^2$ of the vector $x-ty$ in R^n is nonnegative, the inequality $x^2 - 2txy + t^2y^2 \geq 0$ holds for all real numbers t. If $x = 0$ the inequality (0.18) is fulfilled. If $y \neq 0$ we take $t = \frac{xy}{y^2}$. In this case the inequality has the form $x^2 - 2\frac{(xy)^2}{y^2} + \frac{(xy)^2}{y^2} \geq 0$, that is, $x^2y^2 \geq (xy)^2$, which is equivalent to (0.18). ■

Theorem 0.2. *The axiom $M.3°$ holds in the space R^n.*

PROOF. Consider three points X, Y, Z in R^n with position vectors x, y, z. Then

$$\begin{aligned}(XZ)^2 =&(z-x)^2 = (z-y+y-z)^2\\ =&(z-y)^2 + 2(z-y)(y-x) + (y-x)^2\\ \leq&(z-y)^2 + 2|z-y|\,|y-x| + (y-x)^2\\ =&(YZ)^2 + 2YZ\,XY + (XY)^2 = (XY+YZ)^2,\end{aligned}$$

which is equivalent to axiom $M.3°$. ■

The inequality (0.18) implies that $\frac{|xy|}{|x|\,|y|} \leq 1$. So, this expression can be regarded as $\cos\varphi$. The real number φ is called the *angle* between the vectors x and y. Vectors for which $\varphi = \frac{\pi}{2}$ are called *orthogonal.*

If the vector basis e_i is chosen in R^n, the inner product of the vectors $x = x^ie_i$ and $y = y^ie_i$ is equal to the bilinear form

$$xy = e_ie_jx^iy^j = e_{ij}x^iy^j. \tag{0.19}$$

The form $x^2 = e_{ij}x^ix^j$ is a positive definite quadratic form.

If the basis vectors e_i are *orthonormal*, that is, mutually orthogonal and with moduli equal to one, then $e_{ij} = \delta_{ij} (= 1$ if $i = j$ and $=0$ if $i \neq j)$, the form (0.19) becomes

$$xy = \sum_i x^iy^i, \tag{0.20}$$

and the inner square of a vector is equal to the sum of the squares of its coordinates.

The m-planes in R^n are the spaces R^m.

For the history of the geometry of Euclidean n-space see [Ro4, pp. 160–161, 179–180, 247–258, 262–264].

0.5.4. Hyperspheres in Euclidean Space. In the space R^n the equation

$$(x - x_0)^2 = r^2 \tag{0.21}$$

defines a *hypersphere*. The point X_0 with position vector x_0 is the *center* of the hypersphere (0.21), the real number r is its *radius*.

The set of points defined by an inequality $(x - x_0)^2 < r^2$ is called the *interior domain of the hypersphere* (0.21).

0.5.5. Natural Topology in Euclidean Space. In each metric space we can define a *natural topology*, in which the limit points of a set are the points for which the infimum of the distances to points of this set is equal to zero.

The natural topology in R^n can be determined by the countable system of neighborhoods each of which is the interior domain of a hypersphere with rational coordinates of the center and with rational radius. The space R^n endowed with its natural topology is homeomorphic to the topological product of n lines R^1 or fields $\mathbb{R}$.

Since the space R^n can be defined is the space E^n endowed with a metric, the natural topology in R^n coincides with the natural topology in E^n. Therefore, the space E^n endowed with its natural topology is homeomorphic to the topological product of n lines E^1 or fields $\mathbb{R}$.

0.5.6. Pseudo-Euclidean Spaces. If, in the space R^n, we replace axiom $E.5°$ by the axiom

$E.5'$ There are ℓ mutually orthogonal vectors a_a with negative inner squares a_a^2 and $n - \ell$ mutually orthogonal vectors a_u with positive inner squares a_u^2, and each vector a_a is orthogonal to each vector a_u,

then the space is called *pseudo-Euclidean space*, and is denoted by R^n_ℓ; the integer ℓ is called the *index* of this space.

The space R^n_ℓ is a pseudometric space, since it contains points with imaginary distance, noncoincident points with zero distance, and the axiom $M.3°$ is not fulfilled. The angle between two vectors in R^n_ℓ is defined by the same formula as in R^n. In R^2_1 the angle is real only for collinear vectors $y = kx$ ($\varphi = 0$ if $k > 0$, $\varphi = \pi$ if $k < 0$) and for orthogonal vectors ($xy = 0$); in all other cases the angle between two vectors in R^2_1 has the form $i\psi$, $\frac{\pi}{2} \pm i\psi$, or $\pi - i\psi$, where ψ is a real number.

If a vector basis e_i has been chosen in R^n_ℓ, then the inner product of the vectors $x = x^i e_i$ and $y = y^i e_i$ is equal to the bilinear form (0.19). However, the form $x^2 = e_{ij} x^i x^j$ is an indefinite quadratic form. If the basis vectors e_i are *orthonormal*, that is, mutually orthogonal and with moduli equal to 1 or i, the form (0.19) becomes

$$xy = -\sum_a x^a y^a + \sum_u x^u y^u \tag{0.22}$$

and the inner square of a vector is equal to the algebraic sum of the squares of its coordinates.

We will denote the inner product (0.22) in R^n_ℓ by xEy where E is the diagonal $(n \times n)$-matrix with ℓ diagonal entries equal to -1 and the other diagonal entries equal to 1.

The m-planes in the space R^n_ℓ are the spaces R^m_k.

For the history of pseudo-Euclidean geometry see [Ro4, pp. 265–268].

0.5.7. Conformal Space. Each hypersphere in R^n determines an *inversion* transforming a point X in R^n into the point X' on the line X_0X such that

$$X_0X' \cdot X_0X = r^2. \tag{0.23}$$

This transformation is determined for all points in the space except X_0. Therefore, for such transformations to become bijections the space R^n must be supplemented by one point, corresponding under inversion in a hypersphere to its center. This supplemented space is called *conformal* (or *inversive*) *n-space* and is denoted by C^n. The new point is called the *point at infinity.*

The space C^n has an incidence structure, the blocks in it are the *m-spheres*, i.e. intersections of hyperspheres with $(m+1)$-planes. For $m = 1$ we obtain *circles*, for $m = 2$ *spheres*; the hyperspheres are the $(n-1)$-spheres. The m-planes in R^n can be regarded as the m-spheres in C^n passing through the point at infinity. The m-spheres in C^n are the spaces C^m.

For the history of conformal geometry see [Ro4, pp. 114–116, 146-147, 149-151].

0.5.8. Pseudoconformal Spaces. In the spaces R^n_ℓ there are also *hyperspheres*, defined by equations (0.21) with centers at points X_0 with position vectors x_0 and with radii r, but in R^n_ℓ the radius of a hypersphere can be a positive number, zero or a pure imaginary number. Each hypersphere in R^n_ℓ also determines an *inversion*, transforming a point X in the space into the point X' on the line X_0X determined by (0.23). This transformation is defined for all points in the space except X_0 and all points on the hypersphere of radius zero with center at X_0. Therefore, for these transformations to become bijections the space R^n_ℓ must be supplemented by one point corresponding under inversion in a hypersphere to its center and by all points corresponding under the inversion to the points on the sphere of radius zero. This supplemented space is called *pseudoconformal* (or *pseudoinversive*) *n-space* and is denoted by C^n_ℓ. The new points are called the *point at infinity* and *ideal points*, respectively.

The space C^n_ℓ has an incidence structure. The blocks in C^n_ℓ are the *m-spheres*, i.e. intersections of hyperspheres with $(m+1)$-planes. For $m = 1$ they are *circles*, for $m = 2$ *spheres*; the hyperspheres are the $(n-1)$-spheres. The m-planes in R^n_ℓ can be regarded as the m-spheres in C^n_ℓ passing through the point at infinity. The m-spheres in C^n_ℓ are the spaces C^m_k.

The space C^6_3 was considered by É. Cartan in [Car18], the general spaces C^n_ℓ were defined by the author in [Ro1,5].

§0.6. Tensors and Linear Operators

0.6.1. Covectors. In L^n, next to vectors we can define *covectors.* If we are given a linear scalar function of a vector argument, $\varphi(x)$, that is, a function such that

$$\varphi(x+y)=\varphi(x)+\varphi(y), \quad \varphi(\lambda x)=\lambda\varphi(x), \tag{0.24}$$

the value of this function on a vector $x=x^i e_i$ is

$$\varphi(x^i e_i)=\varphi(e_i)x^i=u_i x^i. \tag{0.25}$$

The numbers u_i can be regarded as the coordinates of a new vector u. The transition

$$e_{i'}=A_{i'}^{i} e_i \tag{0.26}$$

from the basis vectors e_i to new basis vectors $e_{i'}$ determines transformations of the coordinates x^i and u_i of the vectors x and u:

$$x^{i'}=A_i^{i'} x^i, \quad u_{i'}=A_{i'}^{i} u_i. \tag{0.27}$$

Here, the matrix $(A_i^{i'})$ is the inverse of the matrix $(A_{i'}^{i})$ of the transformation (0.26). Therefore, the vectors x are called *contravariant vectors* and the vectors u are called *covariant vectors* or *covectors*. The scalar function (0.25) is called *convolution* of a covector u with a vector x and is denoted by ux. The covectors u form a new linear n-space, L^{n*}, called the linear space *conjugate* with the space L^n.

0.6.2. Covectors in Affine and Projective Spaces. For each hyperplane $x=x_0+a_a t^a$ in E^n there is a covector u such that $ua_a=0$ for all $a=1,2,\dots,n-1$. Therefore, this hyperplane is determined by an equation

$$ux+v=0, \tag{0.28}$$

where $v=-ux_0$. Analogously, each hyperplane in P^n is determined by an equation

$$ux=0. \tag{0.29}$$

In the spaces R^n and R_ℓ^n each covector with coordinates u_i is related to a vector with coordinates x^i by $u_i=e_{ij}x^j$, and convolution can be regarded as the inner product of two vectors. Therefore, the equality $ua_a=0$ for a hyperplane in R^n or R_ℓ^n can be regarded as the orthogonality condition for the vector u to all vectors a_a, and the vector u in the hyperplane equation (0.28) can be regarded as a *normal vector* of the hyperplane.

0.6.3. Tensors. Contravariant and covariant vectors are particular cases of *tensors*. If we consider a polylinear scalar function of r vectors and s covectors,

$$\varphi(x,y,\dots,z,u,v,\dots,w)=T_{i_1 i_2\dots i_r}^{j_1 j_2\dots j_s} x^{i_1} y^{i_2}\dots z^{i_r} u_{j_1} v_{j_2}\dots w_{j_s}, \tag{0.30}$$

the numbers $T_{i_1 i_2\dots i_r}^{j_1 j_2\dots j_s}$ are called the coordinates of a *tensor of covariant valency r and of contravariant valency s*. Under the transition (0.26) these coordinates are multiplied by the entries of the r matrices $(A_{i'}^{i})$ and of the s matrices $(A_i^{i'})$.

0.6.4. Linear Operators. In **0.1.6** we have defined *linear transformations* in a linear space L^n, that is, automorphisms and endomorphisms in this space. Such a transformation is determined by a linear vector function of a vector argument, $f(x)$, that is, a function such that

$$f(x+y) = f(x) + f(y), \quad f(\lambda x) = \lambda f(x). \tag{0.31}$$

The value of this function on a vector $x = x^i e_i$ is

$$f(x^i e_i) = f(e_i)x^i = a_i x^i, \tag{0.32}$$

where the a_i are vectors. Therefore, if e^i is a basis in the space L^{n*} of covectors conjugate with the basis e_i in L^n, that is, satisfying the conditions $e_i e^j = \delta_i^j (= 1$ if $i = j$ and $= 0$ if $i \neq j)$, then the coordinates x^i are equal to the convolutions $e^i x$ and (0.32) can be written in the form

$$f(x) = a_i \cdot e^i x. \tag{0.33}$$

Therefore, if we write a vector function $f(x)$ in the form Ax and call A a *linear operator*, we can write a linear operator A as the sum of the n products of the vectors a_i and the covectors e^i:

$$A = a_i \cdot e^i. \tag{0.34}$$

Since the function $y = f(x)$ can be written in coordinate form $y^i = A^i_j x^j$, the coordinates A^i_j, which are the coordinates of a tensor, can also be regarded as the coordinates of the linear operator A. The coordinates A^i_j are the entries of a *square* $(n \times n)$-*matrix* (A^i_j).

Similarly, linear functions $u = f(x)$, $y = f(u)$, and $v = f(u)$ also define linear operators which can be written as $\sum_i a_i \cdot e_i$, $\sum_i a^i \cdot e^i$, and $a^i \cdot e_i$, similar to (0.34). These functions can be written in coordinate form as $u_i = A_{ij}x^j$, $y^i = A^{ij}u_j$, and $v_i = A_i^j u_j$. The coordinates A_{ij}, A^{ij}, and A_i^j, which are the coordinates of tensors, can also be regarded as the coordinates of the corresponding linear operators. The coordinates A_{ij}, A^{ij}, and A_i^j are also the entries of the square $(n \times n)$-matrices (A_{ij}), (A^{ij}), and (A_i^j).

A linear map from a linear space L^n onto or into a linear space L^m is expressed by analogous formulas $y = Ax$ and $y^a = A_i^a x^i$, where (A_i^a) is a *rectangular* $(n \times m)$-*matrix.*

For the history of tensors and linear operators see [Ro4, pp. 310–311, 390–394].

0.6.5. Tensor and Exterior Products of Vectors and Covectors. The products of vectors or covectors a_i, e^i which appear as summands in (0.34) are called *tensor products.* The tensor product $a \cdot b$ of a vector a and a covector b is a linear operator of linear function $y = f(x) = a \cdot bx$. The tensor product $a \cdot b$ is often also written as $a \otimes b$.

Using the tensor product of two vectors, or of two covectors, a and b we can define the *exterior product* of these vectors or covectors:

$$a \wedge b = a \cdot b - b \cdot a. \tag{0.35}$$

In the Euclidean space R^3 the linear operator (0.35) of taking the exterior product $a \wedge b$ of the vectors a and b coincides with the linear operator of linear function $f(x) = [x[ab]]$ where $[ab]$ is the vector product of a and b.

In the Euclidean space R^n the exterior product $a_1 \wedge a_2 \wedge \cdots \wedge a_n$ can be regarded as a scalar equal to the determinant of the matrix (a_i^j) whose entries are the coordinates of the vectors a_i and also equal to the volume of a parallelepiped built on $a_1, a_2, \ldots, a_n$. This scalar is denoted by $[a_1 a_2 \ldots a_n]$ and is called the *skew product* of $a_1, a_2, \ldots, a_n$.

0.6.6. Differentiable Scalar, Vector, and Tensor Fields. The functions $\varphi(x)$ and $f(x)$, where x is the position vector of a point in a domain in E^n, are called *scalar* and *vector fields*, respectively. These functions are said to be *differentiable* if the increments $\Delta\varphi$ and Δf can be represented as sums $adx + edx$ or $Adx + Edx$, where e and E are an infinitesimal covector or linear operator of order equal or less than dx. The covector a and linear operator A are called the *derivatives* of these functions and are denoted by $\frac{d\varphi}{dx}$ and $\frac{df}{dx}$, respectively.

The derivative $\frac{d\varphi}{dx}$ is called the *gradient* of the scalar field $\varphi(x)$.

The derivative of a scalar field is a covector and the derivative of a vector field is a linear operator, or, equivalently, a tensor of contravariant valency one and of covariant valency one.

The derivatives $\frac{d\varphi}{dx}$ and $\frac{df}{dx}$ are also denoted by $\nabla\varphi$ and $\nabla \cdot f$, where ∇ is the differential operator called *nabla*.

The linear operator $\nabla \wedge f = \nabla \cdot f - f \cdot \nabla$ is called the *curl* or *rotation* of the vector field $f(x)$. The inner product ∇f, which is equal to the trace of the matrix of the linear operator $\nabla \cdot f$, is called *the divergence* of the vector field $f(x)$.

Analogously we can define the covariant derivatives of fields of arbitrary tensors $T_{ii_1i_2\ldots i_r}^{j_1j_2\ldots j_s}$. They are denoted by $\nabla_k T_{ii_1i_2\ldots i_r}^{j_1j_2\ldots j_s}$. These derivatives are tensors of contravariant valency r and of covariant valency $s+1$.

§0.7. Riemannian Manifolds and Manifolds with Affine Connections

0.7.1. Topological Manifolds. Topological spaces each point of which has a neighborhood homeomorphic to the space R^n endowed with its natural topology are called *n-dimensional topological manifolds* or *n-manifolds.*

This definition implies that the spaces R^n and E^n and the spaces R_ℓ^n, which are just E^n equipped with a pseudometric, are n-manifolds.

Since the space P^n is just the space E^n supplemented by a hyperplane at infinity, P^n is also an n-manifold.

Since the spaces C^n and C_ℓ^n are just the spaces R^n and R_ℓ^n supplemented by a point at infinity, and, for C_ℓ^n, by an ideal hypercone, the spaces C^n and C_ℓ^n are also n-manifolds.

Since the spaces E^n, R^n, and R_ℓ^n contain infinite sets of integer points on lines without limit points, they are noncompact n-manifolds.

Since a closed bounded set in R^n is compact, the hyperspheres in R^n are compact and, therefore, the space P^n, which is homeomorphic to a bundle of lines in E^{n+1} and, therefore, to a hypersphere in R^{n+1} with identified antipodal points, and the space C^n, which is homeomorphic to a hypersphere in R^{n+1}, are compact n-manifolds; the limit points of sets of integer points on lines in these spaces are the points at infinity on these lines. The spaces C_ℓ^n, which are homeomorphic to hyperquadrics in P^{n+1}, are also compact n-manifolds.

0.7.2. Differentiable Manifolds. A homeomorphism from a domain in an n-manifold onto the space E^n, or onto a domain in this space, determines a coordinate system in this domain in the n-manifold. The coordinates x^i of a point in this domain are equal to the affine coordinates of the corresponding point in E^n.

An n-manifold X^n is called *differentiable* if following two conditions are fulfilled:

$D.1°$ The domains of definition of all coordinate systems cover the whole n-manifold;

$D.2°$ If a point x belongs to the intersection of the domains of definition of two coordinate systems and has in these systems coordinates x^i and $x^{i'}$, then the functions $x^{i'} = f^{i'}(x^i)$ have k-th continuous derivatives and the jacobian $\det\left(\frac{\partial x^{i'}}{\partial x^i}\right)$ is not equal to zero at all points in the intersection.

If all functions $x^{i'} = f^{i'}(x^i)$ are analytic, the manifold X^n is called *analytic*.

0.7.3. Vectors and Tensors in Differentiable Manifolds. Under a coordinate transformation $x^{i'} = f^{i'}(x^i)$ in a differentiable manifold X^n the differentials dx^i of the coordinates are transformed according to the law

$$dx^{i'} = A_i^{i'} dx^i, \qquad A_i^{i'} = \frac{\partial x^{i'}}{\partial x^i}. \tag{0.36}$$

If curves $x^i(t)$ pass through a point x with coordinates x^i, then the derivatives $X^i = \frac{dx^i}{dt}$ at this point are transformed according to the same law (0.36). They can be regarded as coordinates of a vector in E^n. This vector is called a *tangent vector* to the curve $x^i(t)$ at x, and E^n is called the *tangent space* $E^n(x)$ to X^n at this point.

The vectors, covectors, and tensors in the spaces $E^n(x)$ tangent to the manifold X^n, with matrices $(A_i^{i'})$ determined by the formula (0.36), are called the *vectors, covectors, and tensors in the manifold* X^n.

0.7.4. Riemannian and Pseudo-Riemannian Manifolds. If the metric in a Euclidean space R^n, or the pseudometric in a pseudo-Euclidean space R_ℓ^n, can be defined in all the tangent spaces $E^n(x)$ of a differentiable manifold X^n, and if the tensor fields e_{ij} in these spaces, determined by formula (0.19), are continuous and differentiable functions of the coordinates of the points in X^n, then X^n is called a *Riemannian manifold*, denoted by V^n, or a *pseudo-Riemannian manifold*, denoted by V_ℓ^n, respectively.

The tensors e_{ij} are called the *metric tensors* in V^n and V_ℓ^n.

Using the metric tensors in V^n and V_ℓ^n we can define inner products of vectors, moduli of vectors, angles between vectors and between curves, distances between infinitesimally near points, and lengths of finite arcs of curves, shortest curves called *geodesics*, and areas of domains in 2-dimensional surfaces. The distance ds between the points with coordinates x^i and $x^i + dx^i$ is defined by

$$ds^2 = e_{ij} dx^i dx^j. \tag{0.37}$$

Since R^n is a metric space and R_ℓ^n is a pseudometric space, the manifolds V^n and V_ℓ^n also are metric, respectively pseudometric, spaces.

The geodesics are the integral curves of the system of differential equations

$$\frac{d^2x^i}{dt^2} + \Gamma^i_{jk}\frac{dx^j dx^k}{dt\,dt} = 0, \tag{0.38}$$

where the parameter t is proportional to the arc length s of these curves. If e^{ij} is the tensor inverse to the metric tensor e_{ij}, that is, $e_{ij}e^{jk} = \delta^k_i$, then the coefficients Γ^i_{jk} are defined by

$$\Gamma^i_{jk} = \frac{1}{2} e^{ih}\left(\frac{de_{jh}}{dx^k} + \frac{de_{kh}}{dx^j} - \frac{de_{jk}}{dx^h}\right). \tag{0.39}$$

0.7.5. Parallel Displacement. The differential equations (0.38) are particular cases of the differential equations

$$\frac{da^i}{dt} + \Gamma^i_{jk}a^j\frac{dx^k}{dt} = 0 \tag{0.40}$$

for $a^i = \frac{dx^i}{dt}$. The solution of the system of equations (0.40) along a curve $x^i(t)$ is called the *parallel displacement* of the vector a with coordinates a^i along this curve.

Thus, geodesics are curves along which vectors tangent to them are displaced by parallel displacement.

Parallel displacement of vectors preserves moduli of the vectors and the angles between them. Therefore, in V^n and V^n_ℓ parallel displacement is an *isometric mapping* from the tangent spaces $R^n(x)$ and $R^n_\ell(x)$ onto tangent spaces $R^n(x+dx)$ and $R^n_\ell(x + dx)$.

Since parallel displacement also preserves convolutions a^ib_i of covariant and contravariant vectors, parallel displacement of covariant vectors is determined by the differential equations

$$\frac{db_j}{dt} - \Gamma^i_{jk}b_i\frac{dx^k}{dt} = 0. \tag{0.41}$$

Parallel displacement of arbitrary tensors can be defined similarly, using equations more general than (0.40) and (0.41).

0.7.6 Covariant Derivatives and Absolute Differentials. In V^n and V^n_ℓ we can also define covariant derivatives of vectors and tensors by means of parallel displacement: if a vector function $a^i(t)$ is given on a curve $x^i(t)$, then the covariant derivative $\nabla_j a^i$ of this function at a point of this curve is the limit of the quotient of the difference between the coordinates $a^i(t)$ of the vector at the point $x(t)$ and the result of parallel displacement of this vector at the point $x(t+dt)$ to the point $x(t)$, and the increment dx^i of the coordinate x^i between these points. We can similarly define the covariant derivative $\nabla_j b_i$, as well as covariant derivatives of arbitrary tensors.

Note that the covariant derivative of the metric tensor e_{ij} is equal to zero, and that the condition $\nabla_k e_{ij} = 0$ is equivalent to (0.39).

Covariant differentiation is not commutative: the difference $(\nabla_i\nabla_j - \nabla_j\nabla_i)a^k$ is not zero and is equal to

$$(\nabla_i\nabla_j - \nabla_j\nabla_i)a^k = R^k_{ij,h}a^h, \tag{0.42}$$

where $R^k_{ij,h}$ is the *Riemann tensor* or *curvature tensor*

$$R^k_{ij,h} = \frac{\partial \Gamma^k_{jh}}{\partial x^i} - \frac{\partial \Gamma^k_{ih}}{\partial x^j} + \Gamma^k_{jl}\Gamma^l_{ih} - \Gamma^k_{il}\Gamma^l_{jl}. \tag{0.43}$$

The convolution $\nabla_j a^i dx^j$ is called the *absolute differential* of the vector a^i, and is denoted by Da^i.

Absolute differentials of covectors and of arbitrary tensors can be defined similarly.

The geodesic equations (0.40) can be written as $\frac{Dt^i}{dt} = 0$, where t^i is the tangent vector $\frac{dx^i}{dt}$ to the curve.

The modulus $\left|\frac{Dt}{dx}\right|$ of the vector $\frac{Dt^i}{ds}$, where ds is the arc length of the curve, is called the *geodesic curvature* of this curve and is denoted by k_g.

0.7.7. Sectional Curvature. The name "curvature tensor" is connected with the role of this tensor in determining the so-called *sectional curvature* in Riemannian and pseudo-Riemannian manifolds. At each point x in a space V^n or V^n_ℓ and for each 2-dimensional direction through this point the sectional curvature K is defined to be equal to the limit of the quotient of the excess over π of the sum $A+B+C$ of the angles of a geodesic triangle with vertex at x and tangent to the given 2-dimensional direction, to the area S of this triangle, when the other two vertices of this triangle tend to x, i.e.

$$K = \lim_{S\to 0} \frac{A+B+C-\pi}{S}. \tag{0.44}$$

The sectional curvature at x can be expressed in terms of the Riemann tensor at this point for the 2-dimensional direction determined by the differentials dx^i and δx^i, as follows:

$$K = \frac{R_{ij,kl}dx^i dx^k \delta x^j \delta x^l}{(e_{ik}e_{jl} - e_{jk}e_{il})dx^i dx^k \delta x^j \delta x^l}, \tag{0.45}$$

where $R_{ij,kl} = e_{hl}R^h_{ij,k}$.

Formula (0.44) is a consequence of the *Gauss-Bonnet theorem*: if in the space V^n or V^n_ℓ we have a domain D on a 2-dimensional surface bounded by a contour C consisting of N arcs of curves joining at N corners, then the integral of the sectional curvature over the part of the surface bounded by C is equal to

$$\int_D K dS = 2\pi - \int_C k_g ds - \sum_i \Delta_i\varphi, \tag{0.46}$$

where dS is the area element of D, k_g is the geodesic curvature of the arcs of C, ds is the arc length of the curvilinear parts of C, and $\Delta_i\varphi$ are the turning angles of the vector tangent to the contour at the corners.

If C is a geodesic polygon, then $k_g = 0$, and since the $\Delta_i\varphi$ are connected with the interior angles A_i by $\Delta_i\varphi = \pi - A_i$, formula (0.46) takes the form

$$\int_D K dS = \sum_i A_i - (N-2)\pi. \tag{0.47}$$

Therefore, for a geodesic triangle with angles A, B, C we have $\int_D K dS = A + B + C - \pi$, hence, since the integral $\int_D K dS$ is equal to the product of the area S of this triangle by the mean value of K, and when the points B and C tend to A, this mean value tends to the value of K at A, we obtain formula (0.44).

For the history of Riemannian and pseudo-Riemannian geometry see [Ro4, pp. 283–295, 309–314, 316–317].

The spaces R^n and R^n_ℓ are particular cases of the manifolds V^n and V^n_ℓ. Since the metric tensors e_{ij} of these spaces are constant, their derivatives are equal to zero and all Γ^i_{jk}, and, therefore, all $R^k_{ij,h}$, and all sectional curvatures in these spaces also are equal to zero.

0.7.8. Manifolds with Affine Connections. If at the points in a differentiable manifold X^n the tensors e_{ij} are not given but only the quantities Γ^i_{jk}, with the same transformation law under changes of coordinates x^i are given, and if the differential equations $\nabla_k e_{ij} = 0$ defined by means of the Γ^i_{jk} have no solutions having the properties of the metric tensors in the manifold V^n or V^n_ℓ, then the manifold is called a *manifold with affine connection*, and is denoted by A^n.

In V^n and V^n_ℓ the quantities Γ^i_{jk} satisfy the condition $\Gamma^i_{jk} = \Gamma^i_{kj}$. In A^n this condition need not hold. If it does not hold, then the tensor

$$S^i_{jk} = \Gamma^i_{jk} - \Gamma^i_{kj}, \tag{0.48}$$

called the *torsion tensor* is defined in the manifold and is nonzero.

Therefore, manifolds A^n in which $\Gamma^i_{jk} = \Gamma^i_{kj}$ are called *manifolds with torsion-free affine connection.*

Geodesics, covariant derivatives, parallel displacement, and Riemann tensor can be defined in A^n by the same formulas (0.38), (0.40), (0.41), and (0.43) as in V^n and V^n_ℓ.

Parallel displacement in A^n is an *affine mapping* from the tangent spaces $E^n(x)$ onto the tangent spaces $E^n(x + dx)$, i.e. it is a bijection preserving straight lines and quotients of collinear vectors.

The geometric meaning of the tensor S^i_{jk} in the manifolds A^n is as follows: if we perform a parallel displacement from a point x to a nearby point x' for which the infinitesimal segment xx' can be represented by an infinitesimal vector a, and from x' to a nearby point x'' for which the infinitesimal segment $x'x''$ can be represented by an infinitesimal vector b, and from x to a nearby point $\tilde{x}'$ for which the segment $x\tilde{x}'$ can be represented by the vector b, and from $\tilde{x}'$ to a nearby point $\tilde{x}''$ for which the segment $\tilde{x}'\tilde{x}''$ can be represented by the vector a, then the segment $x''\tilde{x}''$ is represented by the vector $\frac{1}{2}S^i_{jk}a^j b^k$ (Fig. 0.1). Therefore, the equality $S^i_{jk} = 0$ is equivalent to the existence of infinitesimal parallelograms.

The geometric meaning of the tensor $R^k_{ij,h}$ in the manifolds A^n is as follows: if we perform a parallel displacement of a vector c from a point x along the closed contour consisting of three infinitesimal segments xx', $x'x''$, and $x''x$, and the segments xx' and xx'' are represented by infinitesimal vectors a and b, then the main part of the difference between c and the vector c' obtained by this parallel displacement is equal to $\frac{1}{2}R^k_{ij,h}a^i b^j c^h$ (Fig. 0.2). Therefore, the equality $R^k_{ij,h} = 0$ is equivalent to the independence of the result of parallel displacement of a vector on the path of

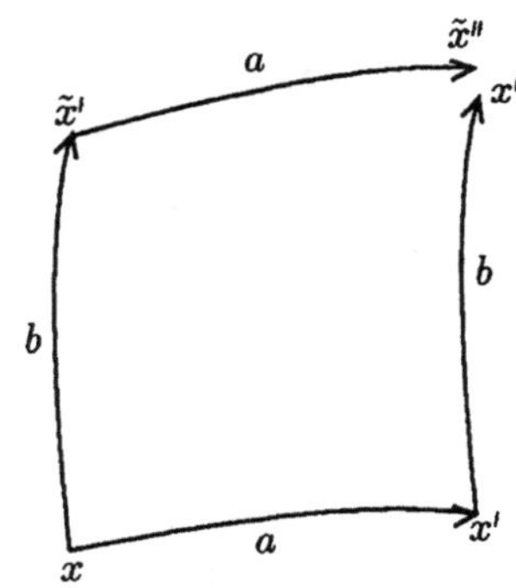

FIG. 0.1

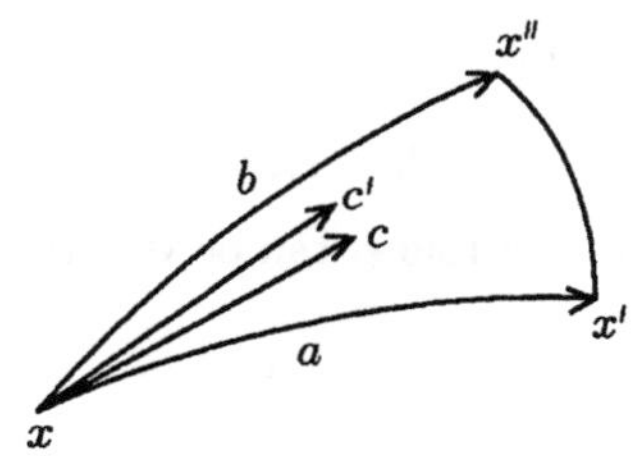

FIG. 0.2

this parallel displacement.

A torsion-free affine connection is determined by its geodesics and their *affine parameters* t for which the differential geodesic equations have the form (0.38): if x and x' are nearby points and x_0 is the middle point between them (if x and x' correspond to values t and t' of the affine parameter, then x_0 corresponds to the value $\frac{1}{2}(t+t')$ of this parameter), then the parallel displacement of vectors at x to x' is generated by the mapping of a neighborhood $U(x)$ onto $U(x')$ consisting of reflections along the geodesics at x and at x_0 (Fig. 0.3).

The term "affine connection" was introduced by H. Weyl [Wey4], the theory of spaces with affine connection was created by É. Cartan [Car2] and J.A. Schouten [Sco]. For the history of geometry of spaces with affine connections see [Ro4, pp. 317–320].

The space E^n is the particular case of a manifold A^n without curvature and torsion. The geodesics in E^n are straight lines, the affine parameter of lines $x = x_0 + at$ is the parameter t. Parallel displacement in E^n is translation, $x' = x + a$. Note that this translation is also the product of the reflections at the point with position vector x and at the point with position vector $\frac{1}{2}(x + x')$.

0.7.9. Arbitrary Frames and Exterior Forms. The affine coordinates in the tangent spaces $E^n(x)$ used by us are determined by the frames of vectors e_i

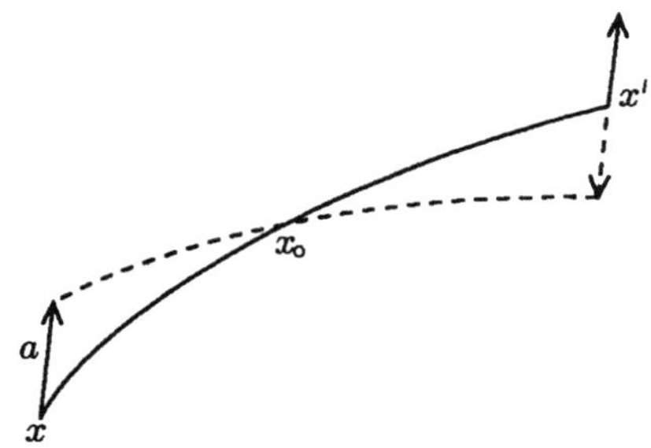

FIG. 0.3

tangent to the coordinate lines of X^n; these vectors can be regarded as derivatives $\frac{\partial x}{\partial u^i}$, and such frames are called *natural frames.*

Now, let us consider frames in $E^n(x)$ consisting of arbitrary vectors e_i. In this case the differential dx of a point in X^n can be written as

$$dx = e_i \omega^i \tag{0.49}$$

and the differentials de_i of the vectors e_i can be written as

$$de_i = e_j \omega_i^j. \tag{0.50}$$

In (0.49) and (0.50), ω^i and ω_i^j are linear differential forms, that is, linear forms of the differentials du^i of the coordinates of the points in X^n; the coefficients of these forms also are functions of the coordinates of the points in X^n. These forms are particular cases of so-called *exterior differential forms*

$$\Omega = a_{i_1 i_2 \ldots i_k} du^{i_1} \wedge du^{i_2} \wedge \ldots \wedge du^{i_k}, \tag{0.51}$$

where $a_{i_1 i_2 \ldots i_k}$ are the coordinates of a skew-symmetric tensor, which are differentiable functions of the coordinates, and the product $du^{i_1} \wedge du^{i_2} \wedge \ldots \wedge du^{i_k}$ is the determinant of the matrix $(d_j x^i)$, where the d_j are differentials in k linearly independent directions.

The operations of exterior multiplication and exterior differentiation are defined on exterior forms. The *exterior product* of the form (0.51) and the form

$$\Psi = b_{j_1 j_2 \ldots j_l} du^{j_1} \wedge du^{j_2} \wedge \ldots \wedge du^{j_l} \tag{0.52}$$

is the form

$$\Omega \wedge \Psi = (-1)^k a_{i_1 i_2 \ldots i_k} b_{j_1 j_2 \ldots j_l} du^{i_1} \wedge du^{i_2} \wedge \ldots \wedge du^{i_k} \wedge du^{j_1} \wedge du^{j_2} \wedge \ldots \wedge du^{j_l}. \tag{0.53}$$

The *exterior differential* of the form (0.51) is the form

$$d\Omega = da_{i_1 i_2 \ldots i_k} \wedge du^{i_1} \wedge du^{i_2} \wedge \ldots \wedge du^{i_k} = \frac{\partial a_{i_1 i_2 \ldots i_k}}{\partial u^j} du^j \wedge du^{i_1} \wedge du^{i_2} \wedge \ldots \wedge du^{i_k}. \tag{0.54}$$

It is easy to check that $d(d\Omega) = 0$.

The exterior differentials of the forms ω^i and ω_i^j appearing in (0.49) and (0.50) give the *structure equations* of the manifolds A^n:

$$d\omega^i = \omega^k \wedge \omega_k^i + \frac{1}{2} S_{hk}^i \omega^h \wedge \omega^k \tag{0.55}$$

and

$$d\omega_j^i = \omega_j^k \wedge \omega_k^i + \frac{1}{2} R_{hk,j}^i \omega^h \wedge \omega^k, \tag{0.56}$$

where the tensors S_{hk}^i and $R_{hk,j}^i$ are given in the same frame as formulas (0.49) and (0.50).

Formulas (0.49) and (0.50) are valid also in the manifolds V^n and V_ℓ^n and in the spaces E^n, R^n, and R_ℓ^n. In R^n, R_ℓ^n, V^n, and V_ℓ^n the vectors e_i of these frames are orthonormal, that is, satisfy the conditions $e_i e_j = \delta_{ij}$ or $-\delta_{ij}$. Differentiation of these conditions gives, for R^n and V^n,

$$\omega_j^i = -\omega_i^j \tag{0.57}$$

and, for R_ℓ^n and V_ℓ^n,

$$\omega_j^i = -e_i^2 e_j^2 \omega_i^j. \tag{0.58}$$

In these spaces and manifolds there are also structure equations (0.55) and (0.56), where $S_{hk}^i = 0$ for V^n and V_ℓ^n, and $S_{hk}^i = R_{hk,j}^i = 0$ for E^n, R^n, and R_ℓ^n.

For the history of the theory of exterior forms see [Ro4, pp. 323–326].

§0.8. Topological Groups and Lie Groups

0.8.1 Topological Groups. A set of elements is said to be a *topological group* if

$TG.1°$ This set is a group (see **0.1.2**);

$TG.2°$ This set is a topological space (see **0.2.1**);

$TG.3°$ The group operations $x \to ax$, $x \to xa$ and $x \to x^{-1}$, where a is a constant element, are continuous mappings from this topological space onto itself.

The group and the topological space participating in the definition of a topological group are called the *basis group* and the *basis space* of this topological group. A topological group is called commutative or noncommutative, additive or multiplicative together with its basis group, and connected or nonconnected, compact or noncompact together with its basis space. A subset of a topological group is called a *topological subgroup* if it is a subgroup of the basis group and a closed subset of the basis space of the topological group.

0.8.2. Lie Groups. The most important for geometry are topological groups that are also *topological n-manifolds*. These topological groups are called n-dimensional *Lie groups*.

Lie groups have the following important property: if the elements x, y, and $z = xy$ belong to a neighborhood $U(e)$ of the neutral group element, there are coordinates x^i, y^i, and z^i of these elements for which the functions

$$z^i = f^i(x^1, x^2, \dots, x^n, y^1, y^2, \dots, y^n)$$

are analytic functions.

Two Lie groups are said to be *locally isomorphic* if their neutral elements have isomorphic neighborhoods.

The set of real numbers and the set of vectors of a linear space L^n are, with respect to addition, additive Lie groups: the 1-dimensional group R and the n-dimensional group R_n, respectively. The group R_n is isomorphic to the direct sum of n groups R.

The set of real numbers without zero is, with respect to multiplication, a multiplicative 1-dimensional Lie group with two connected components; the component, which itself is a group, is isomorphic to the group R.

The quotient group R/Z (Z is the discrete group of integers with respect to addition) is isomorphic to the multiplicative 1-dimensional Lie group T of complex numbers with modulus one. The quotient group R_n/Z_n (Z_n is the direct sum of n groups Z) is isomorphic to the n-dimensional group T_n, called a *torus group*; T_n is isomorphic to the direct product of n groups T.

The set of real $(n \times n)$-matrices is, with respect to addition, an additive n^2-dimensional Lie group, isomorphic to the group R_{n^2}. The set of all nonsingular (invertible) real $(n \times n)$-matrices is, with respect to multiplication, a multiplicative n^2-dimensional Lie group, called the *general linear group* and denoted by GL_n. These matrices represent the automorphisms in the linear space L^n; all real $(n \times n)$-matrices represent the endomorphisms in L^n. The set of all real $(n \times n)$-matrices is, with respect to addition and multiplication, an algebra: $\mathbb{R}_n$.

The subgroup of GL_n consisting of the matrices with determinant 1 is an (n^2-1)-dimensional Lie group, called the *special linear group* and denoted by SL_n.

The subgroup of GL_n consisting of the orthogonal matrices, that is, satisfying the condition

$$U^\top U = I \tag{0.59}$$

(I is the identity matrix, while the matrix $U^\top$ is obtained from U by transposition) is an $\frac{n(n-1)}{2}$-dimensional Lie group, called the *orthogonal group* and denoted by O_n. The determinants of the matrices in this group are equal to ± 1; the subgroup of O_n consisting of the matrices with determinant $+1$ is called the *special orthogonal group* and is denoted by SO_n.

The subgroup of GL_n consisting of the pseudoorthogonal matrices of index ℓ, that is, satisfying the condition

$$U^\top E U = E \tag{0.60}$$

(as in **0.5.6**, the matrix E is a diagonal $(n \times n)$-matrix with ℓ diagonal entries equal to -1 and the other diagonal entries equal to $+1$), is an $\frac{n(n-1)}{2}$-dimensional Lie group called the *pseudoorthogonal group of index* ℓ and denoted by $O_{n,\ell}$. The determinants of matrices in these groups are also equal to ± 1. The subgroup of $O_{n,\ell}$ consisting of the matrices with determinant $+1$ is called the *special pseudoorthogonal group of index* ℓ and is denoted by $SO_{n,\ell}$.

The subgroup of GL_{2n} consisting of the matrices satisfying the condition

$$U^\top J U = J, \tag{0.61}$$

where J is the $(2n \times 2n)$-matrix

$$J = \begin{bmatrix} O & I \\ -I & O \end{bmatrix}, \tag{0.62}$$

is an $\frac{n(n+1)}{2}$-dimensional Lie group, called the *symplectic group* and denoted by Sp_{2n}. The matrices in this group are called *symplectic* $(2n \times 2n)$-matrices.

0.8.3. Solvable, Simple, Semisimple, and Reductive Lie Groups. A Lie group G is said to be *solvable* if it has a system of subgroups $G_1, G_2, \ldots, G_k = \{e\}$ such that G_i is an invariant subgroup (see **0.1.4**) of G_{i-1} ($G_0 = G$), and G_k is the neutral element e in G.

A Lie group is said to be *simple* if it has no invariant subgroups of smaller dimension. A Lie group is said to be *semisimple* if it has no solvable invariant subgroups. Every semisimple Lie group is locally isomorphic to a direct product of *noncommutative* simple Lie groups (therefore, semisimple groups are noncommutative).

A Lie group is said to be *reductive* if it is locally isomorphic to a direct product of simple (not necessarily noncommutative) Lie groups. All compact and commutative Lie groups are reductive.

The groups SL_n, SO_n and $SO_{n,\ell}$ for $n \neq 2$ and 4, and Sp_{2n} are simple and semisimple, that is, noncommutative simple. The groups SO_2 and $SO_{2,1}$, isomorphic to the groups T and R, respectively, are commutative simple, that is, are not semisimple. The groups SO_4 and $SO_{4,2}$ are locally isomorphic to the direct products $SO_3 \otimes SO_3$ and $SO_{3,1} \otimes SO_{3,1}$ and are semisimple but not simple, the group $SO_{4,1}$ is simple and semisimple.

W. Killing [Ki3] and É. Cartan [Car1] proved that there are four infinite series of semisimple simple Lie groups which are not locally isomorphic (A_n for $n \geq 1$, B_n for $n \geq 2$, C_n for $n \geq 3$, and D_n for $n \geq 4$) and 5 exceptional classes of these groups: G_2, F_4, E_6, E_7, E_8.

The groups SL_n are real groups of the series A_n, the groups SO_{2n+1} and $SO_{2n+1,\ell}$ are real groups of the series B_n, the groups Sp_{2n} are real groups of the series C_n, the groups SO_{2n} and $SO_{2n,\ell}$ are real groups of the series D_n. The groups GL_n are reductive.

For the history of the theory of Lie groups see [Ro4, pp. 333–344, 348–363].

0.8.4. Lie Algebras. In the tangent space $E^n(e)$ at the neutral element e in an n-dimensional Lie group we can define the structure of a nonassociative algebra, that is, an *algebra* (see **0.1.7**) not satisfying the condition of associativity of multiplication. The multiplication in this algebra is called *commutation* and is denoted by $[ab]$.

The operations in this algebra are defined as follows: the elements in this algebra can be regarded as tangent vectors to 1-parameter subgroups $x(t)$ of the group with multiplication $x(t_1)x(t_2) = x(t_1 + t_2)$ for $x(0) = e$, the coordinate a^i of a vector a is equal to the derivative $\frac{dx^i}{dt}$. If vectors a and b correspond to subgroups $x(t)$ and $y(t)$, their sum $a + b$ corresponds to the product $x(t)y(t)$ and the coordinate $[ab]^i$ of the *commutator* $[ab]$ is equal to the derivative $\frac{\partial}{\partial t}\frac{\partial}{\partial s}[x(t)y(s)x^{-1}(t)y^{-1}(s)]^i$. This algebra is called the *Lie algebra* tangent to the Lie group G. We denote this algebra by $\mathbb{G}$. In this book all algebras including Lie algebras and the subspaces of these algebras are denoted by capital double letters although often Lie algebras are denoted by small gothic letters.

The commutation is *anticommutative*, that is,

$$[ab] = -[ba], \tag{0.63}$$

and satisfies the *Jacobi identity*

$$[[ab]c] + [[bc]a] + [[ca]b] = 0. \tag{0.64}$$

In Lie algebras the role of formula (0.14) is played by the formula

$$[e_i e_j] = C^k_{ij} e_k. \tag{0.65}$$

The numbers C^k_{ij} are called the *structure constants* of the Lie algebra.

Two locally isomorphic Lie groups have isomorphic Lie algebras.

Sophus Lie, the founder of the theory of Lie groups (see [LiE]) proved that each Lie algebra is the tangent Lie algebra of some Lie group.

The set of vectors in the Euclidean space R^3 with the usual vector multiplication is a Lie algebra. This Lie algebra is the tangent Lie algebra of the group SO_3.

If the Lie group consists of matrices, its tangent Lie algebra consists of matrices of the same algebra, and the vector tangent to the 1-parameter group $U(t)$ is given by the derivative $\frac{dU}{dt}$ for $t = 0$, while $[AB]$ is equal to the difference $AB - BA$.

Therefore, the tangent Lie algebra $\mathbb{GL}_n$ of the group GL_n consists of all $(n \times n)$-matrices.

Since differentiation of the condition $\det U = 1$ gives $\operatorname{Tr} A = 0$, the tangent Lie algebra $\mathbb{SL}_n$ of the group SL_n consists of all $(n \times n)$-matrices with trace zero.

Since differentiation of the conditions (0.59), (0.60) and (0.61) at $t = 0$ gives

$$A^\top + A = 0, \tag{0.66}$$

$$A^\top E + EA = 0, \tag{0.67}$$

and

$$A^\top J + JA = 0, \tag{0.68}$$

the tangent Lie algebras $\mathbb{SO}_n$, $\mathbb{SO}_{n,\ell}$, and $\mathbb{S}p_{2n}$ of the groups $SO_n, SO_{n,\ell}$, and Sp_{2n} consist of all skew-symmetric $(n \times n)$-matrices, of all $(n \times n)$-matrices satisfying the condition (0.67), and of all $(2n \times 2n)$-matrices satisfying the condition (0.68), respectively.

0.8.5. Invariant Riemannian Metrics and Affine Connections in Lie Groups. In every noncommutative Lie group we can define two invariant affine connections with torsion and without curvature: the parallel displacements of vectors from an element a in this group to an element b are generated by the transformations $x \to (ba^{-1})x$ and $x \to x(a^{-1}b)$ of a neighborhood $U(a)$ into a neighborhood $U(b)$. The quantities Γ^k_{ij} for these connections are equal to the structure constants C^k_{ij}, respectively to $-C^k_{ij}$. These connections are without curvature since their parallel displacements do not depend on the paths joining a and b.

In every Lie group we can define an invariant torsion-free affine connection. In the case of noncommutative groups it has a curvature. This affine connection is determined by the geodesics and their affine parameter. In this affine connection the role of geodesics is played by 1-parameter subgroups $x(t)$ and their cosets $ax(t)$ and $x(t)a$. The role of affine parameter on a 1-parametric subgroup $x(t)$ is played by the canonic parameter t of this subgroup, while on a coset the role of affine parameter is played by the parameter generated by the canonic parameter of the subgroup.

If the Lie group is semisimple and compact, the last affine connection is generated by Riemannian metric in this group. The metric tensor of this metric is related to the structure constants of this Lie group by

$$e_{ij} = -C^l_{ih}C^h_{jl} \tag{0.69}$$

and the Riemann tensor of this metric is related to these structure constants by

$$R^h_{ij,k} = \frac{1}{4}C^l_{ik}C^h_{jl}. \tag{0.70}$$

If the Lie group is semisimple and noncompact, the last affine connection is generated by the geometry of a pseudo-Riemannian manifold with tensors e_{ij} and $R^h_{ij,k}$ also determined by formulas (0.69) and (0.70), respectively.

Therefore, a necessary and sufficient condition for semisimplicity of a Lie group is nondegeneracy of the quadratic form

$$-e_{ij}a^ib^j = C^l_{ih}C^h_{jl}a^ib^j, \tag{0.71}$$

which is called the *Killing-Cartan form.* This fact was proved by É. Cartan [Car1]. This form was composed by Cartan from the forms defined by W. Killing [Ki3].

0.8.6. Homogeneous Spaces. A topological space T in which a transitive group G of homeomorphisms acts, that is, for every two points x and y in T there is a homeomorphism f such that $f(x) = y$, is called a *homogeneous space.*

If the group of homeomorphisms in T is not transitive, T can be partitioned into domains on which this group is transitive; such domains are called *orbits.*

The transitive group G is called the *fundamental group* of the homogeneous space T. The subgroup H_x of G preserving a point x in T is called the *stabilizer* of x.

In homogeneous spaces the stabilizers of all points are isomorphic, that is, $H_x = H_y = H$ for all x and y. A homogeneous space with fundamental group G and stabilizer H for its points is in bijection with the quotient space G/H and often is denoted by this quotient space.

Every group G and subgroup H of G that does not contain an invariant subgroup of G determine a homogeneous space G/H.

The fundamental group G of a homogeneous space G/H is called *imprimitive* if this space can be partitioned into classes such that every transformation in G preserves these classes. If the space G/H cannot be partitioned into such classes, G is called *primitive.*

The most important spaces in geometry are homogeneous spaces G/H with G a Lie group.

0.8.7. Symmetric Spaces. Very important in the classes of manifolds V^n, V^n_ℓ and A^n are the *symmetric spaces*, which can be defined as manifolds V^n or V^n_ℓ in which geodesic symmetries in points are isometries, or as manifolds A^n with torsion-free affine connection in which geodesic symmetries in points preserve the affine connection, that is, preserve geodesics and their affine parameters. These spaces can also be defined as V^n, V^n_ℓ, or torsion-free A^n in which

$$\nabla_\ell R^h_{ij,k} = 0. \tag{0.72}$$

All Lie groups, with their invariant torsion-free affine connections, and all semi-simple Lie groups, with their invariant metrics or pseudometrics, as well as all totally geodesic submanifolds of these manifolds, that is submanifolds all whose geodesics are geodesics of the space, are symmetric spaces.

A symmetric space can be realized as a totally geodesic manifold in the Lie group generated by the geodesic symmetries in points of this symmetric space.

Symmetric Riemannian spaces were first defined by P. A. Shirokov [ShP1] and H. Levy [Lev], the theory of symmetric Riemannian spaces and spaces with affine connections was created by Cartan [Car12-15]. Symmetric pseudo-Riemannian spaces were investigated by M. Berger [Ber] and A. S. Fedenko [Fed2]. For the geometry of symmetric spaces see also the book [Hel] by S. Helgason, for the history of this theory see [Ro4, pp. 363-366].

0.8.8. Reductive and k-Symmetric Spaces. There are various generalizations of symmetric spaces. P. K. Rashevsky [Ra] has defined "symmetric spaces with torsion", that is, manifolds with affine connection with torsion, in which covariant derivative of both curvature and torsion tensors are equal to zero. K. Nomizu [Nom] has given another definition of these homogeneous spaces and called them "reductive spaces".

Another generalization of symmetric spaces was proposed by V. I. Vedernikov [Ved1-2]: he replaced in the definition of symmetric homogeneous spaces the involutive automorphism by arbitrary analytic automorphism Φ of the fundamental group G of this space and called these spaces "homogeneous Φ-spaces". Reductive homogenous Φ-spaces, in which k-th power Φ^k of the automorphism Φ is identity, are called *k-symmetric spaces.* These spaces were investigated by N. A. Stepanov [Stp], A. J. Ledger and M. Obata [Led], [LeO], J. A. Wolf and A. Gray [WG], A. S. Fedenko [Fed1], and O. Kowalski [Ko].

0.8.9. Fundamental Groups of the Most Important Homogeneous Spaces. The fundamental group of the affine space E^n is the group of *affine transformations* in this space, that is, the group of bijections in it preserving the straight lines. These transformations have the form

$$x' = Ux + a, \tag{0.73}$$

where x and x' are position vectors of points, a is a vector and U is a nonsingular $(n \times n)$-matrix. For $a = 0$ the transformations (0.73) become

$$x' = Ux, \tag{0.74}$$

and are called *affine rotations.* For $U = I$ they become $x' = x + a$, and are called *translations.* The group (0.73) of affine transformations in E^n is an $n(n+1)$-dimensional Lie group, its subgroup (0.74) is an n^2-dimensional Lie group isomorphic to the group GL_n.

The group of translations in E^n is an n-dimensional commutative invariant subgroup of the group (0.73); it is isomorphic to the group R_n.

An affine transformation (0.73) in E^n for which U has positive, respectively negative, determinant is called an *affine transformation of the first,* respectively *second, kind.*

The fundamental groups of the Euclidean space R^n and the pseudo-Euclidean space R^n_ℓ are the groups of *motions* in these spaces, that is, the groups of isometries in them. These transformations have the form (0.73), where U is an orthogonal matrix in the group O_n for R^n or a pseudoorthogonal matrix in the group $O_{n,\ell}$ for R^n_ℓ. These groups are $\frac{n(n+1)}{2}$-dimensional Lie groups, their subgroups (0.74) are called the groups of *rotations* in these spaces and are isomorphic to the groups O_n and $O_{n,\ell}$, respectively. A motion (0.73) in R^n or R^n_ℓ with a matrix U with determinant $+1$ or -1 is called a *motion of the first,* respectively *second, kind.*

The transformations (0.73) in R^n or R^n_ℓ for which the matrix U is the product of an orthogonal or a pseudoorthogonal matrix by a real number not equal to ± 1, are called *similitudes* in these spaces. If the matrix U of a similitude has the form $U = \lambda I$, the similitude is called a *homothety.* A similitude (0.73) in R^n or R^n_ℓ with a matrix U with positive, respectively negative, determinant is called a *similitude of the first,* respectively *second, kind.*

The fundamental group of the projective space P^n is the group of *collineations* in this space, that is, like the group of affine transformations in E^n, the group of bijections in it preserving the straight lines. These transformations have the form (0.74), where x and x' are vectors in L^{n+1} representing points in P^n and U is a nonsingular $((n+1) \times (n+1))$-matrix. Since the vectors representing points in P^n are determined up to a real multiplier, the group of collineations in P^n is an $n(n+2)$-dimensional Lie group which is locally isomorphic to the group SL_{n+1}; this group is denoted by PSL_{n+1}. Mappings of points in P^n onto hyperplanes and of lines onto pencils of hyperplanes are called *correlations* in P^n and have the form

$$u = Ux. \tag{0.75}$$

The product of two correlations in P^n is a collineation, therefore, the collineations and correlations in P^n form a group, called the group of *projective transformations* in this space.

The fundamental groups of the conformal space C^n and pseudoconformal space C^n_ℓ are the groups of *conformal transformations* in these spaces, that is, the groups of bijections in them preserving all circles. These transformations have the form (0.74), where U is a pseudoorthogonal matrix in the group $O_{n+2,1}$ for C^n or $O_{n+2,\ell+1}$ for C^n_ℓ. These groups are $\frac{(n+1)(n+2)}{2}$-dimensional Lie groups and are denoted by $PO_{n+2,1}$ and $PO_{n+2,\ell+1}$, respectively.

The fundamental groups of the spaces E^n, R^n, R^n_ℓ, P^n, C^n, and C^n_ℓ are transitive. The groups of rotations in R^n and R^n_ℓ are intransitive, their orbits are concentric hyperspheres with a common center. These groups of rotations are transitive on these hyperspheres, but are imprimitive on them, and their classes of primitivity are pairs of antipodal points on these spheres.

The spaces E^n, R^n and R^n_ℓ are symmetric spaces, with an affine connection, a Riemannian metric, and a pseudo-Riemannian pseudometric, respectively. In these spaces the role of geodesics is played by straight lines. In these spaces the symmetry in a point with position vector a has the form

$$x' = 2a - x. \tag{0.76}$$

Such symmetries generate the groups of translations in these spaces.

The groups of similitudes and similitudes-rotations in the spaces R^n and R^n_ℓ have the forms (0.73) and (0.74), where the matrices U are the products of an orthogonal, respectively pseudoorthogonal, matrix by a real number $\neq \pm 1$. Therefore, the matrices in the tangent Lie algebras of the groups of similitudes-rotations in R^n and R^n_ℓ satisfy conditions which differ from (0.66) and (0.67) by replacing 0 at the right hand side by scalar matrices λI. In particular, the matrices in the tangent Lie algebras of the groups of similitudes-rotations in the planes R^2 and R^2_1 have the respective forms

$$\begin{bmatrix} A & B \\ -B & A \end{bmatrix} \quad \text{and} \quad \begin{bmatrix} A & B \\ B & A \end{bmatrix}. \tag{0.77}$$

CHAPTER I

ALGEBRAS AND LIE GROUPS

§1.1. Commutative Associative Algebras

1.1.1. Complex Numbers. In **0.1.7** we have defined n-*algebras*, that is, rings that are linear spaces L^n satisfying the axiom $A.1°$ (0.13). In **0.1.7** we mentioned that the field $\mathbb{C}$ of *complex numbers* is a commutative associative 2-algebra with basis $e_0 = 1$, $e_1 = i$ $(i^2 = -1)$.

Every complex number $x = x_0 + ix_1$ has a *conjugate complex number* $\bar{x} = x_0 - ix_1$, and the following equalities hold:

$$\overline{xy} = \bar{y}\,\bar{x}, \tag{1.1}$$

$$|x|^2 = \bar{x}\,x = x_0^2 + x_1^2, \tag{1.2}$$

$$|xy| = |x|\,|y|. \tag{1.3}$$

The nonnegative real number $|x|$ is called the *modulus* of the complex number x. $|x| = 0$ if and only if $x = 0$. Every complex number $x \neq 0$ has an inverse number

$$x^{-1} = \frac{\bar{x}}{|x|^2}. \tag{1.4}$$

If we regard the modulus $|y - x|$ of the difference $y - x$ between two complex numbers x and y as the *distance* between these numbers, we introduce into $\mathbb{C}$ the metric of the Euclidean plane R^2.

Complex numbers can also be regarded as vectors in R^2, and the *inner product* (x, y) and the *skew product* $[x, y]$ of two vectors x and y are

$$(x, y) = \frac{1}{2}(\bar{x}y + \bar{y}x) \tag{1.5}$$

and

$$[x, y] = \frac{1}{2i}(\bar{x}y - \bar{y}x). \tag{1.6}$$

Therefore, the angle φ between two vectors x and y can be defined by

$$\cos\varphi = \frac{1}{2}\frac{\bar{x}y + \bar{y}x}{|x|\,|y|} \tag{1.7}$$

and

$$\sin\varphi = \frac{1}{2i}\frac{\bar{x}y - \bar{y}x}{|x|\,|y|}. \tag{1.8}$$

Motions of the first and the second kind in R^2 can be expressed by functions of a complex variable as

$$x' = ax + b \quad \text{and} \quad x' = a\bar{x} + b, \tag{1.9}$$

with $|a| = 1$, respectively. The functions (1.9) with $|a| \neq 1$ express *similitudes of the first and the second kind* in R^2, respectively.

For the history of complex numbers and the complex plane see [Ro4, pp.146–147, 163–167, 382].

1.1.2. Split Complex Numbers. The 2-algebra with basis $e_0 = 1$, $e_1 = e\,(e^2 = +1)$ is called the algebra of *split complex numbers* (or *double numbers*) and is denoted by $\mathbb{C}'$.

Every split complex number $x = x_0 + ex_1$ has a *conjugate split complex number* $\bar{x} = x_0 - ex_1$. In this algebra the equalities (1.1) and (1.3) are fulfilled, but (1.2) is replaced by

$$|x|^2 = \bar{x}x = x_0^2 - x_1^2. \tag{1.10}$$

Therefore, the modulus $|x|$ of a split complex number x can be real and pure imaginary and can be equal to 0 for $x \neq 0$. A split complex number x with $|x| \neq 0$ has an inverse number (1.4).

Split complex numbers x with $|x| = 0$ are zero divisors, that is, nonzero numbers with some product equal to 0 (for x with $|x| = 0$, $\bar{x}x = 0$). Therefore, *the algebra* $\mathbb{C}'$ *is not a field.*

The zero divisors

$$e_+ = \frac{1+e}{2} \text{ and } e_- = \frac{1-e}{2} \tag{1.11}$$

are *idempotents*, that is, all powers of them are equal to themselves. Their product is equal to 0.

In **0.1.5** we have defined the *direct sum* of two rings. The direct sum of an n-algebra $\mathbb{A}$ and an m-algebra $\mathbb{B}$ with bases $e_1, e_2, \ldots, e_n$ and $f_1, f_2, \ldots, f_m$ is the $(n+m)$-algebra $\mathbb{A} \oplus \mathbb{B}$ with basis $e_1, e_2, \ldots, e_n$, $f_1, f_2, \ldots, f_m$, where $e_i f_\alpha = f_\alpha e_i = 0$.

Theorem 1.1. *The algebra* $\mathbb{C}'$ *is isomorphic to the direct sum* $\mathbb{R} \oplus \mathbb{R}$.

PROOF. This theorem follows from the properties of the idempotents (1.11). These idempotents can be taken as a basis of the algebra $\mathbb{C}'$, and by the above mentioned properties these elements can also be regarded as the elements 1 of two fields $\mathbb{R}$. The direct sum of these $\mathbb{R}$ is isomorphic to the given algebra. ■

If we regard the modulus $|y - x|$ of the difference $y - x$ between two split complex numbers x and y as the *distance* between these numbers, we introduce into $\mathbb{C}'$ the pseudometric of the pseudo-Euclidean plane R_1^2.

Split complex numbers can also be regarded as vectors in R_1^2, and the *inner product* (x,y) and the *skew product* $[x,y]$ of two vectors x and y are (1.5) and

$$[x,y] = \frac{1}{2e}(\bar{x}y - \bar{y}x), \tag{1.12}$$

respectively.

Therefore, the angle $i\varphi$ (see **0.5.6**) between two vectors x and y in R_1^2 satisfying the conditions $x_0 > 0,\ y_0 > 0,\ x_0^2 > x_1^2,\ y_0^2 > y_1^2$ can be defined by

$$\cos(i\varphi) = \frac{1}{2}\frac{\bar{x}y + \bar{y}x}{|x|\,|y|} \tag{1.13}$$

and

$$\sin(i\varphi) = \frac{i}{2e}\frac{\bar{x}y - \bar{y}x}{|x|\,|y|}. \tag{1.14}$$

Motions of the first and the second kind in R_1^2 can be expressed by functions of a split complex variable as (1.9), with $|a| = 1$. The functions (1.9) with $|a| = i$ express *antimotions of the first and the second kind* in R_1^2. The functions (1.9) with $|a|^2 > 0,\ |a| \neq 1$ express *similitudes of the first*, respectively *the second, kind* in R_1^2.

The functions (1.9) with $|a|^2 < 0,\ |a| \neq i$ express *antisimilitudes of the first, respectively the second, kind* in R_1^2.

Split complex numbers were introduced by W. K. Clifford [Cli2]. For the history of split complex numbers see [Ro4, p. 395].

1.1.3. Dual Numbers. The 2-algebra with basis $e_0 = 1,\ e_1 = \epsilon\,(\epsilon^2 = 0)$ is called the algebra of *dual numbers* and is denoted by $\mathbb{C}^0$. Every dual number $x = x_0 + \epsilon x_1$ has a conjugate dual number $\bar{x} = x_0 - \epsilon x_1$. In this algebra the equalities (1.1) and (1.3) are fulfilled, but (1.2) is replaced by

$$|x|^2 = \bar{x}x = x_0^2. \tag{1.15}$$

Therefore, the modulus $|x|$ of a dual number x is real and can be equal to 0 for $x \neq 0$. A dual number x with $|x| \neq 0$ has an inverse number (1.4). Dual numbers x with $|x| = 0$ have the form ϵx_1 and are zero divisors. Therefore, *the algebra $\mathbb{C}^0$ is not a field.*

Theorem 1.2. *An associative 2-algebra is isomorphic to one of the algebras $\mathbb{C}$, $\mathbb{C}'$ or $\mathbb{C}^0$.*

PROOF. Let us write the elements in the associative 2-algebra as $a + \theta b$. Then θ^2 can be written as $\theta^2 = p + \theta q$ and hence θ satisfies the quadratic equation $\theta^2 - p - \theta q = 0$. The expression $N = 4p - q^2$ can be < 0, > 0 or $= 0$. In these cases the elements

$$i = \frac{1}{\sqrt{-N}}(q - 2\theta), \quad e = \frac{1}{\sqrt{-N}}(q - 2\theta) \ \text{ and } \epsilon = q - 2\theta \tag{1.16}$$

satisfy the conditions $i^2 = -1$, $e^2 = +1$ and $\epsilon^2 = 0$, respectively. ■

If we regard the modulus $|y-x|$ of the difference $y-x$ between two dual numbers x and y as the *distance* between these numbers, this distance, d, between two dual numbers $x = x_0 + \epsilon x_1$ and $y = y_0 + \epsilon y_1$ is equal to 0 if $x_0 = y_0$. Therefore, for such numbers we define a second distance $d' = |y_1 - x_1|$. Thus we introduce into the algebra $\mathbb{C}^0$ the *quasimetric* with two distances d and d'. The plane with this quasimetric is called the *flag plane* and is denoted by F^2. We will consider this plane and other spaces with analogous quasimetrics in Chapter 5.

Motions of the first and the second kind in F^2 can be expressed by functions of a dual variable (1.9) with $|a| = 1$.

Dual numbers, like split complex numbers, were introduced by W. K. Clifford [Cli2]. For the history of dual numbers see [Ro4, p. 395].

1.1.4. Automorphisms in 2-Algebras. In **0.1.4** we have defined *ring automorphisms* as bijections $f(x)$ from a ring onto itself satisfying the two conditions

$$f(x+y) = f(x) + f(y), \tag{1.17}$$

$$f(xy) = f(x)f(y). \tag{1.18}$$

We will first find the ring automorphisms in the 1-algebra, that is, in the field $\mathbb{R}$.

Theorem 1.3. *The only ring automorphism in the field* $\mathbb{R}$ *is the identity,* $f(x) = x$.

PROOF. Applying the condition (1.17) to both sides of the equality $x + 0 = x$, we obtain the equality $f(x) + f(0) = f(x)$, hence $f(0) = 0$.

Applying the condition (1.18) to both sides of the equality $x \cdot 1 = x$, we obtain the equality $f(x) \cdot f(1) = f(x)$, hence $f(1) = 1$.

Since a natural number n can be written as $1 + 1 + \cdots + 1$, we obtain $f(n) = f(1) + f(1) + \cdots + f(1) = 1 + 1 + \cdots + 1 = n$.

Therefore, for a negative integer $x = -n$, since $x+n = 0$, we obtain $f(x)+n = 0$, hence $f(-n) = -n$.

For $x = \frac{1}{m}$, since $x \cdot m = 1$, we obtain $f(x) \cdot m = 1$, hence $f(\frac{1}{m}) = \frac{1}{m}$.

Since a rational number $\frac{n}{m}$ can be written as $\frac{1}{m} + \frac{1}{m} + \cdots + \frac{1}{m}$, we obtain $f(\frac{n}{m}) = f(\frac{1}{m}) + f(\frac{1}{m}) + \cdots + f(\frac{1}{m}) = \frac{1}{m} + \frac{1}{m} + \cdots + \frac{1}{m} = \frac{n}{m}$.

Therefore, for every rational x we obtain $f(x) = x$.

If x is positive, we can write it as $(\sqrt{x})^2$ and, therefore, $f(x) = [f(\sqrt{x})]^2$, that is, $f(x) > 0$ for $x > 0$. So, if $x_1 > x_2$, $f(x_1 - x_2) = f(x_1) - f(x_2) > 0$, that is, $f(x_1) > f(x_2)$.

Now, if the irrational number x lies between the rational numbers $\frac{n}{m}$ and $\frac{n+1}{m}$, $f(x)$ lies between the same rational numbers. Hence, since n can be any natural number, $f(x)$ coincides with x for all real numbers. ■

Theorem 1.4. *The continuous ring automorphisms in the field* $\mathbb{C}$ *or the algebra* $\mathbb{C}'$ *are* $f(x) = x$ *and* $f(x) = \bar{x}$.

The continuous ring automorphisms in the algebra $\mathbb{C}^0$ *have the form*
$f_t(x_0 + \epsilon x_1) = x_0 + \epsilon t x_1$, *where* t *is an arbitrary real number.*

PROOF. As in Theorem 1.3 we prove that $f(x) = x$ for every rational x. The continuity of $f(x)$ then implies $f(x) = x$ for all real x (note that in Theorem 1.3 continuity of the automorphisms was not assumed).

In the case of $\mathbb{C}$ we apply (1.18) to both sides of the equality $i^2 = -1$, to obtain $f(i)^2 = -1$, hence $f(i) = \pm i$.

In the case of $\mathbb{C}'$ we apply (1.18) to both sides of the equality $e^2 = +1$, to obtain $f(e)^2 = +1$, hence $f(e) = \pm 1$ or $\pm e$. Since $f(x)$ is equal to ± 1 only for $x = \pm 1$, we have $f(e) = \pm e$.

In the case of $\mathbb{C}^0$ we apply (1.18) to both sides of the equality $\epsilon^2 = 0$, to obtain $f(\epsilon)^2 = 0$, hence $f(\epsilon) = t\epsilon$.

Since we introduced in the algebra $\mathbb{C}^0$ the quasimetric of the plane F^2, the continuous automorphisms in this algebra preserve the distance d and do not preserve the distance d'. Therefore, the continuous automorphisms $f_t(x)$ in $\mathbb{C}^0$ that do preserve the distance d' are called *metric automorphisms*. These automorphisms occur for $t = \pm 1$, and have the form $f(x) = x$ or $\bar{x}$. ∎

In **0.1.7** we have defined *algebra automorphisms*. Note that the continuous ring automorphisms in the algebras $\mathbb{C}$ and $\mathbb{C}'$ and the metric ring automorphisms in the algebra $\mathbb{C}^0$ are algebra automorphisms in these algebras.

1.1.5. Differentiable and Analytic Functions. It is possible to define differentiable functions in a commutative associative algebra. In **0.6.6** we have defined differentiable vector functions $y = f(x)$ on linear spaces L^n, the derivative of such a function being a linear operator A such that $dy = Adx$. If the space L^n is an n-algebra, a function $y = f(x)$ on this algebra is called differentiable if for any x there exists an element a in this algebra such that $dy = adx$.

Theorem 1.5 (G. SCHEFFERS [Scf]). *Conditions for the differentiability of a function $y = f(x)$ on a commutative associative algebra $\mathbb{A}$ with basis e_i, i.e. when the law of multiplication of basis elements is expressed by (0.14), have the form*

$$\frac{\partial y^i}{\partial x^h} C^h_{jk} = \frac{\partial y^h}{\partial x^k} C^i_{jh}. \tag{1.19}$$

PROOF. If x and y are written in the form $x = x^i e_i$ and $y = y^i e_i$, then the matrix of the operator A such that $dy = Adx$ is the Jacobi matrix $\left(\frac{\partial y^i}{\partial x^j}\right)$. If the relation $dy = Adx$ is determined by the relation $dy = adx$, where $a = a^i e_i$ is an element in the algebra, this matrix can also be written as $(C^i_{jk} a^k)$. Since $\mathbb{A}$ is associative, for any three basis elements e_j, e_k, e_l we have $(e_j e_k) e_l = e_j (e_k e_l)$ and, therefore, $C^h_{jk} C^i_{hl} = C^i_{jh} C^h_{kl}$. If we take the convolution of both sides of this equality with a^k, replace the matrix $(C^i_{jk} a^k)$ by the Jacobi matrix and use the commutativity in $\mathbb{A}$, we obtain (1.19). ∎

For the algebras $\mathbb{C}$, $\mathbb{C}'$ and $\mathbb{C}^0$, whose elements can be written as $x_0 + \theta x_1$, where θ is i, e or ϵ, respectively, the conditions (1.19) can be written in the form

$$\frac{\partial y_0}{\partial x_0} = \frac{\partial y_1}{\partial x_j}, \quad \frac{\partial y_0}{\partial x_1} = \theta^2 \frac{\partial y_1}{\partial x_0}. \tag{1.20}$$

For the field $\mathbb{C}$ the conditions (1.20) coincide with the *Cauchy-Riemann conditions*. For the algebra $\mathbb{C}'$ the conditions (1.20) are equivalent to the conditions $\frac{\partial y_+}{\partial x_-} = \frac{\partial y_-}{\partial x_+} = 0$, where x_+, x_-, y_+, y_- are the coordinates of the split complex numbers x and y in the basis e_+, e_-. In these three cases the Jacobi matrices belong to the Lie algebras of the groups of similitudes-rotations of the planes R^2, R_1^2 or F^2, respectively (in the first two cases this matrix has the form (0.77)).

It is possible to define analytic functions in a commutative associative algebra:

$$y = f(x) = \sum_k a_k x^k. \tag{1.21}$$

For the algebra $\mathbb{C}^0$, setting $a_k = A_k + \epsilon B_k$, the functions $y_0 = F(x_0, x_1)$ and $y_1 = G(x_0, y_1)$, have the form

$$\begin{aligned} y_0 &= \sum_k A_k x_0^k = F(x_0), \\ y_1 &= \sum_k B_k x_0^k + x_1 \sum_k A_k x_0^{k-1} = G(x_0) + x_1 F'(x_0). \end{aligned} \tag{1.22}$$

For them (1.20) for $\theta = \epsilon$ clearly holds.

In particular,

$$e^{x_0 + \epsilon x_1} = e^{x_0}(1 + \epsilon x_1), \tag{1.23}$$

$$\sin(x_0 + \epsilon x_1) = \sin x_0 + \epsilon x_1 \cos x_0, \quad \cos(x_0 + \epsilon x_1) = \cos x_0 - \epsilon x_1 \sin x_0. \tag{1.24}$$

1.1.6. Tensor Products of Commutative Algebras. The *tensor product* of an n-algebra $\mathbb{A}$ and an m-algebra $\mathbb{B}$ with bases $e_1, e_2, \ldots, e_n$ and $f_1, f_2, \ldots, f_m$ is the nm-algebra $\mathbb{A} \otimes \mathbb{B}$ with basis elements $e_1 f_\alpha = f_\alpha e_i$.

The elements in the tensor product $\mathbb{C} \otimes \mathbb{C}$ are called *bicomplex numbers.* Analogously, the elements in the tensor products $\mathbb{C} \otimes \mathbb{C}'$, $\mathbb{C} \otimes \mathbb{C}^0$, $\mathbb{C}' \otimes \mathbb{C}^0$, $\mathbb{C}^0 \otimes \mathbb{C}^0$ are called *dycomplex, duocomplex, dydual,* and *duodual numbers.*

Theorem 1.6. *The algebras $\mathbb{C} \otimes \mathbb{C}$ and $\mathbb{C} \otimes \mathbb{C}'$ are isomorphic to the direct sum $\mathbb{C} \oplus \mathbb{C}$.*

The algebra $\mathbb{C}' \otimes \mathbb{C}'$ is isomorphic to the direct sum $\mathbb{R} \oplus \mathbb{R} \oplus \mathbb{R} \oplus \mathbb{R}$.

The algebra $\mathbb{C}' \otimes \mathbb{C}^0$ is isomorphic to the direct sum $\mathbb{C}^0 \oplus \mathbb{C}^0$.

PROOF. If we denote the basis in the algebra $\mathbb{C} \otimes \mathbb{C}$ by $1, i, I$, with $iI = Ii$, the square $(iI)^2$ is equal to $+1$. Therefore, iI can be regarded as the basis element e in $\mathbb{C}'$. This proves that $\mathbb{C} \otimes \mathbb{C}$ and $\mathbb{C} \otimes \mathbb{C}'$ are isomorphic.

Theorem 1.1 implies that the tensor product of any algebra $\mathbb{A}$ and $\mathbb{C}'$ is isomorphic to the direct sum $\mathbb{A} \oplus \mathbb{A}$, hence follow the remaining assertions of the theorem.■

A bicomplex number $x = x_0 + ix_1 + Ix_3 + iIx_4$ has three *conjugate bicomplex numbers*

$$\left.\begin{aligned} \bar{x} &= x_0 - ix_1 + Ix_3 - iIx_4, \\ \tilde{x} &= x_0 + ix_1 - Ix_3 - iIx_4, \\ \approx\!\!\!\!x &= x_0 - ix_1 - Ix_3 + iIx_4, \end{aligned}\right\} \tag{1.25}$$

and following equalities hold:

$$\left.\begin{aligned} \overline{xy} &= \bar{y}\,\bar{x}, \\ \widetilde{xy} &= \tilde{y}\,\tilde{x}, \\ \widetilde{\overline{xy}} &= \tilde{\bar{y}}\,\tilde{\bar{x}}\,. \end{aligned}\right\} \tag{1.26}$$

The products $\bar{x}x$, $\tilde{x}x$, and $\tilde{\bar{x}}\,x$ are the squares of three kinds of moduli of the bicomplex number x: $\bar{x}x$ is the square of the modulus $|x|$, $\tilde{x}x$ is the square of the modulus $\wr x \wr$, and $\tilde{\bar{x}}\,x$ is the square of the modulus $\wr |x| \wr$. For them the equality (1.3) and two analogous equalities hold.

Analogously we can define conjugate elements and moduli of elements of other tensor products of 2-algebras.

1.1.7. Cyclic, Anticyclic, and Plural Numbers. The algebra $\mathbb{C}'$ is a particular case of the algebra $\mathbb{Z}\mathbf{y}_m$ of *cyclic numbers*

$$x = x_0 + x_1 e + x_2 e^2 + \cdots + x_{m-1} e^{m-1}, \quad e^m = +1. \tag{1.27}$$

The field $\mathbb{C}$ is a particular case of the algebra $\mathbb{Z}\mathbf{y}'_m$ of *anticyclic numbers*

$$x = x_0 + x_1 i + x_2 i^2 + \cdots + x_{m-1} i^{m-1}, \quad i^m = -1. \tag{1.28}$$

The algebra $\mathbb{C}^0$ is a particular case of the algebra $\mathbb{Z}\mathbf{y}^0_m$ of *plural numbers*

$$x = x_0 + x_1 \epsilon + x_2 \epsilon^2 + \cdots + x_{m-1} \epsilon^{m-1}, \quad \epsilon^m = 0. \tag{1.29}$$

The algebra $\mathbb{C}'$ is also a particular case of the algebra $\mathbb{R}^m$ of *m-multiple numbers*

$$x = x_1 e_1 + x_2 e_2 + \cdots + x_m e_m, \quad e_i^2 = +1, \quad e_i e_j = 0, \tag{1.30}$$

which is isomorphic to the direct sum of m fields $\mathbb{R}$.

The algebra $\mathbb{C} \otimes \mathbb{C}$ is a particular case of the algebra $\mathbb{C}^m$ of *m-multiple complex numbers.* An m-multiple complex number has the form (1.30) but all coordinates x_i of elements of this algebra are complex numbers; this algebra is isomorphic to the direct sum of m fields $\mathbb{C}$.

The algebra $\mathbb{Z}\mathbf{y}_3$ is isomorphic to the *triple algebra of A. De Morgan* and to the algebra of *triplets of Ch. Graves*, see [Ro4, pp. 382–384].

The algebras $\mathbb{R}^3$ and $\mathbb{R}^4$ are called the algebras of *triple* and *quadruple numbers*, respectively.

Theorem 1.7. *The tensor products $\mathbb{Z}\mathbf{y}_m \otimes \mathbb{C}$ and $\mathbb{Z}\mathbf{y}'_m \otimes \mathbb{C}$ are isomorphic to the algebra $\mathbb{C}^m$.*

PROOF. That the algebra $\mathbb{Z}\mathbf{y}_m \otimes \mathbb{C}$ is isomorphic to $\mathbb{C}^m$ follows from the fact that the elements

$$\left.\begin{aligned} e_1 &= \frac{1}{m}(1+e+e^2+\cdots+e^{m-1}),\\ e_2 &= \frac{1}{m}(1+\omega_m e+\omega_m^2 e^2+\cdots+\omega_m^{m-1}e^{m-1}),\\ e_3 &= \frac{1}{m}(1+\omega_m^2 e+\omega_m^4 e^2+\cdots+\omega_m^{m-2}e^{m-1}),\\ &\vdots\\ e_m &= \frac{1}{m}(1+\omega_m^{m-1} e+\omega_m^{m-2} e^2+\cdots+\omega_m e^{m-1}) \end{aligned}\right\} \tag{1.31}$$

in this algebra, where ω_m is a primitive m-th root of 1, for instance, $\cos\frac{2\pi}{m}+I\sin\frac{2\pi}{m}$, satisfy the conditions $e_i^2=+1$, $e_ie_j=0$.

If we replace in formulas (1.31) e by $i\omega_{2m}$ we obtain the elements of the algebra $\mathbb{Z}\mathbf{y}'_m\otimes\mathbb{C}$ satisfying the conditions $e_i^2=+1$, $e_ie_j=0$. Hence the algebras $\mathbb{Z}\mathbf{y}_m\otimes\mathbb{C}$ and $\mathbb{C}^m$ are isomorphic. ■

Theorem 1.8. *The algebra $\mathbb{Z}\mathbf{y}'_{2m}$ is isomorphic to the algebra $\mathbb{C}^m$.*
The algebra $\mathbb{Z}\mathbf{y}_{2m}$ is isomorphic to the algebra $\mathbb{C}^{m-1}\oplus\mathbb{C}'$.
The algebras $\mathbb{Z}\mathbf{y}_{2m-1}$ and $\mathbb{Z}\mathbf{y}'_{2m-1}$ are isomorphic to the algebra $\mathbb{C}^{m-1}\oplus\mathbb{R}$.

PROOF. Since the algebras $\mathbb{Z}\mathbf{y}_n\otimes\mathbb{C}$ and $\mathbb{Z}\mathbf{y}'_n\otimes\mathbb{C}$ are isomorphic to the algebra $\mathbb{C}^n$, the algebras $\mathbb{Z}\mathbf{y}_n$ and $\mathbb{Z}\mathbf{y}'_n$ are isomorphic to subalgebras of $\mathbb{C}^n$.

The algebra $\mathbb{Z}\mathbf{y}'_n$ can be represented by the set of linear combinations of basis elements e_i in $\mathbb{C}^n$ with real coefficients. But since $\overline{\omega}_n=\omega_n^{n-1}$ and $\overline{\omega}_n^k=\omega_n^{n-k}$, the pairs of basis elements e_k and e_{n-k-2}, provided $k\neq 1, \frac{n}{2}+1$, are complex conjugates and are basis elements in $\mathbb{C}$, and the basis elements e_i with $i=1, \frac{n}{2}+1$ are real and are basis elements in $\mathbb{R}$. Therefore, since the elements e_i are basis elements in the algebra $\mathbb{R}^n$, the algebra $\mathbb{Z}\mathbf{y}'_n$ is isomorphic to $\mathbb{C}^{m-1}\oplus\mathbb{R}\oplus\mathbb{R}=\mathbb{C}^{m-1}\oplus\mathbb{C}'$ if $n=2m$, and to $\mathbb{C}^{m-1}\oplus\mathbb{R}$ if $n=2m-1$.

For the algebra $\mathbb{Z}\mathbf{y}_{2m}$, since its basis element i can be represented by $e\overline{\omega}_{2m}$, all basis elements in $\mathbb{Z}\mathbf{y}_{2m}$ occur in pairs of complex conjugates and this algebra is isomorphic to $\mathbb{C}^m$. ■

In particular, the algebras $\mathbb{Z}\mathbf{y}_3$ and $\mathbb{Z}\mathbf{y}'_3$ are isomorphic to the direct sum $\mathbb{C}\oplus\mathbb{R}$. If we regard elements $x_1+x_2e+x_3e^2$ in $\mathbb{Z}\mathbf{y}_3$ as points X in the Euclidean space R^3 with coordinates x_1,x_2,x_3, the subalgebras $\mathbb{R}$ and $\mathbb{C}$ of $\mathbb{Z}\mathbf{y}_3$ can be regarded as a line L and a plane P through the point O with zero coordinates. The line L is determined by the real vector $e_1=\frac{1}{3}(1+e+e^2)$, the plane P is determined by the two complex conjugate vectors $e_2=\frac{1}{3}(1+\omega_3e+\omega_3^2e^2)$ and $e_3=\frac{1}{3}(1+\omega_3^2e+\omega_3e^2)$ or by the real vectors $E=\frac{1}{2}(e_2+e_3)=\frac{1}{3}(1-e-e^2)$, and $I=\frac{1}{2i}(e_2-e_3)=\frac{1}{\sqrt{3}}(e-e^2)$. The orthogonal projections from the point X onto the line L and onto the plane P are homomorphisms of the algebra $\mathbb{Z}\mathbf{y}_3$ onto the fields $\mathbb{R}$ and $\mathbb{C}$, respectively (see [Ro4, p. 384]).

1.1.8. Polynomial and Group Algebras. The algebras $\mathbb{Z}\mathbf{y}_m$, $\mathbb{Z}\mathbf{y}'_m$ and $\mathbb{Z}\mathbf{y}^0_m$ are particular cases of *polynomial algebras*, that is, algebras of expressions (1.27) where e satisfies an algebraic equation $P_m(x)=0$ ($P_m(x)$ is a polynomial of degree m). A. A. Albert [Al1, pp. 245–250] has proved that if the polynomial $P_m(x)$ has the form

$$P_m(x) = (x-\alpha_1)^{k_1} \dots (x-\alpha_r)^{k_r} (x-\beta_1)^{\ell_1} \dots (x-\beta_s)^{\ell_s} (x-\overline{\beta}_1)^{\ell_1} \dots (x-\overline{\beta}_s)^{\ell_s}, \quad (1.32)$$

where α_i and β_j are the real and complex roots of $P_m(x)$, respectively ($m = k_1 + \dots + k_r + 2l_1 + \dots + 2l_s$), then the polynomial algebra determined by this polynomial is isomorphic to the direct sum of the r algebras $\mathbb{Z}\mathbf{y}^0_{k_1}, \dots, \mathbb{Z}\mathbf{y}^0_{k_r}$ and the s algebras $\mathbb{Z}\mathbf{y}^0_{\ell_1} \otimes \mathbb{C}, \dots, \mathbb{Z}\mathbf{y}^0_{\ell_s} \otimes \mathbb{C}$. If k_i or ℓ_j is equal to 1, the corresponding algebra $\mathbb{Z}\mathbf{y}^0_{k_i}$ or $\mathbb{Z}\mathbf{y}^0_{\ell_j} \otimes \mathbb{C}$ is replaced by $\mathbb{R}$, respectively $\mathbb{C}$.

The algebras $\mathbb{Z}\mathbf{y}_m$ are particular cases of *group algebras*, that is, algebras whose basis elements form a group (for the algebras $\mathbb{Z}\mathbf{y}_m$ this group is the cyclic group Z_m, hence the name of these algebras).

1.1.9. Simple and Semisimple Algebras. In **0.8.3** we have defined simple and semisimple Lie groups. Analogously, an algebra is called *simple* if it has no ideals distinct from the whole algebra and from its zero element. A direct sum of simple algebras is called a *semisimple algebra.* Semisimple algebras can also be defined as algebras whose *radical* is the zero element. The *radical* is the ideal consisting of nilpotent elements, i.e. elements some power of which is equal to zero. If J is the radical in a non-semisimple algebra $\mathbb{A}$, the quotient algebra $\mathbb{A}/J$ is a semisimple associative algebra.

The criterion of Th. Molien [Mol] for semisimplicity of algebras is that the form (0.71) is nondegenerate, where C^k_{ij} are the structure constants of the algebra. This criterion resembles that for semisimplicity of Lie groups.

K. Weierstrass [Wei] has proved that the only simple commutative associative algebras are the fields $\mathbb{R}$ and $\mathbb{C}$. Therefore, the algebras $\mathbb{C}'$, $\mathbb{C}\otimes\mathbb{C}$, $\mathbb{C}'\otimes\mathbb{C}'$, $\mathbb{R}^n$, $\mathbb{C}^n$, $\mathbb{Z}\mathbf{y}_m$, and $\mathbb{Z}\mathbf{y}'_m$, which are isomorphic to direct sums of these fields, are semisimple.

1.1.10. Quasisimple and r-Quasisimple Algebras. If a semisimple associative algebra $\mathbb{A}$ with basis e_i ($i = 1, 2, \dots, n$) has an involutive automorphism

$$e_a \to e_a, \quad e_u \to -e_u \quad (1.33)$$

fixing basis elements e_a ($a = 1, 2, \dots, m$) and multiplying basis elements e_u ($u = m+1, \dots, n$) by -1, then (0.14), expressing products of basis elements of $\mathbb{A}$, can be rewritten in the form

$$e_a e_b = C^c_{ab} e_c, \quad e_a e_u = C^v_{au} e_v, \quad e_u e_v = C^c_{uv} e_c. \quad (1.34)$$

The sets of elements e_a and e_u are sets of basis elements of two linear subspaces, $\mathbb{B}$ and $\mathbb{E}$, of $\mathbb{A}$, and $\mathbb{A}$ can be written as the direct sum $\mathbb{B} \oplus \mathbb{E}$. The first formula (1.34) shows that $\mathbb{B}$ is a subalgebra of $\mathbb{A}$. The formulas (1.34) are equivalent to the symbolic formulas

$$\mathbb{B}^2 \subset \mathbb{B}, \quad \mathbb{B}\mathbb{E} \subset \mathbb{E}, \quad \mathbb{E}^2 \subset \mathbb{B}. \quad (1.35)$$

If we replace the constants C^c_{uv} by their products by a real number t, $0 < t \le 1$, we obtain algebras which are isomorphic to $\mathbb{A}$, but if we replace these constants by 0, we obtain a new algebra, $\mathbb{A}^0$, for which formulas (0.14) take the form

$$e_a e_b = C^c_{ab} e_c, \quad e_a e_u = C^v_{au} e_v, \quad e_u e_v = 0. \quad (1.36)$$

This new algebra $\mathbb{A}^0$ is called a *quasisimple associative algebra.* This algebra is not semisimple and its radical coincides with the subspace $\mathbb{E}$; the quotient algebra $\mathbb{A}/\mathbb{E}$ is isomorphic to $\mathbb{B}$. The formulas (1.36) are equivalent to the symbolic formulas

$$\mathbb{B}^2 \subset \mathbb{B}, \quad \mathbb{B}\mathbb{E} \subset \mathbb{E}, \quad \mathbb{E}^2 = 0. \tag{1.37}$$

The algebras $\mathbb{C}^0$, $\mathbb{C} \otimes \mathbb{C}^0$, $\mathbb{C}' \otimes \mathbb{C}^0$ are quasisimple. Their radicals consist of the elements ϵx, their subalgebras $\mathbb{B}$ are isomorphic to $\mathbb{R}$, $\mathbb{C}$ and $\mathbb{C}'$, respectively.

The quasisimple algebra $\mathbb{A}^0$ obtained from a semisimple algebra $\mathbb{A} = \mathbb{B} \oplus \mathbb{E}$ can be written as $\mathbb{A} = \mathbb{B} \oplus \epsilon\mathbb{E}$, where ϵ is the basis element of the algebra $\mathbb{C}^0$ of dual numbers.

If we apply this process to a quasisimple associative algebra, we obtain a *biquasisimple associative algebra.* An example of such an algebra is the algebra $\mathbb{C}^0 \otimes \mathbb{C}^0$.

If we apply this process r times to a semisimple associative algebra, we obtain an *r-quasisimple associative algebra.* An example of such an algebra is the tensor product of r algebras $\mathbb{C}^0$.

1.1.11. Hjelmslev Algebras. A non-semisimple associative algebra $\mathbb{A}$ is said to be a *Hjelmslev algebra* if the following conditions hold:

$H.1°$ Each element x in $\mathbb{A}$ that is not a zero divisor has an inverse element x^{-1}.

$H.2°$ The zero divisors form an ideal J such that the quotient algebra $\mathbb{A}/J$ is a field.

$H.3°$ For any two zero divisors, one is the product of the other by a nonzero element.

The term "Hjelmslev algebras" is connected with the Hjelmslev spaces introduced by J. Hjelmslev [Hje]. The term "Hjelmslev algebras" was introduced by W. Klingenberg [Kli].

$\mathbb{C}^0$ and $\mathbb{C} \otimes \mathbb{C}^0$ are Hjelmslev algebras, since for them $\mathbb{A}/J$ is isomorphic to the fields $\mathbb{R}$ and $\mathbb{C}$, respectively. $\mathbb{C}' \otimes \mathbb{C}^0$ and $\mathbb{C}^0 \otimes \mathbb{C}^0$ are not Hjelmslev algebras, since in $\mathbb{C}' \otimes \mathbb{C}^0$ the set of zero divisors is not an ideal, while in $\mathbb{C}^0 \otimes \mathbb{C}^0$, elements ϵ and $\mathcal{E}$ of the respective tensor factors of this algebra are not related by an element a such that $\mathcal{E} = a\epsilon$ or $\epsilon = a\mathcal{E}$.

The algebra $\mathbb{Z}\mathbf{y}^0_m$ of plural numbers is a Hjelmslev algebra. Indeed, 1) its zero divisors have the form $x = x_k\epsilon^k$, where $0 < k < m$, and for an element $x = x_i\epsilon^i$ with $x_0 \neq 0$ there is an inverse element x^{-1}, so condition $H.1°$ is fulfilled. 2) All zero divisors of $\mathbb{Z}\mathbf{y}^0_m$ form an ideal J, and $\mathbb{Z}\mathbf{y}^0_m/J$ is isomorphic to the field $\mathbb{R}$. So, condition $H.2°$ is fulfilled. 3) Any two zero divisors $x = x_k\epsilon^k$, with $r \leq k < m$ and $y = y_l\epsilon^l$, with $s \leq k < m$, can, if $r < s$, be written in the form $x = \epsilon^r z$ and $y = \epsilon^r w$, where z is not a zero divisor and w can be a zero divisor. Therefore $\epsilon^r = xz^{-1}$ and $y = x(z^{-1}w)$, and condition $H.3°$ is fulfilled.

1.1.12. Frobenius Algebras. For all algebras considered by us the dimension of the algebra is equal to the sum of the dimensions of the radical and of the *annihilator* of this radical, that is, the set of elements whose products by all elements in the radical are equal to 0.

In semisimple algebras the radicals consist of 0 only, and their annihilators are algebras themselves.

In the algebras $\mathbb{Z}\mathbf{y}^0_m$ the radical consists of the elements $x = x_k\epsilon^k$ with $0 < k < m$, and its annihilator consists of the elements $a\epsilon^{m-1}$.

In $\mathbb{C}^0 \otimes \mathbb{C}^0$ the radical consists of the elements $x_1\epsilon + x_2\mathcal{E} + x_3\epsilon\mathcal{E}$, and its annihilator consists of the elements $a\epsilon\mathcal{E}$.

It is possible to prove that in algebras with this property there exist covectors ℓ_i such that the bilinear form

$$\Phi(x,y) = C^k_{ij}\ell_k x^i y^j = D_{ij}x^i y^j \tag{1.38}$$

is nondegenerate and has the property of "associativity", that is,

$$\Phi(xy,z) = \Phi(x,yz). \tag{1.39}$$

Algebras having a nondegenerate form $\Phi(x,y)$ with the property of "associativity" are called *Frobenius algebras* (some properties of these algebras were found by G. Frobenius [Fro2] for group algebras, see **1.1.8** ; for more on Frobenius algebras see the book by C. W. Curtis and I. Reiner [CuRe, pp. 413–415]).

The nonsingular matrix D determined by (1.38) is related to the matrices $A_l = (C^i_{kj}a^k)$ and $A_r = (C^i_{jk}a^k)$ by $A_l = D^{-1}A_rD$, which follows from the equality $C^h_{jk}C^i_{h\ell} = C^i_{jh}C^h_{k\ell}$ considered in **1.1.5** if we take the convolution of both sides of this equality with the vector a^k and the covector ℓ_i.

For the semisimple algebras $\mathbb{C}$, $\mathbb{C}'$ and their tensor products, and also for $\mathbb{Z}\mathbf{y}_m$ and $\mathbb{Z}\mathbf{y}'_m$ we can take $\mathrm{Re}(xy)$ as (1.38).

For $\mathbb{C}$ and $\mathbb{C}'$ these forms are

$$\Phi(x,y) = x_0y_0 - x_1y_1, \quad \Phi(x,y) = x_0y_0 + x_1y_1. \tag{1.40}$$

For $\mathbb{C}\otimes\mathbb{C}$ and $\mathbb{C}'\otimes\mathbb{C}'$ these forms are

$$\begin{aligned} \Phi(x,y) &= x_0y_0 - x_1y_1 - x_2y_2 + x_3y_3, \\ \Phi(x,y) &= x_0y_0 + x_1y_1 + x_2y_2 + x_3y_3. \end{aligned} \tag{1.41}$$

For $\mathbb{Z}\mathbf{y}_m$ and $\mathbb{Z}\mathbf{y}'_m$ these forms are

$$\begin{aligned} \Phi(x,y) &= x_0y_0 + x_1y_{m-1} + x_2y_{m-2} + \cdots + x_{m-1}y_1, \\ \Phi(x,y) &= x_0y_0 - x_1y_{m-1} - x_2y_{m-2} - \cdots - +x_{m-1}y_1. \end{aligned} \tag{1.42}$$

For the non-semisimple algebras considered by us the form $\mathrm{Re}(xy)$ is degenerate, and as (1.38) we can take the coefficient of the product xy at one of the basis elements of the radical. For instance, for $\mathbb{C}^0$ we can take the coefficient at ϵ:

$$\Phi(x,y) = x_0y_1 + x_1y_0; \tag{1.43}$$

for $\mathbb{C}^0\otimes\mathbb{C}^0$ that at $\epsilon\mathcal{E}$:

$$\Phi(x,y) = x_0y_3 + x_1y_2 + x_2y_1 + x_3y_0; \tag{1.44}$$

and for $\mathbb{Z}\mathbf{y}^0_m$ that at ϵ^{m-1}:

$$\Phi(x,y) = x_0y_{m-1} + x_1y_{m-2} + x_2y_{m-3} + \cdots + x_{m-1}y_0. \tag{1.45}$$

That these forms have the property of "associativity" (1.39) follows from the associativity of multiplication in the algebras.

§1.2. Noncommutative Associative Algebras

1.2.1. Quaternions. In **0.1.7** we mentioned that the skew field $\mathbb{H}$ of *quaternions* is a noncommutative associative 4-algebra with basis $e_0 = 1$, $e_1 = i$, $e_2 = j$, $e_3 = k$ $(i^2 = j^2 = k^2 - 1,\ ij = -ji = k,\ jk = -kj = i,\ ki = -ik = j)$.

Each quaternion $x = x_0 + ix_1 + jx_2 + kx_3$ has a *conjugate quaternion* $\overline{x} = x_0 - ix_1 - jx_2 - kx_3$, and the following equalities hold: (1.1),

$$|x|^2 = \overline{x}x = x_0^2 + x_1^2 + x_2^2 + x_3^2, \tag{1.46}$$

and (1.3).

The nonnegative real number $|x|$ is called the *modulus* of the quaternion x. The modulus $|x|$ is equal to 0 only if $x = 0$. A quaternion $x \neq 0$ has an inverse element (1.4).

If we regard the modulus $|y-x|$ of the difference $y-x$ between two quaternions x and y as the *distance* between them, we introduce into $\mathbb{H}$ the metric of the Euclidean space R^4.

Quaternions can be also regarded as vectors in R^4, and the *inner product* (x, y) of two vectors x and y has the form (1.5). Therefore, the angle φ between two vectors x and y can be defined by (1.7).

The *motions of the first and the second kind* in R^4 can be expressed by functions of a quaternionic variable:

$$x' = axb + c \text{ and } x' = a\overline{x}b + c, \tag{1.47}$$

with $|a| = |b| = 1$, respectively.

The functions (1.47) with $|a|$ and $|b|$ not equal to 1 express *similitudes of the first,* respectively, *the second, kind* in R^4.

Quaternions were defined by W. R. Hamilton [Ham] as sums of scalars and vectors, defined in the same work. For the history of quaternions see [Ro4, pp. 385–386].

Note that, according to a *theorem of G. Frobenius* [Fro1], an associative division algebra, that is, an associative algebra which is a field or a skew field, is either the field $\mathbb{R}$, the field $\mathbb{C}$, or the skew field $\mathbb{H}$. According to a *theorem of L. S. Pontryagin* [Pon1], these three division algebras are the only connected locally compact topological fields or skew field (defined analogously to topological groups, see **0.8.1**).

The *theorem of A. Hurwitz* [Hur], asserting that the only positive definite quadratic forms whose products are also quadratic forms are binary, quaternary and octonary forms, implies the *theorem of Yu. V. Linnik* [Lin] that the division algebras $\mathbb{C}$ and $\mathbb{H}$ are the only associative algebras with 1 and with involution $x \to \overline{x}$, in which the elements x for which $x = \overline{x}$ are products of 1 by a real number and in which the products $\overline{x}x$ of these numbers are positive definite quadratic forms admitting composition.

1.2.2. Matrices. In **0.8.2** we mentioned the n^2-algebra $\mathbb{R}_n$ of real square $(n \times n)$-*matrices*, which is isomorphic to the algebra of endomorphisms and linear operators in the space L^n. A basis in this algebra is given by the matrices E_i^j with

1 at the intersection of the i-th row and the j-th column and with all other entries equal to 0. The rule of multiplication of basis elements is

$$E_i^j E_k^\ell = \delta_i^\ell E_k^j, \tag{1.48}$$

where δ_k^ℓ is equal to 1 if $k = \ell$ and equal to 0 if $k \neq \ell$. Therefore, the product of the matrices $A = (A_j^i)$ and $B = (B_\ell^k)$ is the matrix C with entries

$$C_j^k = A_h^k B_j^h. \tag{1.49}$$

We can associate with every matrix A a real number, $|A| = \det A$, called its *determinant.* The determinant of the product of two matrices is equal to the product of their determinants,

$$|AB| = |A|\,|B|. \tag{1.50}$$

Any matrix $A = (A_j^i)$ has a *transposed matrix* $A^\top = (A_i^j)$ and an *adjoint matrix* A^*, obtained by transposition from the matrix whose entry $(A^*)_j^i$, is equal to the determinant of the $((n-1) \times (n-1))$-submatrix obtained from A by deleting the row and the column containing the entry A_i^j, multiplied by $(-1)^{i+j}$.

The following properties hold:

$$(AB)^\top = B^\top A^\top, \quad (AB)^* = B^* A^* \tag{1.51}$$

and

$$AA^* = |A| \cdot I, \tag{1.52}$$

where I is the identity (unit) matrix. Therefore, a matrix A with nonzero determinant has an inverse:

$$A^{-1} = \frac{A^*}{|A|}. \tag{1.53}$$

The matrices with nonzero determinant are called *nonsingular matrices*; they correspond to invertible linear operators and form the group GL_n with respect to multiplication.

The matrices with zero determinant are zero divisors; they are called *singular matrices.*

Theorem 1.9. *The algebra $\mathbb{R}_{mn}$ is isomorphic to the tensor product $\mathbb{R}_m \otimes \mathbb{R}_n$ of the algebras $\mathbb{R}_m$ and $\mathbb{R}_n$.*

PROOF. Let us write the indices I, $J = 1, 2, \ldots, mn$ as $i\alpha$, $j\beta$, where i, $j = 1, 2, \ldots, n$, $\alpha, \beta = 1, 2, \ldots, m$. Then the basis elements $E_{j\beta}^{i\alpha}$ in the algebra $\mathbb{R}_{mn}$ can be represented by products $E_j^i E_\beta^\alpha$ of basis elements in the algebras $\mathbb{R}_m$ and $\mathbb{R}_n$, hence the isomorphism between $\mathbb{R}_{mn}$ and the algebra $\mathbb{R}_m \otimes \mathbb{R}_n$. ■

Matrices were defined by A. Cayley [Cay2]. For the history of matrices see [Ro4, pp. 388–394].

Matrices can be defined over any field or algebra. We denote by $\mathbb{A}_n$ the algebra of $(n \times n)$-matrices over an algebra $\mathbb{A}$, that is, with entries that are elements in $\mathbb{A}$. It is obvious that the algebra $\mathbb{A}_n$ can be regarded as the tensor product $\mathbb{A} \otimes \mathbb{R}_n$.

If $\mathbb{A}$ is an m-algebra, then $\mathbb{A}_n$, regarded as an algebra over $\mathbb{R}$, is an mn^2-algebra.

Theorem 1.10. *The algebras $\mathbb{C}$ and $\mathbb{C}'$ are isomorphic to the subalgebras of the algebra $\mathbb{R}_2$ consisting of the (2×2)-matrices having the first, respectively the second, form in (0.77).*

The algebras $\mathbb{C}_n$ and $\mathbb{C}'_n$ are isomorphic to the subalgebras of the algebra $\mathbb{R}_{2n}$ consisting of the $(2n \times 2n)$-matrices having the first, respectively the second, form in (0.77). Here, A and B are matrices in $\mathbb{R}_n$.

PROOF. The real (2×2)-matrices in (0.77) are linear combinations of the basis matrices

$$\begin{bmatrix} 1 & 0 \\ 0 & 1 \end{bmatrix}, \quad \begin{bmatrix} 0 & 1 \\ -1 & 0 \end{bmatrix} \quad \text{and} \quad \begin{bmatrix} 1 & 0 \\ 0 & 1 \end{bmatrix}, \quad \begin{bmatrix} 0 & 1 \\ 1 & 0 \end{bmatrix}, \tag{1.54}$$

respectively, with multiplication rules which coincide with the multiplication rules of the basis elements $1, i$ and $1, e$ in the algebras $\mathbb{C}$ and $\mathbb{C}'$, respectively. Hence the subalgebras consisting of these matrices and the algebras $\mathbb{C}$ and $\mathbb{C}'$ are isomorphic.

The second assertion of the theorem follows since the subalgebra of $\mathbb{R}_{2n}$ consisting of the first, respectively the second, matrices in (0.77), as well as $\mathbb{C}_n$, respectively $\mathbb{C}'_n$, is isomorphic to the tensor product of the subalgebra of the first, respectively the second, (2×2)-matrices (0.77), and of $\mathbb{C}$, respectively $\mathbb{C}'$, by $\mathbb{R}_n$. ■

Theorem 1.11. *The algebra $\mathbb{C}^0$ is isomorphic to the subalgebra of the algebra $\mathbb{R}_2$ consisting of the (2×2)-matrices*

$$\mathcal{A} = \begin{bmatrix} A & B \\ 0 & A \end{bmatrix}. \tag{1.55}$$

The algebra $\mathbb{C}^0_n$ is isomorphic to the subalgebra of the algebra $\mathbb{R}_{2n}$ consisting of the $(2n \times 2n)$-matrices (1.55). Here A and B are matrices in $\mathbb{R}_n$.

PROOF. The real (2×2)-matrices (1.55) are linear combinations of the basis matrices

$$\begin{bmatrix} 1 & 0 \\ 0 & 1 \end{bmatrix}, \quad \begin{bmatrix} 0 & 1 \\ 0 & 0 \end{bmatrix} \tag{1.56}$$

with multiplication rules which coincide with the multiplication rules of the basis elements $1, \epsilon$ in the algebra $\mathbb{C}^0$. Hence the subalgebra consisting of these matrices and the algebra $\mathbb{C}^0$ are isomorphic. The second assertion of the theorem follows from the isomorphism of the subalgebra of $\mathbb{R}_{2n}$ consisting of the matrices (1.55), and of $\mathbb{C}^0_n$, to the tensor product of the subalgebra of (2×2)-matrices (1.55) by $\mathbb{R}_n$. ■

Theorem 1.12. *The skew field $\mathbb{H}$ is isomorphic to the subalgebra of the algebra $\mathbb{C}_2$ consisting of the (2×2)-matrices*

$$\mathcal{A} = \begin{bmatrix} A & B \\ -\overline{B} & \overline{A} \end{bmatrix}. \tag{1.57}$$

The algebra $\mathbb{H}_n$ is isomorphic to the subalgebra of the algebra $\mathbb{C}_{2n}$ consisting of the $(2n \times 2n)$-matrices (1.57). Here, A and B are matrices in $\mathbb{C}_n$.

PROOF. The complex (2×2)-matrices (1.57) are linear combinations with real coefficients of the basis matrices

$$\begin{bmatrix} 1 & 0 \\ 0 & 1 \end{bmatrix}, \quad \begin{bmatrix} 0 & 1 \\ -1 & 0 \end{bmatrix}, \quad \begin{bmatrix} 0 & i \\ i & 0 \end{bmatrix}, \quad \begin{bmatrix} i & 0 \\ 0 & -i \end{bmatrix}, \tag{1.58}$$

whose multiplication rules coincide with the multiplication rules of the basis elements $1, i, j, k$ in $\mathbb{H}$. Hence the subalgebra consisting of these matrices is isomorphic to $\mathbb{H}$.

The second assertion of the theorem follows since the subalgebra of the algebra $\mathbb{C}_{2n}$ consisting of the matrices (1.57), respectively $\mathbb{H}_n$, is isomorphic to the tensor product of the subalgebra of (2×2)-matrices (1.57), respectively $\mathbb{H}$, by $\mathbb{R}_n$. ∎

The determinant of a matrix can be defined only for matrices with entries from a commutative field or algebra. Therefore, in $\mathbb{H}_n$ matrices have no determinants. However, a matrix A in $\mathbb{H}_n$ has a *semideterminant* $|A| = s\det A$, which is equal to the determinant of a $(2n \times 2n)$-matrix in $\mathbb{C}_{2n}$ representing the matrix A.

Therefore, for semideterminants of quaternionic matrices the rule (1.50) is valid. These semideterminants are real, and even positive for nonzero matrices.

E. Study [Stu3] has proved that the semideterminant of a quaternionic matrix $\begin{bmatrix} a & b \\ c & d \end{bmatrix}$, called "nabla-function" by him, is equal to $a\bar{a}d\bar{d} - a\bar{c}d\bar{b} - b\bar{d}c\bar{a} + b\bar{b}c\bar{c}$.

1.2.3. Split Quaternions. The 4-algebra with basis $e_0 = 1$, $e_1 = i$, $e_2 = e$, $e_3 = f$ $(i^2 = -1, e^2 = f^2 = +1, ie = -ei = f, ef = -fe = -i, fi = -if = e)$ is called the algebra of *split quaternions* (or *pseudoquaternions*) and is denoted by $\mathbb{H}'$. Any split quaternion $x = x_0 + ix_1 + ex_2 + fx^3$ has a *conjugate split quaternion* $\overline{x} = x_0 - ix_1 - ex_2 - fx_3$. In this algebra the equalities (1.1) and (1.3) are fulfilled, but equality (1.2) is replaced by

$$|x|^2 = \overline{x}x = x_0^2 + x_1^2 - x_2^2 - x_3^2. \tag{1.59}$$

Therefore, the modulus $|x|$ of a split quaternion, like the modulus of a split complex number, can be real or pure imaginary, and can be equal to 0 for $x \neq 0$.

A split quaternion x with $|x| \neq 0$ has an inverse (1.4).

Split quaternions x with $|x| = 0$ are zero divisors. Therefore, *the algebra $\mathbb{H}'$ is not a division algebra.*

If we regard the modulus $|y - x|$ of the difference $y - x$ between two split quaternions x and y as the *distance* between them, we introduce into $\mathbb{H}'$ the pseudometric of the pseudo-Euclidean space R^4_2.

Split quaternions can also be regarded as vectors in R^4_2, and the *inner product* (x, y) of two vectors x and y has the form (1.5). Therefore, the angle φ between two vectors x and y can be defined by (1.7).

The *motions of the first and the second kind* in R^4_2 can be expressed by functions of a split quaternionic variable (1.47) with $|a| = |b| = 1$ or $|a| = |b| = i$, respectively.

The functions (1.47) with $|a| = 1$, $|b| = i$, respectively $|a| = i$, $|b| = 1$, express *antimotions of the first,* respectively *the second, kind* in R_2^4.

The functions (1.47) with $|a|^2 > 0$, $|b|^2 < 0$, respectively $|a|^2 < 0$, $|b|^2 > 0$, but $|a| \cdot |b| \neq 1$ express *similitudes of the first,* respectively *the second, kind* in R_2^4.

The functions (1.47) with $|a|^2 > 0$, $|b| < 0$, respectively $|a|^2 < 0$, $|b|^2$, but $|a| \cdot |b| \neq i$ express *antisimilitudes of the first,* respectively *the second, kind* in R_2^4.

Theorem 1.13. *The algebra $\mathbb{H}'$ is isomorphic to the algebra $\mathbb{R}_2$.*
The algebra $\mathbb{H}'_n$ is isomorphic to the algebra $\mathbb{R}_{2n}$.

PROOF. The real (2×2)-matrices are linear combinations of the basis matrices

$$\begin{bmatrix} 1 & 0 \\ 0 & 1 \end{bmatrix}, \quad \begin{bmatrix} 0 & 1 \\ -1 & 0 \end{bmatrix}, \quad \begin{bmatrix} 0 & 1 \\ 1 & 0 \end{bmatrix}, \quad \begin{bmatrix} 1 & 0 \\ 0 & -1 \end{bmatrix}, \tag{1.60}$$

whose multiplication rules coincide with the multiplication rules of the basis elements $1, i, e, f$ in $\mathbb{H}'$. Hence the algebra of these (2×2)-matrices is isomorphic to $\mathbb{H}'$.

The second assertion of the theorem follows since the algebra $\mathbb{R}_{2n}$ and the algebra $\mathbb{H}'_n$ are isomorphic to the respective tensor products of $\mathbb{R}_2$ and $\mathbb{H}'$ by $\mathbb{R}_n$. ■

Since the algebra $\mathbb{H}'$ is isomorphic to $\mathbb{R}_2$, (2×2)-matrices are often called *generalized quaternions.* Therefore, É. Cartan [Car5] called $(n \times n)$-matrices *n^2-ions.*

Theorem 1.14. *The algebra $\mathbb{H}'$ is isomorphic to the subalgebra of the algebra $\mathbb{C}_2$ consisting of the (2×2)-matrices*

$$\mathcal{A} = \begin{bmatrix} A & B \\ \overline{B} & \overline{A} \end{bmatrix} \tag{1.61}$$

and to the subalgebra of the algebra $\mathbb{C}'_2$ consisting of the (2×2)-matrices (1.57).

The algebra $\mathbb{H}'_n$ is isomorphic to the subalgebra of the algebra $\mathbb{C}_{2n}$ consisting of the $(2n \times 2n)$-matrices (1.61) and to the subalgebra of the algebra $\mathbb{C}'_{2n}$ consisting of the $(2n \times 2n)$-matrices (1.57). Here A and B are matrices of $\mathbb{C}_n$ of $\mathbb{C}'_n$, respectively.

PROOF. The complex (2×2)-matrices (1.61) are linear combinations with real coefficients of the basis matrices

$$\begin{bmatrix} 1 & 0 \\ 0 & 1 \end{bmatrix}, \quad \begin{bmatrix} i & 0 \\ 0 & -i \end{bmatrix}, \quad \begin{bmatrix} 0 & 1 \\ 1 & 0 \end{bmatrix}, \quad \begin{bmatrix} 0 & i \\ -i & 0 \end{bmatrix}, \tag{1.62}$$

whose multiplication rules coincide with the multiplication rules of the basis elements 1, i, e, f in $\mathbb{H}'$. Hence the subalgebra consisting of these matrices is isomorphic to $\mathbb{H}'$.

The split complex (2×2)-matrices (1.57) are linear combinations with real coefficients of the basis matrices

$$\begin{bmatrix} 1 & 0 \\ 0 & 1 \end{bmatrix}, \quad \begin{bmatrix} 0 & 1 \\ -1 & 0 \end{bmatrix}, \quad \begin{bmatrix} 0 & e \\ e & 0 \end{bmatrix}, \quad \begin{bmatrix} e & 0 \\ 0 & -e \end{bmatrix}, \tag{1.63}$$

whose multiplication rules also coincide with the multiplication rules of the basis elements 1, i, e, f in $\mathbb{H}'$. Hence the subalgebra consisting of these matrices is isomorphic to $\mathbb{H}'$.

The other assertions of the theorem follow since the subalgebra of $\mathbb{C}_{2n}$ consisting of the matrices (1.61), the subalgebra of $\mathbb{C}'_{2n}$ consisting of the matrices (1.57), and the algebra $\mathbb{H}'_n$ are isomorphic to the tensor product of the subalgebra of (2×2)-matrices (1.61) in $\mathbb{C}_2$, of the subalgebra of (2×2)-matrices (1.57) in $\mathbb{C}'_2$, and of $\mathbb{H}'$ by the algebra $\mathbb{R}_n$. ■

A matrix A in $\mathbb{H}'_n$ has a *semideterminant* $|A| = s\det A$, which is equal to the determinant of the matrix in the algebra $\mathbb{R}_{2n}$ representing this matrix.

Therefore, for semideterminants of split quaternionic matrices the rule (1.50) is also valid.

These semideterminants are real and can be positive, negative and, for singular matrices, equal to zero.

Let us mention *Theorem of A. A. Albert* [Al2], which states that the only associative algebras with 1 and with involution $x \to \overline{x}$, in which an element x for which $x = \overline{x}$ is the product of 1 by a real number, and such that products $\overline{x}x$ are products of 1 by nondegenerate quadratic forms admitting composition, are the algebras $\mathbb{C}$, $\mathbb{C}'$, $\mathbb{H}$, and $\mathbb{H}'$.

1.2.4. Simple and Semisimple Algebras. In **1.1.9** we have defined simple and semisimple associative algebras. In the already mentioned book [Mol] by Th. Molien, all complex simple associative algebra were found. All real simple associative algebras were found by É. Cartan in [Car5].

Theorem of Th. Molien. *The only complex simple associative algebras are the algebras* $\mathbb{C}_n$.

Theorem of É. Cartan. *The only real simple associative algebras are the algebras* $\mathbb{R}_n$, $\mathbb{C}_n$ *and* $\mathbb{H}_n$.

Therefore, all complex semisimple associative algebras are direct sums of algebras of complex matrices of various orders, and all real semisimple associative algebras are direct sums of algebras of real, complex or quaternionic matrices of various orders.

For the history of simple associative algebras see [Ro4, pp. 403–404].

1.2.5. Automorphisms in Simple Associative Algebras. In noncommutative associative algebras there are ring automorphisms, that is, bijections from these algebras onto themselves satisfying the conditions (1.17) and (1.18), and algebra automorphisms, that is, linear transformations in the algebras satisfying the same conditions.

Theorem 1.15. *All ring automorphisms in the skew field* $\mathbb{H}$ *are inner automorphisms, i.e. of the form*

$$f(x) = axa^{-1}. \tag{1.64}$$

PROOF. As in Theorem 1.4 (see **1.1.4**) we can prove that $f(x) = x$ for every element x in $\mathbb{H}$ that is the product of the 1 in $\mathbb{H}$ by a rational number. In a noncommutative algebra the center of the algebra, that is, the set of elements x in this algebra commuting with all other elements in it, consists of products of 1 by real numbers, so $f(x) = x$ for all x in the center. Since $\mathbb{H}$ is a noncommutative algebra, this holds for $\mathbb{H}$.

For any element x in $\mathbb{H}$ whose square x^2 is a real number we have $[f(x)]^2 = x^2$. In this case, if $f(x)$ is not real, then x is also not real. Therefore, x and $f(x)$ are linear combinations of i, j, and k. If for such an x, $|x| = 1$, $x^2 = -1$, then $[f(x)]^2$ is also -1 and so $|f(x)| = 1$.

If two such elements x and y, regarded as vectors in R^3, are orthogonal, they satisfy the condition $xy = -yx$. Therefore, $f(x)f(y) = -f(y)f(x)$ and the elements $f(x)$ and $f(y)$, regarded as vectors in R^3, are also orthogonal.

Therefore, every ring automorphism in $\mathbb{H}$ leaves invariant the real axis in $\mathbb{H}$ and determines a motion in the hyperplane orthogonal to this axis and consisting of the elements x such that $\overline{x} = -x$, that is the space R^3. This motion has the form (1.47) and since it preserves 0 and the real axis, $b = a^{-1}$ and $c = 0$ in (1.47). Since it transforms right triplets of mutually orthogonal vectors to right triplets, this motion is a motion of the first kind, and, therefore, a ring automorphism in $\mathbb{H}$ has the form (1.64). ■

Theorem 1.16. *All ring automorphisms in the algebra $\mathbb{R}_n$ are inner automorphisms, i.e. have the form*

$$f(X) = PXP^{-1}. \tag{1.65}$$

PROOF. In the algebra $\mathbb{R}_n$ the role of 1 is played by the identity matrix I, and, since the center of this algebra consists of the "scalar matrices" aI with a a real number, for each automorphism $f(X)$ in this algebra, $f(aI) = aI$. The equation $AX = aX$ has a nonzero solution only if a is an eigenvalue of the matrix A. If we apply to this equality the automorphism $f(X)$, we obtain $f(A)f(X) = af(X)$, and, therefore, a is also an eigenvalue of the matrix $f(A)$, so that the matrix $f(A)$ has the same eigenvalues as A. Therefore, for a matrix X in $\mathbb{R}_n$ with n distinct eigenvalues, the matrix $f(X)$ has the same n distinct eigenvalues, and under transition from the vector basis of eigenvectors of X to the vector basis of eigenvectors of $f(X)$ by a linear operator with matrix P, the matrices X and $f(X)$ are related by (1.65). Since each matrix in $\mathbb{R}_n$ can be obtained from a matrix with n distinct eigenvalues by a limit process, the relation (1.65) is valid for all matrices in $\mathbb{R}_n$. Therefore the ring automorphism $f(X)$ in $\mathbb{R}_n$ is inner, i.e. has the form (1.65). ■

Theorem 1.17. *The continuous ring automorphisms in the algebra $\mathbb{C}_n$ are the inner automorphisms (1.65) and the transformations*

$$f(X) = P\overline{X}P^{-1}. \tag{1.66}$$

This theorem follows from theorems 1.4 and 1.16.

Theorem 1.18. *All ring automorphisms in the algebra $\mathbb{H}_n$ are inner automorphisms, i.e. of the form (1.65).*

This theorem follows from theorems 1.14 and 1.15.

1.2.6. Semiquaternions, Split Semiquaternions, and $\frac{1}{4}$-Quaternions. The 4-algebra with basis $e_0 = 1$, $e_1 = i$, $e_2 = \epsilon$, $e_3 = \eta$ $(i^2 = -1$, $\epsilon^2 = \eta^2 = 0$, $i\epsilon = -\epsilon i = \eta$, $\eta i = -i\eta = \epsilon$, $\epsilon\eta = \eta\epsilon = 0)$ is called the algebra of *semiquaternions* and is denoted by $\mathbb{H}^0$. Each semiquaternion $x = x_0 + ix_1 + \epsilon x_2 + \eta x_3$ has a *conjugate* *semiquaternion* $\overline{x} = x_0 - ix_1 - \epsilon x_2 - \eta x_3$. In this algebra the equalities (1.1), (1.2), and (1.3) hold. Therefore, the modulus $|x|$ of a semiquaternion, like the modulus of a dual number x, is real and can be equal to 0 for $x \neq 0$. A semiquaternion x for which $|x| \neq 0$ has an inverse element (1.4). A semiquaternion x for which $|x| = 0$ has the form $\epsilon x_2 + \eta x_3$ and is a zero divisor.

The 4-algebra with basis $e_0 = 1$, $e_1 = e$, $e_2 = \epsilon$, $e_3 = \eta$ $(e^2 = +1, \epsilon^2 = \eta^2 = 0, e\epsilon = -\epsilon e = \eta, \eta e = -e\eta = -\epsilon, \epsilon\eta = \eta\epsilon = 0)$ is called the algebra of *split semiquaternions*, or *semipseudoquaternions*, and is denoted by $\mathbb{H}'^0$. Each split semiquaternion, $x = x_0 + ix_1 + \epsilon x_2 + \eta x_3$ has a *conjugate split semiquaternion* $\overline{x} = x_0 - ix_1 - \epsilon x_2 - \eta x_3$. In this algebra the equalities (1.1), (1.10), and (1.3) hold. Therefore, the modulus $|x|$ of a split semiquaternion x can be real or pure imaginary and can be equal to 0 for $x \neq 0$. A split semiquaternion x for which $|x| \neq 0$ has an inverse element (1.4). A split semiquaternion x for which $|x| = 0$ has the form $\epsilon x_2 + \eta x_3$ and is a zero divisor, but not all zero divisors of this algebra have this form.

The 4-algebra with basis $e_0 = 1$, $e_1 = \epsilon$, $e_2 = \eta$, $e_3 = \omega$ $(\epsilon^2 = \eta^2 = \omega^2 = 0, \epsilon\eta = -\eta\epsilon = \omega, \epsilon\omega = \omega\epsilon = \eta\omega = \omega\eta = 0)$ is called the algebra of $\frac{1}{4}$*-quaternions* and is denoted by $\mathbb{H}^{00}$. Each $\frac{1}{4}$-quaternion $x = x_0 + \epsilon x_1 + \eta x_2 + \omega x_3$ has a *conjugate* $\frac{1}{4}$*-quaternion,* $\overline{x} = x_0 - \epsilon x_1 - \eta x_2 - \omega x_3$. In this algebra the equalities (1.1), (1.15), and (1.3) hold. Therefore, the modulus $|x|$ of a $\frac{1}{4}$-quaternion x is real and can be equal to 0 for $|x| \neq 0$. A $\frac{1}{4}$-quaternion x for which $|x| \neq 0$ has an inverse element (1.4). A $\frac{1}{4}$-quaternion x for which $|x| = 0$ has the form $\epsilon x_1 + \eta x_2 + \omega x_3$ and is a zero divisor.

Theorem 1.19. *The algebra $\mathbb{H}^0$ is isomorphic to the subalgebra of the algebra $\mathbb{C}_2$ consisting of the (2×2)-matrices*

$$\mathcal{A} = \begin{bmatrix} A & B \\ 0 & \overline{A} \end{bmatrix} \tag{1.67}$$

and to the subalgebra of the algebra $\mathbb{C}_2^0$ consisting of the (2×2)-matrices (1.55).

The algebra $\mathbb{H}_n^0$ is isomorphic to the subalgebra of the algebra $\mathbb{C}_{2n}$ consisting of the $(2n \times 2n)$-matrices (1.67) and to the subalgebra of the algebra $\mathbb{C}_{2n}^0$ consisting of the $(2n \times 2n)$-matrices (1.55). Here A and B are matrices in $\mathbb{C}_n$ and $\mathbb{C}_n^0$, respectively.

The algebra $\mathbb{H}'^0$ is isomorphic to the subalgebra of the algebra $\mathbb{C}'_2$ consisting of the (2×2)-matrices (1.67) and to the subalgebra of the algebra $\mathbb{C}_2^0$ consisting of the (2×2)-matrices (1.61).

The algebra $\mathbb{H}'^0_n$ is isomorphic to the subalgebra of the algebra $\mathbb{C}'_{2n}$ consisting of the $(2n \times 2n)$-matrices (1.67) and to the subalgebra of the algebra $\mathbb{C}_{2n}^0$ consisting of the $(2n \times 2n)$-matrices (1.61). Here A and B are matrices in $\mathbb{C}'_n$ and $\mathbb{C}_n^0$, respectively.

The algebra $\mathbb{H}^{00}$ is isomorphic to the subalgebra of the algebra $\mathbb{C}_2^0$ consisting of the (2×2)-matrices (1.67).

The algebra $\mathbb{H}_n^{00}$ is isomorphic to the subalgebra of the algebra $\mathbb{C}_{2n}^0$ consisting of the $(2n \times 2n)$-matrices (1.67). Here A and B are matrices in $\mathbb{C}_n^0$,respectively.

The proof is analogous to the proof of Theorem 1.14.

1.2.7. Tensor Products of Algebras. The tensor product $\mathbb{A} \otimes \mathbb{B}$ of two algebras $\mathbb{A}$ and $\mathbb{B}$, has been defined in **1.1.6**. Now we consider the tensor product $\mathbb{A} \otimes \mathbb{B}$ when one or both of $\mathbb{A}$ and $\mathbb{B}$ are noncommutative.

The elements in the tensor product $\mathbb{C} \otimes \mathbb{H}$ are called *biquaternions*. Analogously, the elements in the tensor products $\mathbb{C}' \otimes \mathbb{H}$ and $\mathbb{C}^0 \otimes \mathbb{H}$ are called *dyquaternions* and *duoquaternions*, respectively.

The elements in the tensor product $\mathbb{H} \otimes \mathbb{H}$ are called *quaterquaternions*. In other tensor products the word "quaternion" (or its part "quater") must be replaced by "pseudoquaternion", "semiquaternion", or "pseudosemiquaternion" (or "pseudoquater", "semiquater", or "pseudosemiquater").

Theorem 1.20. *The algebras $\mathbb{C} \otimes \mathbb{H}$ and $\mathbb{C} \otimes \mathbb{H}'$ are isomorphic to the algebra $\mathbb{C}_2$.*

PROOF. The second assertion follows from the isomorphism of the algebras $\mathbb{H}'$ and $\mathbb{R}_2$, the first follows since the multiplication rules of the elements $1, i, jI, kI$ in the algebra $\mathbb{C} \otimes \mathbb{H}$ and of the basis elements $1, i, e, f$ in the algebra $\mathbb{H}'$ coincide. ■

Theorem 1.21. *The algebra $\mathbb{H} \otimes \mathbb{H}$ is isomorphic to the algebra $\mathbb{R}_4$.*

PROOF. This isomorphism follows from the isomorphism of the tensor products $\mathbb{H} \otimes \mathbb{H}$ and $\mathbb{H}' \otimes \mathbb{H}'$, which holds since the multiplication rules of the elements $1, i, jI, kI, I, iI, -j, -k, iJ, -J, -kK, jK, iK, -K, kJ, -jJ$ in the algebra $\mathbb{H} \otimes \mathbb{H}$ and of the basis elements $1, i, e, f, I, iI, eI, fI, E, iE, eE, fE, F, iF, eF, fF$ in the algebra $\mathbb{H}' \otimes \mathbb{H}'$,which is isomorphic to $\mathbb{R}_2 \otimes \mathbb{R}_2 = \mathbb{R}_4$, coincide. ■

Each biquaternion or quaterquaternion x has three conjugate biquaternions or quaterquaternions: $\bar{x}$, in which the basis elements i, j, k are multiplied by -1, $\tilde{x}$, in which the basis elements I, J, K are multiplied by -1, and $\tilde{\bar{x}}$ in which the basis elements i, j, k, I, J, K are multiplied by -1. For biquaternions all conditions (1.26) are fulfilled, for quaterquaternions only the third condition in (1.26) is fulfilled.

The products $\bar{x}x$, $\tilde{x}x$, and $\tilde{\bar{x}}\, x$ are called the squares of the moduli $|x|$, $\wr x \wr$, and $\wr|x|\wr$. For them the equality (1.3) and analogous equalities are fulfilled.

Conjugate elements and moduli of elements of other tensor products are defined analogously.

1.2.8. Alternions. Complex numbers and quaternions are particular cases of *alternions* or *Clifford numbers*. The algebra $\mathbb{A}_n$ of alternions, or the *Clifford algebra* (do not confuse this with the algebra of $(n \times n)$-matrices over an algebra $\mathbb{A}$!), is the algebra with as basis the 2^{n-1} elements

$$1, e_1, e_2, \dots, e_{n-1}, e_{i_1 i_2 \dots i_k} = e_{i_1} e_{i_2} \dots e_{i_k} \tag{1.68}$$

where these elements satisfy the conditions

$$e_i e_j = -e_j e_i \tag{1.69}$$

and

$$e_i^2 = -1. \tag{1.70}$$

The algebra $\mathbb{A}_1$ has unique basis element 1 and coincides with the field $\mathbb{R}$. The algebra $\mathbb{A}_2$ has 2 basis elements, 1 and e_1, and is isomorphic to the field $\mathbb{C}$. The algebra $\mathbb{A}_3$ has 4 basis elements, $1, e_1, e_2, e_{12}$, and is isomorphic to the division algebra $\mathbb{H}$.

The basis element e_{123} in the algebra $\mathbb{A}_4$ commutes with all basis elements in this algebra and, since $e_{123}^2 = +1$, the algebra $\mathbb{A}_4$ is isomorphic to the tensor product $\mathbb{C}' \otimes \mathbb{H}$, that is, to the direct sum $\mathbb{H} \oplus \mathbb{H}$.

All indices of basis elements $e_{i_1 i_2 \ldots i_k}$ are different. Therefore, we can write these elements in the "normal form" where $i_1 < i_2 < \cdots < i_k$.

Any element in $\mathbb{A}_n$ can be written in the form

$$\begin{aligned} x &= x_0 + x^i e_i + x^{ij} e_{ij} + \cdots + x^{i_1 i_2 \ldots i_k} e_{i_1 i_2 \ldots i_k} + \cdots \\ &\quad + x^{i_1 i_2 \ldots i_{n-1}} e_{i_1 i_2, \ldots i_{n-1}} \\ &= \sum_k x^{i_1 i_2 \ldots i_k} e_{i_1 i_2 \ldots i_k}, \end{aligned} \tag{1.71}$$

in which all elements $e_{i_1 i_2 \ldots i_k}$ are written in normal form.

Each alternion (1.71) has a *conjugate alternion*,

$$\overline{x} = \sum_k (-1)^{k(k+1)/2} x^{i_1 i_2 \ldots i_k} e_{i_1 i_2 \ldots i_k}. \tag{1.72}$$

As in $\mathbb{C}$ and $\mathbb{H}$, for it condition (1.3) holds. This condition follows from the equality

$$(-1)^{k(k+1)/2} e_{i_1 i_2 \ldots i_k} = (-1)^k e_{i_k} e_{i_{k-1}} \ldots e_{i_2} e_{i_1}. \tag{1.73}$$

The coordinate x_0 of x is called the *real part* of x and is denoted by Re x.

The real part of the product $\overline{x}y$ is equal to

$$\operatorname{Re}(\overline{x}y) = \sum_k x^{i_1 i_2 \ldots i_k} y^{i_1 i_2 \ldots i_k}. \tag{1.74}$$

The real nonnegative number $|x|$ whose square is equal to

$$|x|^2 = \operatorname{Re}(\overline{x}x) = \sum_k (x^{i_1 i_2 \ldots i_k})^2 \tag{1.75}$$

is called the *modulus* of the alternion x. The modulus $|x|$ is equal to 0 only if $x = 0$.

If we regard the modulus $|y - x|$ of the difference $y - x$ between two alternions x and y as the *distance* between them, we introduce into $\mathbb{A}_n$ the metric of the Euclidean space $R^{2^{n-1}}$.

Alternions can also be regarded as vectors in $R^{2^{n-1}}$, and the *inner product* (x, y) of two vectors x and y is equal to Re $(\overline{x}y)$.

Theorem 1.22. *The algebra $\mathbb{A}_n$ is isomorphic to the subalgebra of the algebra $\mathbb{A}_{n+1}$ consisting of all linear combinations of the basis elements in $\mathbb{A}_{n+1}$ with an even number of indices.*

PROOF. If we add to each basis element in the expression (1.71) which has an odd number of indices the index n, we obtain an element of the subalgebra of $\mathbb{A}_{n+1}$ and it is easy to check that this subalgebra is isomorphic to the algebra $\mathbb{A}_n$. ■

In particular, $1, e_{12}$, is a basis in $\mathbb{A}_2 = \mathbb{C}$, $1, e_{12}, e_{13}, e_{23}$ is a basis in $\mathbb{A}_3 = \mathbb{H}$, and $1, e_{12}, e_{13}, e_{23}, e_{14}, e_{24}, e_{34}, e_{1234}$ is a basis in $\mathbb{A}_4 = \mathbb{H} \otimes \mathbb{C}'$.

Theorem 1.23. *The algebra $\mathbb{A}_{8m+1}$ is isomorphic to the algebra $\mathbb{R}_{2^{4m}}$.*
The algebra $\mathbb{A}_{8m+2}$ is isomorphic to the algebra $\mathbb{C}_{2^{4m}}$.
The algebra $\mathbb{A}_{8m+3}$ is isomorphic to the algebra $\mathbb{H}_{2^{4m}}$.
The algebra $\mathbb{A}_{8m+4}$ is isomorphic to the algebra $\mathbb{H}_{2^{4m}} \oplus \mathbb{H}_{2^{4m}}$.
The algebra $\mathbb{A}_{8m+5}$ is isomorphic to the algebra $\mathbb{H}_{2^{4m+1}}$.
The algebra $\mathbb{A}_{8m+6}$ is isomorphic to the algebra $\mathbb{C}_{2^{4m+2}}$.
The algebra $\mathbb{A}_{8m+7}$ is isomorphic to the algebra $\mathbb{R}_{2^{4m+3}}$.
The algebra $\mathbb{A}_{8m+8}$ is isomorphic to the algebra $\mathbb{R}_{2^{4m+3}} \oplus \mathbb{R}_{2^{4m+3}}$.

PROOF. The center of $\mathbb{A}_n$ can contain only the two basis elements 1 and $e_{12\ldots n-1}$, since for $0 < k < n-1$ we have $e_{12\ldots k}e_{k,k+1} = -e_{k,k+1}e_{12\ldots k}$. But $e_{12\ldots n-1}$ belongs to the center of $\mathbb{A}_n$ only if n is even, since

$$e_{12\ldots n-1}e_i = -(-1)^{n-i-1}e_{12\ldots i-1,i+1\ldots n-1}, e_ie_{12\ldots n-1} = -(-1)^{i-1}e_{12\ldots i-1,i+i\ldots n-1}$$

and the numbers $n-i-1$ and $i-1$ are simultaneously even or odd only if n is even. Therefore, the center of $\mathbb{A}_n$ contains only 1 if n is odd, and only 1 and $e_{12\ldots n-1}$ if n is even.

Since $e^2_{12\ldots 2m-1} = (-1)^m$ the transition $\mathbb{A}_{4m+1} \to \mathbb{A}_{4m+2}$ comes down to tensor multiplication by $\mathbb{C}$ and the transition $\mathbb{A}_{4m+3} \to \mathbb{A}_{4m+4}$ comes down to tensor multiplication by $\mathbb{C}' = \mathbb{R} \oplus \mathbb{R}$.

Since the elements $1, e_{12\ldots 4m+1}, e_{12\ldots 4m+2}, e_{4m+1,4m+2}$ form a basis in $\mathbb{H}$ and commute with all other basis elements in $\mathbb{A}_{4m+3}$, the transition $\mathbb{A}_{4m+1} \to \mathbb{A}_{4m+3}$ comes down to tensor multiplication by $\mathbb{H}$.

Since the elements $1, e_{12\ldots 4m+3}, e_{12\ldots 4m+4}, e_{4m+3,4m+4}$ form a basis in $\mathbb{H}'$ and commute with all other basis elements of $\mathbb{A}_{4m+5}$, the transition $\mathbb{A}_{4m+3} \to \mathbb{A}_{4m+5}$ comes down to tensor multiplication by $\mathbb{H}' = \mathbb{R}_2$.

Therefore, the algebra $\mathbb{A}_5$ is isomorphic to $\mathbb{H} \otimes \mathbb{H}' = \mathbb{H} \otimes \mathbb{R}_2 = \mathbb{H}_2$, the algebra $\mathbb{A}_6$ is isomorphic to $\mathbb{H}_2 \otimes \mathbb{C} = \mathbb{H}'_2 \otimes \mathbb{C} = \mathbb{R}_4 \otimes \mathbb{C} = \mathbb{C}_4$, the algebra $\mathbb{A}_7$ is isomorphic to $\mathbb{H}_2 \otimes \mathbb{H} = \mathbb{R}_2 \otimes (\mathbb{H} \otimes \mathbb{H}) = \mathbb{R}_2 \otimes \mathbb{R}_4 = \mathbb{R}_8$, and the algebra $\mathbb{A}_8$ is isomorphic to $\mathbb{R}_8 \otimes \mathbb{C}' = \mathbb{R}_8 \otimes (\mathbb{R} \oplus \mathbb{R}) = \mathbb{R}_8 \oplus \mathbb{R}_8$.

The assertion of the theorem is obtained by complete mathematical induction. ■

This theorem was formulated without proof by W. K. Clifford in the same paper [Cli3] in which "Clifford numbers" were defined. For a complete proof see [Ro1, pp. 452–457].

Theorem 1.23 shows that the algebras $\mathbb{A}_{2m+1}$ are isomorphic to algebras of real or quaternionic matrices, that is, are simple algebras, while the algebras $\mathbb{A}_{2m}$ are isomorphic to algebras of complex matrices or to direct sums of algebras of real or quaternionic matrices, that is, are simple or semisimple algebras.

1.2.9. Pseudoalternions. Split complex numbers and split quaternions are particular cases of *pseudoalternions* or *pseudo-Clifford numbers.* The algebra $\mathbb{A}_{n,\ell}$ of pseudoalternions, or the *pseudo-Clifford algebra,* is the algebra with basis consisting of the 2^{n-1} elements (1.68) satisfying the conditions (1.69) and

$$e_i^2 = -\epsilon_i, \tag{1.76}$$

where $\epsilon_i = \pm 1$, and ℓ of them are equal to -1 and $n-1-\ell$ of them are equal to $+1$.

The algebra $\mathbb{A}_{2,1}$ has 2 basis elements, 1 and e_1, and is isomorphic to the algebra $\mathbb{C}'$. The algebras $\mathbb{A}_{3,1}$ and $\mathbb{A}_{3,2}$ have 4 basis elements, $1, e_1, e_2, e_{12}$, and are isomorphic to the algebra $\mathbb{H}'$.

The basis element e_{123} in the algebras $\mathbb{A}_{4,1}, \mathbb{A}_{4,2}$, and $\mathbb{A}_{4,3}$ commutes with all other basis elements in these algebras and, since in $\mathbb{A}_{4,1}$ and $\mathbb{A}_{4,3}$ we have $e_{123}^2 = -1$ and in $\mathbb{A}_{4,2}$ we have $e_{123}^2 = +1$, the algebras $\mathbb{A}_{4,1}$ and $\mathbb{A}_{4,3}$ are isomorphic to the tensor product $\mathbb{C} \otimes \mathbb{H}$, that is, to the algebra $\mathbb{C}_2$, and the algebra $\mathbb{A}_{4,2}$ is isomorphic to the tensor product $\mathbb{C}' \otimes \mathbb{H}$, that is, to the direct sum $\mathbb{H} \oplus \mathbb{H}$.

The pseudoalternions in $\mathbb{A}_{n,\ell}$ can be written in the same form (1.71), as alternions in $\mathbb{A}_n$, in which all elements $e_{i_1 i_2 \ldots i_k}$ are written in normal form.

Each pseudoalternion (1.71) has a *conjugate pseudoalternion* (1.72). As in $\mathbb{C}'$ and $\mathbb{H}'$ condition (1.3) is fulfilled. This condition follows from (1.73), as for alternions.

The coordinate x_0 of a pseudoalternion x is called the *real part* of x and is denoted by Re x. The real part of the product $\bar{x}y$ is equal to

$$\mathrm{Re}(\bar{x}y) = \sum_k \epsilon_{i_1}\epsilon_{i_2}\ldots\epsilon_{i_k} x^{i_1 i_2 \ldots i_k} y^{i_1 i_2 \ldots i_k}. \tag{1.77}$$

The real or pure imaginary number $|x|$ whose square is equal to

$$|x|^2 = \mathrm{Re}(\bar{x}x) = \sum_k \epsilon_{i_1}\epsilon_{i_2}\ldots\epsilon_{i_k} (x^{i_1 i_2 \ldots i_k})^2 \tag{1.78}$$

is called the *modulus* of the pseudoalternion x. The modulus $|x|$ can be equal to 0 for $x \neq 0$.

If we regard the modulus $|y-x|$ of the difference $y-x$ between two pseudoalternions x and y as the *distance* between them, we introduce into $\mathbb{A}_{n,\ell}$ the pseudometric of the pseudo-Euclidean space $R_L^{2^{n-1}}$.

Pseudoalternions can also be regarded as vectors in $R_L^{2^{n-1}}$, and the *inner product* (x, y) of the vectors x and y is equal to Re $(\bar{x}y)$.

For the algebras $\mathbb{A}_{n,\ell}$ a theorem analogous to Theorem 1.22 for $\mathbb{A}_n$ holds, that is, the algebra $\mathbb{A}_{n,\ell}$ is isomorphic to the subalgebra of the algebra $\mathbb{A}_{n+1,\ell}$ consisting of all linear combinations of the basis elements in $\mathbb{A}_{n+1,\ell}$ with an even number of indices.

Theorem 1.24. *The algebras $\mathbb{A}_{8m+1,4k}$ and $\mathbb{A}_{8m+1,4k+1}$ are isomorphic to the algebra $\mathbb{R}_{2^{4m}}$.*

The algebras $\mathbb{A}_{8m+1,4k+2}$ and $\mathbb{A}_{8m+1,4k+3}$ are isomorphic to the algebra $\mathbb{H}_{2^{4m-1}}$.

The algebras $\mathbb{A}_{8m+2,4k}$ and $\mathbb{A}_{8m+2,4k+2}$ are isomorphic to the algebra $\mathbb{C}_{2^{4m}}$.

The algebra $\mathbb{A}_{8m+2,4k+1}$ is isomorphic to the algebra $\mathbb{R}_{2^{4m}} \oplus \mathbb{R}_2$.

The algebra $\mathbb{A}_{8m+2,4k+3}$ *is isomorphic to the algebra* $\mathbb{H}_{2^{4m-1}} \oplus \mathbb{H}_{2^{4m-1}}$.
The algebras $\mathbb{A}_{8m+3,4k}$ *and* $\mathbb{A}_{8m+3,4k+3}$ *are isomorphic to the algebra* $\mathbb{H}_{2^{4m}}$.
The algebras $\mathbb{A}_{8m+3,4k+1}$ *and* $\mathbb{A}_{8m+3,4k+2}$ *are isomorphic to the algebra* $\mathbb{R}_{2^{4m+1}}$.
The algebra $\mathbb{A}_{8m+4,4k}$ *is isomorphic to the algebra* $\mathbb{H}_{2^{4m}} \oplus \mathbb{H}_{2^{4m}}$.
The algebras $\mathbb{A}_{8m+4,4k+1}$ *and* $\mathbb{A}_{8m+4,4k+3}$ *are isomorphic to the algebra* $\mathbb{C}_{2^{4m+1}}$.
The algebra $\mathbb{A}_{8m+4,4k+2}$ *is isomorphic to the algebra* $\mathbb{R}_{2^{4m+1}} \oplus \mathbb{R}_{2^{4m+1}}$.
The algebras $\mathbb{A}_{8m+5,4k}$ *and* $\mathbb{A}_{8m+5,4k+1}$ *are isomorphic to the algebra* $\mathbb{H}_{2^{4m+1}}$.
The algebras $\mathbb{A}_{8m+5,4k+2}$ *and* $\mathbb{A}_{8m+5,4k+3}$ *are isomorphic to the algebra* $\mathbb{R}_{2^{4m+2}}$.
The algebras $\mathbb{A}_{8m+6,4k}$ *and* $\mathbb{A}_{8m+6,4k+2}$ *are isomorphic to the algebra* $\mathbb{C}_{2^{4m+2}}$.
The algebra $\mathbb{A}_{8m+6,4k+1}$ *is isomorphic to the algebra* $\mathbb{H}_{2^{4m+1}} \oplus \mathbb{H}_{2^{4m+1}}$.
The algebra $\mathbb{A}_{8m+6,4k+3}$ *is isomorphic to the algebra* $\mathbb{R}_{2^{4m+2}} \oplus \mathbb{R}_{2^{4m+2}}$.
The algebras $\mathbb{A}_{8m+7,4k}$ *and* $\mathbb{A}_{8m+7,4k+3}$ *are isomorphic to the algebra* $\mathbb{R}_{2^{4m+3}}$.
The algebras $\mathbb{A}_{8m+7,4k+1}$ *and* $\mathbb{A}_{8m+7,4k+2}$ *are isomorphic to the algebra* $\mathbb{H}_{2^{4m+2}}$.
The algebra $\mathbb{A}_{8m+8,4k}$ *is isomorphic to the algebra* $\mathbb{R}_{2^{4m+3}} \oplus \mathbb{R}_{2^{4m+3}}$.
The algebras $\mathbb{A}_{8m+8,4k+1}$ *and* $\mathbb{A}_{8m+8,4k+3}$ *are isomorphic to the algebra* $\mathbb{C}_{2^{4m+3}}$.
The algebra $\mathbb{A}_{8m+8,4k+2}$ *is isomorphic to the algebra* $\mathbb{H}_{2^{4m+2}} \oplus \mathbb{H}_{2^{4m+2}}$.

The proof is analogous to the proof of theorem 1.23.

This theorem was formulated without a proof by É. Cartan in the paper [Car6]; see [Ro1, pp. 452–458] for a proof.

Theorem 1.24 shows that the algebras $A_{n,\ell}$, like the algebras $\mathbb{A}_n$, are simple or semisimple algebras.

1.2.10. Split Alternions and Bialternions. The algebras $\mathbb{A}_{2,1}$ and $\mathbb{A}_{3,2}$ of split complex numbers and split quaternions are particular cases of the important class of algebras $\mathbb{A}_{n,\ell}$ for $n = 2\ell$ or $n = 2\ell - 1$. These algebras are called algebras of *split alternions.*

Theorem 1.24 implies that the algebra $\mathbb{A}_{2m+1,m+1}$ of split alternions of odd order is isomorphic to the algebra $\mathbb{R}_{2^m}$, and the algebra $\mathbb{A}_{2m+2,m+1}$ of split alternions of even order is isomorphic to the algebra $\mathbb{R}_{2^m} \oplus \mathbb{R}_{2^m}$.

The algebras $\mathbb{C} \otimes \mathbb{C}$ and $\mathbb{C} \otimes \mathbb{H}$ of bicomplex numbers and biquaternions are particular cases of the algebras $\mathbb{C} \otimes \mathbb{A}_n$ of *bialternions.* It is clear that the algebra $\mathbb{C} \otimes \mathbb{A}_{n,l}$ is isomorphic to the algebra $\mathbb{C} \otimes \mathbb{A}_n$. Therefore, theorems 1.23 and 1.24 imply the *theorem of R. Brauer and H. Weyl* [BrWe]: the algebra $\mathbb{A}_{2m+1} \otimes \mathbb{C}$ of bialternions of odd order is isomorphic to the algebra $\mathbb{C}_{2^m}$, and the algebra $\mathbb{A}_{2m+2} \otimes \mathbb{C}$ of bialternions of even order is isomorphic to the algebra $\mathbb{C}_{2^m} \oplus \mathbb{C}_{2^m}$. Brauer and Weyl proved this theorem independently of theorems 1.23 and 1.24.

1.2.11. Quasisimple and r-Quasisimple Noncommutative Algebras. In **1.1.10** we have defined quasisimple and r-quasisimple associative algebras. If we apply the process of transition from a semisimple to a quasisimple algebra, as described in **1.1.10**, to the algebra $\mathbb{H}$ and its involutive automorphism $x \to ixi^{-1}$, we obtain the algebra $\mathbb{H}^0$ of semiquaternions. We obtain the same algebra if we apply the transition process to the algebra $\mathbb{H}'$ and its involutive automorphism $x \to exe$. If we apply the transition process to the same algebra and its involutive automorphism $x \to ixi^{-1}$ we obtain the algebra $\mathbb{H}'^0$ of split semiquaternions.

If we apply the transition process to the algebra $\mathbb{R}_n$ and its involutive automorphism

$$\begin{bmatrix} A & B \\ C & D \end{bmatrix} \to \begin{bmatrix} A & -B \\ -C & D \end{bmatrix}, \tag{1.79}$$

where the submatrices A, B, C, D are an $(m \times m)$-matrix, an $(m \times (n-m))$-matrix, an $((n-m) \times m)$-matrix and an $((n-m) \times (n-m))$-matrix, respectively, we obtain the algebra isomorphic to the subalgebra of the algebra $\mathbb{C}_n^0$ consisting of the dual matrices

$$\mathcal{A} = \begin{bmatrix} A & \epsilon B \\ \epsilon C & D \end{bmatrix}. \tag{1.80}$$

We call this new algebra the algebra of *$(n \times n)$-quasimatrices*, and denote it by $\mathbb{R}_n^m$.

Theorem 1.25. *The algebra $\mathbb{H}'^0$ is isomorphic to the algebra $\mathbb{R}_2^1$.*
The algebra $\mathbb{H}_n'^0$ is isomorphic to the algebra $\mathbb{R}_{2n}^n$.

PROOF. A (2×2)-quasimatrix is a linear combination of the basis quasimatrices

$$\begin{bmatrix} 1 & 0 \\ 0 & 1 \end{bmatrix}, \quad \begin{bmatrix} 0 & \epsilon \\ -\epsilon & 0 \end{bmatrix}, \quad \begin{bmatrix} 0 & \epsilon \\ \epsilon & 0 \end{bmatrix}, \quad \begin{bmatrix} 1 & 0 \\ 0 & -1 \end{bmatrix}, \tag{1.81}$$

whose multiplication rules coincide with the multiplication rules of the basis elements $1, \eta, \epsilon, e$, of the algebra $\mathbb{H}'^0$. This implies that $\mathbb{H}'^0$ and the algebra consisting of these quasimatrices are isomorphic.

The second assertion of the theorem follows from the isomorphism of the algebra $\mathbb{R}_{2n}^n$ and of the algebra $\mathbb{H}_n'^0$ to the tensor products of the algebras $\mathbb{R}_2^1$ and $\mathbb{H}'^0$, respectively, by the algebra $\mathbb{R}_n$. ∎

If we apply this transition process to the algebra $\mathbb{A}_n$ and its involutive automorphism generated by the transformation $e_i \to \pm e_i$ of one-index basis elements, where there are k signs $-$ and $n-1-k$ signs $+$, we obtain the algebra $\mathbb{A}_n^k$ of *quasialternions*.

If we apply this transition process to the algebra $\mathbb{A}_{n,\ell}$ and its involutive automorphism generated by the transformation $e_i \to \pm e_i$ of one-index basis elements, where k_0, respectively k_1, of the elements $e_a(e_a^2 = +1)$, respectively $e_u(e_u^2 = -1)$, appear with a sign $-$, respectively a sign $+$, and $\ell - k_0$, respectively $n - \ell - k_1$, of the elements e_a, respectively e_u, appear with a sign $+$, respectively a sign $-$, we obtain the algebra $\mathbb{A}_{n,\ell}^{k_0 k_1}$ of *quasipseudoalternions*.

It is easy to check that the algebra $\mathbb{A}_2^1$ is isomorphic to the algebra $\mathbb{C}^0$, the algebras $\mathbb{A}_3^1$ and $\mathbb{A}_{3,1}^{01}$ are isomorphic to the algebra $\mathbb{H}^0$, and the algebra $\mathbb{A}_{3,1}^{10}$ is isomorphic to the algebra $\mathbb{H}'^0$.

If we apply this transition process to the algebras $\mathbb{R}_n, \mathbb{A}_n$, and $\mathbb{A}_{n,\ell}$ r times we obtain the r-quasisimple algebras of *r-quasimatrices*, *r-quasialternions*, and *r-quasipseudoalternions*, respectively.

1.2.12. Grassmann Algebras. The algebra of $(n-1)$-quasialternions, that is, the 2^{n-1}-algebra with basis (1.68) satisfying the conditions (1.69) and $e_i^2 = 0$, is called a *Grassmann algebra*, since the products $e_{i_1} e_{i_2} \ldots e_{i_k}$ of one-index basis elements in this algebra can be regarded as exterior products $e_{i_1} \wedge e_{i_2} \wedge \ldots \wedge e_{i_k}$ of vectors (see (0.35)), which were introduced by H. Grassmann [Gr] (see [Ro4, p. 394]).

For $n = 2$ the Grassmann algebra coincides with the algebra $\mathbb{C}^0$ of dual numbers, and for $n = 3$ with the algebra $\mathbb{H}^{00}$ of $\frac{1}{4}$-quaternions.

1.2.13. Frobenius Algebras. In **1.1.12** we have defined associative Frobenius algebras. As for semisimple commutative algebras, for semisimple noncommutative algebras we can take the form $\Phi(x, y)$ to be $\mathrm{Re}(xy)$. For instance, for $\mathbb{H}$ and $\mathbb{H}'$ we then have

$$\Phi(x,y) = x_0y_0 - x_1y_1 - x_2y_2 - x_3y_3, \quad \Phi(x,y) = x_0y_0 - x_1y_1 + x_2y_2 + x_3y_3. \quad (1.82)$$

As for non-semisimple commutative algebras, for the non-semisimple noncommutative algebras we can take the form $\Phi(x, y)$ to be the coefficient of the product xy at a basic element of the radical.

For instance, for $\mathbb{H}^0$ and $\mathbb{H}'^0$ we can take the coefficients at ϵ:

$$\Phi(x,y) = x_0y_2 + x_2y_0 - x_1y_3 + x_3y_1, \quad \Phi(x,y) = x_0y_2 + x_2y_0 + x_1y_3 - x_3y_1. \quad (1.83)$$

§1.3. Alternative Algebras

1.3.1. Octonions. Another generalization of the algebra $\mathbb{H}$ of quaternions is the nonassociative 8-algebra $\mathbb{O}$ of *octonions*, or *octaves*, or *Cayley numbers*, with basis $e_0 = 1, e_1 = i, e_2 = j, e_3 = k, e_4 = \ell, e_5 = p, e_6 = q, e_7 = r$, for which the subalgebras with bases

$$1,i,j,k; \quad 1,i,\ell,p; \quad 1,i,q,r; \quad 1,j,q,\ell; \quad 1,j,p,r; \quad 1,k,p,q; \quad 1,k,\ell,r \quad (1.84)$$

are isomorphic to the algebra $\mathbb{H}$ of quaternions. The bases (1.84) can be symbolically represented as in Fig. 1.1, which was proposed by H. Freudenthal in [Fr1].

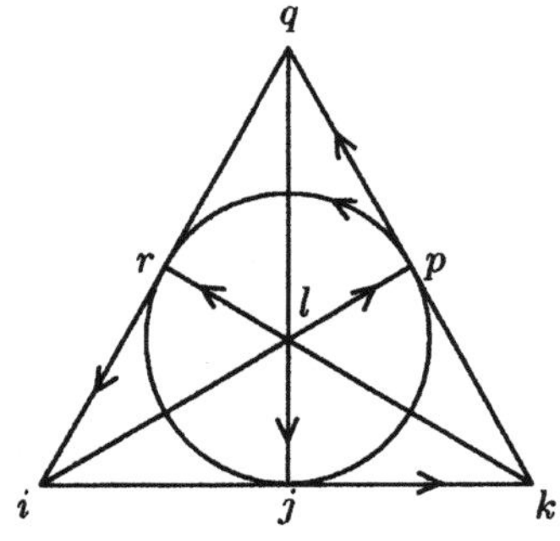

FIG. 1.1

The algebra $\mathbb{O}$ is not associative, since $(ij)\ell = k\ell = r$, $i(j\ell) = -iq = -r$. But it has the property of *alternativity*, that is, any two elements in it generate an associative subalgebra isomorphic to $\mathbb{R}, \mathbb{C}$ or $\mathbb{H}$.

1 and each basis element e_i $(i > 0)$ generate a subalgebra isomorphic to $\mathbb{C}$, 1 and any two basis elements e_i and e_j $(i, j > 0)$ generate an algebra with basis (1.84) isomorphic to $\mathbb{H}$.

In particular, any two octonions x and y satisfy the condition

$$\overline{x}(xy) = (\overline{x}x)y, \quad (yx)\overline{x} = y(x\overline{x}). \tag{1.85}$$

Each octonion $x = x_0 + ix_1 + jx_2 + kx_3 + lx_4 + px_5 + qx_6 + rx_7$ has a *conjugate octonion* $\overline{x} = x_0 - ix_1 - jx_2 - kx_3 - lx_4 - px_5 - qx_6 - rx_7$ and the following equalities are fulfilled: (1.1),

$$|x|^2 = \overline{x}x = x_0^2 + x_1^2 + x_2^2 + x_3^2 + x_4^2 + x_5^2 + x_6^2 + x_7^2, \tag{1.86}$$

and (1.3).

The nonnegative real number $|x|$ is called the *modulus* of the octonion x. The modulus $|x|$ is equal to 0 only if $x = 0$. An octonion $x \neq 0$ has an inverse element (1.4). Therefore, the algebra $\mathbb{O}$ is a division algebra.

If we regard the modulus $|x - y|$ of the difference $y - x$ between two octonions x and y as the *distance* between them, we introduce into $\mathbb{O}$ the metric of the Euclidean space R^8.

Octonions can also be regarded as vectors in the space R^8, and the *inner product* (x, y) of two vectors x and y has the form (1.5). Therefore, the angle φ between two vectors x and y can be defined by formula (1.7).

Octonions were defined by A. Cayley [Cay1], hence the name "Cayley numbers". For the history of octonions see [Ro4, pp. 386–388].

In **1.2.1** we have mentioned the *theorem of A. Hurwitz*, stating that the only positive definite quadratic forms whose products also are quadratic forms are binary, quaternary, and octonary forms. Yu. V. Linnik has proved in [Lin] that this theorem implies that the only algebras with 1 and with involution $x \to \overline{x}$, and in which the elements x for which $x = \overline{x}$ are products of 1 by real numbers and for which the products $\overline{x}x$ are positive definite quadratic forms admitting composition, are the division algebras $\mathbb{C}, \mathbb{H}$, and $\mathbb{O}$.

If we replace in the *theorem of L. S. Pontryagin*, mentioned also in **1.2.1**, the condition of associativity by the condition of alternativity, we find that $\mathbb{O}$ is the only topological alternative division algebra.

1.3.2. Split Octonions and Bioctonions. The 8-algebra with basis $e_0 = 1, e_1 = i, e_2 = j, e_3 = k, e_4 = e, e_5 = f, e_6 = g, e_7 = h$ for which the subalgebra with basis $1, i, j, k$ is isomorphic to the algebra $\mathbb{H}$ of quaternions and the algebras with bases

$$1, i, e, f; \quad 1, i, g, h; \quad 1, j, f, h; \quad 1, j, g, e; \quad 1, k, e, h; \quad 1, k, f, g \tag{1.87}$$

are isomorphic to the algebra $\mathbb{H}'$ of split quaternions, is called the algebra of *split octonions*, or *split octaves*, or *pseudooctonions*, or *Cayley-Dickson numbers*, and is denoted by $\mathbb{O}'$.

The algebra $\mathbb{O}'$ is not associative, since $(ij)e = ke = h$, $i(je) = -ig = -h$. But it has the property of *alternativity*, that is, any two elements in it generate an associative subalgebra isomorphic to an algebra $\mathbb{R}, \mathbb{C}, \mathbb{C}', \mathbb{C}^0, \mathbb{H}, \mathbb{H}', \mathbb{H}^0, \mathbb{H}'^0$, or $\mathbb{H}^{00}$.

1 and i generate a subalgebra isomorphic to $\mathbb{C}$, 1 and e generate a subalgebra isomorphic to $\mathbb{C}'$, 1 and $i+e$ generate a subalgebra isomorphic to $\mathbb{C}^0$, the subalgebras

with bases $1, i, j, k$ and $1, i, e, f$ are isomorphic to $\mathbb{H}$ and $\mathbb{H}'$, the subalgebras with bases $1, i, j+e, k+f$ and $1, e, i+h, -f-k$ are isomorphic to $\mathbb{H}^0$ and $\mathbb{H}'^0$, the subalgebra with basis $1, \frac{i+h}{2}, \frac{j+e}{2}, \frac{k+f}{2}$ is isomorphic to $\mathbb{H}^{00}$. Therefore, any two split octonions satisfy the condition (1.85).

Like $\mathbb{C}$ and $\mathbb{C}'$, which have as common complex form the algebra $\mathbb{C} \otimes \mathbb{C}$ of bicomplex numbers, and like $\mathbb{H}$ and $\mathbb{H}'$, which have as common complex form the algebra $\mathbb{C} \otimes \mathbb{H}$ of biquaternions, the algebras $\mathbb{O}$ and $\mathbb{O}'$ have a common complex form: the algebra $\mathbb{C} \otimes \mathbb{O}$ of *bioctonions*. Like the algebra $\mathbb{C}'$, which is isomorphic to the subalgebra of $\mathbb{C} \otimes \mathbb{C}$ with basis $1, iI$, and the algebra $\mathbb{H}'$, which is isomorphic to the subalgebra of $\mathbb{C} \otimes \mathbb{H}$ with basis $1, i, jI, kI$, the algebra $\mathbb{O}'$ is isomorphic to the subalgebra of $\mathbb{C} \otimes \mathbb{O}$ with basis $1, i, j, k, \ell I, pI, qI, rI$, the last four of which play the role of basis elements e, f, g, h.

Each split octonion $x = x_0 + ix_1 + jx_2 + kx_3 + ex_4 + fx_5 + gx_6 + hx_7$ has a *conjugate split octonion*, $\overline{x} = x_0 - ix_1 - jx_2 - kx_3 - ex_4 - fx_5 - gx_6 - hx_7$, and the following equalities are fulfilled: (1.1),

$$|x|^2 = \overline{x}x = x_0^2 + x_1^2 + x_2^2 + x_3^2 - x_4^2 - x_5^2 - x_6^2 - x_7^2, \tag{1.88}$$

and (1.3).

Therefore, the modulus $|x|$ of a split octonion x, like the modulus of a split complex number or split quaternion, can be real or pure imaginary and can be equal to 0 for $x \neq 0$. A split octonion x for which $|x| \neq 0$ has an inverse element (1.4).

If we regard the modulus $|y-x|$ of the difference $y-x$ between two split octonions x and y as the *distance* between them, we introduce into $\mathbb{O}'$ the pseudometric of the pseudo-Euclidean space R_4^8.

Split octonions can also be regarded as vectors in R_4^8, and the *inner product* (x, y) of two vectors x and y has the form (1.5). Therefore, the angle φ between two vectors x and y can be defined by formula (1.7).

In **1.2.3** we have mentioned the *theorem of A. Albert* [Al2], stating that the only associative algebras with 1 and with involution $x \to \overline{x}$, in which the elements x for which $x = \overline{x}$ are products of 1 by real numbers and for which the products $\overline{x}x$ are nondegenerate quadratic forms admitting composition, are the algebras $\mathbb{C}, \mathbb{C}', \mathbb{H}$ and $\mathbb{H}'$. If in this theorem we drop the condition of associativity, we obtain the algebras $\mathbb{C}, \mathbb{C}', \mathbb{H}, \mathbb{H}', \mathbb{O}$, and $\mathbb{O}'$.

Split octonions were defined by L. E. Dickson [Dic], hence the name "Cayley-Dickson numbers". For the history of split octonions see [Ro4, pp. 411–412].

1.3.3. Zorn Vector-Matrices. M. Zorn [Zo] has defined the algebra of *real vector-matrices*

$$X = \begin{bmatrix} \xi & x \\ y & \eta \end{bmatrix}, \tag{1.89}$$

where ξ, η are real numbers and x, y are vectors in R^3. The sum of the vector-matrix (1.89) and the vector-matrix $Y = \begin{bmatrix} \zeta & z \\ w & \omega \end{bmatrix}$ is the vector-matrix $X+Y$ with as entries the numbers $\xi + \zeta$ and $\eta + \omega$ and the vectors $x+z$ and $y+w$. The product of these vector-matrices is the vector-matrix XY with as entries the numbers $\xi\zeta - xw$ and $\eta\omega - yz$ and the vectors $\xi z + \omega x + [yw]$ and $\zeta y + \eta w + [xz]$ (xy is the inner product of the vectors x and y, $[xy]$ is the vector product of these vectors). If we replace real numbers and vectors in the definition of Zorn vector-matrices by complex numbers and vectors, we obtain *complex Zorn vector-matrices.*

Theorem 1.26. *The algebra of real Zorn vector-matrices is isomorphic to the algebra $\mathbb{O}'$ of split octonions.*

The algebra of complex Zorn vector-matrices is isomorphic to the algebra $\mathbb{C}\otimes\mathbb{O}$ of bioctonions.

The algebra of complex Zorn vector-matrices satisfying the conditions $\xi = \overline{\eta}, x = \overline{y}$ is isomorphic to the algebra $\mathbb{O}$ of octonions.

PROOF. Real Zorn vector-matrices are linear combinations of the basis vector-matrices

$$\begin{bmatrix}1 & 0\\ 0 & 1\end{bmatrix}, \begin{bmatrix}0 & e_1\\ e_1 & 0\end{bmatrix}, \begin{bmatrix}0 & e_2\\ e_2 & 0\end{bmatrix}, \begin{bmatrix}0 & e_3\\ e_3 & 0\end{bmatrix},$$
$$\begin{bmatrix}1 & 0\\ 0 & -1\end{bmatrix}, \begin{bmatrix}0 & -e_1\\ e_1 & 0\end{bmatrix}, \begin{bmatrix}0 & e_2\\ -e_2 & 0\end{bmatrix}, \begin{bmatrix}0 & -e_3\\ e_3 & 0\end{bmatrix}, \tag{1.90}$$

whose multiplication rules coincide with the multiplication rules of the basis elements $1, i, j, k, e, f, g, h$ of the algebra $\mathbb{O}'$. Hence we obtain the isomorphism of the algebra of real Zorn vector-matrices and the algebra $\mathbb{O}'$.

The second assertion can be proved analogously.

The third assertion follows since the algebra $\mathbb{O}$ is isomorphic to the subalgebra of $\mathbb{C}\otimes\mathbb{O}$ with basis $1, i, j, k, eI, fI, gI, hI$, therefore, $\mathbb{O}$ is isomorphic to the subalgebra of the algebra of complex vector-matrices with basis consisting of the first four vector-matrices (1.90) and of the products of the last four vector-matrices (1.90) by the imaginary unit I. This implies the conditions $\xi = \overline{\eta}, x = \overline{y}$. ■

1.3.4. Simple Alternative Algebras. Like associative algebras, alternative algebras contain *ideals*, and alternative algebras that have no ideals different from 0 and the algebra itself are called *simple alternative algebras.*

Theorem of M. Zorn [Zo]. *The only simple alternative algebras are the algebras $\mathbb{O}, \mathbb{O}'$, and $\mathbb{C}\otimes\mathbb{O}$.*

1.3.5. Automorphisms in Algebras of Octonions and Split Octonions. Like associative algebras, alternative algebras have ring and algebra automorphisms.

Theorem 1.27. *The ring automorphisms in each of the algebras $\mathbb{O}$ and $\mathbb{O}'$ form groups which are 14-dimensional subgroups of the groups SO_7 and $SO_{7,3}$, respectively.*

PROOF. Like in Theorem 1.15 (see **1.2.5**) we can prove that for any x in $\mathbb{O}$ and $\mathbb{O}'$ that is the product of the 1 in the algebra by a real number, $f(x) = x$. In $\mathbb{O}$ and $\mathbb{O}'$, for each element x whose square x^2 is a real number, $f(x)$ is such that $[f(x)]^2 = x^2$. In this case, if $f(x)$ is real, x is real. Therefore, if $f(x)$ is not real, x and $f(x)$ are linear combinations of the basis elements $i, j, \dots, r$ in $\mathbb{O}$ or $i, j, \dots, h$ in $\mathbb{O}'$.

We first consider $\mathbb{O}$. For an x such that $|x| = 1$, $x^2 = -1$, we have $[f(x)]^2 = -1$, and so $|f(x)| = 1$. If two such elements x and y, regarded as vectors in R^7, are orthogonal, they satisfy the condition $xy = -yx$. Therefore, $f(x)f(y) = -f(y)f(x)$ and the elements $f(x)$ and $f(y)$, regarded as vectors in R^7, are also orthogonal.

Therefore, each ring automorphism of $\mathbb{O}$ leaves invariant the real axis of this algebra and determines a motion in R^7 consisting of the octonions satisfying the condition $\overline{x} = -x$. Since this motion preserves 0 and the orientation of orthonormal frames, it is a rotation of the first kind; the group of all these rotations is isomorphic to SO_7. These rotations can be written in orthogonal coordinates as

$$\left.\begin{aligned}
'x^1 &= x^1, \\
'x^2 &= x^2 \cos \varphi_0 + x^3 \sin \varphi_0, \\
'x^3 &= -x^2 \sin \varphi_0 + x^3 \cos \varphi_0, \\
'x^4 &= x^4 \cos \varphi_1 + x^5 \sin \varphi_1, \\
'x^5 &= -x^4 \sin \varphi_1 + x^5 \cos \varphi_1, \\
'x^6 &= x^6 \cos \varphi_2 + x^7 \sin \varphi_2, \\
'x^7 &= -x^6 \sin \varphi_2 + x^7 \cos \varphi_2.
\end{aligned}\right\} \tag{1.91}$$

Since the automorphisms in $\mathbb{O}$ preserve its associative subalgebras, if an automorphism $f(x)$ in $\mathbb{O}$ leaves invariant the basis element i, it preserves the subalgebras with basis elements $1, i, j, k; 1, i, \ell, p; 1, i, q, r$ and it maps the basis elements in $\mathbb{O}$ as follows:

$$\begin{aligned}
f(i) &= i, \\
f(j) &= j \cos \varphi_0 + k \sin \varphi_0, \\
f(k) &= -j \sin \varphi_0 + k \cos \varphi_0, \\
f(\ell) &= \ell \cos \varphi_1 + p \sin \varphi_1, \\
f(p) &= -\ell \sin \varphi_1 + p \cos \varphi_1, \\
f(q) &= q \cos \varphi_2 + r \sin \varphi_2, \\
f(r) &= -q \sin \varphi_2 + r \cos \varphi_2.
\end{aligned}$$

Since $j = q\ell$, we obtain

$$\begin{aligned}
f(j) =& j \cos \varphi_0 + k \sin \varphi_0, \\
f(q)f(\ell) =& (q \cos \varphi_2 + r \sin \varphi_2)(\ell \cos \varphi_1 + p \sin \varphi_1) \\
=& j(\cos \varphi_1 \cos \varphi_2 - \sin \varphi_1 \sin \varphi_2) \\
& - k(\cos \varphi_1 \sin \varphi_2 + \sin \varphi_1 \cos \varphi_2) \\
=& j \cos(\varphi_1 + \varphi_2) - k \sin(\varphi_1 + \varphi_2).
\end{aligned}$$

Therefore,

$$\varphi_0 + \varphi_1 + \varphi_2 = 0. \tag{1.92}$$

An automorphism in $\mathbb{O}$ is characterized by: an element in $\mathbb{O}$, for which we can take the basis element i; the planes of the basis elements j and k, ℓ and p, and q and r, in which the rotations over the angles φ_0, φ_1, and φ_2 are realized; and by two of these three angles. Instead of i we can take an octonion x satisfying $x = -\overline{x}$ and $|x| = 1$. The dimension of the set of such octonions is 6. The plane of the vectors j and k is determined by one element j, for which we can take an octonion y satisfying the same two conditions plus the condition of orthogonality of y and i, that is $iy+yi = 0$; if we have fixed the elements i and j, then $k = ij$. However, for j we can take any linear combination $j\cos\alpha + k\sin\alpha$, and so the dimension of the set of planes of the basis elements j and k is 4. Analogously we find that the dimension of the set of planes of the basis elements ℓ and p is 2. If the basis element i and the planes of the basis elements j, k and l, p are chosen, the plane of the basis elements q and r is determined as the orthogonal complement of the 5-plane containing i, j, k, ℓ, p. Therefore, the dimension of the group of automorphisms in $\mathbb{O}$ is $6+4+2+2 = 14$.

Since in $\mathbb{O}'$ the pseudometric of R^8_4 is defined, we can similarly prove that the automorphisms in $\mathbb{O}'$ are represented by rotations of the first kind in R^7_3, that is, by matrices in the group $SO_{7,3}$. These rotations can be written in orthogonal coordinates in the form (1.91) or in the form

$$\left.\begin{aligned} 'x^1 &= x^1 \cosh\varphi_0 + x^6 \sinh\varphi_0 \\ 'x^6 &= x^1, \sinh\varphi_0 + x^6 \cos\varphi_0, \\ 'x^2 &= x^2, \cosh\varphi_1 + x^5 \sinh\varphi_1, \\ 'x^5 &= x^2 \sinh\varphi_1 + x^5 \cosh\varphi_1, \\ 'x^3 &= x^3 \cosh\varphi_2 + x^4 \sinh\varphi_2, \\ 'x^4 &= x^3 \sinh\varphi_2 + x^4 \cosh\varphi_2, \\ 'x^7 &= x^7. \end{aligned}\right\} \tag{1.93}$$

In the first case formula (1.92) is proved analogously to the case of automorphisms in $\mathbb{O}$. In the second case the automorphism in $\mathbb{O}'$ maps the basis elements in $\mathbb{O}'$ as follows:

$$\begin{aligned} f(i) &= i\cosh\varphi_0 + g\sinh\varphi_0, \\ f(g) &= i\sinh\varphi_0 + g\cosh\varphi_0, \\ f(j) &= j\cosh\varphi_1 + f\sinh\varphi_1, \\ f(f) &= j\sinh\varphi_1 + f\cosh\varphi_1, \\ f(k) &= k\cosh\varphi_2 + e\sinh\varphi_2, \\ f(e) &= k\sinh\varphi_2 + e\cosh\varphi_2, \\ f(h) &= h. \end{aligned}$$

Since $i = jk$, we obtain

$$\begin{aligned} f(i) =& i\cosh\varphi_0 + g\sinh\varphi_0, \\ f(j)f(k) =& (j\cosh\varphi_1 + f\sinh\varphi_1)(k\cosh\varphi_2 + e\sinh\varphi_2) \\ =& i\cosh\varphi_1\cosh\varphi_2 + \sinh\varphi_1\sinh\varphi_2) \\ & - g(\cosh\varphi_1\sinh\varphi_2 + \sinh\varphi_1\cosh\varphi_2) \\ =& i\cosh(\varphi_1+\varphi_2) - g\sinh(\varphi_1+\varphi_2). \end{aligned}$$

Therefore, again we obtain formula (1.92).

The dimension of the group of automorphisms in $\mathbb{O}'$, which is calculated analogously to the dimension of the group of automorphisms in $\mathbb{O}$, is also 14. ∎

1.3.6. Tensor Products. In **1.3.2** we have defined the algebra $\mathbb{C}\otimes\mathbb{O}$ of bioctonions. Analogously we can define the tensor products $\mathbb{C}'\otimes\mathbb{O}$ and $\mathbb{C}^0\otimes\mathbb{O}$ of *dyoctonions* and *duooctonions*. The first of these algebras is a semisimple alternative algebra, the second is a quasisimple alternative algebra obtained from the simple alternative algebra $\mathbb{C}\otimes\mathbb{O}$ by means of the transition process described in **1.1.10** and corresponding to the involutive automorphism $x \to \tilde{x}$.

Analogously we can define the tensor products $\mathbb{H}\otimes\mathbb{O}$ and $\mathbb{O}\otimes\mathbb{O}$ of *quateroctonions* and *octooctonions*, as well as the tensor products obtained from these algebras by means of replacing one or both factors by the corresponding split algebras.

These algebras are not alternative; in general, neither two elements nor one element of these algebras can generate an associative algebra. For instance, if the element x of these algebras has the form $x = iI + jJ + 1$, then $(xx)x = 3iI + 3jJ + 1 + 2rK$, $x(xx) = 3iI + 3jJ + 1 - 2rK$, and $(xx)x - x(xx) = 4rK$.

1.3.7. Semioctonions, Split Semioctonions, $\frac{1}{4}$-Octonions, Split $\frac{1}{4}$-Octonions, $\frac{1}{8}$-Octonions. In $\mathbb{O}$ and $\mathbb{O}'$ there are the following involutive algebra automorphisms

$$\begin{aligned} x = x_0 + ix_1 + jx_2 + kx_3 + lx_4 + px_5 + qx_6 + rx_7 \to \\ \to \hat{x} = x_0 + ix_1 + jx_2 + kx_3 - lx_4 - px_5 - qx_6 - rx_7, \end{aligned} \tag{1.94}$$

$$\begin{aligned} x = x_0 + ix_1 + jx_2 + kx_3 + ex_4 + fx_5 + gx_6 + hx_7 \to \\ \to \hat{x} = x_0 + ix_1 + jx_2 + kx_3 - ex_4 - fx_5 - gx_6 - hx_7, \end{aligned} \tag{1.95}$$

$$\begin{aligned} x = x_0 + ix_1 + jx_2 + kx_3 + ex_4 + fx_5 + gx_6 + hx_7 \to \\ \to \check{x} = x_0 + ix_1 - jx_2 - kx_3 + ex_4 + fx_5 - gx_6 - hx_7. \end{aligned} \tag{1.96}$$

The quasisimple alternative algebra obtained from $\mathbb{O}$ by means of the transition process described in **1.1.10** and corresponding to the involutive automorphism (1.94) is called the algebra of *semioctonions* and is denoted by $\mathbb{O}^0$. This algebra has basis elements $1, i, j, k, \epsilon, \vartheta, \kappa, \lambda$ and is isomorphic to the subalgebra of the algebra $\mathbb{C}^0\otimes\mathbb{O}$ with basis elements $1, i, j, k, \ell\mathcal{E}, p\mathcal{E}, q\mathcal{E}, r\mathcal{E}$.

The same algebra can be obtained from $\mathbb{O}'$ by means of the same transition process but corresponding to the involutive automorphism (1.95); it is isomorphic to the subalgebra of the algebra $\mathbb{C}^0\otimes\mathbb{O}'$ with basis elements $1, i, j, k, e\mathcal{E}, f\mathcal{E}, g\mathcal{E}, h\mathcal{E}$.

The quasisimple alternative algebra obtained from $\mathbb{O}'$ by means of the same transition process and corresponding to the involutive automorphism (1.96) is called the algebra of *split semioctonions* and is denoted by $\mathbb{O}'^0$. This algebra has basis elements $1, i, e, f, \epsilon, \vartheta, \zeta, \mu$ and is isomorphic to the subalgebra of the algebra $\mathbb{C}^0\otimes\mathbb{O}'$ with basis elements $1, i, e, f, j\mathcal{E}, k\mathcal{E}, g\mathcal{E}, h\mathcal{E}$.

Each semioctonion $x = x_0 + ix_1 + jx_2 + kx_3 + \epsilon x_4 + \vartheta x_5 + \kappa x_6 + \lambda x_7$ or split semioctonion $x = x_0 + ix_1 + ex_2 + fx_3 + \epsilon x_4 + \vartheta x_5 + \zeta x_6 + \mu x_7$ has a *conjugate*,

$\overline{x}$, which differs from x in that the coordinates $x_1, x_2, \ldots, x_7$ are multiplied by -1. The following equalities are fulfilled: (1.1), (1.46) for $\mathbb{O}^0$ and (1.59) for $\mathbb{O}'^0$, and (1.3).

Using the same transition processes we can obtain from $\mathbb{O}^0$ and $\mathbb{O}'^0$ the biquasisimple alternative algebras $\mathbb{O}^{00}$ and $\mathbb{O}'^{00}$ of $\frac{1}{4}$*-octonions* and *split* $\frac{1}{4}$*-octonions* and the 3-quasisimple alternative algebra $\mathbb{O}^{000}$ of $\frac{1}{8}$*-octonions.* As in $\mathbb{O}, \mathbb{O}', \mathbb{O}^0, \mathbb{O}'^0$, in these algebras each element x has a conjugate element, $\overline{x}$, and the equalities (1.1), (1.2) for $\mathbb{O}^{00}$, (1.10) for $\mathbb{O}'^{00}$, and (1.15) for $\mathbb{O}^{000}$, and (1.3) are fulfilled.

§1.4. Lie Algebras and Lie Groups

1.4.1. Real and Complex Lie Groups and Algebras. In **0.8.2** we have defined real Lie groups, in **0.8.3** - solvable, simple, semisimple, and reductive real Lie groups, and in **0.8.4** - real Lie algebras tangent to real Lie groups.

If in these definitions we replace real coordinates of elements by complex ones, we obtain *complex Lie groups and algebras.* Since in real r-dimensional Lie groups the functions $z^i = f^i(x^1, x^2, \ldots, x^r; y^1, y^2, \ldots, y^r)$ of coordinates of products $z = xy$ can always be assumed to be analytic functions, each real Lie group has a complex form determined by the same formulas.

The complex form of the real Lie group G is denoted by $\mathbb{C}G$; the complex form of the real Lie algebra $\mathbb{G}$ is denoted by $\mathbb{C} \otimes \mathbb{G}$. In particular, the complex forms of the general linear group GL_n, the special linear group SL_n, the orthogonal or pseudoorthogonal group O_n and $O_{n,\ell}$, and the symplectic group Sp_{2n} are the complex groups $\mathbb{C}GL_n$, $\mathbb{C}SL_n$, $\mathbb{C}O_n$, and $\mathbb{C}Sp_{2n}$, respectively. The last three groups are determined by the same formulas $\det U = 1$, (0.59), and (0.61) as for the corresponding real groups, and the complex forms of the tangent Lie algebras of these groups are the complex algebra $\mathbb{C}_n$, its linear subspace determined by the condition $\operatorname{Tr} A = 0$ and the Lie algebras determined by equations (0.66) and (0.68).

1.4.2. Split Complex and Dual Lie Groups. If in the definition of the complex Lie group $\mathbb{C}G$ we replace complex coordinates by coordinates from $\mathbb{C}'$ or $\mathbb{C}^0$, we obtain the *split complex* or the *dual Lie group* $\mathbb{C}'G$ or $\mathbb{C}^0G$, respectively.

Theorem 1.28. *The split complex Lie group $\mathbb{C}'G$ is isomorphic to the direct product $G \otimes G$.*

The dual Lie group $\mathbb{C}^0G$ is isomorphic to the direct product $G \otimes \mathbb{G}$ of the group G by its tangent Lie algebra $\mathbb{G}$.

PROOF. The assertions of this theorem follow from the properties of analytic functions of split complex and dual variables $f(x) = \sum_k a_k x^k$ with real coefficients a_k, mentioned in **1.1.5**, i.e.

$$f(x_+e_+ + x_-e_-) = f(x_+)e_+ + f(x_-)e_-$$

and

$$f(x_0 + \epsilon x_1) = f(x_0) + x_1 f'(x_0).$$

Analytic functions of several split complex or dual variables have analogous properties:

$$\begin{aligned}&z^i_+e_+ + z^i_-e_-\\&=f^i(x^1_+e_+ + x^1_-e_-, \dots, x^n_+e_+ + x^n_-e_-;\ y^1_+e_+ + y^1_-e_-, \dots, y^n_+e_+ + y^n_-e_-)\\&=f^i(x^1_+, \dots, x^n_+;\ y^1_+, \dots, y^n_+)e_+ + f^i(x^1_-, \dots, x^n_-;\ y^1_-, \dots, y^n_-)e_-,\end{aligned}\tag{1.97}$$

and

$$\begin{aligned}z^i_0 + \epsilon z^i_1 =&f^i(x^1_0 + \epsilon x^1_1, \dots, x^n_0 + \epsilon x^n_1;\ y^1_0 + \epsilon y^1_1, \dots, y^n_0 + \epsilon y^n_1)\\=&f^i(x^1_0, \dots, x^n_0; y^1_0, \dots, y^n_0)\\&+\epsilon\left(x^1_1\frac{\partial f^i}{\partial x^1_0} + \dots + x^n_1\frac{\partial f^i}{\partial x^n_0} + y^1_1\frac{\partial f^i}{\partial y^1_0} + \dots + y^n_1\frac{\partial f^i}{\partial y^n_0}\right).\end{aligned}\tag{1.98}$$

From (1.97) we find that the split complex Lie group $\mathbb{C}'G$ with coordinates $x^i = x^i_+e_+ + x^i_-e_-$ is isomorphic to the direct product of two isomorphic real Lie groups G with coordinates x^i_+ and x^i_-.

Comparison of formula (1.98) with the formula

$$\frac{dz^i_0}{dt} = \frac{dx^1_0}{dt}\frac{\partial f^i}{\partial x^1_0} + \dots + \frac{dx^n_0}{dt}\frac{\partial f^i}{\partial x^n_0} + \frac{dy^1_0}{dt}\frac{\partial f^i}{\partial y^1_0} + \dots + \frac{dy^n_0}{dt}\frac{\partial f^i}{\partial y^n_0}$$

shows that the coordinates x^i_1, y^i_1, z^i_1 can be viewed as derivates $\frac{dx^i_0}{dt}, \frac{dy^1_0}{dt}, \frac{dz^i_0}{dt}$, that is, as coordinates of elements in $\mathbb{G}$. Therefore, the dual Lie group $\mathbb{C}^0G$ with coordinates $x^i = x^i_0 + \epsilon x^i_1$ is isomorphic to the direct product of the group G with coordinates x^i_0 and x^i_1 by its tangent Lie algebra $\mathbb{G}$. ■

1.4.3. Root Vectors of Simple Complex Lie Groups. Consider a noncommutative simple complex Lie group of the infinite series A_n, B_n, C_n, or D_n. For noncommutative semisimple Lie groups the Killing-Cartan quadratic form (0.71) is nondegenerate. In the case of complex Lie groups this form can be regarded as the square of the modulus of a vector in a complex Euclidean space, the tangent Lie algebra of this group.

With each vector A in this Lie algebra we can associate the linear operator $\hat{A}$ of the linear transformation $X \to \hat{A}X = [AX]$ with matrix $(A^i_j) = (C^i_{jk}A^k)$. Therefore the Killing-Cartan form (0.71) for the vector A can be written as

$$|A|^2 = -C^l_{ki}C^k_{ij}A^iA^j = -A^l_kA^k_l = -\mathrm{Tr}(\hat{A}^2).\tag{1.99}$$

Since $[AA] = 0$, the operator $\hat{A}$ has a zero eigenvalue. The multiplicity of this zero eigenvalue is different for different vectors A in the tangent Lie algebra of a semisimple Lie group. The minimal multiplicity of the zero eigenvalue of all vectors A is called the *rank* of the semisimple Lie group.

The vectors A in the tangent Lie algebra of a semisimple Lie group for which this multiplicity is minimal are called *regular vectors.*

If A is a regular vector, the set of all vectors X for which $[AX] = 0$ is a subalgebra of the Lie algebra, called the *Cartan subalgebra.* The dimension of this subalgebra is equal to the rank of the Lie group. The subgroup of the Lie group corresponding to this subalgebra is called the *Cartan subgroup.* This subgroup consists of all group

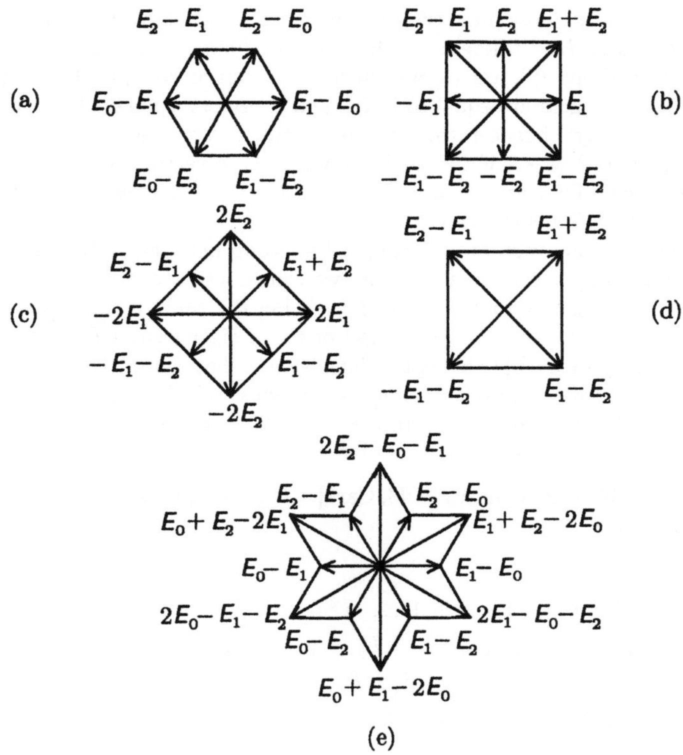

FIG. 1.2

elements that commute with all elements in the one-parameter subgroup tangent to the regular vector A in the Lie algebra.

W. Killing [Kil3] proved that the operators $\hat{A}$ corresponding to all vectors $A = a^i E_i$ of such a subalgebra have common eigenvectors and the corresponding eigenvalues are linear forms αA with A vectors in the Cartan subalgebra. Since the eigenvalues of a linear operator $\hat{A}$ are the roots of the characteristic equation det $(\hat{A} - \lambda I) = 0$ of this operator, these linear forms are called *roots* or *root forms* of the Lie group. The covectors α determining these linear forms are called *root vectors* of the semisimple Lie group.

Since a Cartan subalgebra endowed with the Euclidean geometry in the Lie algebra is also a Euclidean space, the covectors in this subalgebra can be regarded as vectors.

The root vectors of a simple Lie group form the *root system* of this group. The classification of simple Lie groups can be reduced to the classification of their root systems. All root vectors are linear combinations of basis vectors E_i in the Euclidean space $\mathbb{C}R^n$, or in the space $\mathbb{C}R^{n+1}$ containing this Euclidean space as a hyperplane, with real coefficients. Therefore these vectors are vectors in the real

Euclidean space R^n, whose dimension n is equal to the rank of the group.

The quotients $2\alpha\beta/\alpha^2$ where $\alpha\beta$ is the inner products of the root vectors α and β in the metric of this R^n, are always integers and can be equal to 0, ± 1, ± 2, and ± 3 only. These integers are equal to 0 if α and β are orthogonal. They are equal to 1 or -1 if the angle between these vectors is equal to $\pi/3$ or $2\pi/3$, respectively. They are equal to 2 or -2 if the angle between these vectors is equal to $\pi/4$ or $3\pi/4$, respectively. They are equal to 3 or -3 if the angle between these vectors is equal to $\pi/6$ or $5\pi/6$, respectively.

The system of root vectors of a simple Lie group determines the tangent Lie algebra of this group, since in this algebra we can choose a vector basis consisting of n linearly independent root vectors H (n being the rank of the group), which are vectors in the Cartan subalgebra, and of $r - n$ eigenvectors E_α of the linear transformations $X \to [HX]$ (r being the dimension of the group). The commutators of these vectors are determined by following relations: $[H_iH_j] = 0$, $[HE_\alpha] = \alpha A E_\alpha$ (αA is the root form corresponding to the root α), $[E_\alpha E_{-\alpha}] = \alpha$, $[E_\alpha E_\beta] = E_{\alpha+\beta}$ if $\alpha + \beta$ is a root vector and $= 0$ if $\alpha + \beta$ is not a root vector.

Theorem of Killing-Cartan [Kil3, Car1]. *The root systems of the simple complex Lie groups are:*

$$\left.\begin{array}{ll}
A_n: & E_i - E_j,\ i \neq j,\ i,j = 0,\ 1, \dots, n. \\
B_n: & \pm E_i,\ i = 1,2,\dots,n, \pm E_i \pm E_j,\ i,j = 1,2,\dots,n. \\
C_n: & \pm 2E_i,\ i = 1,2,\dots,n,\ \pm E_i \pm E_j,\ i,j = 1,2,\dots,n. \\
D_n: & \pm E_i \pm E_j,\ i < j,\ i,j = 1,2,\dots,n. \\
G_2: & E_i - E_j \pm \left(\sum_i E_i - 3E_j\right),\ i,j = 0,1,2. \\
F_4: & \pm E_i,\ \pm E_i \pm E_j, \frac{1}{2}(\pm E_1 \pm E_2 \pm E_3 \pm E_4),\ i,j = 1,2,3,4. \\
E_6: & E_i - E_j,\ \pm\sqrt{2}E_0,\ \pm\left(\frac{\sqrt{2}}{2}E_0 + \frac{1}{2}\sum_i E_i - E_j - E_k - E_l\right), \\
 & i,j,k,l = 1,2,3,4,5,6. \\
E_7: & E_i - E_j, \pm\left(\frac{1}{2}\sum_i E_i - E_h - E_j - E_k - E_l\right), \\
 & h,i,j,k,l = 1,2,3,4,5,6,7. \\
E_8: & \pm E_i \pm E_j,\ \pm\left(\frac{1}{2}\sum_i E_i - E_j\right),\ \pm\left(\frac{1}{2}\sum_i E_i - E_k - E_l\right), \\
 & i,j,k,l = 1,2,3,4,5,6,7,8.
\end{array}\right\} \qquad (1.100)$$

The complex dimension r of a complex simple Lie group is equal to the sum of number of vectors in the root system of this group and to its rank n.

For the group A_n this dimension r is equal to $n(n+2)$, for the groups B_n and

C_n it is equal to $n(2n+1)$, for the group D_n it is equal to $n(2n-1)$, for the groups G_2, F_4, E_6, E_7, and E_8 it is equal to 14, 52, 78, 133, and 248, respectively.

Note that the dimension r is always a multiple of the rank n, and the integer

$$h = \frac{r}{n} - 1 = \frac{r-n}{n} \tag{1.101}$$

is called the *Coxeter number* of the simple Lie group. For the groups A_n, B_n and C_n, D_n, G_2, F_4 and E_6, E_7, and E_8 these numbers are equal to $n+1$, $2n$, $2n-2$, 6, 12, 18, and 30, respectively.

The root systems of the groups A_1, B_1, C_1 consist of the two vectors $\pm E_1$, the root systems of the groups A_2, B_2, C_2, D_2, and G_2 are depicted in Fig. 1.2a,b,c,d,e, respectively.

Theorem 1.29. *The groups $\mathbb{C}SL_{n+1}$, $\mathbb{C}O_{2n+1}$, $\mathbb{C}Sp_{2n}$ and $\mathbb{C}O_{2n}$ are complex simple Lie groups of the classes A_n, B_n, C_n, and D_n, respectively.*

PROOF. For the proof it is sufficient to calculate the root systems of these groups and to convince ourselves that these root systems coincide with the root systems of the groups A_n, B_n, C_n, and D_n, respectively.

The matrices of the group $\mathbb{C}SL_{n+1}$ in the basic case can be reduced to diagonal form with diagonal entries $u_0, u_1, \ldots, u_n$ whose product is equal to 1. The tangent Lie algebra of this group consists of the matrices in the algebra $\mathbb{C}_{n+1}$ with zero trace. In the basic case these matrices can also be reduced to diagonal form with diagonal entries $a_0, a_1, \ldots, a_n$ whose sum is equal to 0. If we denote by E_{ij} the matrix with entry at the intersection of the i-th row and the j-th column equal to 1 and with all other entries equal to 0, a regular element X in a Cartan subalgebra of this Lie algebra can be written as $X^k E_{kk}$.

Therefore, the commutator $[XE_{ij}]$ can be written as $[XE_{ij}] = (X^k E_{kk})E_{ij} - E_{ij}(X^k E_{kk})$. But formula (1.48) can be written as $E_{ij}E_{kl} = \delta_{jk}E_{il}$. Therefore, $[XE_{ij}] = X^i E_{ij} - X^j E_{ij} = (X^i - X^j)E_{ij}$. We see that all vectors E_{ij} for $i \neq j$ are eigenvectors of the linear operator $\widehat{X}$ and the corresponding eigenvalues are the linear forms $X^i - X^j$. If we denote the vectors E_{kk} by E_k, the covectors of these forms, that is, the root vectors, are the vectors $E_i - E_j$, and we obtain the first line in (1.100), that is, the root system of $\mathbb{C}SL_{n+1}$ coincides with the root system of A_n.

The group $\mathbb{C}O_{2n+1}$ is the subgroup of $\mathbb{C}SL_{2n+1}$ preserving the hypercone

$$x^1 x^{n+1} + x^2 x^{n+2} + \cdots + x^n x^{2n} + (x^{2n+1})^2 = 0. \tag{1.102}$$

The matrices in this group in the basic case can be reduced to diagonal form with diagonal entries $u_1, u_2, \ldots, u_n, u_1^{-1}, u_2^{-1}, \ldots, u_n^{-1}, 1$. The tangent Lie algebra of this group consists of the matrices in $\mathbb{C}_{2n+1}$ that, in the basic case, can also be reduced to diagonal form with diagonal entries $a_0, a_1, \ldots, a_n, -a_1, -a_2, \ldots, -a_n, 0$. A regular element X in a Cartan subalgebra of this Lie algebra can be reduced to the form $X^k E_{kk} - X^k E_{n+k,n+k}$. Therefore, the commutators $[XE_{IJ}]$ $(I, J = 1, 2, \ldots, 2n+1)$ can be written as

$$\left.\begin{aligned}
&[XE_{ij}] = (X^kE_{kk})E_{ij} - E_{ij}(X^kE_{kk}) = (X^i - X^j)E_{ij},\\
&[XE_{i,n+j}] = (X^kE_{kk})E_{i,n+j} - E_{i,n+j}(-X^{n+k}E_{n+k,n+k})\\
&\qquad = (X^i + X^j)E_{i,n+j},\\
&[XE_{n+i,j}] = (-X^kE_{n+k,n+k})E_{n+i,j} - E_{n+i,j}(X^kE_{kk})\\
&\qquad = (-X^i - X^j)E_{n+i,j},\\
&[XE_{n+i,n+j}] = (-X^kE_{n+k,n+k})E_{n+i,n+j} -\\
&E_{n+i,n+j}(-X^kE_{n+k,n+k}) = (-X^i + X^j)E_{n+i,j},\\
&[XE_{i,2n+1}] = (X^kE_{kk})E_{i,2n+1} = X^iE_{i,2n+1},\\
&[XE_{n+i,2n+1}] = (-X^kE_{n+k,n+k})E_{n+i,2n+1} = -X^iE_{n+i,2n+1}.
\end{aligned}\right\} \tag{1.103}$$

We see that all vectors E_{IJ} for $I \neq J$ are eigenvectors of the linear operator $\widehat{X}$ and the corresponding eigenvalues are the linear forms $\pm X^i \pm X^j$ and $\pm X^i$. The covectors of these forms, that is, the root vectors, are the vectors $\pm E_i \pm E_j$ and $\pm E_i$, and we obtain the second line in (1.100), that is, the root system of $\mathbb{C}O_{2n+1}$ coincides with the root system of B_n.

The group $\mathbb{C}O_{2n}$ is the subgroup of the group $\mathbb{C}SL_{2n}$ preserving the hypercone

$$x^1x^{n+1} + x^2x^{n+2} + \cdots + x^nx^{2n} = 0. \tag{1.104}$$

In the basic case the matrices in this group can be reduced to diagonal form with diagonal entries $u_1, u_2, \ldots, u_n, u_1^{-1}, u_2^{-1}, \ldots, u_n^{-1}$. The tangent Lie algebra of this group consists of the matrices in the algebra $\mathbb{C}_{2n}$ that, in the basic case, can also be reduced to diagonal form with diagonal entries $a_0, a_1, \ldots, a_n, -a_1, -a_2, \ldots, -a_n$. A regular element X in a Cartan subalgebra of this Lie algebra can be reduced to the same form $X^kE_{kk} - X^kE_{n+k,n+k}$, as for the group B_n. Therefore, the commutators $[XE_{IJ}]$ $(I, J = 1, 2, \ldots, 2n)$ can be written as the first four lines in (1.103). Therefore, all vectors E_{IJ} with $I \neq J$ are also eigenvectors of the linear operator $\widehat{X}$ and the corresponding eigenvalues are the linear forms $\pm X^i \pm X^j$. The covectors of these forms, that is, the root vectors, are the vectors $\pm E_i \pm E_j$, and we obtain the fourth line in (1.100), that is, the root system of $\mathbb{C}O_{2n}$ coincides with the root system of D_n.

Note that here the matrices of the Lie algebras of the groups $\mathbb{C}O_{2n+1}$ and $\mathbb{C}O_{2n}$ satisfy the condition (0.67) where E is not a diagonal matrix with diagonal entries ± 1, as in **0.5.6** and **0.8.2**, but for $\mathbb{C}O_{2n}$ is the $(2n \times 2n)$-matrix

$$E = \begin{bmatrix} 0 & I \\ I & 0 \end{bmatrix} \tag{1.105}$$

while for $\mathbb{C}O_{2n+1}$ it is the $((2n+1) \times (2n+1))$-matrix obtained from the $(2n \times 2n)$-matrix (1.105) by adjoining an $(2n+1)$-th row and $(2n+1)$-th column consisting of zeros (the equations (1.102) and (1.104) can also be written as $x^{\top}Ex = 0$). If the matrix E has this form and we write in (0.67) the matrix A as

$$A = \begin{bmatrix} A_1^1 & A_2^1 \\ A_1^2 & A_2^2 \end{bmatrix}, \tag{1.106}$$

or similarly with adjoined $(2n+1)$-th row and $(2n+1)$-th column consisting of zeros, the condition (0.67) can be written as

$$(A_1^1)^{\top} = -A_2^2, \quad (A_2^1)^{\top} = -A_2^1, \quad (A_1^2)^{\top} = -A_1^2. \tag{1.107}$$

The group $\mathbb{C}Sp_{2n}$ is the subgroup of $\mathbb{C}SL_{2n}$ satisfying the condition (0.61) where J is the $(2n \times 2n)$-matrix (0.62). The matrices in the tangent Lie algebra of this group are the matrices in $\mathbb{C}_{2n}$ satisfying the condition (0.68). If in (0.68) we write the matrix A in the form (1.106), this condition can be written as

$$(A_1^1)^{\top} = -A_2^2, \quad (A_2^1)^{\top} = A_2^1, \quad (A_1^2)^{\top} = A_1^2. \tag{1.108}$$

In the basic case the matrices in this group can be reduced to diagonal form with diagonal entries $u_1, u_2, \dots, u_n, u_1^{-1}, u_2^{-1}, \dots, u_n^{-1}$ and the matrices (1.106) of the tangent Lie algebra of this group can be reduced to diagonal form with diagonal entries $a_0, a_1, \dots, a_n, -a_1, -a_2, \dots, -a_n$. A regular element X in a Cartan subalgebra of this Lie algebra can be reduced to the same form, $X^k E_{kk} - X^k E_{n+k,n+k}$, as for B_n and D_n. Therefore, the commutators $[XE_{IJ}]$ $(I, J = 1, 2, \dots, 2n)$ can be written as the first four lines in (1.103). Therefore, all vectors E_{IJ} with $I \neq J$ are also eigenvectors of the linear operator $\widehat{X}$ and the corresponding eigenvalues are the linear forms $\pm X^i \pm X^j$. However, for the group $\mathbb{C}O_{2n}$ the formulas (1.107) show that the submatrices A_2^1 and A_1^2 in (1.106) are *skew symmetric*, and, therefore, do not contain diagonal entries and, therefore, in the linear forms $\pm X^i \pm X^j$ we cannot have i equal to j. For the group $\mathbb{C}Sp_{2n}$ the formulas (1.108) show that the submatrices A_2^1 and A_1^2 in (1.106) are *symmetric*, and, therefore, do contain diagonal entries and in the linear forms $\pm X^i \pm X^{ji}$ we can have i equal to j. In that case these forms can be written as $2X^i$. The covectors of these forms, that is, the root vectors, are the vectors $\pm E_i \pm E_j$ and $2E_i$, and we obtain the third line in (1.100), that is, the root system of the group $\mathbb{C}Sp_{2n}$ coincides with the root system of C_n. ■

1.4.4. Dynkin Diagrams. The vectors in the root systems (1.100) can be written as linear combinations of n *fundamental root vectors.* For these vectors we can choose *simple root vectors*, which are defined as follows: if we introduce in the set of vectors a^iE_i the *order correlation*: $a^iE_i > b^iE_i$ if the first coordinate $a^i \neq b^i$ is greater than b^i, we call a root vector *simple* if it is positive (i.e. greater than the zero vector) and cannot be written as a sum of other positive vectors. If we single out from the root vectors (1.100) the simple ones, we obtain

Theorem of E.Dynkin [Dy1]. *The systems of simple root vectors of the simple*

complex Lie groups are:

$$\left.\begin{aligned}
A_n :&\alpha_1 = E_0 - E_1, \alpha_2 = E_1 - E_2, \dots, \alpha_n = E_{n-1} - E_n.\\
B_n :&\alpha_1 = E_1 - E_2, \alpha_2 = E_2 - E_3, \dots, \alpha_{n-1} = E_{n-1} - E_n,\\
&\alpha_n = E_n.\\
C_n :&\alpha_1 = E_1 - E_2, \alpha_2 = E_2 - E_3, \dots, \alpha_{n-1} = E_{n-1} - E_n,\\
&\alpha_n = 2E_n.\\
D_n :&\alpha_1 = E_1 - E_2, \alpha_2 = E_2 - E_3, \dots, \alpha_{n-1} = E_{n-1} - E_n,\\
&\alpha_n = E_{n+1} + E_n.\\
G_2 :&\alpha_1 = E_0 - E_1 \ \alpha_2 = E_1 + E_2 - 2E_0.\\
F_4 :&\alpha_1 = E_2 - E_3, \alpha_2 = E_3 - E_4, \alpha_3 = E_4,\\
&\alpha_4 = \frac{1}{2}(E_1 - E_2 - E_3 - E_4).\\
E_6 :&\alpha_1 = E_1 - E_2, \alpha_2 = E_2 - E_3,\ \alpha_3 = E_3 - E_4,\ \alpha_4 = E_4 - E_5,\\
&\alpha_5 = E_5 - E_6,\ \alpha_6 = \frac{\sqrt{2}}{2}E_0 - \frac{1}{2}(E_1 + E_2 + E_3 - E_4 - E_5 - E_6).\\
E_7 :&\alpha_1 = E_1 - E_2,\ \alpha_2 = E_2 - E_3, \alpha_3 = E_3 - E_4, \alpha_4 = E_4 - E_5,\\
&\alpha_5 = E_5 - E_6, \alpha_6 = E_6 - E_7,\\
&\alpha_7 = \frac{1}{2}(E_0 - E_1 - E_2 - E_3 + E_4 + E_5 + E_6 + E_7).\\
E_8 :&\alpha_1 = E_2 - E_3,\ \alpha_2 = E_3 - E_4,\ \alpha_3 = E_4 - E_5,\ \alpha_4 = E_5 - E_6,\\
&\alpha_5 = E_6 - E_7,\ \alpha_6 = E_7 - E_8,\\
&\alpha_7 = \frac{1}{2}(E_1 - E_2 - E_3 - E_4 - E_5 - E_6 - E_7 + E_8),\ \alpha_8 = E_7 + E_8.
\end{aligned}\right\} \tag{1.109}$$

The systems of simple root vectors can be represented by introduced in [Dy1] *Dynkin diagrams* whose dots represent simple roots, and dots not joined if the root vectors are orthogonal and joined by a simple, double or triple line if the angle between the root vectors is equal to $2\pi/3$, $3\pi/4$, or $5\pi/6$, respectively. The Dynkin diagrams of the simple complex Lie groups are given in Fig. 1.3.

1.4.5. Cartan Matrices. The Dynkin diagrams are equivalent to the *Cartan matrices* with entries a_{ij} equal to the integers $2\alpha_i\alpha_j/\alpha_i^2$. The Cartan matrices of the simple complex Lie groups of the infinite series A_n, B_n, C_n, D_n, are

$$\begin{gathered}
\begin{pmatrix} 2 & -1 & 0 & \dots & 0 & 0\\ -1 & 2 & -1 & \dots & 0 & 0\\ 0 & -1 & 2 & \dots & 0 & 0\\ \dots & \dots & \dots & \dots & \dots & \dots\\ 0 & 0 & 0 & \dots & 2 & -1\\ 0 & 0 & 0 & \dots & -1 & 2 \end{pmatrix},
\begin{pmatrix} 2 & -1 & 0 & \dots & 0 & 0\\ -1 & 2 & -1 & \dots & 0 & 0\\ 0 & -1 & 2 & \dots & 0 & 0\\ \dots & \dots & \dots & \dots & \dots & \dots\\ 0 & 0 & 0 & \dots & 2 & -2\\ 0 & 0 & 0 & \dots & -1 & 2 \end{pmatrix},\\
\begin{pmatrix} 2 & -1 & 0 & \dots & 0 & 0\\ -1 & 2 & -1 & \dots & 0 & 0\\ 0 & -1 & 2 & \dots & 0 & 0\\ \dots & \dots & \dots & \dots & \dots & \dots\\ 0 & 0 & 0 & \dots & 2 & -1\\ 0 & 0 & 0 & \dots & -2 & 2 \end{pmatrix},
\begin{pmatrix} 2 & -1 & \dots & 0 & 0 & 0 & 0\\ -1 & 2 & \dots & 0 & 0 & 0 & 0\\ \dots & \dots & \dots & \dots & \dots & \dots & \dots\\ 0 & 0 & \dots & -1 & 2 & -1 & -1\\ 0 & 0 & \dots & 0 & -1 & 2 & 0\\ 0 & 0 & \dots & 0 & -1 & 0 & 2 \end{pmatrix}.
\end{gathered} \tag{1.110}$$

The Cartan matrices of the simple complex Lie groups of the five exceptional classes G_2, F_4, E_6, E_7, E_8 are

$$\begin{pmatrix} 2 & -1 \\ 3 & 2 \end{pmatrix}, \begin{pmatrix} 2 & -1 & 0 & 0 \\ -1 & 2 & -2 & 0 \\ 0 & -1 & 2 & -1 \\ 0 & 0 & -1 & 2 \end{pmatrix}, \begin{pmatrix} 2 & 0 & -1 & 0 & 0 & 0 \\ 0 & 2 & 0 & -1 & 0 & 0 \\ -1 & 0 & 2 & -1 & 0 & 0 \\ 0 & -1 & -1 & 2 & -1 & 0 \\ 0 & 0 & 0 & -1 & 2 & -1 \\ 0 & 0 & 0 & 0 & -1 & 2 \end{pmatrix},$$
$$\begin{pmatrix} 2 & 0 & -1 & 0 & 0 & 0 & 0 \\ 0 & 2 & 0 & -1 & 0 & 0 & 0 \\ -1 & 0 & 2 & -1 & 0 & 0 & 0 \\ 0 & -1 & -1 & 2 & -1 & 0 & 0 \\ 0 & 0 & 0 & -1 & 2 & -1 & 0 \\ 0 & 0 & 0 & 0 & -1 & 2 & -1 \\ 0 & 0 & 0 & 0 & 0 & -1 & 2 \end{pmatrix}, \begin{pmatrix} 2 & 0 & -1 & 0 & 0 & 0 & 0 & 0 \\ 0 & 2 & 0 & -1 & 0 & 0 & 0 & 0 \\ -1 & 0 & 2 & -1 & 0 & 0 & 0 & 0 \\ 0 & -1 & -1 & 2 & -1 & 0 & 0 & 0 \\ 0 & 0 & 0 & -1 & 2 & -1 & 0 & 0 \\ 0 & 0 & 0 & 0 & -1 & 2 & -1 & 0 \\ 0 & 0 & 0 & 0 & 0 & -1 & 2 & -1 \\ 0 & 0 & 0 & 0 & 0 & 0 & -1 & 2 \end{pmatrix}. \tag{1.111}$$

Since the Dynkin diagrams and the Cartan matrices of the simple complex Lie groups determine the systems of simple root vectors of these groups, and since these systems determine the complete root systems of these groups, the Dynkin diagrams and the Cartan matrices of the simple complex Lie groups determine the tangent Lie algebras of these groups and, therefore, these groups themselves up to local isomorphism.

1.4.6. Isomorphisms of Simple Complex Lie Groups. Since the Dynkin diagrams of the simple complex Lie groups A_1, B_1, and C_1 consist of a dot only, the Dynkin diagrams of the simple complex Lie groups B_2 and C_2 are similar, as are the Dynkin diagrams of the simple complex Lie groups A_3 and D_3 (Fig. 1.4a,b,c), these simple complex Lie are locally isomorphic. Since the Dynkin diagram of the semisimple complex Lie group D_2 consists of two separate dots, that is, of two Dynkin diagrams of the group A_1 (Fig. 1.4d), this semisimple complex Lie group is locally isomorphic to the direct product $A_1 \otimes A_1$.

1.4.7. Spherical Weyl Groups for Simple Lie Groups. Symmetry of the root system of a simple complex Lie group implies the existence of a group of rotations in the space R^n preserving this system. Analogous to the Galois group of an algebraic equation, for a simple Lie group, the complete group of these transformations is called the *Galois group*. The subgroup of this group consisting of the rotations generated by the reflections in hyperplanes through the common initial point of all root vectors orthogonal to these vectors, is called the *spherical Weyl group*.

Both the Galois group Γ and the spherical Weyl group W are finite; W is an invariant subgroup of Γ. These groups coincide for all simple Lie groups except A_n, D_n, and E_6. For A_n, D_n for $n \neq 4$, and E_6 the quotient group Γ/W consists of 2 elements, for D_4 this quotient group consists of $3! = 6$ elements. This fact is connected with the symmetries of the Dynkin diagrams of these groups: for A_n, D_n for $n \neq 4$, and E_6 these diagrams have two-fold symmetries, for A_n the root vectors α_i are symmetric to the root vectors α_{n+1-i}, for D_n the root vector α_{n-1} is

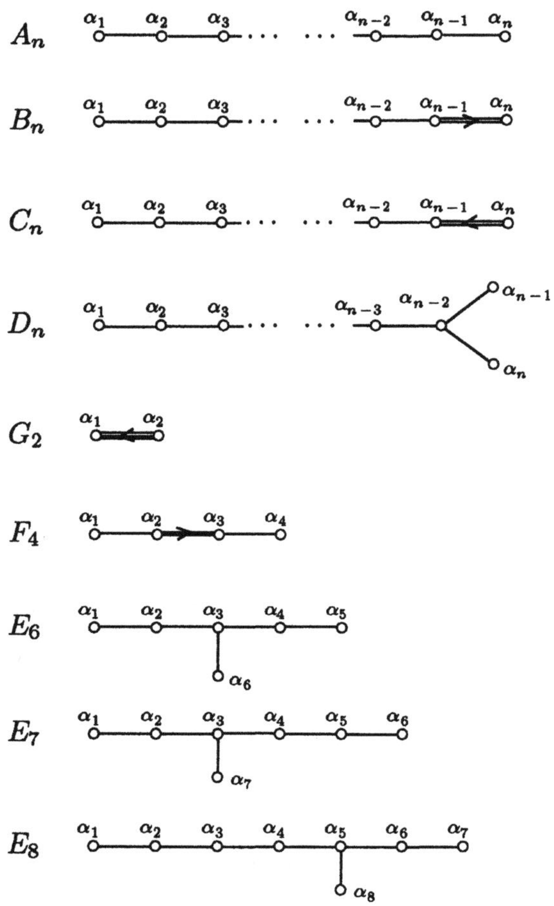

FIG. 1.3

symmetric to the root vector α_n, for E_6 the root vectors α_1 and α_2 are symmetric to the root vectors α_5 and α_4, respectively. The Dynkin diagram for D_4 has a three-fold symmetry of the root vectors α_1, α_3, and α_4 (Fig. 1.5).

The two-fold symmetries of the Dynkin diagrams for A_n, D_n, and E_6 determine *duality principles* in the spaces with these groups of transformations. The first of these principles is the classical duality principle of projective geometry, the second and third of these principles were discovered by É. Cartan in [Car11]. The three-fold symmetry of the Dynkin diagram of the group D_4 determines the *triality principle* in the corresponding space also discovered by É. Cartan in [Car11].

The spherical Weyl groups for simple Lie groups, as all finite groups of rotations of Euclidean space generated by reflections in hyperplanes, can be characterized by

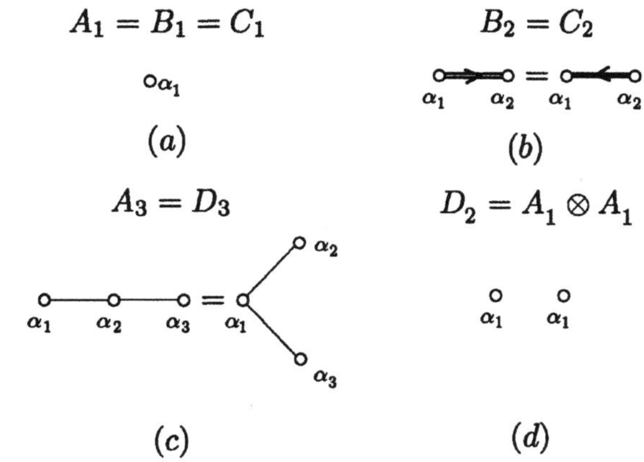

FIG. 1.4

the *Coxeter diagrams* introduced by H. S. M. Coxeter in [Cox3].

These diagrams are very similar to Dynkin diagrams and consist of dots representing the reflections which generate the group and lines joining these dots. The dots are not joined if the hyperplanes are orthogonal, are joined with a line without a number if the angle between the corresponding hyperplanes is equal to $\pi/3$, and are joined with a line with a number $n > 3$ if this angle is equal to π/n. The notations of these groups are also similar to the notations for the simple Lie groups. The Coxeter diagrams of finite groups of rotations generated by reflections are given in the Fig. 1.6.

Note that the finite group A_n is the group of symmetries of a regular n-dimensional *simplex*.

The finite group $B_n = C_n$ is the group of symmetries of an n-dimensional *cube* or *cocube* (which corresponds to a cube according to the projective duality principle, for $n = 3$ it is an octahedron).

The finite group D_n is the group of symmetries of an n-dimensional *semicube*, obtained from a cube by selecting of an vertex on each edge and rejecting of the other vertex on this edge.

The finite groups $G_2(r)$, H_3, F_4, and H_4 are the groups of symmetries of a regular plane r-gon, a regular 3-dimensional *icosahedron* and *dodecahedron*, a regular 4-dimensional *bicube* (obtained from a cube by adding to its 16 vertices the 8 vertices obtained by reflection of its center of symmetry in its 8 faces), and a regular 4-dimensional polyhedra with 120 and 600 faces, respectively.

The 24 vertices of a bicube in R^4 determining the finite group F_4 can be represented by the quaternions

$$\pm 1, \pm i, \pm j, \pm k, \frac{1}{2}(\pm 1 \pm i \pm j \pm k). \tag{1.112}$$

The group E_8 is the group of symmetries of a polyhedron in R^8 with 240 vertices, representable by the octonions

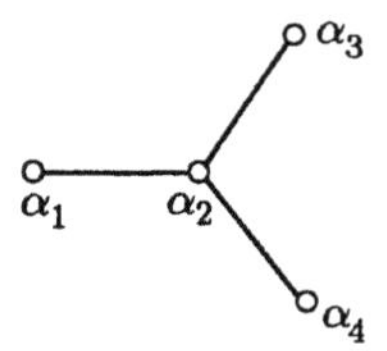

FIG. 1.5

$$\begin{gathered}
\pm 1, \pm i, \pm j, \pm k, \pm \ell, \pm p, \pm q, \pm r, \quad \frac{1}{2}(\pm 1 \pm j \pm k \pm p), \\
\frac{1}{2}(\pm 1 \pm k \pm i \pm q), \quad \frac{1}{2}(\pm 1 \pm i \pm j \pm r), \\
\frac{1}{2}(\pm 1 \pm p \pm q \pm r), \quad \frac{1}{2}(\pm 1 \pm i \pm \ell \pm p), \\
\frac{1}{2}(\pm 1 \pm j \pm \ell \pm q), \quad \frac{1}{2}(\pm 1 \pm k \pm \ell \pm r), \\
\frac{1}{2}(\pm i \pm \ell \pm q \pm r), \quad \frac{1}{2}(\pm j \pm \ell \pm r \pm p), \\
\frac{1}{2}(\pm k \pm \ell \pm p \pm q), \quad \frac{1}{2}(\pm i \pm j \pm k \pm \ell), \\
\frac{1}{2}(\pm j \pm k \pm q \pm r), \quad \frac{1}{2}(\pm k \pm i \pm r \pm p), \quad \frac{1}{2}(\pm i \pm j \pm p \pm q)
\end{gathered} \tag{1.113}$$

(see [Cox8]).

The finite groups E_6 and E_7 are isomorphic to the groups of collineations of the projective space P^3 and the plane P^2 preserving a *smooth cubic surface with* 27 *rectilinear generators* and a *quartic curve with* 28 *double tangents*, respectively.

The spherical Weyl group of a simple Lie group is the finite group of rotations generated by the reflections and has the same notation as the corresponding simple Lie group. The spherical Weyl group for the group G_2 is the finite group $G_2(6)$.

1.4.8. Coxeter Transformations. The transformations in spherical Weyl groups which are products, in any order, of all n reflections in hyperplanes orthogonal to all simple root vectors are especially important. These transformations, investigated by Coxeter [Cox4–5] are called *Coxeter transformations.*

The eigenvalues of the matrices of these transformations have the form $e^{2\pi i a_i/h}$, where h is the *Coxeter number* (1.101) of the simple Lie group and the numbers a_i, called the *exponents* of this group, are

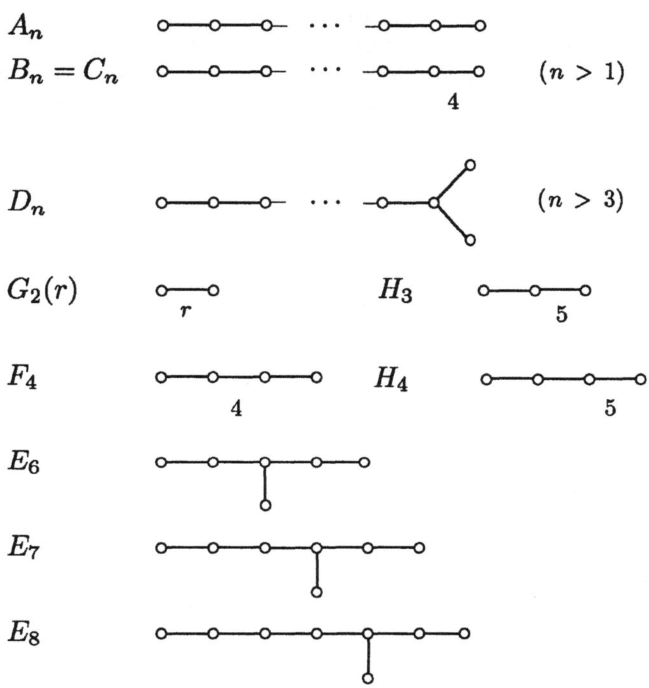

FIG. 1.6

$$\left.\begin{array}{ll}\text{for} & A_n: \quad 1,2,\ldots,n-1,n,\\ \text{for} & B_n \text{ and } C_n: \quad 1,3,5,\ldots,2n-1\\ \text{for} & D_n \text{ for odd } n: \quad 1,3,5,\ldots,n-2,n-1,n,\ldots,2n-3,\\ & \qquad \text{for even } n: \quad 1,3,5,\ldots,n-3,n-1,n-1,n+1,\ldots,2n-3,\\ \text{for} & G_2: \quad 1,5,\\ \text{for} & F_4: \quad 1,5,7,11,\\ \text{for} & E_6: \quad 1,4,5,7,8,11,\\ \text{for} & E_7: \quad 1,5,7,9,11,13,17,\\ \text{for} & E_8: \quad 1,7,11,13,17,19,23,29.\end{array}\right\} \quad (1.114)$$

Note that for simple Lie groups the sum of the exponents symmetric with respect to the middle of their row is equal to the Coxeter number h of this group.

The exponents of simple Lie groups were found by E. B. Dynkin [Dy2] (see also [KOV, p. 3]); the term was introduced by Coxeter [Cox5] and Kostant [Kos].

1.4.9. Extended Dynkin Diagrams. If the minimal root vector of a simple complex Lie group is $\mu = -\sum_i m_i\alpha_i$ and we add a dot representing this vector to the Dynkin diagram for this group, we obtain the *extended Dynkin diagram* for this group. For all simple complex groups these diagrams are given in Fig. 1.7 (over each dot of the Dynkin diagram the corresponding number m_i is written, whereas over the dot representing a minimal root vector the number 1 is written).

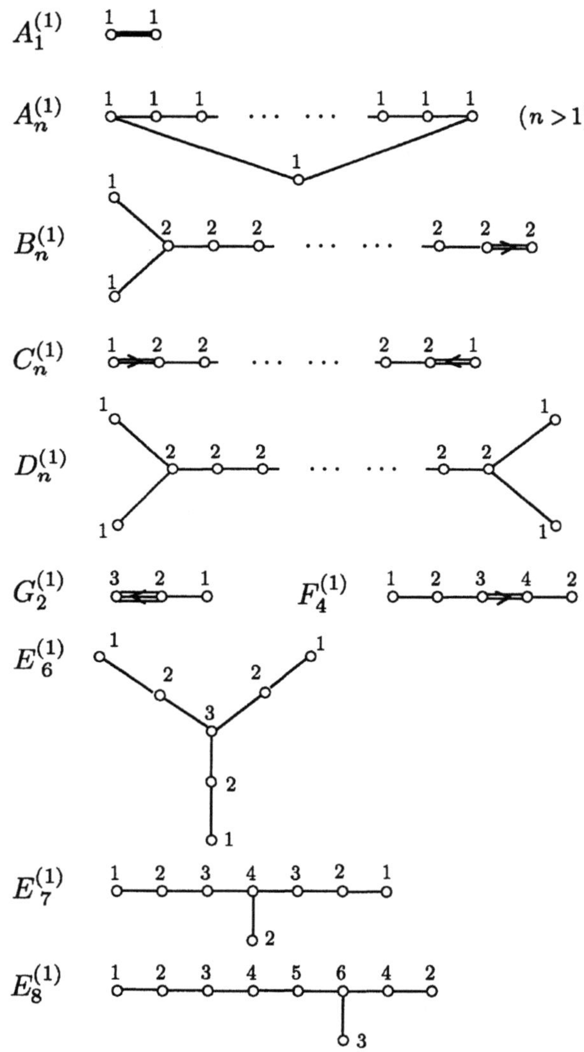

FIG. 1.7

Extended Dynkin diagrams also have symmetries. The diagram $A_n^{(1)}$ for $n > 1$ can be regarded as a circle with $n+1$ dots on its circumference dividing it into

$n+1$ equal arcs, therefore, this diagram admits a group of rotations isomorphic to the cyclic group Z_{n+1}. The diagrams $A_1^{(1)}$, $B_n^{(1)}$, $C_n^{(1)}$, and $E_7^{(1)}$ have two-fold symmetries. The diagram $D_n^{(1)}$ has four-fold symmetry. The diagram $E_6^{(1)}$ can be regarded as a figure with three-fold symmetry. The numbers determined by these symmetries for the simple Lie groups are: $n+1$ for A_n, 2 for B_n, C_n, and E_7, 4 for D_n, 1 for G_2, F_4, and E_8, and 3 for E_6. These numbers are equal to the determinants of the Cartan matrices of these groups and are called the *coefficients of connectedness* of these groups. These numbers determine important topological and algebraic properties of the compact real forms of these Lie groups.

1.4.10. Affine Weyl Groups for Simple Complex Lie Groups. Besides the spherical Weyl group of rotations in R^n with any simple complex Lie group there is also related a group of motions in this space. This group, discovered by Coxeter [Cox4], is generated by the reflections in the same hyperplanes as the spherical Weyl groups and by the reflection in the hyperplane through the terminal point of the minimal root vector orthogonal to this vector. This group is called the *affine Weyl group* for the simple complex Lie group. The affine Weyl group for a simple Lie group, as all discrete groups of motions in Euclidean space generated by reflections in hyperplanes, can be characterized by *Coxeter diagrams* also introduced by Coxeter in [Cox3]. Such a diagram is very similar to an extended Dynkin diagram, and also consists of dots representing reflections generating the group and of lines joining these dots. These dots are not joined, or are joined by a line without number of with a number n,in the same cases as for the spherical Weyl diagram. The notations of these groups are also similar to the notations of the extended Dynkin diagrams of the simple Lie groups. The Coxeter diagrams of the discrete groups of motions generated by reflections are given in Fig.1.8.

Note that although there are finite groups generated by reflections that are

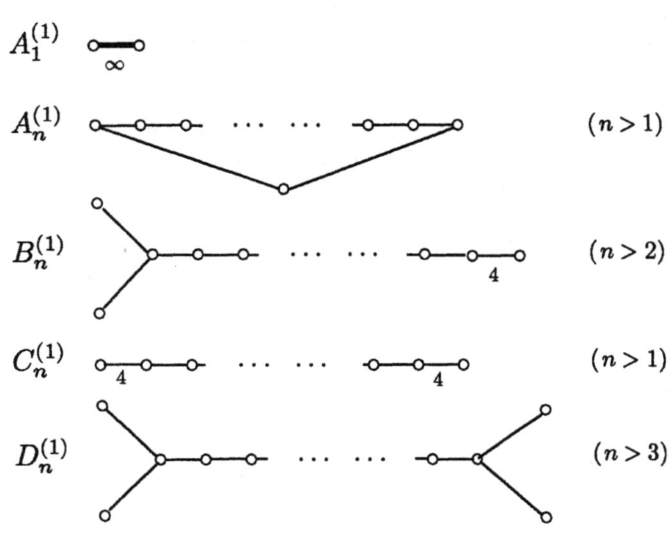

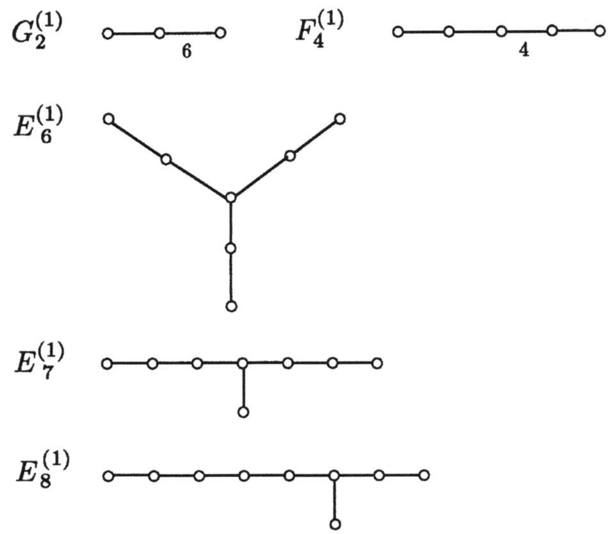

FIG. 1.8

not spherical Weyl groups of simple Lie groups, each discrete infinite group generated by reflections is the affine Weyl group of a simple Lie group.

1.4.11. Root Vectors of Compact Simple Real Lie Groups. The root forms and root vectors of simple real Lie groups are defined similarly as for simple complex Lie groups. For a compact simple real Lie group the Killing-Cartan form (0.71), which can also be written as (1.99), determines a real Riemannian metric in the Lie group and a real Euclidean metric in its tangent Lie algebra. The Cartan subalgebras of compact simple Lie groups of rank n, with metric induced by the Euclidean metric in the Lie algebra, are real spaces R^n. For compact simple real Lie groups all root forms are pure imaginary, as are all root vectors. Therefore, if the root vectors of the compact simple real Lie groups are written as $i\alpha$, for these groups formulas (1.100) and (1.109) are the same as for the simple complex groups. Therefore for compact simple real Lie groups the diagrams of root vectors and the Dynkin diagrams of simple root vectors are the same as for simple complex groups (see Fig. 1.2 and 1.3, respectively).

Analogously to Theorem 1.29 we can prove

Theorem 1.30. *The group $\mathbb{C}SU_{n+1}$ of complex special unitary $((n+1)\times(n+1))$-matrices, the group O_{2n+1} of real orthogonal $((2n+1)\times(2n+1))$-matrices, the intersection of the groups $\mathbb{C}SU_{2n}$ and $\mathbb{C}Sp_{2n}$ of complex $(2n\times 2n)$-matrices, and the group O_{2n} of real $(2n\times 2n)$-matrices are compact real groups of the classes A_n, B_n, C_n, and D_n, respectively.*

PROOF. The *complex special unitary group* $\mathbb{C}SU_n$ is the subgroup of the complex general linear group $\mathbb{C}GL_n$ consisting of the *unitary matrices*, i.e. the matrices satisfying the condition

$$\overline{U}^{\top} U = I. \tag{1.115}$$

The compactness of the groups O_n and $\mathbb{C}U_n$ of matrices $U = (U_i^j)$, where $U_i^j = V_i^j + iW_i^j$ for $\mathbb{C}U_n$ can be proved as follows: the condition (0.59) of orthogonality and the condition (1.115) of unitarity can be written as

$$\sum_k U_i^k U_j^k = \delta_{ij}, \quad \sum_k \overline{U}_i^k U_j^k = \sum_k V_i^k V_j^k + \sum_k W_i^k W_j^k = \delta_{ij}. \tag{1.116}$$

Summation of these expressions gives for $i = j$,

$$\sum_i \sum_k (U_i^k)^2 = n, \quad \sum_i \sum_k \overline{U}_i^k U_j^k = \sum_i \sum_k (V_i^k)^2 + \sum_i \sum_k (W_i^k)^2 = n. \tag{1.117}$$

Therefore, the entries of the matrices in O_n can be regarded as coordinates of points on a hypersphere of radius $\sqrt{n}$ in R^{n^2} and the entries of the matrices in $\mathbb{C}U_n$ can be regarded as coordinates of points on a hypersphere of the same radius in R^{2n^2}. Since these hyperspheres are compact manifolds and the groups O_n and $\mathbb{C}U_n$ can be regarded as intersections of these hyperspheres with the algebraic varieties (1.117), these groups are also compact. ■

The matrices in $\mathbb{C}SU_{n+1}$ always can be reduced to diagonal form with diagonal entries $u_0, u_1, \ldots, u_n$, having product equal to 1. From condition (1.115) it follows that the complex numbers u_i satisfy the condition $\overline{u}_i u_i = |u_i|^2 = 1$, that is, $u_i = e^{ia_i}$. Therefore, the tangent Lie algebra of this group consists of the complex $((n+1) \times (n+1))$-matrices with zero trace that can be reduced to diagonal form with pure imaginary diagonal entries $ia_0, ia_1, \ldots, ia_n$. The root forms and root vectors of this group have the same form as the root forms and root vectors of the group $\mathbb{C}SL_{n+1}$. The real dimensions of compact simple Lie groups are equal to the complex dimensions of the corresponding complex simple groups.

The matrices in the real groups O_{2n+1} and O_{2n}, which are intersections of $\mathbb{C}O_{2n+1}$ with $\mathbb{C}SU_{2n+1}$ and $\mathbb{C}O_{2n}$ with $\mathbb{C}SU_{2n}$ can also be reduced to diagonal form. The diagonal entries are equal to 1 or to complex numbers u_i satisfying the condition $|u_i| = 1$, that is $u_i = e^{ia_i}$. Therefore, the tangent Lie algebras of these groups consist of the complex matrices that can be reduced to diagonal form with diagonal entries equal to 0 or pure imaginary numbers $\pm ia_i$. The root forms and root vectors of these groups have the same form as the root forms and root vectors of $\mathbb{C}O_{2n+1}$ and $\mathbb{C}O_{2n}$.

The matrices in the intersection of the groups $\mathbb{C}SU_{2n}$ and $\mathbb{C}Sp_{2n}$ can also be reduced to diagonal form. The diagonal entries are complex numbers u_i satisfying the condition $|u_i| = 1$, that is, $u_i = e^{ia_i}$. Therefore, the tangent Lie algebra of this group consists of the complex matrices that can be reduced to diagonal form with diagonal entries equal to pure imaginary numbers $\pm ia_i$. The root forms and root vectors of this group have the same form as the root forms and root vectors of $\mathbb{C}Sp_{2n}$.

Theorem 1.31 (É. CARTAN [Car 13]). *The intersection of the groups* $\mathbb{C}SU_{2n}$ *and* $\mathbb{C}Sp_{2n}$ *is isomorphic to the group* $\mathbb{H}U_n$ *of quaternionic unitary* $(n\times n)$*-matrices, that is, the matrices in the group* $\mathbb{H}GL_n$ *satisfying the condition (1.115).*

PROOF. The compactness of the group $\mathbb{H}U_n$ of matrices $U = (U_i^j)$ with $U_i^j = V_i^j + iW_i^j + jX_i^j + kY_i^j$ can be proved as follows: the condition (1.115) of unitarity can be written in the form

$$\sum_k \overline{U}_i^k U_j^k = \sum_k V_i^k V_j^k + \sum_k W_i^k W_j^k + \sum_k X_i^k X_j^k + \sum_k Y_i^k Y_j^k = \delta_{ij}. \tag{1.118}$$

Summation of these expressions for gives $i = j$

$$\sum_i \sum_k \overline{U}_i^k U_i^k = \sum_i \sum_k (V_i^k)^2 + \sum_i \sum_k (W_i^k)^2 + \sum_i \sum_k (X_i^k)^2 + \sum_i \sum_k (Y_i^k)^2 = n. \tag{1.119}$$

Therefore, the entries of the matrices in the group $\mathbb{H}U_n$ can be regarded as coordinates of points on the hypersphere of radius $\sqrt{n}$ in R^{4n^2}. Since this hypersphere is a compact manifold and the group $\mathbb{H}U_n$ can be regarded as the intersection of this hypersphere with algebraic variety (1.119), this group is also compact.

In theorem 1.12 we have seen that the algebra $\mathbb{H}_n$ is isomorphic to the subalgebra of $\mathbb{C}_{2n}$ consisting of the matrices (1.57) where A and B are matrices in $\mathbb{C}_n$. We write quaternions z in the form $x + jy$ with x and y complex numbers, and similarly write quaternionic matrices U in the analogous form $V + jW$ with V and W complex matrices of the same order. If we replace in the quaternionic condition (1.115) the matrix U by $V + jW$, we obtain two conditions for the complex matrices V and W; these cobditions coincide with the conditions (0.61) and (1.115) for complex $(2n \times 2n)$-matrices. ■

Note that equality to 1 of the determinant of a complex symplectic $(2n \times 2n)$-matrix is equivalent to equality to 1 of the semideterminant of the corresponding quaternionic matrix.

Since differentiation of the conditions (1.115) gives for $t = 0$

$$\overline{A}^{\top} + A = 0, \tag{1.120}$$

the tangent Lie algebras $\mathbb{CU}_n$ and $\mathbb{HU}_n$ of the groups $\mathbb{C}U_n$ and $\mathbb{H}U_n$ consist of all complex, respectively quaternionic, skew-Hermitian $(n \times n)$-matrices, and the tangent Lie algebra $\mathbb{CSU}_n$ of the group $\mathbb{C}SU_n$ consists of the complex $(n \times n))$-matrices with zero trace.

Therefore, since the diagonal entries of complex and quaternionic skew-Hermitian matrices satisfy the condition $x = -\overline{x}$, the dimension of the group $\mathbb{C}SU_{n+1}$ is equal to $n + 2 \cdot \frac{n(n+1)}{2} = n(n+2)$, and the dimension of the group $\mathbb{H}U_n$ is equal to $3n + 4 \cdot \frac{n(n-1)}{2} = n(2n+1)$. These dimensions are equal to the dimension of the real simple Lie groups of the class A_n, respectively C_n.

Since a real skew-symmetric $(N \times N)$-matrix depends on $\frac{N(N-1)}{2}$ parameters, the dimensions of O_{2n+1} and O_{2n} are equal to $n(2n+1)$, respectively $n(2n-1)$. These dimensions are equal to the dimension of the real simple Lie groups of the class B_n, respectively D_n.

Theorem 1.32. *The group of automorphisms in the division algebra* $\mathbb{O}$ *of octonions is the compact simple real Lie group* G_2.

PROOF. In theorem 1.27 we have seen that the group of automorphisms in $\mathbb{O}$ is a 14-dimensional subgroup of SO_7. Therefore, this group is compact.

The matrices in the tangent Lie algebra of this group are the skew-symmetric (7×7)-matrices that can be reduced to diagonal form with diagonal entries ia_0, ia_1, ia_2, $-ia_0$, $-ia_1$, $-ia_2, 0$ satisfying the condition

$$ia_0 + ia_1 + ia_2 = 0. \tag{1.121}$$

The root forms of SO_7 are $\pm iX_i \pm iX_j$, $\pm iX_i$ $(i = 0, 1, 2)$. By (1.121) the forms $\pm iX_i$ can be expressed in the terms of the forms $\pm(iX_i + iX_j)$. Therefore, the root forms of this group are $iX_0 - iX_1$, $iX_0 - iX_2$, $iX_1 - iX_0$, $iX_1 - iX_2$, $iX_2 - iX_0$, $iX_2 - iX_1$ and $iX_0 + iX_1$, $iX_0 + iX_2$, $iX_1 + iX_2$, $-iX_0 - iX_1$, $-iX_0 - iX_2$, $-iX_1 - iX_2$. The first six root forms determine root vectors $E_i - E_j$ which are located in the plane $x^0 + x^1 + x^2 = 0$, the last six root forms and forms $\pm iX_i$ equivalent to them can be replaced by linear combinations, $\pm(X_i + X_j - 2X_k)$ $(i \neq j \neq k \neq i)$ of them, which determine root vectors $\pm(E_i + E_j - 2E_k)$ located in the same plane. These vectors are vectors of the 5th line in (1.100), that is, the root vectors of the group of automorphisms in $\mathbb{O}$ coincide with the root vectors of the simple group G_2. ■

Theorem of E. B. Vinberg [Vin]. . *The tangent Lie algebras of the compact simple Lie groups* F_4, E_6, E_7, *and* E_8 *are direct sums of the linear spaces of skew-Hermitian* (3×3)*-matrices whose entries are elements in the algebras* $\mathbb{O}$, $\mathbb{C} \otimes \mathbb{O}$, $\mathbb{H} \otimes \mathbb{O}$, *and* $\mathbb{O} \otimes \mathbb{O}$, *respectively, with zero traces and of the tangent Lie algebras of the groups of ring automorphisms in these algebras. In the second case we must take continuous automorphisms.*

The condition of skew-Hermiticy is $a_{ij} = \overline{a}_{ji}$ *for octonionic matrices and* $a_{ij} = \tilde{\overline{a}}_{ji}$ *for matrices with entries from the tensor products. In the first two cases the tangent Lie algebras are the Lie algebra of the compact real simple Lie group* G_2, *in the third and fourth cases they are the direct sum of this Lie algebra with the Lie algebras of the compact simple groups* $A_1 = B_1 = C_1$, *respectively* G_2.

In these direct sums the commutator $[XY]$ *of two elements* X *and* Y *is determined as follows:*

(1) *if* X *and* Y *are elements in the direct sum of the tangent Lie algebras of the groups of automorphisms in a tensor product of two division algebras, and, therefore, are elements in the tangent Lie algebra of the group of automorphisms in this tensor product,* $[XY]$ *is the usual commutator of two elements in this Lie algebra;*

(2) *if* X *is an element in the tangent Lie algebra of the group of automorphisms in the tensor product, also called the Lie algebra of infinitesimal automorphisms in the tensor product, and* Y *is a* (3×3)*-matrix with entries from this tensor product,* $[XY]$ *is the matrix with as entries the results of the action of the infinitesimal automorphism* X *on the corresponding entries of* Y.

If we define the operations

$$D_{a,b}c = [a[bc]] - [b[ac]] + [ab]c]], \quad \text{where } [uv] = uv - vu, \tag{1.122}$$

$$D_{a_1 \otimes b_1, a_2 \otimes b_2} = (b_1, b_2) D_{a_1, a_2} + (a_1, a_2) D_{b_1, b_2}, \tag{1.123}$$

where (a_1, a_2) *and* (b_1, b_2) *are the inner products of the elements* a_1, a_2 *and* b_1, b_2 *in algebras* $\mathbb{C}$, $\mathbb{H}$, *and* $\mathbb{O}$ *in their metrics of* R^2, R^4, *and* R^8, *respectively,*

(3) *if* $X = (x_{ij})$ *and* $Y = (y_{ij})$ *are* (3×3)*-matrices,*

$$[XY] = XY - YX - \frac{1}{3}\mathrm{Tr}\,(XY - YX) + \frac{1}{3}\sum_i \sum_j D_{x_{ij}, y_{ij}}. \tag{1.124}$$

The dimensions of these Lie algebras and groups are equal to $3 \cdot 8 + 2 \cdot 7 + 14 = 52$, $3{\cdot}16+2{\cdot}8+14 = 78$, $3{\cdot}32+2{\cdot}10+14+3 = 133$, and $3{\cdot}64+2{\cdot}14+14+14 = 248$, respectively. These dimensions coincide with the dimensions of the compact real simple groups F_4, E_6, E_7, and E_8, respectively.

Note that a Cartan subalgebra of the Lie algebra for $\mathbb{O}$ consists of the diagonal (3×3)-matrices with diagonal entries of the form $a+ib$ and of the diagonal matrices of a Cartan subalgebra of the tangent Lie algebra of G_2.

For $\mathbb{C}\otimes\mathbb{O}$ a Cartan subalgebra of the Lie algebra consists of the diagonal (3×3)-matrices with diagonal entries of the form $a + ib + cI + diI$ $(iI = Ii)$ and of the diagonal matrices of a Cartan subalgebra of the tangent Lie algebra of G_2.

For $\mathbb{H}\otimes\mathbb{O}$ a Cartan subalgebra of the Lie algebra consists of the diagonal (3×3)-matrices with the diagonal entries of the same form and of diagonal matrices of a Cartan subalgebras of the tangent Lie algebras of G_2 and $A_1 = B_1 = C_1$.

For $\mathbb{O}\otimes\mathbb{O}$ a Cartan subalgebra of the Lie algebra consists of the diagonal (3×3)-matrices with diagonal entries of the same form and of the diagonal matrices of two Cartan subalgebras of G_2.

Therefore, the dimensions of these Cartan subalgebras are equal to $2 + 2 = 4$, $4+2 = 6$, $4+2+1 = 7$, and $4+2+2 = 8$, respectively. These dimensions coincide with the ranks of the compact real simple groups F_4, E_6, E_7, and E_8, respectively.

In **1.4.9** we have defined the *coefficient of connectedness* of a simple Lie group determined by its extended Dynkin diagram, which is equal to the determinant of its Cartan matrix. This coefficient is equal to the *number of elements of the center of the simply connected compact real Lie group* with the same tangent Lie algebra. This coefficient is also equal to the *number of elements of the Poincaré group* (also called *connectedness group*) *of the adjoint group of the Lie group* with the same tangent Lie algebra (that is, the group of transformations $x \to axa^{-1}$).

1.4.12. Root Vectors of Noncompact Simple Real Lie Groups. A noncompact semisimple real Lie group, which has the same complex form as a compact semisimple real Lie group, is determined by this compact group and by an involutive automorphism in this group. If a compact semisimple real Lie group G and its involutive automorphism J are given, this automorphism determines an involutive automorphism (1.33) in the tangent Lie algebra $\mathbb{G}$ of G. In this basis formula (0.65) for $\mathbb{G}$ can be written as

$$[e_a e_b] = C^c_{ab} e_c, \quad [e_a e_u] = C^v_{au} e_v, \quad [e_u e_v] = C^c_{uv} e_c. \tag{1.125}$$

Here e_a and e_u are basis elements in two linear subspaces $\mathbb{K}$ and $\mathbb{E}$ of $\mathbb{G}$ which are direct summands, that is, $\mathbb{G} = \mathbb{K} \oplus \mathbb{E}$.

The first formula in (1.125) shows that $\mathbb{K}$ is a subalgebra of $\mathbb{G}$. Formulas (1.125) are equivalent to the symbolic formulas

$$[\mathbb{K}\mathbb{K}] \subset \mathbb{K}, \quad [\mathbb{K}\mathbb{E}] \subset \mathbb{E}, \quad [\mathbb{E}\mathbb{E}] \subset \mathbb{K}. \tag{1.126}$$

If we multiply the basis elements in $\mathbb{E}$ by the imaginary unit i, we obtain a new Lie algebra, $\mathbb{G}'$, in which (0.65) has the form

$$[e_a e_b] = C^c_{ab} e_c, \quad [e_a e_u] = C^v_{au} e_v, \quad [e_u e_v] = -C^c_{uv} e_c. \tag{1.127}$$

This new Lie algebra has the same complex form $\mathbb{C}\otimes\mathbb{G}$ as the tangent Lie algebra $\mathbb{G}$ of the compact Lie group G. The Lie algebra $\mathbb{G}'$ determines a noncompact semisimple Lie group G' with the same complex form $\mathbb{C}G$ as G. The algorithm for passing from a compact semisimple group G to a noncompact semisimple group G' was proposed by Cartan in [Car15] and is called the *Cartan algorithm.*

Root forms and root vectors of noncompact simple real Lie groups are defined similarly to root forms and the root vectors of complex simple Lie groups. For a noncompact simple real Lie group the Killing-Cartan form (0.71), which can also be written as (1.99), determines a real pseudo-Riemannian pseudometric in the Lie group and a real pseudo-Euclidean pseudometric in its tangent Lie algebra.

A Cartan subalgebra of a noncompact simple real Lie group of rank n, endowed with its pseudometric, is a real space R^n_ℓ. The forms (0.71) and (1.99) can be reduced to an algebraic sum of P positive and N negative squares; the difference $\delta = N - P$ is called the *character* of the noncompact simple real Lie group.

For a noncompact simple real Lie group the root forms can be real, pure imaginary, and conjugate imaginary. So can the root vectors. A noncompact simple real Lie group all root forms and root vectors of which are real is called a *split* simple real Lie group. A noncompact simple Lie group all root forms and root vectors of which are real or conjugate imaginary is called a *semisplit* simple Lie group.

1.4.13. Satake Diagrams. The vectors in the root systems of noncompact simple real Lie groups can be represented by the *Satake diagrams* [Sat], which are versions of the Dynkin diagrams. In them, real root vectors are represented by white dots, pure imaginary root vectors are imaged by black dots, and conjugate imaginary root vectors are represented by white dots joined by a double arrow. Therefore, the Satake diagram of a compact real simple Lie group contains black dots only, the Satake diagram of a split or semisplit real simple Lie group contains white dots only.

The Satake diagrams of the noncompact simple real Lie groups are given in Fig. 1.9, for each diagram Cartan's notation of the corresponding group and its character δ are given. Note that character δ of any split simple Lie group is equal to its rank n.

1.4.14. Isomorphisms of Simple Real Lie Groups. The isomorphisms of the simple complex Lie groups determine isomorphisms of the corresponding simple real Lie groups. The Satake diagrams of isomorphic simple real Lie groups, like the Dynkin diagrams of isomorphic simple complex Lie groups, are similar. The Satake diagrams of the respective isomorphic real simple groups $A_1 = B_1 = C_1$, $B_2 = C_2$, $A_3 = D_3$ are given in Fig. 1.10a,b,c; the Satake diagrams of the respective isomorphic real semisimple groups $D_2 = A_1 \otimes A_1$ and $\mathbb{C}A_1$ are given in Fig. 1.10d. In these figures we represent the Satake diagrams of not only noncompact groups, as in Fig. 1.9, but also those of the compact groups which we denote by A0, B0, C0, and D0.

AI
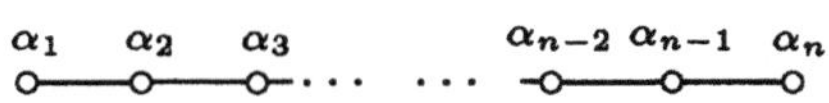

$\delta = n$

AII
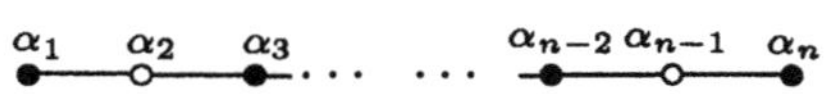

$\delta = -n-2$

AIII

$\delta = 4l(n-l+1) - n(n+2)$

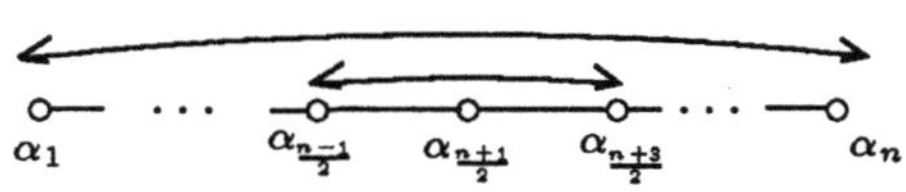

$\delta = 1$

AIV
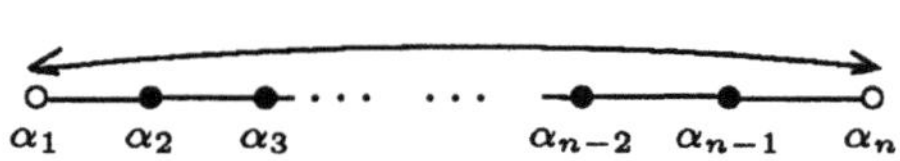

$\delta = 2n - n^2$

BI

$\delta = 2l(2n-l+1) - n(2n+1)$

BII 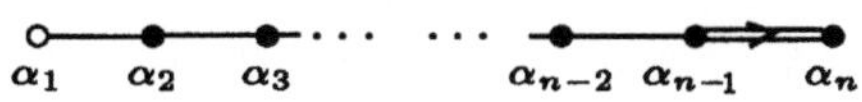

$\delta=3n-n^2$

CI 

$\delta=n$

CII 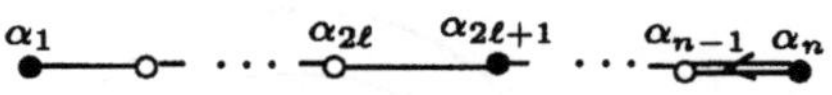

$\delta=8l(n-1)-n(2n+1)$

DI 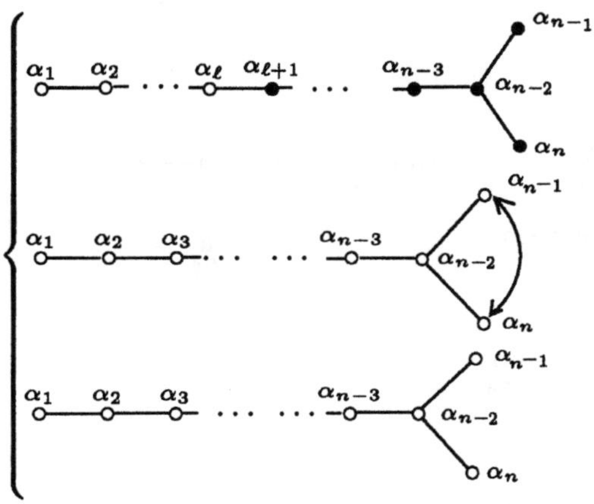

$\delta=2l(2n-1)-n(2n-1)$

$\delta=2(n^2-1)-n(2n-1)$

$\delta=n$

DII 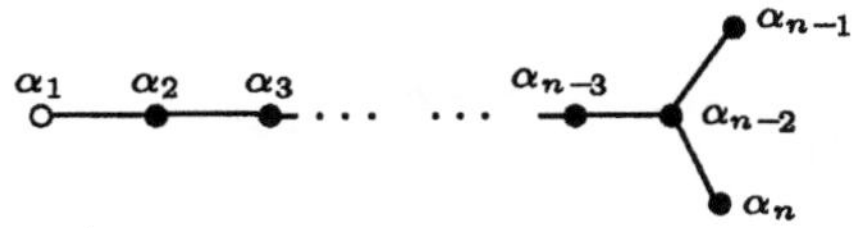

$\delta=5n-2n^2-2$

DIII 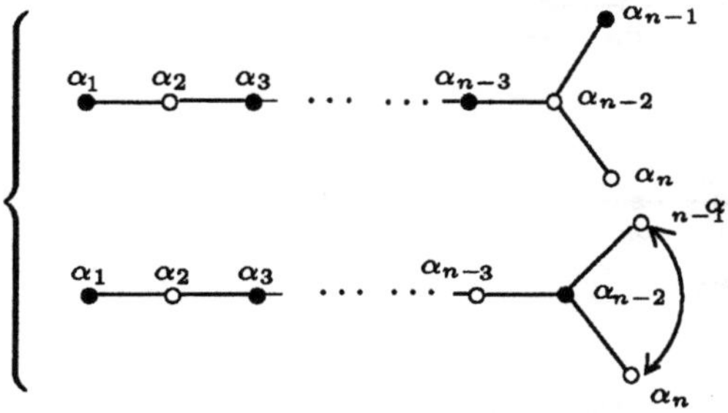

$\delta=-n$

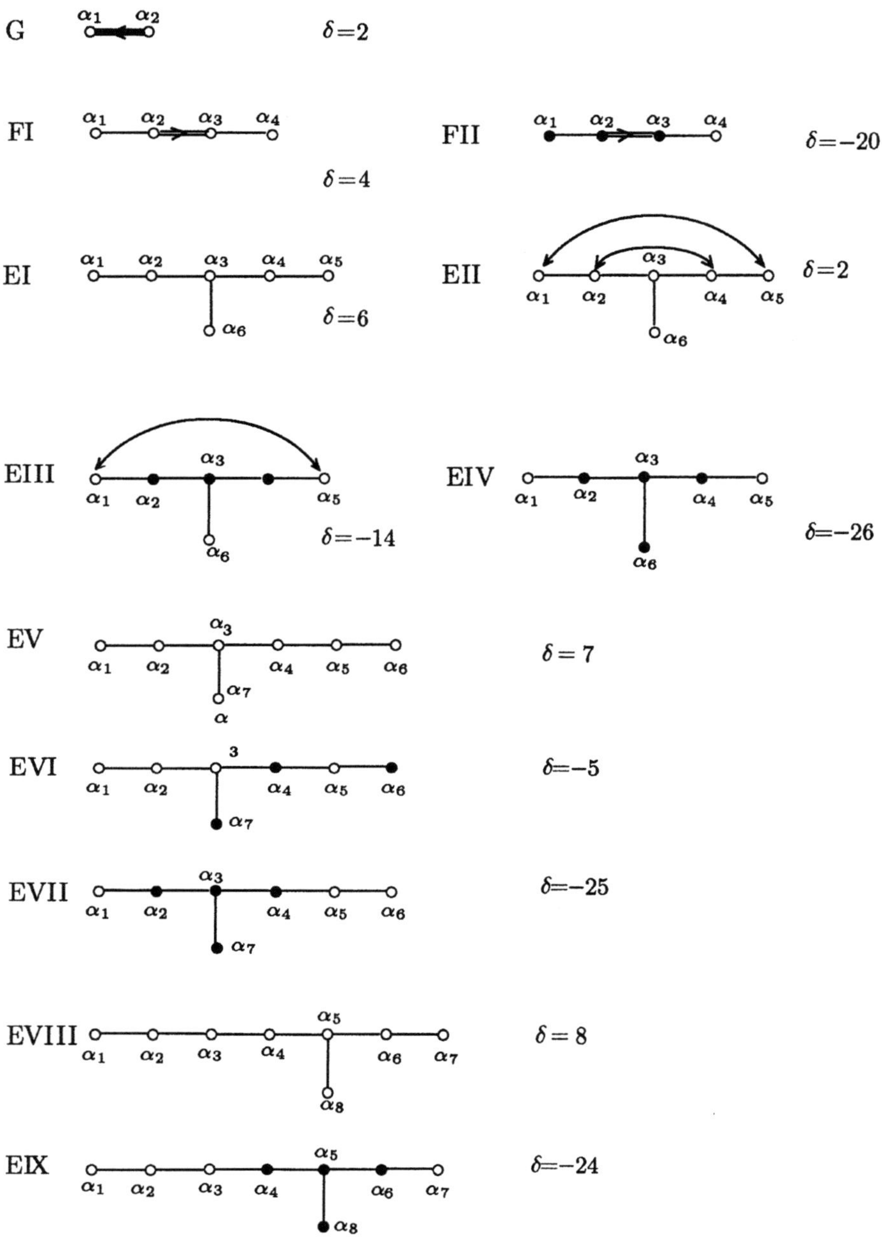
G
α_1 α_2
$\delta=2$
FI
α_1 α_2 α_3 α_4
$\delta=4$
FII
α_1 α_2 α_3 α_4
$\delta=-20$
EI
α_1 α_2 α_3 α_4 α_5 α_6
$\delta=6$
EII
α_1 α_2 α_3 α_4 α_5 α_6
$\delta=2$
EIII
α_1 α_2 α_3 α_5 α_6
$\delta=-14$
EIV
α_1 α_2 α_3 α_4 α_5 α_6
$\delta=-26$
EV
α_1 α_2 α_3 α_4 α_5 α_6 α_7
$\delta=7$
EVI
α_1 α_2 α 3 α_4 α_5 α_6 α_7
$\delta=-5$
EVII
α_1 α_2 α_3 α_4 α_5 α_6 α_7
$\delta=-25$
EVIII
α_1 α_2 α_3 α_4 α_5 α_6 α_7 α_8
$\delta=8$
EIX
α_1 α_2 α_3 α_4 α_5 α_6 α_7 α_8
$\delta=-24$

FIG. 1.9

1.4.15. Topology of Lie Groups. Let us mention some theorems concerning the topological structure of real Lie groups.

Theorem of É Cartan. [Car16]. *A real Lie group is homeomorphic to the topological product of its maximal compact subgroup by a Euclidean space.*

Therefore, the topological structure of any real Lie group is determined by the topological structure of its maximal compact subgroup.

In **1.4.11** we have mentioned that the number of elements in the Poincaré group of a compact simple real Lie group coincides with its coefficient of connectedness determined by its Cartan matrix and its extended Dynkin diagram.

Other topological invariant of compact simple real Lie groups, the *Betti numbers* b_i, are related with their Weyl groups. These topological invariants of topological manifolds, introduced by B. Riemann [Ri1] and E. Betti [Bet], were called "Betti numbers" by H. Poincaré [Poi3]. Therefore, Cartan called the polynomial $P(t) = \sum_k b_k t^{n-k}$ the *Poincaré polynomial.*

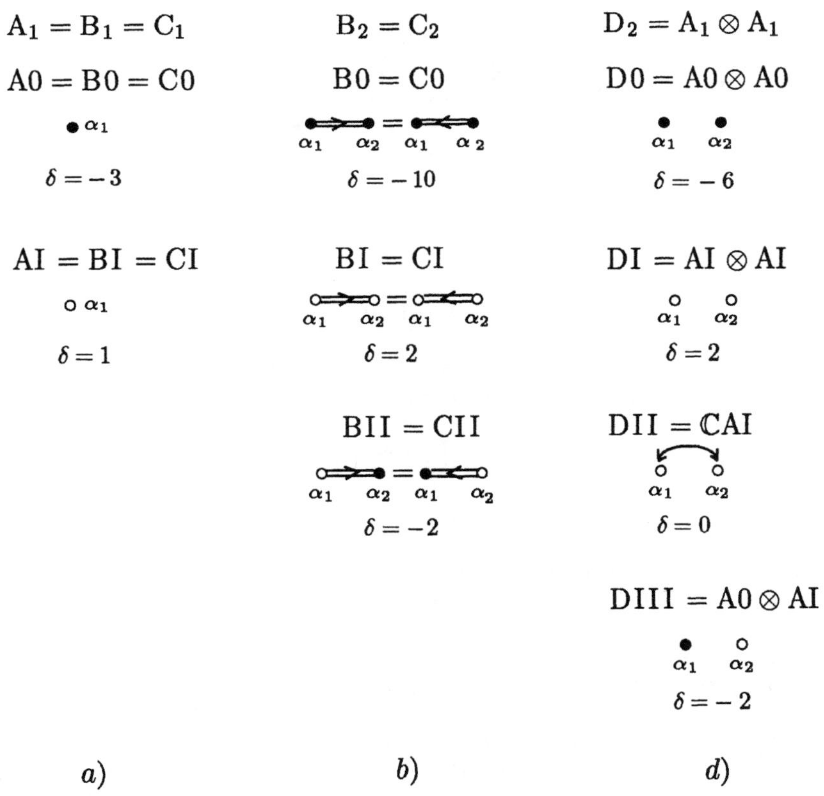

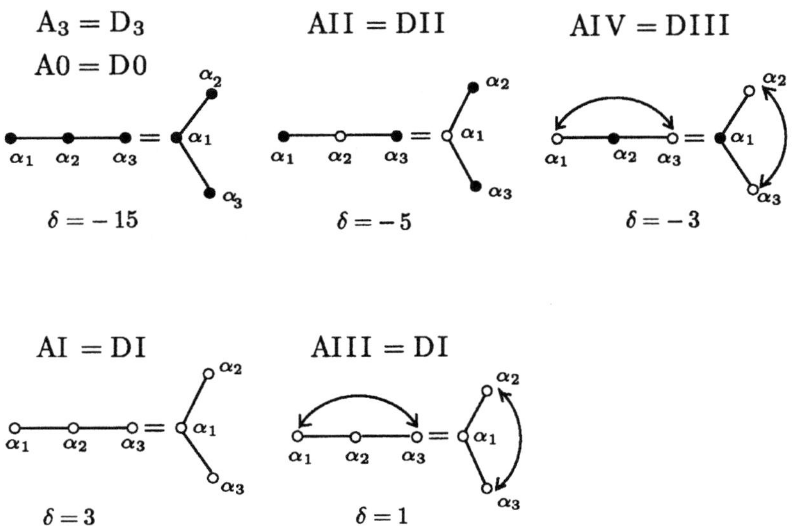

c)

FIG.1.10

Theorem of C. Chevalley. [Chv]. *The Poincaré polynomial of a simply connected compact simple real Lie group is*

$$P(t) = \prod_i (t^{2a_i+1} + 1), \tag{1.128}$$

where a_i are the exponents (1.114) *of the simple Lie group.*

Since formula (1.114) implies that $a_i + a_{n-i+1} = h = \frac{r-n}{n}$, we have $\sum_i a_i = \frac{nh}{2} = \frac{r-n}{2}$ and $\sum_i (2a_i + 1) = r - n + n = r$, that is, the degree of the polynomial (1.128) is equal to the dimension r of the group.

In particular, for the group $A_1 = B_1 = C_1$ we have $P(t) = t^3 + 1$, for the group A_2 we have $P(t) = (t^3+1)(t^5+1) = t^8+t^5+t^3+1$, for the group $A_3 = D_3$ we have $P(t) = (t^3+1)(t^5+1)(t^7+1) = t^{15}+t^{12}+t^{10}+t^8+t^7+t^5+t^3+t^1$, for the group $B_2 = C_2$ we have $P(t) = (t^3+1)(t^7+1) = t^{10}+t^7+t^3+1$, for the groups B_3 and C_3 we have $P(t) = (t^3+1)(t^7+1)(t^{11}+1) = t^{21}+t^{18}+t^{14}+t^{11}+t^{10}+t^7+t^3+1$, for the group D_2 we have $P(t) = (t^3+1)^2 = t^6+2t^3+1$, for the group G_2 we have $P(t) = (t^3+1)(t^{11}+1) = t^{14}+t^{11}+t^3+1$.

1.4.16. Symmetric and Reductive Spaces. In **0.8.7** and **0.8 .8** we have defined symmetric and reductive homogeneous spaces with Lie groups of transformations. Since symmetric space G/K with fundamental group G is determined by an involutive automorphism in this group formula (0.65) for the tangent Lie algebra $\mathbb{G}$ of this group can be written as (1.125) equivalent to formulas (1.126).

Reductive spaces according to the definition of Nomizu [Nom] are homogeneous spaces G/K satisfying only two first conditions (1.126). Therefore formula (0.65) for the tangent Lie algebra of G can be written as the formula obtained from (1.125) by replacing third part of this formula by $[e_u e_v] = C^a_{uv} e_a + C^w_{uv} e_w$. It is easy to check that tensor $-C^w_{uv}$ is the torsion tensor in this reductive space [KoN].

1.4.17. Finite Groups of Lie Type. Besides the spherical Weyl groups, another class of finite groups is connected with Lie groups: the *simple finite groups of Lie type*. These groups are the groups of automorphisms in Lie algebras over the finite fields $\mathbb{F}_q$, defined by the same structure constants as the complex and split real Lie groups of the same type. If a real or complex Lie algebra is the tangent Lie algebra of a simple Lie group A_n, B_n C_n, D_n, G_2, F_4, E_6, E_7, or E_8, the corresponding finite group is denoted by $\mathbb{F}_q A_n$, $\mathbb{F}_q B_n$, $\mathbb{F}_q C_n$, $\mathbb{F}_q D_n$, $\mathbb{F}_q G_2$, $\mathbb{F}_q F_4$, $\mathbb{F}_q E_6$, $\mathbb{F}_q E_7$, or $\mathbb{F}_q E_8$. The first four classes of these groups also are denoted by $\mathbb{F}_q SL_{n+1}$, $\mathbb{F}_q SO_{2n+1}$, $\mathbb{F}_q Sp_{2n}$, and $\mathbb{F}_q SO_{2n}$. All these groups except $\mathbb{F}_q SO_N$ are simple finite groups, that is, they have no invariant subgroup; the groups $\mathbb{F}_q SO_N$ have simple invariant subgroups, denoted by $\mathbb{F}_q S\Omega_N$. Note that C. Chevalley [Chv] proved that the number of elements in such a group G (that is, the*cardinality* of this group) is equal to

$$\text{card}\, G = \frac{1}{u} q^N \prod_i (q^{a_i+1} - 1), \tag{1.129}$$

where u is the number of elements in the center of the finite group of automorphisms in the Lie algebra over $\mathbb{F}_q$, N is the number of positive root vectors of the corresponding simple Lie group, and a_i are the same exponents (1.114) of this group as in (1.128). Formula (1.129) shows a deep connection between the theories of simple real and complex Lie groups on the one hand and of finite groups of Lie type on the other hand.

§1.5. Jordan and Elastic Algebras

1.5.1. Commutative Jordan Algebras. An other important class of nonassociative algebras is that of the *commutative Jordan algebras*, that is, algebras with commutative multiplication $X \cdot Y = Y \cdot X$ in which a condition of associativity is replaced by the condition

$$(X \cdot (Y \cdot Y)) \cdot Y = (X \cdot Y) \cdot (Y \cdot Y). \tag{1.130}$$

We denote $X \cdot X$ by $X^{\cdot 2}$, $X \cdot (X \cdot X)$ by $X^{\cdot 3}$, $X \cdot (X \cdot (X \cdot X))$ by $X^{\cdot 4}$ and the analogous product of n factors by $X^{\cdot n}$. Commutativity of the product $X \cdot Y$ implies the relations $X^{\cdot 3} = X \cdot X^{\cdot 2}$, $X^{\cdot 4} = X \cdot X^{\cdot 3}$, and so on. If, in (1.130), $X = Y$, we obtain $X^{\cdot 4} = X^{\cdot 2} \cdot X^{\cdot 2}$. Analogously, we can prove that $X^{\cdot (m+n)} = X^{\cdot m} \cdot X^{\cdot n}$ for any m and n.

An associative algebra $\mathbb{A}$ with multiplication XY gives rise to a Jordan algebra, $\mathbb{J}(\mathbb{A})$, with multiplication

$$X \cdot Y = \frac{1}{2}(XY + YX). \tag{1.131}$$

Theorem 1.33. *The subalgebras of the algebras* $\mathbb{J}(\mathbb{R}_n)$, $\mathbb{J}(\mathbb{C}_n)$, *and* $\mathbb{J}(\mathbb{C}'_n)$ *consisting of the symmetric matrices* ($X = X^\top$) *and the subalgebras of the algebras* $\mathbb{J}(\mathbb{C}_n)$ $\mathbb{J}(\mathbb{C}'_n)$, $\mathbb{J}(\mathbb{H}_n)$, *and* $\mathbb{J}(\mathbb{H}'_n)$ *consisting of the Hermitian symmetric matrices* ($X = \overline{X}^\top$) *are Jordan algebras.*

We denote these Jordan algebras by $\mathbb{J}\mathbf{s}(\mathbb{R}_n)$, $\mathbb{J}\mathbf{s}(\mathbb{C}_n)$, $\mathbb{J}\mathbf{s}(\mathbb{C}'_n)$, $\mathbb{J}\bar{\mathbf{s}}(\mathbb{C}_n)$, $\mathbb{J}\bar{\mathbf{s}}(\mathbb{C}'_n)$, $\mathbb{J}\bar{\mathbf{s}}(\mathbb{H}_n)$, *and* $\mathbb{J}\bar{\mathbf{s}}(\mathbb{H}'_n)$, *respectively.*

PROOF. It is sufficient to prove that these sets of matrices are closed under the operations $X \to X^\top$ and $X \to \overline{X}^\top$. For the first operation this follows from the equality

$$(X \cdot Y)^\top = \frac{1}{2}(Y^\top X^\top + X^\top Y^\top) = \frac{1}{2}(XY + YX) = X \cdot Y.$$

For the second operation the proof is analogous. ∎

If A is a nonsingular matrix in one of the algebras $\mathbb{R}_n$, $\mathbb{C}_n$, $\mathbb{C}'_n$, $\mathbb{H}_n$ or $\mathbb{H}'_n$, we can define the Jordan algebras $\mathbb{J}\mathbf{a}(\mathbb{R}_n)$, $\mathbb{J}\mathbf{a}(\mathbb{C}_n)$, $\mathbb{J}\mathbf{a}(\mathbb{C}'_n)$, $\mathbb{J}\bar{\mathbf{a}}(\mathbb{C}_n)$, $\mathbb{J}\bar{\mathbf{a}}(\mathbb{C}'_n)$, $\mathbb{J}\bar{\mathbf{a}}(\mathbb{H}_n)$, and $\mathbb{J}\bar{\mathbf{a}}(\mathbb{H}'_n)$ of them, with multiplication

$$X \cdot Y = \frac{1}{2}(XAY + YAX) \quad \text{or} X \cdot Y = \frac{1}{2}(\bar{X}AY + \bar{Y}AX), \tag{1.132}$$

respectively.

If A is the diagonal $(n \times n)$-matrix E with ℓ diagonal entries equal to -1 and $n - \ell$ diagonal entries equal to 1, defined in **0.5.6**, we will replace the letter $\mathbf{a}$ in the denotation of the Jordan algebras by $\mathbf{s}_\ell$. Note that the algebras $\mathbb{J}\mathbf{s}_\ell(\mathbb{R}_n)$, $\mathbb{J}\mathbf{s}_\ell(\mathbb{C}'_n)$, $\mathbb{J}\bar{\mathbf{s}}_\ell(\mathbb{C}_n)$, and $\mathbb{J}\bar{\mathbf{s}}_\ell(\mathbb{H}_n)$ are not isomorphic to $\mathbb{J}\mathbf{s}(\mathbb{R}_n)$, $\mathbb{J}\mathbf{s}(\mathbb{C}'_n)$, $\mathbb{J}\bar{\mathbf{s}}(\mathbb{C}_n)$, and $\mathbb{J}\bar{\mathbf{s}}(\mathbb{H}_n)$, respectively. However, $\mathbb{J}\mathbf{s}_\ell(\mathbb{C}_n)$ is isomorhic to $\mathbb{J}\mathbf{s}(\mathbb{C}_n)$, $\mathbb{J}\bar{\mathbf{s}}_\ell(\mathbb{C}'_n)$, is isomorphic to $\mathbb{J}\bar{\mathbf{s}}(\mathbb{C}'_n)$, and $\mathbb{J}\bar{\mathbf{s}}_\ell(\mathbb{H}'_n)$ is is isomorphic to $\mathbb{J}\bar{\mathbf{s}}(\mathbb{H}'_n)$.

Theorem 1.34. *The algebras* $\mathbb{J}\mathbf{j}(\mathbb{R}_{2n})$, $\mathbb{J}\mathbf{j}(\mathbb{C}_{2n})$, *and* $\mathbb{J}\mathbf{j}(\mathbb{C}'_{2n})$, *that is, the algebras* $\mathbb{J}\mathbf{a}(\mathbb{R}_{2n})$, $\mathbb{J}\mathbf{a}(\mathbb{C}_{2n})$ *and* $\mathbb{J}\mathbf{a}(\mathbb{C}'_{2n})$ *for which the matrix* A *is the* $(2n \times 2n)$*-matrix* J *of the form* (0.62), *are isomorphic to the algebras of skew symmetric matrices in the algebras* $\mathbb{R}_{2n}$, $\mathbb{C}_{2n}$ *and* $\mathbb{C}'_{2n}$ *with multiplication* (1.132), *where* $A = J$. *The algebras* $\mathbb{J}\bar{\mathbf{j}}(\mathbb{C}_{2n})$, $\mathbb{J}\bar{\mathbf{j}}(\mathbb{C}'_{2n})$, $\mathbb{J}\bar{\mathbf{j}}(\mathbb{H}_{2n})$, *and* $\mathbb{J}\bar{\mathbf{j}}(\mathbb{H}'_{2n})$, *that is, the algebras* $\mathbb{J}\bar{\mathbf{a}}(\mathbb{C}_{2n})$, $\mathbb{J}\bar{\mathbf{a}}(\mathbb{C}'_{2n})$ $\mathbb{J}\bar{\mathbf{a}}(\mathbb{H}_{2n})$ *and* $\mathbb{J}\bar{\mathbf{a}}(\mathbb{H}'_{2n})$ *for which the matrix* A *is the* $(2n \times 2n)$*-matrix* J, *are isomorphic to the algebras of skew Hermitian matrices in the algebras* $\mathbb{C}_{2n}$, $\mathbb{C}'_{2n}$, $\mathbb{H}_{2n}$ *and* $\mathbb{H}'_{2n}$, *with the same multiplication* (1.132), *where* $A = J$.

PROOF. It is sufficient to prove that these sets of matrices are closed under the operations $X \to -X^\top$ and $X \to -\overline{X}^\top$. For the first operation,

$$(X \cdot Y)^\top = \frac{1}{2}(Y^\top J^\top X^\top + X^\top J^\top Y^\top) = -\frac{1}{2}(XJY + YJX) = -X \cdot Y.$$

For the second operation the proof is analogous. ∎

If the matrix A in the algebra $\mathbb{H}_n$ is the diagonal $(n \times n)$-matrix iI, we replace the letter **a** in the notation of the Jordan algebra by **i**.

The Jordan algebras $\mathbb{J}(\mathbb{A}_n)$, $\mathbb{J}(\mathbb{A}_{n,\ell})$, and $\mathbb{J}(\mathbb{C} \otimes \mathbb{A}_n)$ are the sets of alternions, pseudoalternions, and bialternions in the algebras $\mathbb{A}_n$, $\mathbb{A}_{n,\ell}$, and $\mathbb{C} \otimes \mathbb{A}_n$ which have the form $x = x_0 + x^i e_i$ and with the multiplication (1.131). If we denote the expression $x^i e_i$ by x_1 and regard it as a vector in the space R^{n-1}, R_ℓ^{n-1}, or $\mathbb{C}R^{n-1}$, respectively, then the product $x \cdot y$ of two elements in these algebras has the form

$$x \cdot y = x_0 y_0 - x_1 y_1 + x_0 y_1 + y_0 x_1, \tag{1.133}$$

where $x_1 y_1$ is the inner product of the two vectors x_1 and y_1 (x_0 and y_0 are real, respectively complex, numbers).

There are also Jordan algebras $\mathbb{J}\bar{\mathbb{s}}(\mathbb{O}_3)$, $\mathbb{J}_3(\mathbb{O}'_3)$ and $\mathbb{J}\bar{\mathbb{s}}(\mathbb{C} \otimes \mathbb{O}_3)$. They consist of the octonionic, split octonionic, and bioctonionic Hermitian symmetric (3×3)-matrices with multiplication (1.131).

As for associative, alternative, and Lie algebras, for commutative Jordan algebras we can define simplicity and semisimplicity. The real commutative Jordan algebras in which for any element X in this algebra the trace of the matrix of the regular linear representation of the element $X^{\cdot 2}$ is a positive definite quadratic form in the coordinates of X, are called *compact commutative Jordan algebras* since this property is analogous to the property of being the tangent Lie algebra of a compact Lie group (tangent Lie algebras of compact Lie groups are often called "compact Lie algebras").

Theorem of A. A. Albert and G. K. Kalish [Al1], [Kal]. *The complex simple commutative Jordan algebras are:* $\mathbb{J}(\mathbb{C}_n)$, $\mathbb{J}\mathbb{s}(\mathbb{C}_n)$, $\mathbb{J}_\mathbb{j}(\mathbb{C}_{2n})$, $\mathbb{J}(\mathbb{C}\otimes\mathbb{A}_n)$, $\mathbb{J}\bar{\mathbb{s}}(\mathbb{C}\otimes \mathbb{O}_3)$

These classes of simple complex commutative Jordan algebras are denoted A, B, C, D, and E, respectively. The same notations are also used for real simple commutative Jordan algebras with corresponding complex forms.

Theorem of F. D. Jacobson and N. Jacobson [JaJa]. *The compact real simple commutative Jordan algebras are:* $\mathbb{J}\bar{\mathbb{s}}(\mathbb{C}_n)$, $\mathbb{J}\mathbb{s}(\mathbb{R}_n)$, $\mathbb{J}\bar{\mathbb{s}}(\mathbb{H}_n)$, $\mathbb{J}(\mathbb{A}_n)$, $\mathbb{J}\bar{\mathbb{s}}(\mathbb{O}_3)$.

The noncompact real simple commutative Jordan algebras are: A) $\mathbb{J}(\mathbb{R}_n)$, $\mathbb{J}\bar{\mathbb{s}}_\ell(\mathbb{C}_n)$, $\mathbb{J}(\mathbb{H}_{n/2})$; *B)* $\mathbb{J}\bar{\mathbb{s}}_\ell(\mathbb{R}_n)$, $\mathbb{J}\bar{\mathbb{i}}(\mathbb{H}_{n/2})$; *C)* $\mathbb{J}\mathbb{j}(\mathbb{R}_{2n})$, $\mathbb{J}\bar{\mathbb{s}}_\ell(\mathbb{H}_n)$; , *D)* $\mathbb{J}(\mathbb{A}_{n,l})$; *E)* $\mathbb{J}\bar{\mathbb{s}}_1(\mathbb{O}_3)$, $\mathbb{J}(\mathbb{O}'_3)$.

Theorem of N. Jacobson [Ja]. *Let $\mathbb{A}$ be a simple associative algebra. Each automorphism in the commutative Jordan algebra $\mathbb{J}(\mathbb{A})$ coincides with an automorphism or antiautomorphism in the algebra $\mathbb{A}$.*

Therefore, since each ring automorphism in the algebras $\mathbb{R}_n$ and $\mathbb{H}_n$ is inner, and each ring antiautomorphism in these algebras, that is, a bijection $f(x)$ in these algebras that satisfies the two conditions (1.17) and $f(xy) = f(y)f(x)$, can be written as the product of a ring automorphism in these algebras and the transposition $X \to X^\top$, each automorphism in the commutative Jordan algebras $\mathbb{J}(\mathbb{R}_n)$ and $\mathbb{J}(\mathbb{H}_n)$ has the form

$$X \to UXU^{-1} \tag{1.134}$$

or

$$X \to UX^{\top}U^{-1}. \tag{1.135}$$

Since each continuous ring automorphism in the algebra $\mathbb{C}_n$ is inner, and each continuous ring antiautomorphism in this algebra can be written as the product of a continuous ring automorphism in this algebra and the transposition $X \to X^{\top}$, each continuous automorphism in the commutative Jordan algebra $\mathbb{J}(\mathbb{C}_n)$ has the form (1.134), (1.135),

$$X \to U\overline{X}U^{-1} \tag{1.136}$$

or

$$X \to U\overline{X}^{\top}U^{-1} \tag{1.137}$$

For $\mathbb{J}\mathbf{s}(R_n)$, $\mathbb{J}\mathbf{s}_\ell(R_n)$, $\mathbb{J}\mathbf{j}(\mathbb{R}_{2n})$, $\mathbb{J}\overline{\mathbf{s}}(\mathbb{C}_n)$, $\mathbb{J}\overline{\mathbf{s}}_\ell(\mathbb{C}_n)$, $\mathbb{J}\overline{\mathbf{s}}_\ell(\mathbb{H}_n)$, $\mathbb{J}\overline{\mathbf{s}}_\ell(\mathbb{H}_n)$, $\mathbb{J}\overline{\mathbf{i}}(\mathbb{H}_n)$ the transformations (1.135) can be reduced to the transformations (1.134) and the transformations (1.137) can be reduced to the transformations (1.136). If the transformation (1.134) maps the set of matrices satisfying the condition $X^{\top}AX = A$ to itself, we obtain

$$(UXU^{-1})^{\top}A(UXU^{-1}) = (U^{-1})^{\top}X^{\top}U^{\top}AUXU^{-1} = A,$$

that is, the matrix U must satisfy the condition

$$U^{\top}AU = A. \tag{1.138}$$

Analogously, if the transformation (1.134) maps the set of matrices satisfying the condition $\overline{X}^{\top}AX = A$ to itself, we find that U must satisfy the condition

$$\overline{U}^{\top}AU = A. \tag{1.139}$$

Therefore, if the transformation (1.134) is an automorphism in $\mathbb{J}\mathbf{s}(\mathbb{R}_n)$, $\mathbb{J}\overline{\mathbf{s}}_\ell(\mathbb{R}_n)$, $\mathbb{J}\overline{\mathbf{s}}(\mathbb{C}_n)$, $\mathbb{J}\overline{\mathbf{s}}_\ell(\mathbb{C}_n)$, $\mathbb{J}\overline{\mathbf{s}}(\mathbb{H}_n)$, or $\mathbb{J}\overline{\mathbf{s}}_\ell(\mathbb{H}_n)$, then U must be a real orthogonal, real pseudoorthogonal, complex unitary, complex pseudounitary, quaternionic unitary, or quaternionic pseudounitary matrix, respectively. Analogously, if the transformation (1.134) is an automorphism in $\mathbb{J}\mathbf{j}(\mathbb{R}_{2n})$ or $\mathbb{J}\overline{\mathbf{i}}(\mathbb{H}_n)$, then U must be a real, respectively quaternionic, symplectic matrix.

In commutative Jordan algebras the automorphisms (1.136) and (1.138) are particular cases of *linear transformations* in these algebras. For instance, such occurs in $\mathbb{J}\mathbf{s}(\mathbb{R}_n)$, $\mathbb{J}\mathbf{s}_\ell(\mathbb{R}_n)$, $\mathbb{J}\mathbf{j}(\mathbb{R}_{2n})$, $\mathbb{J}\overline{\mathbf{s}}(\mathbb{C}_n)$, $\mathbb{J}\overline{\mathbf{s}}_\ell(\mathbb{C}_n)$, $\mathbb{J}\overline{\mathbf{s}}(\mathbb{H}_n)$, $\mathbb{J}\overline{\mathbf{s}}_\ell(\mathbb{H}_n)$, and $\mathbb{J}\overline{\mathbf{i}}(\mathbb{H}_n)$, where the matrix U need not satisfy the condition (1.138) or (1.139).

In commutative Jordan algebras we also can define *fractional linear transformations*

$$X' = (AX + B)(CX + D)^{-1}, \tag{1.140}$$

where A, B, C, D are matrices in corresponding matrix algebra.

1.5.2. Elastic Algebras. The commutative Jordan algebras are particular cases of algebras belonging to a large class of algebras containing also noncommutative algebras. This class consists of the unital algebras (i.e. algebras with 1) satisfying the conditions

$$(X \cdot Y) \cdot X = X \cdot (Y \cdot X), \tag{1.141}$$

$$(X \cdot Y) \cdot (X \cdot X) = X \cdot (Y \cdot (X \cdot X)), \tag{1.142}$$

and

$$(X \cdot (X \cdot X) \cdot Y) = (X \cdot X) \cdot (X \cdot Y). \tag{1.143}$$

For commutative algebras the condition (1.141) is automatically satisfied, and the condition (1.142) can be written as (1.130). Therefore, these commutative algebras are commutative Jordan algebras. In general, these algebras are called *general Jordan algebras* or *elastic algebras.*

The conditions (1.141), (1.142), and (1.143) are satisfied for associative algebras, and, since in these conditions only two elements participate, also for alternative algebras. All three conditions (1.141), (1.142), and (1.143) are satisfied for algebras with an anticommutative multiplication, in particular for Lie algebras. Therefore, associative, alternative, Lie, and commutative Jordan algebras are particular cases of elastic algebras. The class of elastic algebras was defined by I. L. Kantor [Kan1].

1.5.3. Algebras of Rectangular Matrices. The sets of real, complex, or quaternionic rectangular $(m \times n)$-matrices with the multiplication

$$X \cdot Y = XA^{\mathsf{T}}Y, \tag{1.144}$$

where A is an $(n \times m)$-matrix, also form elastic algebras since for such matrices both sides in (1.141) are equal to $XAYAX$, both sides in (1.142) are equal to $XAYAXAX$, and both sides in (1.143) are equal to $XAXAXAY$. The role of unit in the algebras of matrices $X = (x_a^u)$ is played by the matrix $I = (\delta_a^u)$. These algebras are denoted by $\mathbb{R}_{m,n}$, $\mathbb{C}_{m,n}$, and $\mathbb{H}_{m,n}$, respectively.

1.5.4. Freudenthal Algebras. H. Freudenthal [Fr2] has defined a class of elastic algebras which we call the *Freudenthal algebras* and denote by $\mathbf{Fr}(\mathbb{R}_3)$, $\mathbf{Fr}(\mathbb{C}_3)$, $\mathbf{Fr}(\mathbb{H}_3)$, and $\mathbf{Fr}(\mathbb{O}_3)$. These algebras can be defined as algebras of vector-matrices (1.89) where ξ and η are real numbers and x and y are elements in the commutative Jordan algebras $\mathbb{J}\mathrm{s}(\mathbb{R}_3)$, $\mathbb{J}\bar{\mathrm{s}}(\mathbb{C}_3)$, $\mathbb{J}\bar{\mathrm{s}}(\mathbb{H}_3)$, and $\mathbb{J}\bar{\mathrm{s}}(\mathbb{O}_3)$, respectively. The sum and product of two such vector-matrices X and Y are defined as for Zorn vector-matrices (see **1.3.3**). Here the role of the inner product xy is played by the trace $\mathrm{Tr}(x \cdot y)$ of the product of the elements x and y in the corresponding Jordan algebra, and the role of the vector product $[xy]$ is played by the element

$$x \times y = x \cdot y - \frac{1}{2}x - \frac{1}{2}y + \frac{1}{2}(I - \mathrm{Tr}\ (x \cdot y) \cdot I).$$

Note that while dimension of the division algebra $\mathbb{R}$, $\mathbb{C}$, $\mathbb{H}$, or $\mathbb{O}$ is equal to $r = 1, 2, 4, 8$, the dimension of the Jordan algebra $\mathbb{J}\mathrm{s}(\mathbb{R}_3)$, $\mathbb{J}\bar{\mathrm{s}}(\mathbb{C}_3)$, $\mathbb{J}\bar{\mathrm{s}}(\mathbb{H}_3)$, or $\mathbb{J}\bar{\mathrm{s}}(\mathbb{O}_3)$ is equal to $3r + 3 = 6, 9, 15$, or 27, and the dimension of the Freudenthal algebra

$\mathbb{F}\mathbf{r}(\mathbb{R}_3)$ $\mathbb{F}\mathbf{r}(\mathbb{C}_3)$, $\mathbb{F}\mathbf{r}(\mathbb{H}_3)$, or $\mathbb{F}\mathbf{r}(\mathbb{O}_3)$ is equal to $2(3r+3)+2 = 6r+8 = 14, 20, 32$, or 56.

Note that in Freudenthal algebras $X(XX) = (XX)X$, which is not true in the tensor products $\mathbb{H}\otimes\mathbb{O}$ and $\mathbb{O}\otimes\mathbb{O}$. In particular, for the vector-matrix X with entries ξ, η, x, y the entries in the vector-matrix $X(XX) = (XX)X$ are $\xi^3 + 2(\xi+\eta)xy$, $\eta^3 + 2(\xi+\eta)xy$, $(\xi^2 + \xi\eta + \eta^2 + xy)x$, and $(\xi^2 + \xi\eta + \eta^2 + xy)y$.

We can similarly define the Freudenthal algebras $\mathbb{F}\mathbf{r}(\mathbb{C}'_3)$, $\mathbb{F}\mathbf{r}(\mathbb{H}'_3)$, and $\mathbb{F}\mathbf{r}(\mathbb{O}'_3)$.

1.5.5. Ternars. A linear space with a trilinear operation

$$w = < x, y, z > \tag{1.145}$$

is called a *ternar*. If one of the elements of the trilinear operation $< x, y, z >$ is fixed, we obtain in this ternar the bilinear operation of an *algebra*, for instance, the operation

$$x \cdot y = < x, a, y > .$$

If a Lie algebra $\mathbb{G}$ and an involutive automorphism (1.33) in it are given, the formula (0.65) for this Lie algebra can be written as (1.125), this Lie algebra can be represented as the direct sum $\mathbb{K}\oplus\mathbb{E}$, and (1.125) is equivalent to (1.126). Therefore, the subspace $\mathbb{E}$ can be regarded as a ternar with the trilinear operation

$$< x, y, z > = [[xy]z]. \tag{1.146}$$

This ternar is called a *Lie ternar*.

The operation (1.146) satisfies the conditions

$$< x, y, z > = - < y, x, z > \tag{1.147}$$

and

$$< x, y, z > + < y, z, x > + < z, x, y > = 0. \tag{1.148}$$

Note that the homogeneous space G/K with fundamental group G and stabilizer K, with tangent Lie algebras $\mathbb{G}$ and $\mathbb{K}$ can be regarded as a *symmetric space* (see **0.8.7**) with tangent space $\mathbb{E}$, and thus we can associate with each symmetric space a Lie ternar.

A ternar satisfying the condition (1.148) can be obtained also from an elastic algebra having an involution $x \to \overline{x}$. Here, the trilinear operation (1.145) is defined by

$$< x, y, z > = x(\overline{y}z) + z(\overline{y}x) - y(\overline{x}z). \tag{1.149}$$

For a commutative Jordan algebra with multiplication $X \cdot Y$ the operation (1.145) can be defined by

$$< X, Y, Z > = X \cdot (Y \cdot Z) + (Y \cdot X) \cdot Z - Y \cdot (X \cdot Z). \tag{1.150}$$

In this case the operation (1.145) satisfies the condition

$$< x, y, z >=< z, y, x > \tag{1.151}$$

and the ternar is called a *Jordan ternar.*

For the algebra of rectangular matrices with multiplication XAY the operation (1.145) is defined by

$$< X, Y, Z >= XY^{\top}Z + ZY^{\top}X - YX^{\top}Z. \tag{1.152}$$

For the Freudenthal algebra with multiplication XY the ternar operation (1.145) can be defined by

$$< X, Y, Z >= X(\overline{Y}Z) + Z(\overline{Y}X) - Y(\overline{X}Z), \tag{1.153}$$

where $X \to \overline{X}$ is the involution in this algebra consisting of interchanging the diagonal entries ξ and η of the vector-matrices in this algebra.

§1.6. Linear Representations of Simple Lie Groups

1.6.1. Linear Representation of Lie Groups. A homomorphic map from a Lie group into the group GL_N or $\mathbb{C}GL_N$ is called a *linear representation* of this Lie group. A linear representation of a Lie group G determines a linear representation of its tangent Lie algebra $\mathbb{G}$ into corresponding Lie algebra $\mathbb{GL}_n$ or $\mathbb{CGL}_n$.

A linear representation of a group G is called *reducible* if in the space L^N or $\mathbb{C}L^N$ of this representation there is a subspace that is invariant under this representation. A linear representation φ of a group G is called *totally reducible* if the space L^N or $\mathbb{C}L^N$ of this representation is a direct sum of invariant subspaces. In this case, if we have chosen basis vectors in these invariant subspaces, the matrix of any linear representation has the form

$$A = \begin{bmatrix} A_1 & 0 & \dots & 0 \\ 0 & A_2 & \dots & 0 \\ \vdots & \vdots & \ddots & \vdots \\ 0 & 0 & \dots & A_k \end{bmatrix}, \tag{1.154}$$

where $A_1, A_2, \dots, A_k$ are the matrices of the linear representations $\varphi_1, \varphi_2, \dots, \varphi_k$ of the group in these invariant subspaces. In this case the linear representation φ is called the direct sum of the linear representations $\varphi_1, \varphi_2, \dots, \varphi_k$ and is denoted by $\varphi_1 \oplus \varphi_2 \oplus \dots \oplus \varphi_k$.

A linear representation of a Lie group is called *irreducible* if it cannot be written as the direct sum of other linear representations.

The linear representation of the Lie algebra $\mathbb{G}$ by the matrices of the linear transformations $X' = [AX]$ and the corresponding linear representation of the Lie group G are called *adjoint representations* of the algebra $\mathbb{G}$ and group G.

1.6.2. Linear Representation of Semisimple Lie Groups. In his theory of linear representations of semisimple complex Lie groups, É. Cartan [Car7] has proved that all such linear representations are totally reducible. Therefore, this theory can be reduced to the theory of linear representations of simple complex Lie groups. In **1.4.3** we have defined Cartan subgroups and subalgebras of simple complex Lie groups, and root systems, which consist of linear combinations of simple root vectors in Cartan subalgebras with integer coefficients, and have considered the adjoint representations of these groups in their Cartan subalgebras. As for adjoint representations, for general linear representations of these groups, the linear operators $\widehat{A}$ representing regular vectors A of the Lie algebra also have common eigenvectors, and the eigenvalues λ of these operators corresponding to different vectors A are linear forms, called the *weights* or *weight forms* of the linear representation of the simple Lie group. The covectors π determining these linear forms are called the *weight vectors* of this linear representation; here these covectors can also be regarded as vectors.

Like a root vector, a weight vector is a linear combination of simple root vectors in a Cartan subalgebra with the geometry of Euclidean space $\mathbb{C}R^n$, or in the space $\mathbb{C}R^{n+1}$ containing this Euclidean space as a hyperplane, with rational coefficients. Therefore, a weight vector, can also be regarded as a vector in real Euclidean space R^n of dimension n equal to the rank of the group. The relations "greater" and "smaller" are defined for weight vectors like for root vectors..

Therefore, among the weight vectors of any linear representation of a simple complex Lie group there is a maximal weight vector, the corresponding weight is called the *dominant weight* of this representation.

Theorem of É. Cartan [Car7]. *The dominant weight of an irreducible linear representation of a simple Lie group determines this representation up to equivalence, that is, if the dominant weights of two irreducible linear representations of a simple Lie group coincide, the matrices of these representations are the matrices of the same linear operators.*

The irreducible linear representations of a simple complex Lie group are called *fundamental representations* of this group.

Theorem of H. Weyl [Wey5]. *The weight vectors π^i of the dominant weights of the fundamental representations of a simple complex Lie group are related with the fundamental root vectors α_i of this group by*

$$\alpha_i = a_{ij}\pi^j, \tag{1.155}$$

where (a_{ij}) is the Cartan matrix (1.110) or (1.111) of this group.

Therefore, if we denote the vectors $2\alpha_i/\alpha_i^2$ by π_i, the vectors π^i form a basis dual to the basis π_i and can be written as

$$\pi^i = A^{ij}\alpha_j, \tag{1.156}$$

where (A^{ij}) is the inverse matrix of the Cartan matrix (a_{ij}). For the classical simple Lie groups A_n, B_n, C_n, and D_n the matrices (A^{ij}) have the form

$$\left.\begin{array}{c}
\begin{pmatrix}
\frac{n}{n+1} & \frac{n-1}{n+1} & \frac{n-2}{n+1} & \cdots & \frac{3}{n+1} & \frac{2}{n+1} & \frac{1}{n+1} \\
\frac{n-1}{n+1} & \frac{2n-2}{n+1} & \frac{2n-2}{n+1} & \cdots & \frac{6}{n+1} & \frac{4}{n+1} & \frac{2}{n+1} \\
\frac{n-2}{n+1} & \frac{2n-4}{n+1} & \frac{2n-6}{n+1} & \cdots & \frac{9}{n+1} & \frac{6}{n+1} & \frac{3}{n+1} \\
\cdots & \cdots & \cdots & \cdots & \cdots & \cdots & \cdots \\
\frac{2}{n+1} & \frac{4}{n+1} & \frac{6}{n+1} & \cdots & \frac{2n-4}{n+1} & \frac{2n-2}{n+1} & \frac{n-1}{n+1} \\
\frac{1}{n+1} & \frac{2}{n+1} & \frac{3}{n+1} & \cdots & \frac{n-1}{n+1} & \frac{n-1}{n+1} & \frac{n}{n+1}
\end{pmatrix},
\begin{pmatrix}
1 & 1 & 1 & \ldots & 1 & 1 \\
1 & 2 & 2 & \ldots & 2 & 2 \\
1 & 2 & 3 & \ldots & 3 & 3 \\
\cdots & \cdots & \cdots & \cdots & \cdots & \cdots \\
1 & 2 & 3 & \ldots & n-1 & n-1 \\
\frac{1}{2} & 1 & \frac{3}{2} & \ldots & \frac{n-1}{2} & \frac{n}{2}
\end{pmatrix}, \\
\begin{pmatrix}
1 & 1 & 1 & \ldots & 1 & \frac{1}{2} \\
1 & 2 & 2 & \ldots & 2 & 1 \\
1 & 2 & 3 & \ldots & 3 & \frac{3}{2} \\
\cdots & \cdots & \cdots & \cdots & \cdots & \cdots \\
1 & 2 & 3 & \ldots & n-1 & \frac{n-1}{2} \\
1 & 2 & 3 & \ldots & n-1 & \frac{n-1}{2}
\end{pmatrix},
\begin{pmatrix}
1 & 1 & 1 & \ldots & 1 & \frac{1}{2} & \frac{1}{2} \\
1 & 2 & 2 & \ldots & 2 & 1 & 1 \\
1 & 2 & 3 & \ldots & 3 & \frac{3}{2} & \frac{3}{2} \\
\cdots & \cdots & \cdots & \cdots & \cdots & \cdots & \cdots \\
1 & 2 & 3 & \ldots & n-2 & \frac{n-2}{2} & \frac{n-2}{2} \\
\frac{1}{2} & 1 & \frac{3}{2} & \ldots & \frac{n-2}{2} & \frac{n}{4} & \frac{n-2}{4} \\
\frac{1}{2} & 1 & \frac{3}{2} & \ldots & \frac{n-2}{2} & \frac{n-2}{4} & \frac{n}{4}
\end{pmatrix}.
\end{array}\right\} \tag{1.157}$$

For the exceptional simple Lie groups G_2, F_4, E_6, E_7, and E_8 the matrices A^{ij} have the form

$$\left.\begin{array}{c}
\begin{pmatrix} 2 & 1 \\ 3 & 2 \end{pmatrix},
\begin{pmatrix} 2 & 3 & 4 & 2 \\ 3 & 6 & 8 & 4 \\ 2 & 4 & 6 & 3 \\ 1 & 2 & 3 & 2 \end{pmatrix},
\begin{pmatrix}
\frac{4}{3} & 1 & \frac{5}{3} & 2 & \frac{4}{3} & \frac{2}{3} \\
1 & 2 & 2 & 3 & 2 & 1 \\
\frac{5}{3} & 2 & \frac{10}{3} & 4 & \frac{8}{3} & \frac{4}{3} \\
2 & 3 & 4 & 6 & 4 & 2 \\
\frac{4}{3} & 2 & \frac{8}{3} & 4 & \frac{10}{3} & \frac{5}{3} \\
\frac{2}{3} & 1 & \frac{4}{3} & 2 & \frac{5}{3} & \frac{4}{3}
\end{pmatrix}, \\
\begin{pmatrix}
2 & 2 & 3 & 4 & 3 & 2 & 1 \\
2 & \frac{7}{2} & 4 & 6 & \frac{9}{2} & 3 & \frac{3}{2} \\
3 & 4 & 6 & 8 & 6 & 4 & 2 \\
4 & 6 & 8 & 12 & 9 & 6 & 3 \\
3 & \frac{9}{2} & 6 & 9 & \frac{15}{2} & 5 & \frac{5}{2} \\
2 & 3 & 4 & 6 & 5 & 4 & 2 \\
1 & \frac{3}{2} & 2 & 3 & \frac{5}{2} & 2 & \frac{3}{2}
\end{pmatrix},
\begin{pmatrix}
4 & 5 & 7 & 10 & 8 & 6 & 4 & 2 \\
5 & 8 & 10 & 15 & 12 & 9 & 6 & 3 \\
7 & 10 & 14 & 20 & 16 & 12 & 8 & 4 \\
10 & 15 & 20 & 30 & 24 & 18 & 12 & 6 \\
8 & 12 & 16 & 24 & 20 & 15 & 10 & 5 \\
6 & 9 & 12 & 18 & 15 & 12 & 8 & 4 \\
4 & 6 & 8 & 12 & 10 & 8 & 6 & 3 \\
2 & 3 & 4 & 6 & 5 & 4 & 3 & 2
\end{pmatrix}.
\end{array}\right\} \tag{1.158}$$

The commutative groups RL and WL consisting of the linear combinations of the root, respectively weight, vectors of a simple complex Lie group are called the *root lattice* and the *weight lattice* of this group, respectively.

The group RL is a subgroup of WL, and for a simple complex Lie group G the quotient group WL/RL is isomorphic to the *connectedness group* of the adjoint group of the compact simple real Lie group whose complex form is G (see **1.4.11**); the number of elements of this group coincides with the index of connectedness of G.

Formulas (1.155) and (1.156) show that between the fundamental root vectors α_i of a simple complex Lie group and the weight vectors π^i of the dominant weights of the fundamental linear representations of this group there is a correspondence, and the number of fundamental representations, like the number of fundamental root vectors, is equal to the rank of this group. Therefore, we can denote the fundamental linear representation of a simple complex Lie group corresponding to the fundamental root vector α_i by φ_i with the same index i.

Any linear representation of a simple Lie group can be built from fundamental ones. If we are given a linear representation φ of a Lie group G in the linear space $\mathbb{C}L^N$, then any *tensor* (see **0.6.3**) also determines a linear representation of this group in a linear space with dimension equal to the number of linearly independent coordinates of this tensor. In the case of symmetric, or skew symmetric tensors $T_{i_1 i_2 \dots i_k}$ we will denote these representations by $\varphi^{(k)}$, respectively $\varphi^{[k]}$. The last representation is called the *k-th exterior power* of the representation φ. The dimension of this representation is equal to the number of combinations C_N^k. If the first k maximal weight vectors are $\pi > \pi_1 > \dots > \pi_{k-1}$, then the weight vector of the dominant weight of the representation $\varphi^{[k]}$ is $\pi + \pi_1 + \pi_2 + \dots + \pi_{k-1}$.

If two linear representations φ and ψ of the group G in spaces $\mathbb{C}L^N$ and $\mathbb{C}L^M$ with coordinates of vectors x^I $(I = 1, 2, \dots, N)$ and y^J $(J = 1, 2, \dots, M)$ are given, the products $x^i y^J$, which can be regarded as coordinates of vectors in the space $\mathbb{C}L^{MN}$, form a linear representation of G in $\mathbb{C}L^{MN}$. This representation is called the *Kronecker product* of φ and ψ and is denoted by $\varphi \otimes \psi$. If the weight vectors of the dominant vectors of the representations φ and ψ are π and ρ, the weight vector of the dominant weight of $\varphi \otimes \psi$ is $\pi + \rho$.

If the weight vector π of the dominant weight of a representation of a simple complex Lie group G is $\pi = \sum_i m_i \alpha_i$, this representation can be represented by the Dynkin diagram of G (Fig. 1.3) with mark m_i at the dot representing the root vector α_i.

H. Weyl [Wey5] has proved that the dimension N of the linear representation with dominant weight represented by the weight vector π is equal

$$N = \prod_{\alpha > 0} \frac{(\pi + \sigma)\alpha}{\sigma\alpha}, \tag{1.159}$$

where α runs over all positive root vectors and σ is half the sum of these vectors.

For the fundamental representations of simple complex Lie groups formula (1.159) gives the numbers shown in Table 1.1.

Table 1.1

A_n	φ_1 and φ_n $n+1$, φ_2 and φ_{n-1} $C_{n+1}^2, \dots, \varphi_i$ and φ_{n-i+1} C_{n+1}^i
B_n	φ_1 $2n+1$, φ_2 $C_{2n+1}^2, \dots, \varphi_i$ $C_{2n+1}^i, \dots, \varphi_{n-1}$ $C_{2n=1}^{n-1}$, φ_n 2^n
C_n	φ_1 $2n$, φ_2 $C_{2n}^2 - 1, \dots, \varphi_i$ $C_{2n}^i - C_{2n}^{i-2}, \dots, \varphi_n$ $C_{2n}^n - C_{2n}^{n-2}$
D_n	φ_1 $2n$, φ_2 $C_{2n}^2, \dots, \varphi_i$ $C_{2n}^i, \dots, \varphi_{n-2}$ C_{2n}^{n-2}, φ_{n-1} and φ_n 2^{n-1}
G_2	φ_1 7, φ_2 14
F_4	φ_1 52, φ_2 1274, φ_3 273, φ_4 26
E_6	φ_1 and φ_5 27, φ_2 and φ_4 351, φ_3 2925, φ_6 78
E_7	φ_1 133, φ_2 8645, φ_3 365750, φ_4 27664, φ_5 1539, φ_6 56, φ_7 912
E_8	φ_1 248, φ_2 30380, φ_3 2450240, φ_4 146325270, φ_5 6899079264, φ_6 6696000, φ_7 3875, φ_8 147250

Table 1.1 shows that the dimensions of the spaces of the representations φ_1 of A_n, B_n, C_n, and D_n are equal to the orders of the matrices in the groups $\mathbb{C}SL_{n+1}$,

$\mathbb{C}SO_{2n+1}$, $\mathbb{C}Sp_{2n}$, and $\mathbb{C}SO_{2n}$, that is, these representations are formed by matrices of these groups themselves.

This table shows that the dimension of the space of the representation φ_i of A_n, B_n, and D_n ($i < n$ for B_n and $i < n-1$ for D_n) is equal to the number of combinations C_N^i where N is the dimension of the linear space of φ_i, that is, this representation is the i-th exterior power of the representation φ_1 of these groups. The representation φ_n of B_n and the representations φ_{n-1} and φ_n of D_n are *spinor representations* of these groups. We will consider these representations in **1.6.3.** The dimensions of the spaces of the representations φ_i of C_n for $i > 1$ are equal to the differences $C_{2n}^i - C_{2n-2}^i$, since for $i > 1$ the i-th exterior power of φ_1 is reducible and is the direct sum of two representations one of which is φ_i.

Table 1.1 shows also that the dimensions of the spaces of the representations φ_1 of G_2, F_4, E_7, and E_8, and of the representation φ_6 of E_6, are equal to the dimensions of these groups, that is, these representations are adjoint representations of these groups. The adjoint representations of the groups A_n, B_n, C_n, and D_n are the representations $\varphi_1 \otimes \varphi_n, \varphi_2, \varphi_1$, and φ_2, respectively. Note that the dots in the Dynkin diagrams of these groups corresponding to adjoint representations are the dots related with the complementary dots in the extended Dynkin diagrams of these groups (see Fig. 1.7).

The theory of linear representations of semisimple real Lie groups built by É. Cartan in [Car9] is analogous to the theory of linear representations of semisimple complex Lie groups built by him in [Car7]; the real dimensions of the spaces of representations of simple real Lie groups are equal to the complex dimensions of the spaces of representations of simple complex Lie groups .

1.6.3. Spinor Representations. The spinor representations of the compact real groups SO_n can be defined as follows. Consider the algebra $\mathbb{A}_n$ of alternions (see **1.2.8**), which can be represented by the alternions in the algebra $\mathbb{A}_{n+1}$ with even number of indices (see Theorem 1.22). If a is an element in this algebra $\mathbb{A}_n$ and $x = x^i e_i$ is an element in $\mathbb{A}_{n+1}$, in general the product $\bar{a}xa$ is not a linear combination of the elements e_i. The group of those elements a satisfying the condition

$$\bar{a}a = 1 \tag{1.160}$$

while for all $x = x^i e_i$ in $\mathbb{A}_{n+1}$ the product $\bar{a}xa$ is a linear combination of the elements e_i, that is,

$$'x = \bar{a}xa = {'x^i} e_i \tag{1.161}$$

is called the *spinor group* of the group SO_n, and is denoted by $\overline{SO}_n$.

Theorem of R. Lipschitz [Lip]. *The spinor group $\overline{SO}_n$ is homomorphic to the group SO_n, and the kernel of this homomorphism consists of the two alternions 1 and -1.*

The coordinates $a^{i_1 i_2 \cdots i_{2k}}$ of the alternions in the algebra $\mathbb{A}_n$, as represented by elements in the algebra $\mathbb{A}_{n+1}$ with an even number of indices satisfying the conditions (1.160) and (1.161) are related by

$$\left.\begin{array}{r}a_0 a^{i_1 i_2 i_3 i_4} = 3!! a^{[i_1 i_{2_a} i_3 i_4]},\\ a_0 a^{i_1 i_2 i_3 i_4 i_5 i_6} = 5!! a^{[i_1 i_{2_a} i_3 i_4 i_5 i_6]}\\ \cdots\cdots\cdots\cdots\cdots\cdots\\ a_0 a^{i_1 i_2 \cdots i_{2k}} = (2k-1)!! a^{[i_1 i_{2_a} i_3 i_4 \cdots i_{2k}]},\\ \cdots\cdots\cdots\cdots\cdots\cdots\end{array}\right\} \tag{1.162}$$

and

$$\sum_k (a^{i_1 i_2 \cdots i_{2k}})^2 = 1. \tag{1.163}$$

In the equations (1.162) the expression $(2k-1)!!$ denotes the product of all odd integers from 1 to $2k-1$ and $[\quad]$ is the symbol of alternation, that is, if $T^{i_1 i_2 \cdots i_m}$ is a general tensor of valency m, then $T^{[i_1 i_2 \cdots i_m]}$ is a skew symmetric tensor of the same valency whose coordinates are equal to the products of $\frac{1}{m!}$ by the algebraic sum of all coordinates $T^{j_1 j_2 \cdots j_m}$ obtained from $T^{i_1 i_2 \cdots i_m}$ over all permutations of the indices $i_1 i_2, \ldots, i_m$, taken with the sign $+$ if the permutation is even and with the sign $-$ if this permutation is odd.

The equations (1.162) are obtained from the condition that all coordinates of the alternion $\bar{a}a$, except its coordinate at 1, and all coordinates of the alternion $\bar{a}xa$, except its coordinate at e_i, are equal to zero. The equation (1.163) follows from the condition (1.160).

If we introduce in $\mathbb{A}_n$ the metric of the Euclidean space $R^{2^{n-1}}$, with the distance between the alternions a and b equal to the modulus $|b-a|$ of their difference, (1.162) is the equation of a hypercone, which is the intersection of hypercones of second order, and the equation (1.163) is the equation of a hypersphere. The intersection of the hypercone (1.162) with the hypersphere (1.163) is an algebraic variety, called the *Lipschitzian*. We denote this variety by Ω_n. Since the equations (1.162) and (1.163) make it possible to express all coordinates $a^{i_1 i_2 \cdots i_{2k}}$ of an alternion a in this group in terms of its coordinates $a^{ij} = -a^{ji}$, the dimension of the Lipschitzian Ω_n and of the spinor group $\overline{SO}_n$ is equal to $\frac{n(n-1)}{2}$, that is, this dimension is equal to the dimension of the group SO_n.

If we similarly introduce the metric of the space R^{2^n} in the algebra $\mathbb{A}_{n+1}$, the subspace of this algebra generated by the elements e_i has been endowed with the metric of the space R^n. If x and y are elements in this subspace, their inner product (x, y) is equal to $-\frac{1}{2}(xy+yx)$ and the squares of their moduli $|x|$ and $|y|$ are equal to $-x^2$ and $-y^2$, respectively. Therefore, the product $(\bar{a}\bar{x}a)(\bar{a}xa)$ is equal to $|x|^2$ and, if $x' = \bar{a}xa$ is also an element in this subspace, $|x'| = |x|$. Since the transformation (1.161) is a linear transformation, $'x^i = A^i_j x^j$, and $|x'| = |x|$, this transformation is a rotation in R^n and the matrix (A^i_j) is an orthogonal matrix in O_n.

Let us prove that each orthogonal matrix in SO_n can be written in the form (1.161). Since each rotation of the first kind in R_n is the product of an even number of reflections in hyperplanes through the center of a rotation, let us consider such a reflection. If $(n, x) = 0$ with $|n| = 1$ is the equation of a hyperplane in R^n through the point O, the reflection in this hyperplane is $x' = -nxn$, since if $x = x_1 + x_2$, where x_1 is collinear with n and x_2 is orthogonal to it, $x_1 n = n x_1$ and $x_2 n = -n x_2$, then $x'_1 = -n x_1 n = -x_1 n^2 = x_1$, $x'_2 = -n x_2 n = x_2 n^2 = -x_2$. Therefore, each rotation of the first kind in R^n can be represented as the product of an even number of transformations (1.161), where the role of alternion a is played by the product $n_1, n_2, \dots, n_{2k}$, that is, a transformation (1.161) with $a = n_1 n_2 \dots n_{2k}$ and $\bar{a} = n_{2k} n_{2k-1} \dots n_1$.

Therefore, the groups SO_n and $\overline{SO}_n$ are homomorphic. It is evident that the alternions a and $-a$ determine the same rotation in R^n and that the kernel of this homomorphism consists of the alternions 1 and -1.

For the groups SO_2 and SO_3, since the algebras $\mathbb{A}_2$ and $\mathbb{A}_3$ are isomorphic to $\mathbb{C}$ and $\mathbb{H}$, respectively, the spinor groups are determined by the single equation (1.163), that is, by the condition $|a| = 1$, and the transformations (1.161) are the transformations in $\mathbb{C}$ and $\mathbb{H}$ where $|a| = 1$, that is, $x' = e^{it}x$ and (1.161) in $\mathbb{H}$, respectively.

For the group SO_4, since the algebra $\mathbb{A}_4$ is isomorphic to the direct sum $\mathbb{H} \oplus \mathbb{H}$, the spinor group is determined by the two equations (1.162) and (1.163), which can be written as

$$a_0 a^{i_1 i_2 i_3 i_4} - a^{i_1 i_2} a^{i_3 i_4} + a^{i_1 i_3} a^{i_2 i_4} + a^{i_1 i_4} a^{i_2 i_3} = 0, \tag{1.164}$$

$$\begin{aligned} &a_0^2 + (a^{i_1 i_2})^2 + (a^{i_1 i_3})^2 + (a^{i_1 i_4})^2 + (a^{i_2 i_3})^2 + \\ &+ (a^{i_2 i_4})^2 + (a^{i_3 i_4})^2 + (a^{i_1 i_2 i_3 i_4})^2 = 1. \end{aligned} \tag{1.165}$$

The group SO_4 can also be represented by the quaternionic transformations $x' = axb$ with $|a| = |b| = 1$, that is, by transformations (1.47) of the first kind with $c = 0$. The spinor group $\overline{SO}_4$ can also be represented by the direct product of the groups of quaternions a and b with modulus one.

For the group SO_5, since the algebra $\mathbb{A}_5$ is isomorphic to the algebra $\mathbb{H}_2$, the spinor group is determined by the five equations (1.162) and the single equation (1.163). This spinor group is isomorphic to the group $\mathbb{H}U_2$ of unitary matrices in the algebra $\mathbb{H}_2$.

For the group SO_6, since the algebra $\mathbb{A}_6$ is isomorphic to the algebra $\mathbb{C}_4$, the spinor group is determined by the sixteen equations (1.162) and the single equation (1.163). This spinor group is isomorphic to the group $\mathbb{C}SU_4$ of unimodular unitary matrices in the algebra $\mathbb{C}_4$.

For the group SO_7, since the algebra $\mathbb{A}_7$ is isomorphic to the algebra $\mathbb{R}_8$, the spinor group is determined by the 42 equations (1.162) and the single equation (1.163).

For the group SO_8, since the algebra $\mathbb{A}_8$ is isomorphic to the direct sum $\mathbb{R}_8 \oplus \mathbb{R}_8$, the spinor group is determined by the 99 equations (1.162) and the single equation (1.163).

We similarly define the spinor group $\overline{SO}_{n,\ell}$ of the group $SO_{n,\ell}$ of pseudoorthogonal matrices. This group is defined by means of the algebra $\mathbb{A}_{n,\ell}$ of pseudoalternions (see **1.2.9**) which can be represented by the pseudoalternions in the algebra $\mathbb{A}_{n+1,\ell}$ with an even number of indices (similar to the algebra $\mathbb{A}_n$, see Theorem 1.22).

If a is an element in the algebra $\mathbb{A}_{n,\ell}$ and $x = x^i e_i$ is an element in the algebra $\mathbb{A}_{n+1,\ell}$, the spinor group $\overline{SO}_{n,\ell}$ consists of the elements in $\mathbb{A}_{n,\ell}$ satisfying the condition

$$\bar{a}a = \pm 1, \tag{1.166}$$

and such that the product $\bar{a}xa$ is a linear combination of the elements e_i, that is, (1.161) is fulfilled.

As for the groups $\overline{SO}_n$, we can prove for $\overline{SO}_{n,\ell}$ a theorem analogous to the theorem of Lipschitz: the spinor group $\overline{SO}_{n,\ell}$ is homomorphic to the group $SO_{n,\ell}$, the kernel of this homomorphism consists of the two pseudoalternions 1 and -1, and the coordinates $a^{i_1 i_2 \dots i_{2k}}$ of the pseudoalternions in $\mathbb{A}_{n,\ell}$, represented by elements in $\mathbb{A}_{n+1,\ell}$ with an even number of indices and satisfying the conditions (1.166) and (1.161), are related by equations (1.162) and

$$\sum_k \epsilon_{i_1} \epsilon_{i_2} \dots \epsilon_{i_{2k}} (a^{i_1 i_2 \dots i_{2k}})^2 = 1. \tag{1.167}$$

If we introduce in the algebras $\mathbb{A}_{n,\ell}$ and $\mathbb{A}_{n+1,\ell}$ the pseudometrics of the pseudo-Euclidean spaces $R_L^{2^{n-1}}$ and $R_L^{2^n}$ with distance between two pseudoalternions a and b equal to the modulus $|b-a|$ of their difference, we have thereby introduced, in the subspace of $\mathbb{A}_{n+1}$ generated by the elements e_i, the pseudometric of the space R^n_ℓ, and the transformations (1.161) are rotations in this space.

For the groups $SO_{2,1}$ and $SO_{3,1}$, since the algebras $\mathbb{A}_{2,1}$ and $\mathbb{A}_{3,1}$ are isomorphic to $\mathbb{C}'$ and $\mathbb{H}'$, respectively, the spinor groups are determined by the single equation (1.167), that is, by the condition $|a|^2 = \pm 1$, and the transformations (1.161) are the transformations in $\mathbb{C}'$ and $\mathbb{H}'$ with $|a|^2 = \pm 1$, that is, $x' = \pm(\cosh t + e \sinh t)x$ or $'itx = \pm(\sinh t + e\cosh t)x$, and (1.166), respectively. Since the algebra $\mathbb{H}'$ is isomorphic to the algebra $\mathbb{R}_2$ and split quaternions with modulus one correspond under this isomorphism to real unimodular matrices, the spinor group $\overline{SO}_{3,1}$ is isomorphic to the group SL_2.

For the groups $SO_{4,1}$ and $SO_{4,2}$, since the algebras $\mathbb{A}_{4,1}$ and $\mathbb{A}_{4,2}$ are isomorphic to the algebra $\mathbb{C}_2$ and to the direct sum $\mathbb{R}_2 \oplus \mathbb{R}_2$ (or to $\mathbb{C}'_2$), respectively, the spinor group is determined by the two equations (1.162) and (1.163), which can be written as (1.164) and

$$\begin{aligned} a_0^2 - (a^{i_1 i_2})^2 - (a^{i_1 i_3})^2 - (a^{i_1 i_4})^2 + (a^{i_2 i_3})^2 + \\ + (a^{i_2 i_4})^2 + (a^{i_3 i_4})^2 - (a^{i_1 i_2 i_3 i_4})^2 = 1 \end{aligned} \tag{1.168}$$

and

$$\begin{aligned} a_0^2 + (a^{i_1 i_2})^2 - (a^{i_1 i_3})^2 - (a^{i_1 i_4})^2 - (a^{i_2 i_3}) - \\ - (a^{i_2 i_4})^2 + (a^{i_3 i_4})^2 + (a^{i_1 i_2 i_3 i_4})^2 = 1, \end{aligned} \tag{1.169}$$

respectively. These spinor groups are isomorphic to the groups $\mathbb{C}SL_2$ and $SL_2 \otimes SL_2 = \mathbb{C}'SL_2$.

The group $SO_{4,2}$ can also be represented by the split quaternionic transformations $x' = axb$, with $|a|^2 = |b|^2 = \pm 1$. The spinor group $\overline{SO}_{4,2}$ can also be represented by the direct product of the groups of quaternions a and b with modulus 1 or i.

For the groups $SO_{5,1}$ and $SO_{5,2}$, since the algebras $\mathbb{A}_{5,1}$ and $\mathbb{A}_{5,2}$ are isomorphic to the algebras $\mathbb{H}_2$ and $\mathbb{R}_4$, respectively, the spinor groups are isomorphic to the group $\mathbb{H}U_{2,1}$ of pseudounitary matrices in $\mathbb{H}_2$ and to the group Sp_4 of symplectic matrices in $\mathbb{R}_4$.

For the groups $SO_{6,1}, SO_{6,2}$, and $SO_{6,3}$, since the algebras $\mathbb{A}_{6,1}, \mathbb{A}_{6,2}$, and $\mathbb{A}_{6,3}$ are isomorphic to $\mathbb{H}_2 \oplus \mathbb{H}_2, \mathbb{C}_4$, and $\mathbb{C}'_4 = \mathbb{R}_4 \oplus \mathbb{R}_4$, respectively, the spinor groups are isomorphic to the group $\mathbb{H}SL_2$ of unimodular matrices in the algebra $\mathbb{H}_2$, to the group $\mathbb{C}SU_{4,1}$ of unimodular pseudounitary matrices in $\mathbb{C}_4$, and to the group SL_4 of unimodular matrices in $\mathbb{R}_4$, respectively.

Since the algebras $\mathbb{A}_n$ and $\mathbb{A}_{n,\ell}$ are isomorphic to algebras of real, complex, or quaternionic matrices or to direct sums of pairs of isomorphic algebras of such matrices, the vectors in the linear spaces on which the linear transformations having these matrices act are called *spinors*, and such linear representations of SO_n and $SO_{n,\ell}$ are called *spinor representations* of these groups.

If the algebras $\mathbb{A}_{2m}$ and $\mathbb{A}_{2m,\ell}$ are isomorphic to direct sums of algebras of matrices, the vectors in the linear spaces on which the linear transformations having these matrices act are called *semispinors.*

In particular, the spinors in $SO_{3,1}$ are vectors in L^2, the spinors in $SO_{4,1}$ are vectors in $\mathbb{C}L^2$, the spinors in $SO_{4,2}$ are vectors in $\mathbb{C}' L^2$, and if the vectors of the last spaces are written as $x = x_+e_+ + x_-e_-$, the real vectors x_+ and x_- are semispinors in $SO_{4,2}$.

1.6.4. Fundamental and Parabolic Figures in Spaces with Simple Fundamental Groups. In **0.8.3** we have defined solvable Lie groups. In each simple Lie group there is a maximal solvable Lie group. Since the meaning of these subgroups was found by Armand Borel, these subgroups are called *Borel subgroups.* The Borel subgroup of the group SL_n consists of *triangular matrices*, that is, of matrices all subdiagonal entries of which are equal to zero. The invariant subgroup G_1 of this group consists of the matrices all diagonal entries of which are equal to 1, the subgroup G_i of this group consists of the matrices in G_1 all entries of which in the $i-1$ diagonal above the main diagonal are equal to zero (in this case the number k in the definition of a solvable group is equal to n).

In the geometry of homogeneous spaces with as fundamental group a simple Lie group a very important role is played by the subgroups containing the Borel subgroup. These subgroups are called *parabolic subgroups.* The subalgebras of the tangent Lie algebra of a simple Lie group corresponding to parabolic subgroups are called *parabolic subalgebras.*

Theorem of I. L. Kantor [Kan1-2]. *The parabolic subalgebras of the tangent Lie algebras of the simple Lie groups can be determined as follows. For each simple root vector α_i of the simple Lie group G and for any set of simple root vectors $\alpha_{i_1}, \alpha_{i_2}, \ldots, \alpha_{i_k}$ of G there is a decomposition of the tangent Lie algebra $\mathbb{G}$ of G as a direct sum of linear subspaces:*

$$\mathbb{G} = \mathbb{G}_{-h} \oplus \mathbb{G}_{-h+1} \oplus \cdots \oplus \mathbb{G}_{-1} \oplus \mathbb{G}_0 \oplus \mathbb{G}_1 \oplus \cdots \oplus \mathbb{G}_{h-1} \oplus \mathbb{G}_h, \tag{1.170}$$

where the subspace $\mathbb{G}_\lambda$ is the set of linear combinations of the eigenvectors of the linear transformations $X \to [HX]$ in $\mathbb{G}$ (see **1.4.3***) whose eigenvalues are root* *forms determined by root vectors which are linear combinations of the simple root vectors with coefficient at the simple root vector α_i equal to λ or with sum of the coefficients at the simple root vectors $\alpha_{i_1}, \alpha_{i_2}, \ldots, \alpha_{i_k}$ equal to λ. The subspace $\mathbb{G}_0$ contains the Cartan subalgebra of the Lie algebra $\mathbb{G}$. Then the direct sum $\mathbb{P} = \mathbb{G}_0 \oplus \mathbb{G}_1 \oplus \cdots \oplus \mathbb{G}_{h-1} \oplus \mathbb{G}_h$ is a parabolic subalgebra, and each parabolic subalgebra of the tangent Lie algebra of a simple Lie group can be obtained in this way.*

If we denote the subgroup of the simple Lie group G corresponding to the subalgebra $\mathbb{P}$ by P, the direct sum $\mathbb{G}_{-h} \oplus \mathbb{G}_{-h+1} \oplus \cdots \oplus \mathbb{G}_{-1}$ can be regarded as the tangent space of the homogeneous space G/P.

If we consider a homogeneous space $M = G/K$ with fundamental group G, the parabolic subgroup P is a stabilizer of a geometric figure in M, called *a parabolic figure* in M. Homogeneous spaces G/P with P parabolic subgroups of simple Lie groups G are called *parabolic spaces.*

Parabolic figures determined by one simple root vector are called *fundamental figures*; fundamental figures determined by a simple root vectors α_i are called *α_i-figures.* Parabolic figures determined by a set of simple root vectors $\alpha_{i_1}, \alpha_{i_2}, \ldots, \alpha_{i_k}$ are called *$(\alpha_{i_1}, \alpha_{i_2}, \ldots, \alpha_{i_k})$-figures.*

α_i-figure and α_j-figure that are parts of (α_i, α_j)-figures are called *incident fundamental figures.* Analogously we can define *incident general parabolic figures.*

Note that *Jacques Tits*, who created the theory of parabolic figures and spaces in [Ti1, Ti2, Ti3], called fundamental figures "fundamental elements", parabolic spaces – "R-spaces", α_i-figures and $(\alpha_{i_1}, \alpha_{i_2}, \ldots, \alpha_{i_k})$-figures – "$\alpha_i$-planes" and "$(\alpha_{i_1}, \alpha_{i_2}, \ldots, \alpha_{i_k})$-planes", respectively. The terms "parabolic spaces" and "parabolic figures" were introduced by the author, M. P. Zamakhovsky, and T. A. Timoshenko (Stepashko) [RZT].

We denote the parabolic spaces G/P whose stabilizers P are determined by one simple root α_i or by a set of simple roots $\alpha_{i_1}, \alpha_{i_2}, \ldots, \alpha_n$ by $[\alpha_i]$ and $[\alpha_{i_1} \alpha_{i_2}, \ldots, \alpha_n]$, respectively.

The dimensions of these parabolic spaces, dim $[\alpha_i]$ and dim $[\alpha_1, \alpha_2, \ldots, \alpha_n]$, are equal to the sums of the dimensions of the subspaces $\mathbb{G}_{-h}, \mathbb{G}_{-h+1}, \ldots, \mathbb{G}_{-1}$ whose direct sum $\mathbb{G}_{-h} \oplus \mathbb{G}_{-h+1} \oplus \cdots \oplus \mathbb{G}_{-1}$ can be regarded as tangent spaces to these parabolic spaces.

Theorem 1.35. *The dimension* $\dim_j(\alpha_i)$ *of the manifold of α_i-figures which are incident with some α_j-figure is equal to*

$$\dim_j(\alpha_i) = \dim[\alpha_i, \alpha_j] - \dim[\alpha_j]. \tag{1.171}$$

<u>Proof</u>. This formula follows from the fact that the manifold of (α_i, α_j)-figures is foliated on manifolds of α_i-figures which are incident with an α_j-figure, while each such manifold is related to one α_j-figure. ■

Theorem 1.36. *The fundamental α_i-figures in the projective space P^n are $(i-1)$-planes, in particular, points for $i = 1$, lines for $i = 2$, planes for $i = 3$, hyperplanes for $i = n$.*

<u>Proof</u>. In **1.4.3** we have seen that for the group SL_{n+1} the role of eigenvectors of the linear transformations $X \to [HX]$ in its tangent Lie algebra is played by the

matrices E_{ij} $(i \neq j)$, and since the root vectors of this group corresponding to E_{ij} can be written in the form $E_i - E_j$ $(i \neq j,\ i, j = 0, 1, \dots, n)$ and the simple root vectors of this group have the form $\alpha_i = E_{i-1} - E_i$, the root vectors of this group corresponding to the eigenvectors E_{ij} are linear combinations $\alpha_{i+1} + \alpha_{i+2} + \dots + \alpha_j$ for $i < j$ and linear combinations $-\alpha_{i+1} - \alpha_{i+2} - \dots - \alpha_j$ for $i > j$. Therefore, the decomposition (1.170) of the Lie algebra $\mathbb{SL}_{n+1}$ corresponding to the simple root vector α_i has the form

$$\begin{bmatrix} A_1^1 & A_2^1 \\ A_1^2 & A_2^2 \end{bmatrix} = \begin{bmatrix} 0 & 0 \\ A_1^2 & 0 \end{bmatrix} + \begin{bmatrix} A_1^1 & 0 \\ 0 & A_2^2 \end{bmatrix} + \begin{bmatrix} 0 & A_2^1 \\ 0 & 0 \end{bmatrix}, \tag{1.172}$$

where A_1^1 is an $((i+1) \times (i+1))$-matrix, A_2^1 is an $((n-i) \times (i+1))$-matrix, A_1^2 is an $((i+1) \times (n-i))$-matrix, and A_2^2 is an $((n-i) \times (n-i))$-matrix. Therefore, the subalgebra $\mathbb{P}$ consists of the matrices

$$\mathcal{A} = \begin{bmatrix} A_1^1 & A_2^1 \\ 0 & A_2^2 \end{bmatrix}. \tag{1.173}$$

It is clear that all these subgroups contain the Borel subgroup of the group SL_{n+1}, which consists of the upper triangular matrices.

Since the subgroup of the group of collineations of the real projective space P^n, represented by the matrices in SL_{n+1} corresponding to the matrices in subalgebra (1.173) of the Lie algebra $\mathbb{SL}_{n+1}$, and, therefore, itself consisting of matrices of such a form, is the stabilizer of the $(i-1)$-plane $E_0 E_1 \dots E_{i-1}$, the fundamental α_i-figures of P^n are $(i-1)$-planes. ■

We can similarly prove that an $(\alpha_{i_1}, \alpha_{i_2}, \dots, \alpha_{i_k})$-figure in the space P^n is a *flag* consisting of incident $(i_1 - 1)$-plane, $(i_2 - 1)$-plane, ..., $(i_k - 1)$-plane.

In particular, $(\alpha_1, \alpha_2, \dots, \alpha_n)$-figures in P^n are called *complete flags.* For this reason general parabolic spaces are sometimes called *flag manifolds* (see, for instance, [Wo2]).

Theorem of Tits [Ti1]. *The fundamental linear representations φ_i of all simple Lie groups coincide with the linear coordinate transformations of the α_i-figures.*

In particular, the projective coordinates x^i of points, the tangential coordinates u_i of hyperplanes, the *Plücker coordinates*

$$p^{ij} = x^i y^j - y^i x^j \tag{1.174}$$

of lines, and the *Grassmann coordinates*

$$p^{i_0 i_1 \dots i_m} = (m+1)! x^{[i_0} y^{i_1} \dots z^{i_m]} \tag{1.175}$$

of m-planes $(m = 2, \dots, n-1)$ coincide with the coordinates of the vectors in the spaces of the fundamental representations of the group A_n (see **1.6.2**).

We will consider fundamental and general parabolic figures of homogeneous spaces with other fundamental groups in all later chapters.

Theorem 1.37 (I. L. KANTOR [Kan2–3]). *The tangent Lie algebra of a simple Lie group having a decomposition (1.170) has the property that the commutator* $[x, y]$ *of two elements* x *and* y *in this algebra, belonging to subspaces* $\mathbb{G}_\lambda$ *and* $\mathbb{G}_\mu$, *respectively, belongs to* $\mathbb{G}_{\lambda+\mu}$. *In the subspaces* $\mathbb{G}_{-\lambda}$ *of the decomposition (1.170) we can introduce ternary and algebra structures.*

PROOF. The first assertion of this theorem can be checked immediately. This assertion can be symbolically written as

$$[\mathbb{G}_\lambda \mathbb{G}_\mu] \subset \mathbb{G}_{\lambda+\mu}. \tag{1.176}$$

Let us determine in the tangent Lie algebra $\mathbb{G}$ of a simple Lie group G an involutive automorphism σ such that every eigenvector of a linear transformation $X \to [HX]$ in this Lie algebra which corresponds to the root form α is mapped to the eigenvector of this linear transformation which corresponds to the root form $-\alpha$, while in the Cartan subalgebra of this Lie algebra every root vector α is mapped to the root vector $-\alpha$. This automorphism maps a linear subspace $\mathbb{G}_\lambda$ to the subspace $\mathbb{G}_{-\lambda}$. In every subspace $\mathbb{G}_\lambda$ we can define a ternary structure by the formula

$$< t, x, y >= [[\sigma(t), x], y]. \tag{1.177}$$

If one of elements x, y, z in formula (1.177) is fixed, we obtain an algebra structure. ■

Note that this way for obtaining an algebra structure in $\mathbb{G}_\lambda$ is not the only way. In the later chapters we will see that for the most important cases the algebra structures introduced in $\mathbb{G}_\lambda$ are *elastic* (see **1.5.2**).

For $h = 1$ this theorem can be made more precise, as follows:

Theorem 1.38 (KANTOR-KOECHER [Kan1, Koe]). *For* $h = 1$, *in the subspace* $\mathbb{G}_{-1}$ *of the decomposition (1.170) we can introduce a Jordan algebra structure, and each Jordan algebra can be imbedded into a Lie algebra as a subspace* $\mathbb{G}_{-1}$.

PROOF. The Jordan algebra multiplication $x \cdot y$ in $\mathbb{G}_{-1}$ can be defined by formula (1.177) with $\sigma(t) = a$ a fixed element in $\mathbb{G}_1$. The Lie group G determined by the Lie algebra $\mathbb{G} = \mathbb{G}_{-1} \oplus \mathbb{G}_0 \oplus \mathbb{G}_1$ is the group of fractional linear transformations (1.140) in the Jordan algebra. ■

Theorem 1.39. *For* $h = 1$, *two fundamental figures with isomorphic stabilizers determine a point in a symmetric space.*

PROOF. Denote the subalgebra $\mathbb{G}_0$ by $\mathbb{K}$ and the subspace $\mathbb{G}_{-1} \oplus \mathbb{G}_1$ by $\mathbb{E}$. Then, by (1.176) $\mathbb{K}$ and $\mathbb{E}$ satisfy condition (1.126) and the groups G and K corresponding to Lie algebras $\mathbb{G}$ and $\mathbb{K}$ determine a symmetric space G/K and K is stabilizer of a point of this space. ■

Theorem 1.39 can be generalized to arbitrary parabolic spaces. M. P. Zamakhovsky [RoZ] proved that two arbitrary parabolic figures with isomorphic stabilizers determine a point of a generic symmetric or k-symmetric space (see **0.8.8**).

If the tangent Lie algebra $\mathbb{G}$ of a simple Lie group G has the form (1.170), let us denote the direct sum $\mathbb{G}_{-h} \oplus \mathbb{G}_0 \oplus \mathbb{G}_h$ by $\mathbb{G}^0$; $\mathbb{G}^0$ is a subalgebra of $\mathbb{G}$. The subalgebras $\mathbb{G}_0$ and $\mathbb{G}^0$ are the tangent algebras of the subgroups G_0 and G^0 in G. Then the homogeneous spaces G/G_0 and G/G^0, called *biparabolic* and *dyparabolic* spaces, respectively, are $(h+1)$-symmetric and h-symmetric spaces, respectively.

Theorem 1.40. *If the decomposition (1.170) is determined by a simple root α_i then h in (1.170) is equal to the coefficient m_i of decomposition $\sum_i m_i\alpha_i$ at α_i.*

The assertion of this theorem can be immediately verified.

M.P. Zamakhovsky [RoZ] has proved that for s-symmetric spaces G/G_0 and G/G^0 of interior type (that is, defined by interior automorphism in G) determined by an α_i-figure $s = m_i + 1$ and $s = m_i$, respectively.

The stabilizers of points in the spaces G/G^0 are semisimple Lie groups whose Dynkin diagrams are obtained from extended Dynkin diagram of G (see Fig. 1.7) by removal of simple root α_i. The stabilizers of points in the spaces G/G_0 are locally isomorphic to direct products of semisimple Lie groups whose Dynkin diagrams are obtained from usual Dynkin diagram of G (see Fig. 1.3) by removal of simple root α_i and the commutative group T. For symmetric spaces with exceptional fundamental groups these stabilizers coincide with subgroups mentioned in Theorems 7.25–29; note that for these groups numbers m_i coincide with the numbers of last entries of tables (7.28–32), respectively.

The exterior automorphisms of simple Lie groups exist only for the groups A_n, D_n, and E_6, that is for the groups with symmetries of Dynkin diagrams (see fig. 1.3 and 1.5); these automorphisms are connected with duality and triality principles in corresponding spaces.

CHAPTER II

AFFINE AND PROJECTIVE GEOMETRIES

§2.1. Affine Geometries

2.1.1. Linear Spaces and Modules. In **0.1.6** we have defined the real linear spaces L^n. If in this definition we replace the field $\mathbb{R}$ of real numbers by the field $\mathbb{C}$ of complex numbers or by the skew field $\mathbb{H}$ of quaternions, we obtain the complex linear space $\mathbb{C}L^n$ or quaternionic linear space $\mathbb{H}L^n$. We have mentioned the space $\mathbb{C}L^n$ in **1.6.1**. Since $\mathbb{H}$ is noncommutative, in the products (0.7) of a vector a and a quaternionic scalar λ, the scalar must always be at the right of the vector.

If, in the definition of the linear spaces L^n, $\mathbb{C}L^n$, and $\mathbb{H}L^n$, we replace the division algebras $\mathbb{R}, \mathbb{C}$, and $\mathbb{H}$ by an algebra $\mathbb{A}$ with zero divisors, we obtain a *module* over the algebra $\mathbb{A}$. If the algebra $\mathbb{A}$ is noncommutative, in the products (0.7) of a vector a and a scalar λ, which is an element in $\mathbb{A}$, the scalar must always be at the right of the vector.

Since $\mathbb{A}$ contains zero divisors, a module over $\mathbb{A}$ contains *singular vectors* a, i.e. such that the product $a\lambda$ by a zero divisor λ can be equal to zero.

Nonsingular vectors in a module are also called *regular vectors*. The product of a nonsingular vector by a zero divisor is a singular vector, since if a is a nonsingular vector and λ is a zero divisor, with product $\lambda\mu$ by another zero divisor μ equal to zero, the product $(a\lambda)\mu = a(\lambda\mu)$ is the zero vector.

In modules over algebras, linear dependence and independence of vectors and coordinates of vectors are defined as in linear spaces. Since the products of all coordinates of a singular vector by a zero divisor are equal to zero, all coordinates of a singular vector are zero divisors (but there are nonsingular vectors with all coordinates zero divisors, for instance, the vector $\{e_+, e_-\}$ in the module $\mathbb{C}' L^2$).

Theorem 2.1. *Any system of vectors in a module over an algebra including a singular vector is linearly dependent.*

<u>PROOF</u>. If the system consists of $a_1, a_2, \ldots, a_k, b$, where the vectors $a_1 a_2, \ldots, a_k$ are nonsingular linearly independent and b is a singular vector with $b\lambda = 0$, then the linear combination $a_1 \cdot 0 + a_2 \cdot 0 + \cdots + a_k \cdot 0 + b\lambda = 0$. ■

A module is called an *n-dimensional free module* if besides the axioms V I, V II and V III of n-dimensional linear space (see **0.1.6**) it also satisfies the axiom

V III. 3° Any vector in the module is a linear combination of n linearly independent vectors.

We denote such a module by $\mathbb{A}L^n$.

The Jordan algebra $\mathbb{R}_{m,n}$ (see **1.5.3**) is a non-free module over the algebra $\mathbb{R}_n$ if n is not a multiple of m, and is a p-dimensional free module $\mathbb{R}_n L^p$ if $n = pm$. In this case the role of basis vectors is played by a set of p $(m \times n)$-matrices consisting of $p-1$ zero $(m \times m)$-matrices and one identity $(m \times m)$-matrix.

2.1.2. Real and Complex Interpretations of Free Modules over Algebras. If $\mathbb{A}$ is a real m-algebra with basis elements e_α, the linear space or free module $\mathbb{A}L^n$ admits a real interpretation in the space L^{mn}. In this interpretation each vector $x = \{x^i\}$ in $\mathbb{A}L^n$ with coordinates $x^i = x^{i\alpha}e_\alpha$ is interpreted as the vector $\hat{x} = \{x^{i\alpha}\}$ in L^{mn}. In this interpretation the sum $x+y$ in $\mathbb{A}L^n$ is interpreted as the vector $\hat{x}+\hat{y}$ in L^{mn} and the product $z = x\lambda$ in $\mathbb{A}L^n$ is interpreted as the vector

$$\hat{z} = \{z^{i\alpha}\} = \{C^{\alpha}_{\beta\gamma}\lambda^\gamma x^{i\beta}\}. \tag{2.1}$$

Therefore, the linear space L^{mn}, with multiplication by elements in $\mathbb{A}$ defined in accordance with the rule (2.1) gives an interpretation of the space or module $\mathbb{A}L^n$. In this case the space L^{mn} is called a space with *algebra structure.* If the algebra $\mathbb{A}$ is $\mathbb{C}, \mathbb{C}', \mathbb{C}^0, \mathbb{H}, \mathbb{H}', \mathbb{H}^0, \mathbb{H}'^0, \mathbb{R}_n$, etc., the algebra structure is called a *complex, split complex, dual, quaternionic, split quaternionic, semiquaternionic, split semiquaternionic, matrix, etc., structure*, respectively.

Since each scalar λ in $\mathbb{A}$ is interpreted as a linear operator $\Lambda = (C^{\alpha}_{\beta\gamma}\lambda^\gamma)$ in L^{mn}, the operators $J, E, \mathcal{E}$ corresponding to the elements i, e, and ϵ in $\mathbb{C}, \mathbb{C}'$, and $\mathbb{C}^0$ are called the *operators of the complex, split complex, and dual structure*, respectively.

Particular cases of these interpretations are the interpretations of the linear space $\mathbb{C}L^n$ and the free modules $\mathbb{C}'L^n$ and $\mathbb{C}^0L^n$ in the space L^{2n} and the interpretations of the linear space $\mathbb{H}L^n$ and the free modules $\mathbb{H}'L^n, \mathbb{H}^0L^n$, and $\mathbb{H}'^0L^n$ in the space L^{4n}.

There are real interpretations of free modules also in real linear spaces of dimension lower than mn. The split complex free module $\mathbb{C}'L^n$ admits an interpretation in the space L^n: a vector $x = x_+e_+ + x_-e_-$ in $\mathbb{C}'L^n$ with coordinates $x^i = x^i_+e_+ + x^i_-e_-$ is interpreted as two vectors x_+ and x_- with coordinates x^i_+ and x^i_- in L^n. The product of the vector x by a split complex number $\lambda = \lambda_+e_+ + \lambda_-e_-$ is interpreted as the product of the vector x_+ by λ_+ and of the vector x_- by λ_-. The singular vectors are interpreted as the pairs of vectors $\{x_+, 0\}$ and $\{0, x_-\}$.

The matrix free module $\mathbb{R}_mL^n$ admits an interpretation in the space L^{mn}: a vector x with coordinates $x^{\beta i}_\alpha$ is interpreted as m vectors x_α with coordinates $x^{\beta i}_\alpha$ for fixed α. The product of the vector x by a matrix $\lambda = (\lambda^\alpha_\beta)$ is interpreted as the linear combination $y_\beta = x_\alpha\lambda^\alpha_\beta$. The singular vectors are interpreted as linearly depended vectors. Since the algebra $\mathbb{H}'$ is isomorphic to the algebra $\mathbb{R}_2$, for $m = 2$ this interpretation can also be regarded as an interpretation of the free module $\mathbb{H}'L^n$ in the space L^{2n}.

Since the algebras $\mathbb{H}, \mathbb{H}'$, and $\mathbb{H}^0$ can be represented by the complex (2×2)-matrices (1.57), (1.61), and (1.67), the linear space $\mathbb{H}L^n$ and the free modules $\mathbb{H}'L^n$ and $\mathbb{H}^0L^n$ can be interpreted as the complex linear space $\mathbb{C}L^{2n}$. Since the algebra $\mathbb{H}'^0$ can be represented by the split complex (2×2)-matrices (1.67), the free module $\mathbb{H}'^0L^n$ can be also interpreted as the split complex free module $\mathbb{C}'L^{2n}$.

2.1.3. Affine Spaces. In **0.4.2** we have defined the real affine space E^n. If we replace in this definition the real linear space L^n by a linear space or free module $\mathbb{A}L^n$, we obtain the *affine space* $\mathbb{A}E^n$ over the algebra $\mathbb{A}$.

In the space $\mathbb{A}E^n$ the *affine coordinates* of points *straight lines, planes, m-planes,* and *hyperplanes* are defined as in E^n. As in E^n, m-planes in $\mathbb{A}E^n$ are determined by m linearly independent vectors $a_1, a_2, \ldots, a_m$ by the equations (0.15). If u is a covector such that all convolutions $ua_1 = ua_2 = \cdots = ua_{n-1} = 0$ for the vectors determining a hyperplane, and if we put $ux_0 = u_0$, the equation of this hyperplane can be written as $ux + u_0 = 0$ or, in coordinate form, as

$$u_i x^i + u_0 = 0. \tag{2.2}$$

2.1.4. Adjacent Points and Lines. Points X and Y in $\mathbb{A}E^n$ such that the vector $\overrightarrow{XY}$ is a singular vector are called *adjacent points.*

Theorem 2.2. *Through two adjacent points there passes more than one straight line.*

PROOF. Since in formula (0.15) the vectors $a_1, a_2, \ldots, a_m$ must be linearly independent, all these vectors must be nonsingular. Therefore, for $m = 1$, that is, for a straight line, this formula has the form $x = x_0 + at$ and the vector a must also be nonsingular. Since a singular vector b can be represented as the product $a\delta$ of various non-collinear nonsingular vectors a by various zero divisors δ, through two points X and Y in $\mathbb{A}E^n$ such that $\overrightarrow{XY}$ is a singular vector, different lines $x = x_0 + at$ with various direction vectors a pass. ■

Two different lines which pass through the same two points are called *adjacent lines.* The different lines passing through two adjacent points are adjacent lines.

Two straight lines in $\mathbb{A}E^n$ which have no common points but can be transformed by a translation $x' = x + a$ into adjacent lines are said to be *diverging lines.*

Theorem 2.3. *Any two straight lines in the affine plane $\mathbb{A}E^2$ can be in five cases of possible positions: they can meet in one point, be parallel, coincide, be adjacent, or be diverging.*

PROOF. For $m = 2$ formula (2.2) has the form

$$u_1 x^1 + u_2 x^2 + u_0 = 0. \tag{2.3}$$

Consider two lines: (2.3) and

$$v_1 x^1 + v_2 x^2 + v_0 = 0, \tag{2.4}$$

and find a common point of these lines. If the coefficient u_2 is not zero or a zero divisor, the coordinate x^2 is equal to

$$x^2 = -u_2^{-1} u_1 x^1 - u_2^{-1} u_0.$$

If the point $x(x^1, x^2)$ is a common point of the lines (2.3) and (2.4), the coordinate x^2 satisfies the equation (2.4), that is,

$$(v_1 - v_2 u_2^{-1} u_1) x^1 + (v_0 - v_2 u_2^{-1} u_0) = 0. \tag{2.5}$$

If the element $v_1 - v_2 u_2^{-1} u_1$ is not zero or a zero divisor, there is a single value

$$x^1 = (v_1 - v_2u_2^{-1}u_1)^{-1}(v_0 - v_2u_2^{-1}u_0)$$

and the lines meet in the single point $x(x^1, x^2)$.

If both elements $v_1 - v_2u_2^{-1}u_1$ and $v_0 - v_2u_2^{-1}u_0$ are equal to zero, equation (2.5) is satisfied by any value of x^1 and any point of the line (2.3) is a point of the line (2.4), that is, these two lines coincide.

If the element $v_1 - v_2u_2^{-1}u_1$ is zero, but the element $v_0 - v_2u_2^{-1}u_0$ is not zero, the lines have no common points, but the translation which transforms the coefficient v_0 into $v_2u_2^{-1}u_0$ transforms the line (2.4) into the line (2.3); therefore, these lines are parallel. These are the possible cases when the algebra $\mathbb{A}$ is a division algebra.

If the algebra $\mathbb{A}$ has zero divisors there are two more possibilities. If the element $v_1 - v_2u_2^{-1}u_1$ is a zero divisor and the element $v_0 - v_2u_2^{-1}u_0$ is zero or a zero divisor such that the equation (2.5) is satisfied for some but not by all values of x^1, the lines are adjacent.

If the element $v_1 - v_2u_2^{-1}u_1$ is a zero divisor and the element $v_0 - v_2u_2^{-1}u_0$ is not zero or is a zero divisor such that the equation (2.5) is satisfied by no value of x^1, the lines have no common point but can be transformed by a translation into adjacent lines, that is, the lines are diverging. ■

Note that if $\mathbb{A}$ is a Hjelmslev algebra (see **1.1.11**), the space $\mathbb{A}E^n$ is called a *Hjelmslev affine space.* Such spaces have adjacent points and lines and if the algebra $\mathbb{A}$ is replaced by its quotient algebra $\mathbb{A}/\mathbb{J}$, where $\mathbb{J}$ is its ideal consisting of zero divisors, the space $\mathbb{A}E^n$ becomes an affine space $\mathbb{F}E^n$ over the division algebra $\mathbb{F} = \mathbb{A}/\mathbb{J}$ and, therefore, has no adjacent points and lines. Hjelmslev spaces first were considered by J. Hjelmslev [Hje]. Examples of Hjelmslev affine spaces are the dual and semiquaternionic affine spaces $\mathbb{C}^0E^n$ and $\mathbb{H}^0E^n$.

2.1.5. Affine Theorem of Pappus. If the algebra $\mathbb{A}$ is commutative, in $\mathbb{A}E^n$ the *affine theorem of Pappus* holds:

Theorem 2.4. *If in a hexagon which is inscribed in a pair of meeting lines there are two pairs of parallel opposite sides, then the sides in the third pair of opposite sides are also parallel.*

PROOF. Consider the hexagon $ABCDEF$ with vertices A, C, E on the line OA and vertices B, D, F on the line OB, with sides BC and EF parallel, and sides AB and DE parallel (Fig. 2.1). Consider the vectors $x = \overrightarrow{OA}$, $y = \overrightarrow{OF}$, $x\alpha = \overrightarrow{OE}$ and $y\beta = \overrightarrow{OB}$. Since the lines EF and BC are parallel, the vector $\overrightarrow{OC}$ is $x\alpha\beta$; since the lines AB and ED are parallel, the vector $\overrightarrow{OD}$ is $y\beta\alpha$. Therefore, if $\alpha\beta = \beta\alpha$ the lines AF and CD also are parallel. ■

A similar theorem is valid also for all hexagons which are inscribed in a pair of parallel lines.

Note that the affine theorem of Pappus holds for all hexagons with parallel opposite sides inscribed in a pair of lines only if the algebra $\mathbb{A}$ is a division algebra; if $\mathbb{A}$ has zero divisors this theorem is only generically valid.

2.1.6. Affine Theorem of Desargues. If the algebra $\mathbb{A}$ is associative, in $\mathbb{A}E^n$ the *affine theorem of Desargues* holds:

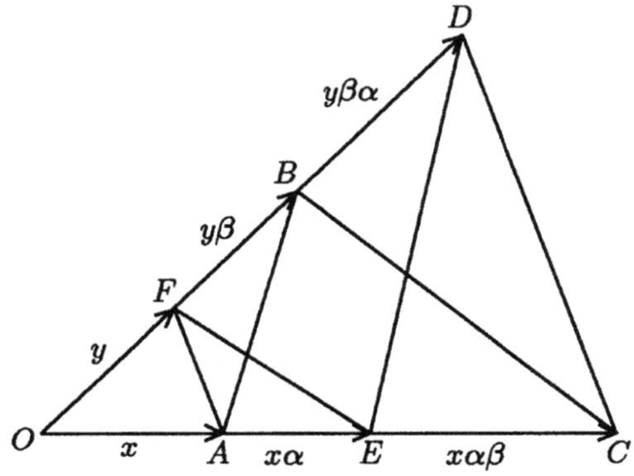

FIG. 2.1

Theorem 2.5. *If, for two triangles, the sides in each of the three pairs of corresponding sides are parallel, then the lines joining the corresponding vertices are parallel or meet at a point.*

If the lines joining corresponding vertices of two triangles are parallel or meet at a point, and the sides in each of two pairs of corresponding sides of the triangles are parallel, then the sides in the third pair of corresponding sides of these triangles are also parallel.

PROOF. Consider the three lines OA, OB, and OC, the vectors $\overrightarrow{OA} = x, \overrightarrow{OB} = y$, and $\overrightarrow{OC} = z$, points A', B', C' on these lines, and the vectors $\overrightarrow{OA'} = x\alpha, \overrightarrow{OB'} = y\beta, \overrightarrow{OC'} = z\gamma$ (Fig. 2.2). If the sides $AB, A'B'$, and $BC, B'C'$ of the triangles ABC and $A'B'C'$ are parallel, the vectors $y - x$ and $y\beta - x\alpha$ are collinear and $\alpha = \beta$. The vectors $z - y$ and $x\gamma - y\beta$ are also collinear, and $\beta = \gamma$. Therefore, the vectors $x - z$ and $x\alpha - z\gamma$ are collinear, and the sides CA and $C'A'$ of the triangles ABC and $A'B'C'$ are parallel. Since the vectors $x\alpha - y, x\alpha\beta - y\beta, x\alpha(\beta\gamma) - y\beta\gamma$, and $x(\alpha\beta)\gamma - y\beta\gamma$ must be collinear, the last two of them can be equal only if $(\alpha\beta)\gamma = \alpha(\beta\gamma)$, so this theorem is valid only if the algebra $\mathbb{A}$ is associative.

The converse assertion can be proved similarly. ■

When the lines AA', BB' and CC' are parallel, that is, the point O is at infinity, this theorem is called the *affine small theorem of Desargues.* In this case the triangle $A'B'C'$ is obtained from the triangle ABC by a translation $x' = x + a$.

Note that the affine theorem of Desargues holds for all pairs of triangles with parallel corresponding sides whose corresponding vertices are on three lines meeting at one point or being parallel only if $\mathbb{A}$ is a division algebra; if $\mathbb{A}$ has zero divisors, this theorem is only generically valid.

2.1.7. Real Interpretations of Affine Spaces over Algebras. The real and complex interpretations of the linear spaces and free modules $\mathbb{A}L^n$ give real and complex interpretations of the affine spaces $\mathbb{A}E^n$. If the free module $\mathbb{A}L^n$ admits a real interpretation as the space L^{mn}, the affine space $\mathbb{A}E^n$ admits a real interpretation as the space E^{mn}; the interpretation of the vectors in the

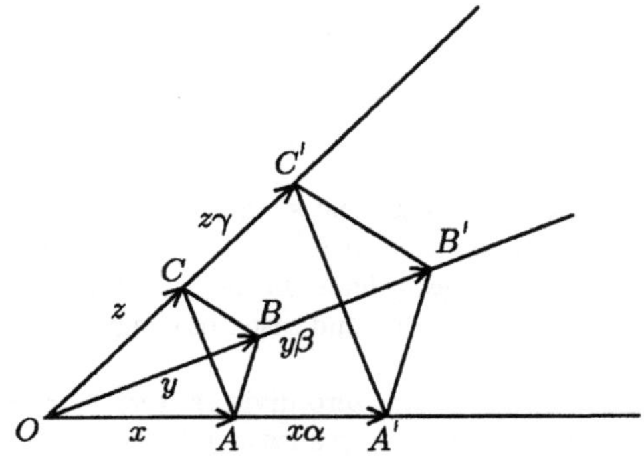

FIG. 2.2

affine space is the same as that of the vectors in $\mathbb{A}L^n$ in the space L^{mn}. In this interpretation lines and p-planes in $\mathbb{A}E^n$ are interpreted as m-planes and mp-planes in E^{mn}, respectively. Adjacent points in $\mathbb{A}E^n$ are interpreted as points through which more than one m-plane, interpreting a line, passes; m-planes which interprete non-adjacent lines meet in one point; m-planes which interprete adjacent lines meet in a line, in a plane or in an m'-plane ($m' < m$).

An m-plane in E^{mn} interpreting a line in $\mathbb{A}E^n$ cuts out from the hyperplane at infinity of E^{mn}, which is the projective space P^{mn-1}, an $(m-1)$-plane. The system of these $(m-1)$-planes is such that through each generic point in P^{mn-1} only one $(m-1)$-plane of this system passes. Systems of $(m-1)$-planes in P^{mn-1} for which this is true are called *congruences*; the points through which more than one $(m-1)$-plane of a congruence passes are called *foci* of the congruence; the set of the foci of a congruence is called the *focal variety* of the congruence. These foci are the points at infinity of the lines in L^{mn} whose direction vectors interpret singular vectors in $\mathbb{A}E^n$.

Theorem 2.6. *For the space $\mathbb{C}'E^n$ we have $m = 2$ and the planes in $\mathbb{E}^{2n}$, interpreting the lines in $\mathbb{C}'E^n$, cut out from the hyperplane at infinity, which is P^{2n-1}, a congruence of lines which intersect two real $(n-1)$-planes.*

For the space $\mathbb{C}E^n$ we have $m = 2$ and the planes in $\mathbb{E}^{2n}$, interpreting the lines in $\mathbb{C}E^n$, cut out from the hyperplane at infinity a congruence of lines which join conjugate imaginary points in two conjugate imaginary $(n-1)$-planes.

PROOF. These interpretations can be obtained from the representation of the complex, or split complex, coordinates X^i of the points in $\mathbb{C}E^n$, or $\mathbb{C}'E^n$, in the form $X^i = x^{2i-1} + ix^{2i}$, or $X^i = x^{2i-1} + ex^{2i}$. The planes in $\mathbb{E}^{2n}$ interpreting the lines in $\mathbb{C}E^n$ or $\mathbb{C}'E^n$ are defined by vectors with coordinates X^i and $iX^i = -x^{2i} + ix^{2i-1}$, and $eX^i = x^{2i} + ex^{2i-1}$, respectively. Therefore, for $\mathbb{C}E^n$ the lines which are cut out by these planes from the hyperplane at infinity, join the points x with coordinates x^{2i-1} and x^{2i} with the points x' with coordinates

$$'x^{2i-1} = -x^{2i}, \quad 'x^{2i} = x^{2i-1}, \tag{2.6}$$

while for $\mathbb{C}'E^n$ they join the points x with coordinates x^{2i-1} and x^{2i} with the points x' with coordinates

$$'x^{2i-1} = x^{2i}, \quad 'x^{2i} = x^{2i-1}. \tag{2.7}$$

These systems of lines are congruences. Since these congruences of lines are defined by systems of linear equations $a_{ij}p^{ij} = 0$ in the Plücker coordinates (1.174), they are called *linear congruences of lines.* In the first case this congruence is called an *elliptic linear congruence*, in the second case this congruence is called a *hyperbolic linear congruence.*

For $\mathbb{C}'E^n$ the coordinates X^i are zero divisors if $x^{2i-1} = \pm x^{2i}$. Therefore, in this case the points in the real $(n-1)$-planes at infinity with these coordinates are the points at infinity of the real lines in E^{2n} whose direction vectors are vectors interpreting the singular vectors x_+e_+ and x_-e_- in $\mathbb{C}'E^n$. These $(n-1)$-planes form the focal variety of the hyperbolic linear congruence.

For $\mathbb{C}E^n$ the points in the imaginary $(n-1)$-planes at infinity are the points at infinity of the lines in $\mathbb{C}E^{2n}$ whose direction vectors are vectors interpreting the singular vectors x_+e_+ and x_-e_- in $(\mathbb{C} \oplus \mathbb{C})E^n$. These imaginary $(n-1)$-planes form the imaginary focal variety of the elliptic linear congruence. ■

2.1.8. Applications of Segreans. We define the *Segreans* $\Sigma_{m,n}$ as the algebraic varieties in P^{mn+m+n} whose order and dimension are equal to $C^n_{n+m} = C^m_{n+m}$ and determined by the parametric equations

$$z^{i\alpha} = x^i y^\alpha \quad (i = 0, 1, \dots, m, \; \alpha = 0, 1, \dots, n). \tag{2.8}$$

A Segrean $\Sigma_{m,n}$ has m-planar and n-planar generators $y^\alpha = \text{const}$ and $x^i = \text{const}$. The equations of a Segrean in projective coordinates are:

$$z^{i\alpha} z^{j\beta} = z^{i\beta} z^{j\alpha}. \tag{2.9}$$

The Segreans $\Sigma_{1,1}$ are ruled quadrics in P^3, general Segreans were defined by Corrado Segre [SeC].

Theorem 2.7. *For the space $\mathbb{H}'E^n$ we have $m = 4$ and the 4-planes in E^{4n}, interpreting the lines in $\mathbb{H}'E^n$, cut out from the hyperplane at infinity, which is P^{4n-1}, a congruence of 3-planes whose focal variety is the Segrean $\Sigma_{1,2n-1}$.*

For the space $\mathbb{H}E^n$ we have $m = 4$ and the 4-planes in E^{4n}, interpreting the lines in $\mathbb{H}E^n$, cut out from the hyperplane at infinity, which is P^{4n-1}, a congruence of 3-planes whose foci are imaginary and whose focal variety is the imaginary Segrean $\Sigma_{1,2n-1}$.

PROOF. Let us represent the split quaternionic coordinates x^i of the points in the space $\mathbb{H}'E^n$ by (2×2)-matrices

$$X^i = \begin{bmatrix} z^{0,2i-1} & z^{1,2i-1} \\ z^{0,2i} & z^{1,2i} \end{bmatrix} \tag{2.10}$$

of the algebra $\mathbb{R}_2$. If the vector $x = \{x^i\}$ is singular, both columns of the $(2 \times 2n)$-matrix X, which consists of the (2×2)-matrices X^i $(i = 1, 2, \dots, n)$, are proportional. Therefore, for all I and J $\quad (I, J = 1, 2, \dots, 2n)$

$$z^{0I} z^{1J} = z^{0J} z^{1I}. \tag{2.11}$$

Equation (2.11) implies that all second order minors of the matrix X vanish. It coincides with equation (2.9) for the Segrean $\Sigma_{1,2n-1}$ in P^{4n-1}.

For the space $\mathbb{H}E^n$ the proof is analogous, but in this case we must consider complex matrices (2.10), which represent quaternions x^i. ∎

Theorem 2.8 (V. A. DOBROMYSLOV [Dob]). *For the space $\mathbb{R}_p E^n$ we have $m = p^2$ and the p^2-planes in $E^{p^2 n}$, interpreting the lines in $\mathbb{R}_p E^n$, cut out from the hyperplane at infinity, which is $P^{p^2 n-1}$, a congruence of (p^2-1)-planes whose focal variety is formed by the $(p-2)$-planes that intersect the Segrean $\Sigma_{p-1,pn-1}$ at $p-1$ points.*

PROOF. The vectors $x = \{x^i\}$ in $\mathbb{R}_p E^n$ can be represented by $(p \times pn)$-matrices X consisting of the $(p\times p)$-matrices x^i $\quad (i = 1, 2, \ldots, n)$. A singular vector $x = \{x^i\}$ in this space whose coordinates x^i are matrices of rank $\leq r$ is said to be a *singular vector of rank* r. The entries of the matrices X which represent singular vectors of rank r satisfy the condition of equality to zero of all minors of rank $r+1$ of the matrix X.

For $r = 1$ this condition coincides with equation (2.7) of the Segrean $\Sigma_{p-1,pn-1}$.

For arbitrary r this condition coincides with the equation of the variety formed by the $(r-1)$-planes that intersect the Segrean $\Sigma_{p-1,pn-1}$ at r points. ∎

2.1.9. Other Real Interpretations of Affine Spaces over Algebras. The interpretation of the free module $\mathbb{C}' L^n$ in the space L^n implies

Theorem 2.9. *The space $\mathbb{C}' E^n$ admits an interpretation in the space E^n: a point x in $\mathbb{C}' E^n$ with coordinates $x^i = x^i_+ e_+ + x^i_- e_-$ is interpreted as two points, x_+ and x_-, with coordinates x^i_+ and x^i_-.*

Adjacent points in $\mathbb{C}' E^n$ are interpreted as pairs of points with coinciding points, adjacent lines in $\mathbb{C}' E^n$ are interpreted as pairs of lines with coinciding lines, diverging lines in $\mathbb{C}' E^n$ are interpreted as pairs of lines one of which is meeting and the second is parallel.

The interpretation of the free module $\mathbb{R}_m L^n$ in the space L^{mn} implies

Theorem 2.10. *The space $\mathbb{R}_m E^n$ admits an interpretation in the space E^{mn}: a point x in $\mathbb{R}_m E^n$ with coordinates $(x^{\beta i}_\alpha)$ is interpreted as m points x_α, with coordinates $x^{\beta i}_\alpha$.*

Adjacent points in $\mathbb{R}_m E^n$ are interpreted as points x_α and y_α such that the vectors $\overrightarrow{x_\alpha y_\alpha}$ are linearly dependent. Lines and p-planes in $\mathbb{R}_m E^n$ are interpreted as m-planes and pm-planes in E^{mn}, respectively.

Theorem 2.11. *The points in the space $\mathbb{H}' E^n$ can be interpreted as planes in $\mathbb{C}E^{2n}$ through a point which cut out from the hyperplane at infinity, which is $\mathbb{C}P^{2n-1}$, a congruence of lines joining the points x with coordinates x^{2i-1} and x^{2i} with the points with coordinates*

$$'x^{2i-1} = \bar{x}^{2i}, \quad 'x^{2i} = \bar{x}^{2i-1}. \tag{2.12}$$

The points in the space $\mathbb{H}E^n$ can be interpreted as planes in $\mathbb{C}E^{2n}$ through a point which cut from the hyperplane at infinity, which is $\mathbb{C}P^{2n-1}$, a congruence of lines joining the points x with coordinates x^{2i-1} and x^{2i} with the points with coordinates

$$'x^{2i-1} = \bar{x}^{2i}, \quad 'x^{2i} = -\bar{x}^{2i-1}. \tag{2.13}$$

These congruences are called the hyperbolic and the elliptic Hermitian linear congruences in the space $\mathbb{C}P^{2n-1}$, respectively.

PROOF. These interpretations can be obtained from the representations of the quaternionic or split quaternionic coordinates X^i of the points in $\mathbb{H}E^n$ and $\mathbb{H}'E^n$ by complex matrices (1.57) and (1.61) where $A = x^{2i-1}$ and $B = -\bar{x}^{2i}$ in the case of $\mathbb{H}E^n$ and $B = \bar{x}^{2i}$ in the case of $\mathbb{H}'E^n$.

Planes in $\mathbb{C}E^{2n}$, interpreting lines in $\mathbb{H}E^n$ and $\mathbb{H}'E^n$, are determined by vectors with coordinates X^i and $X^i i$ and $X^i e$, respectively. Therefore, the lines which are cut out by these planes from the hyperplane at infinity join, in the case of $\mathbb{H}E^n$, the points x with coordinates x^{2i-1} and x^{2i} with the points x' with coordinates (2.13) and, in the case of $\mathbb{C}'E^n$, the points x with coordinates x^{2i-1} and x^{2i} with the points x' with coordinates (2.12).

These systems of lines are elliptic and hyperbolic Hermitian linear congruences in $\mathbb{C}P^{2n-1}$, respectively. ■

§2.2. Projective Geometries

2.2.1. Projective Spaces. In **0.4.3** we have defined real projective space P^n. If we replace in this definition the field $\mathbb{R}$ by an algebra $\mathbb{A}$, we obtain the projective space $\mathbb{A}P^n$ over the algebra $\mathbb{A}$.

The points in $\mathbb{A}P^n$, like the points in P^n, can be represented by the nonzero and nonsingular vectors in the space $\mathbb{A}E^{n+1}$, which are called the *analytic points* in $\mathbb{A}P^n$ or *position vectors* of points in this space..

The space $\mathbb{A}P^n$ can be regarded as an extension of the space $\mathbb{A}E^n$, which is a hyperplane in $\mathbb{A}E^{n+1}$, by new points corresponding to lines in $\mathbb{A}E^{n+1}$ through a point S outside this hyperplane. The new points corresponding to lines parallel to this hyperplane are called *points at infinity,* the new points corresponding to lines diverging with lines in this hyperplane are called *ideal points.*

The *projective coordinates* of points, *straight lines*, *planes*, *m-planes*, and *hyperplanes* in $\mathbb{A}P^n$ are defined as in P^n; m-planes in $\mathbb{A}P^n$ are determined by $m+1$ linearly independent vectors $a_0, a_1, \ldots, a_m$ in $\mathbb{A}E^{n+1}$ using equations (0.16). If u is a covector such that all convolutions $ux_0 = ua_1 = \cdots = ua_n = 0$, the equation of this hyperplane can be written in the form $ux = 0$ or, in coordinate form,

$$u_i x^i = 0. \tag{2.14}$$

2.2.2. Topology of Projective Spaces. The topology of affine spaces is very simple: the space E^n endowed with this topology, which coincides with the natural topology of the space R^n (see **0.5.5**), is homeomorphic to the topological product of n lines E^1, the space $\mathbb{A}E^n$ is homeomorphic to the topological product of n lines $\mathbb{A}E^1$, and since the line $\mathbb{A}E^1$ can be identified with the m-algebra $\mathbb{A}$ and this algebra is homeomorphic to E^m, the space $\mathbb{A}E^n$ is homeomorphic to the space E^{mn}.

The topology of projective spaces is much complicated. The simplest topological structure is that of a projective space $\mathbb{F}P^n$ over a division algebra $\mathbb{F}$.

Theorem 2.12. *The projective space* $\mathbb{F}P^n$ *can be written as the union*

$$\mathbb{F}P^n = \mathbb{F}E^n \cup \mathbb{F}E^{n-1} \cup \mathbb{F}E^{n-2} \cup \cdots \cup \mathbb{F}E^2 \cup \mathbb{F}E^1 \cup \mathbb{F}E^0, \tag{2.15}$$

where $\mathbb{F}E^0$ *is a point.*

PROOF. This theorem follows from that fact that for a division algebra $\mathbb{F}$ the space $\mathbb{F}P^n$ can be written as the union $\mathbb{F}P^n = \mathbb{F}E^n \cup \mathbb{F}P^{n-1}$ of the affine space $\mathbb{F}E^n$ and the hyperplane at infinity of $\mathbb{F}P^n$, which is $\mathbb{F}P^{n-1}$. ■

Since the summands of the union (2.15), which are homeomorphic to affine spaces, are called *cells* of $\mathbb{F}P^n$, the decomposition (2.15) is called the *cellular decomposition* of $\mathbb{F}P^n$.

If $\mathbb{F}$ is the real field $\mathbb{R}$, formula (2.15) takes the form

$$P^n = E^n \cup E^{n-1} \cup E^{n-2} \cup \cdots \cup E^2 \cup E^1 \cup E^0. \tag{2.16}$$

If the cellular decomposition of an n-manifold M^n has the form

$$M^n = a_0 E^n \cup a_1 E^{n-1} \cup a_{n-2} E^{n-2} \cup \cdots \cup a_{n-2} E^2 \cup a_1 E^1 \cup a_n E^0, \tag{2.17}$$

where the $a_0, a_1, \ldots, a_{n-1}, a_n$ are integers, the polynomial

$$C(t) = a_0 t^n + a_1 t^{n-1} + a_2 t^{n-2} + \cdots + a_{n-2} t^2 + a_{n-1} t^1 + a_n \tag{2.18}$$

is called the *cellular polynomial* of M^n. Therefore, the cellular polynomial of P^n is

$$C(t) = t^n + t^{n-1} + t^{n-2} + \cdots + t^2 + t + 1 = \frac{t^{n+1} - 1}{t - 1} \tag{2.19}$$

and, if the dimension of the division algebra $\mathbb{F}$ is equal to m, the cellular polynomial of $\mathbb{F}P^n$ is

$$C(t) = t^{nm} + t^{(n-1)m} + t^{(n-2)m} + \cdots + t^{2m} + t^m + 1 = \frac{t^{(n+1)m} - 1}{t^m - 1}. \tag{2.20}$$

Note that the cellular polynomial (2.18) of a manifold M^n coincides with the Poincaré polynomial $\sum_k b_k t^{n-k}$ (see **1.4.15**) only for manifolds whose cellular polynomial contains no terms at powers with differences equal to 1. If the nonzero coefficients of the polynomial $C(t)$ of a manifold M^{mn} are $a_0, a_m, a_{2m}, \ldots, a_{(n-2)m}$, $a_{(n-1)m}$, a_{nm}, the Poincaré polynomial of this manifold has the form

$$P(t) = a_0 t^{nm} + a_1 t^{(n-1)m} + a_2 t^{(n-2)m} + \cdots + a_{(n-2)m} t^2 + a_{(n-1)m} t + a_{nm}. \tag{2.21}$$

In particular, the Poincaré polynomials of $\mathbb{C}P^n$ and $\mathbb{H}P^n$ coincide with the cellular polynomials of these spaces and are

$$P(t) = t^{2n} + t^{2n-2} + t^{2n-4} + \cdots + t^4 + t^2 + 1 \tag{2.22}$$

and

$$P(t) = t^{4n} + t^{4n-4} + t^{4n-8} + \cdots + t^8 + t^4 + 1, \tag{2.23}$$

respectively.

Let us consider the projective line $\mathbb{F}P^1$ over a division algebra $\mathbb{F}$ of dimension m.

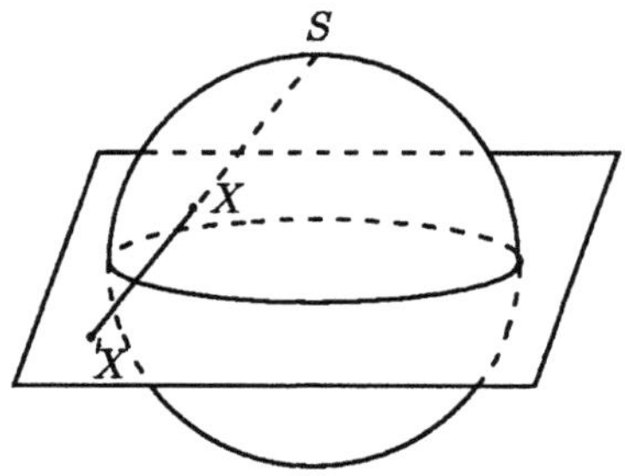

FIG. 2.3

Theorem 2.13. *The projective line $\mathbb{F}P^1$ over a division m-algebra $\mathbb{F}$ is homeomorphic to the hypersphere Σ^m in the Euclidean space R^{m+1}.*

PROOF. Formula (2.15) for $\mathbb{F}P^1$ gives $\mathbb{F}P^1 = \mathbb{F}E^1 \cup \mathbb{F}E^0$. Since $\mathbb{F}E^1$ can be identified with the division algebra $\mathbb{F}$ and since $\mathbb{F}$ is homeomorphic to E^m, the line $\mathbb{F}P^1$ is homeomorphic to the union of E^m and a point.

Introduce into $\mathbb{F}$ the metric of the space R^m, identify this space R^m with the hyperplane $x^{m+1} = 0$ in R^{m+1}, and consider the stereographic projection of this hyperplane onto the hypersphere $x^2 = 1$ in R^{m+1} from the pole S of this hypersphere with coordinates $0, 0, \dots, 0, 1$ (Fig. 2.3). This projection establishes a homeomorphism between the hypersphere without S and the hyperplane. Therefore, the hypersphere can be regarded as the union of the space R^m or E^m and a point, and, therefore, it is homeomorphic to the line $\mathbb{F}P^1$. ■

So, the line P^1 is homeomorphic to a circle in R^2, the line $\mathbb{C}P^1$ is homeomorphic to a sphere in R^3 (the so-called Riemann sphere), and the line $\mathbb{H}P^1$ is homeomorphic to a hypersphere in R^5.

Since the hypersphere Σ^m is homeomorphic to the conformal space C^m, the lines $P^1, \mathbb{C}P^1$, and $\mathbb{H}P^1$ are homeomorphic to the line C^1, the plane C^2, and the space C^4, respectively.

If $\mathbb{A}$ is an n-algebra, the space $\mathbb{A}P^n$ is an nm-manifold. This manifold is differentiable, and its tangent spaces have the structure of the corresponding affine space $\mathbb{A}E^n$. The real or imaginary singular vectors in the tangent spaces of $\mathbb{C}P^n$ and $\mathbb{C}'P^n$ determine on the hyperplanes at infinity of these spaces pairs of conjugate imaginary or real $(n-1)$-planes.

The real or imaginary singular vectors in the tangent spaces $\mathbb{H}P^n$ and $\mathbb{H}'P^n$ determine on the hyperplanes at infinity of these spaces imaginary or real Segreans $\Sigma_{1,2n-1}$. The singular vectors in the tangent space of $\mathbb{R}_m P^n$ determine on the hyperplane at infinity of this space a Segrean $\Sigma_{m-1,mn-1}$ (see Theorems 2.8).

Theorem 2.14. *The space P^n is non-orientable if n is even, and orientable if n is odd.*

PROOF. Since the points in P^n can be represented by the lines in a bundle of lines in E^{n+1} through one point, P^n is homeomorphic to a hypersphere in R^{n+1} with identified antipodal points. The theorem follows from the fact that the reflection in

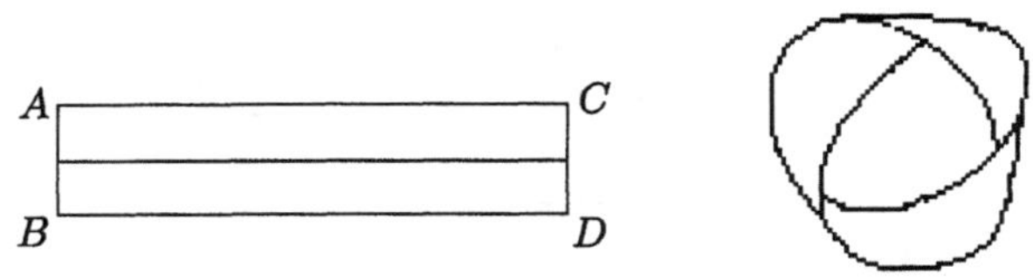

a) b)

FIG. 2.4

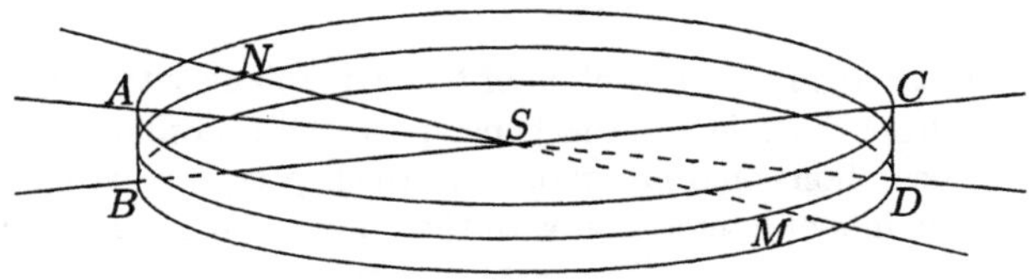

FIG. 2.5

the center of a hypersphere maps an oriented vector frame on the hypersphere to a vector frame with the same orientation if the dimension of the hypersphere is odd, and to a vector frame with opposite orientation if the dimension of the hypersphere is even. ∎

Since the plane P^2 is non-orientable, it is one-sided. It is homeomorphic to the one-sided *Möbius strip* (Fig. 2.4) whose boundary is glued together along a circle.

The sphere in R^3 is homeomorphic to the surface of a cylinder of finite height with the same center of symmetry (Fig. 2.5); the identification of antipodal points of this surface identifies the top of the cylinder with its bottom, and maps its lateral surface into a Möbius strip.

A projective space $\mathbb{A}P^n$ over an algebra $\mathbb{A}$ with zero divisors can be represented as a union containing not only cells, i.e. $\mathbb{A}E^m$, as in formula (2.15), but also ideal points of this space. In **2.2.6**, in Theorem 2.19 , we will prove that the projective space $\mathbb{R}_{m+1}P^n$ admits an interpretation as the Grassmann manifold $Gr^P_{mn+n+m,m}$ of m-planes in the space P^{mn+n+m}, and, therefore, it is homeomorphic to this manifold. In **2.2.6** we will also give a cellular decomposition of this manifold.

Theorem 2.15. *The split complex projective space $\mathbb{C}' P^n$ is homeomorphic to the topological product $P^n \times P^n$.*

The dual projective space $\mathbb{C}^0 P^n$ is homeomorphic to the topological product $P^n \times E^n$.

PROOF. The proof is analogous to the proof of Theorem 1.28 (see **1.4.2**). The last assertion follows from the fact that the tangent space of P^n can be regarded to be E^n. ∎

If in the m-algebra $\mathbb{A}$ we introduce the pseudometric of the pseudo-Euclidean space R^m_L, we can prove that the projective line $\mathbb{A}P^1$ is homeomorphic to the hypersphere in R^{m+1}_L. The space R^m_L can be regarded as the hyperplane $x^{m+1} = 0$

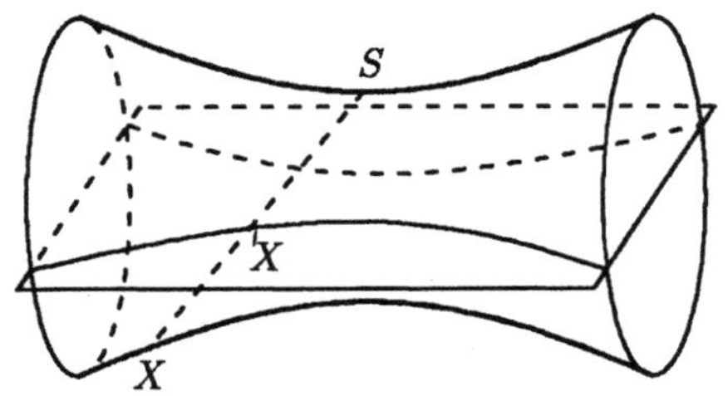

FIG. 2.6

in R_L^{m+1}, and the stereographic projection of this hyperplane onto the hypersphere $xEx = 1$ in R_L^{m+1} from the pole S of this hypersphere with coordinates $0, 0, \ldots, 0, 1$ (Fig. 2.6) establishes a homeomorphism between the hypersphere without the point S and without its points of intersection with the tangent hyperplane to the hypersphere at S and the hyperplane. The point S itself corresponds to the point at infinity of $\mathbb{A}P^1$, while the points of the intersection correspond to ideal points of $\mathbb{A}P^1$ provided that the space R_L^{m+1} is completed to the space P^{m+1}.

Since we have introduced in the algebras $\mathbb{C}'$ and $\mathbb{H}'$ the pseudometrics of the plane R_1^2 and of the space R_2^4, these lines are homeomorphic to a sphere of real radius in R_1^3 and to a hypersphere of real radius in R_2^5, respectively, provided that R_1^3 and R_2^5 are completed to the projective spaces P^3 and P^5. Therefore, these extended sphere and hypersphere are a ruled quadric in P^3 and a hyperquadric with planar generators in P^5, respectively.

Since Theorem 2.15 states that $\mathbb{C}'P^n$ is homeomorphic to $P^n \times P^n$, the line $\mathbb{C}'P^1$ is homeomorphic to the topological product $P^1 \times P^1$, which is also homeomorphic to a ruled quadric in P^3.

Since the line $\mathbb{H}'P^1$ coincides with the line $\mathbb{R}_2P^1$, this line is homeomorphic to the manifold $Gr_{3,1}^P$ of lines in P^3, which, as we will see in **2.2.6**, is homeomorphic to a hyperquadric with planar generators in P^5.

Therefore, the lines $\mathbb{C}'P^1$ and $\mathbb{H}'P^1$ are homeomorphic, respectively, to the pseudoconformal plane C_1^2 and to the pseudoconformal space C_2^4.

Note that all the above mentioned projective spaces over simple and semisimple algebras are compact, but that the space $\mathbb{C}^0P^n$ over the non-semisimple algebra $\mathbb{C}^0$ is noncompact.

2.2.3. Adjacent Points and Lines. Under completion of $\mathbb{A}E^2$ to $\mathbb{A}P^2$ the adjacent points and lines in $\mathbb{A}E^2$ become adjacent points and lines in $\mathbb{A}P^2$; diverging lines in $\mathbb{A}E^2$ also become adjacent lines, with as common points ideal points. Therefore, if $\mathbb{A}$ is an algebra with zero divisors, there are only three possible situations for two lines in $\mathbb{A}P^2$: intersection (at one point), adjacence, and coincidence.

Adjacent points in projective spaces over rings with zero divisors were considered by D. Barbilian [Ba] and W. Klingenberg [Kli].

2.2.4. Projective Theorems of Pappus and Desargues. In **2.1.5** and **2.1.6** we have stated the affine theorems of Pappus and Desargues. If we complete the affine plane $\mathbb{A}E^2$ to the projective plane $\mathbb{A}P^2$, we arrive to the *projective theorems of Pappus and Desargues*:

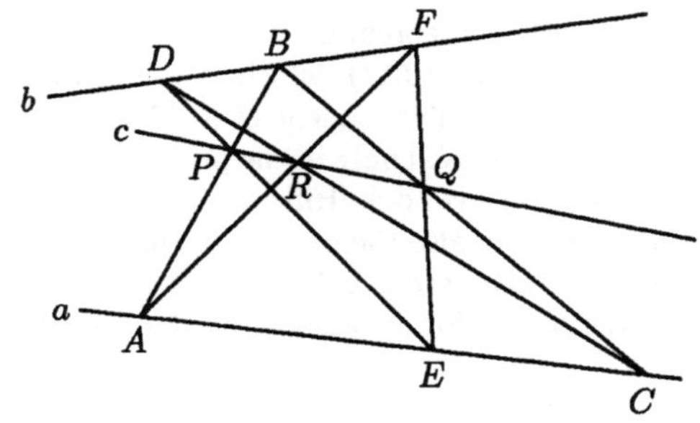

FIG. 2.7

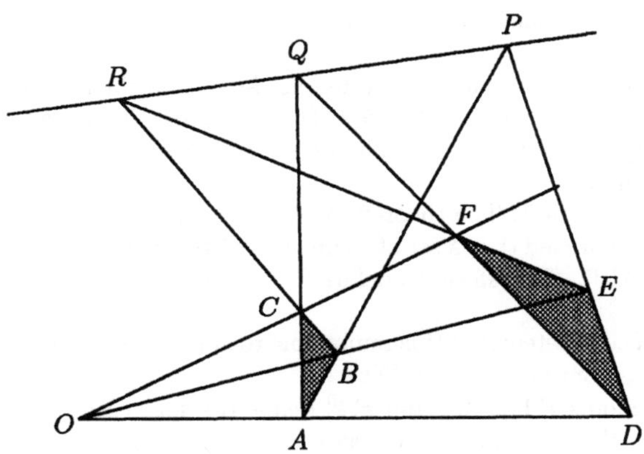

FIG. 2.8

Theorem 2.16. *Opposite sides of a hexagon inscribed in a pair of meeting lines meet at points lying on a single line (Fig. 2.7).*

Theorem 2.17. *If the sides in all three pairs of corresponding sides of two triangles meet at points lying on a line, the lines joining corresponding vertices of these triangles meet at one point.*

If the lines joining the corresponding vertices of two triangles meet at one point, all sides in the three pairs of corresponding sides of the triangles meet at points lying on a line (Fig. 2.8).

Note that the projective theorem of Pappus is a particular case of the *projective theorem of Pascal*: the opposite sides of a hexagon inscribed in a conic meet at points lying on a line.

Note also that the projective theorems of Pappus and Desargues hold for all hexagons as described above and for all pairs of triangles as described above if the algebra $\mathbb{A}$ is a division algebra; if $\mathbb{A}$ has zero divisors, these theorems are only generically valid.

Note also, that the projective theorem of Desargues in projective spaces $\mathbb{A}P^n$ for $n > 2$ follows from the relation (0.17), which is valid in all projective spaces, but that for $n = 2$ this theorem is independent of (0.17). Therefore the projective theorem of Desargues can be non-valid only in the projective planes $\mathbb{A}P^2$. This fact was established by D. Hilbert in his book [Hil].

Since the affine, and, therefore, also the projective theorem of Desargues is valid only if the algebra $\mathbb{A}$ is associative (see **2.1.6**), this theorem is possible in $\mathbb{A}P^2$ only for nonassociative algebras $\mathbb{A}$, and for nonassociative $\mathbb{A}$ projective spaces $\mathbb{A}P^n$ with $n > 2$ do not exist.

2.2.5. Axioms of Kolmogorov. A. N. Kolmogorov [Kol] proved that projective spaces P^n, and $\mathbb{C}P^n$, can be defined as sets of elements of n kinds, called m-planes ($m = 0, 1, 2, \ldots, n-1$), among which there are relations of incidence satisfying the following system of axioms:

$P.1^0$ The set of all m-planes for any fixed $m = 0, 1, 2, \ldots, n-1$ is a compact connected topological space.

$P.2^0$ Each m-plane and l-plane have a unique d-plane as their intersection and a unique s-plane as their sum (see **0.4.3**). The numbers l, m, d, s are related by (0.17), and the maps from the set of m-planes and l-planes onto their intersections and sums are continuous.

$P.3^0$ ("Axiom of Pappus") The theorem of Pappus is valid.

A. N. Kolmogorov proved that a set of elements satisfying the axioms $P.1^0 - P.3^0$ can only be P^n or $\mathbb{C}P^n$. Therefore, to define $\mathbb{C}P^n$ we only have to add the axiom of completeness:

$P.4^0$ ("Axiom of completeness") It is impossible to add to the space new elements for which all axioms continue to be fulfilled.

For $n \neq 2$ the axioms $P.1^0, P.2^0$, and $P.4^0$ define the space $\mathbb{H}P^n$. To define the plane $\mathbb{H}P^2$ to these three axioms it is necessary to add the axiom:

$P.5^0$ ("Axiom of Desargues") The theorem of Desargues is valid.

2.2.6. Real and Complex Interpretations of Projective Spaces over Algebras. The real and complex interpretations of the affine spaces $\mathbb{A}E^n$ determine real and complex interpretations of the projective spaces $\mathbb{A}P^n$.

The interpretation of $\mathbb{C}'E^n$ in E^n, which is established in Theorem 2.9, implies

Theorem 2.18. *The space $\mathbb{C}'P^n$ admits an interpretation in the space P^n: a point x in $\mathbb{C}'P^n$ with coordinates $x^i = x^i_+ e_+ + x^i_- e_-$ is interpreted as two points, x_+ and x_-, with coordinates x^i_+ and x^i_-.*

Adjacent points in $\mathbb{C}'P^n$ are interpreted as pairs of points with coinciding points, adjacent lines in $\mathbb{C}'E^n$ are interpreted as pairs of lines with coinciding lines.

The first assertion of Theorem 2.15, regarding the fact that $\mathbb{C}'P^n$ is homeomorphic to the topological product $P^n \times P^n$, can also be obtained from this interpretation.

The interpretation of $\mathbb{R}_m E^n$ in E^{mn}, which is established in Theorem 2.10, implies

Theorem 2.19 (M. Javadov [Jav2], J. Thas [Th]). *The space $\mathbb{R}_{m+1}P^n$ admits an interpretation in the space P^{mn+n+m}: a point x in $\mathbb{R}_{m+1}P^n$ with coordinates $(x^{\beta i}_\alpha)$ is interpreted as an m-plane through the points x_α with coordinates $x^{\beta i}_\alpha$.*

PROOF. This theorem is based on the fact that the products of coordinates $(x_\alpha^{\beta i})$ by a matrix (k_β^γ) are the coordinates y_α^γ of a point y_α in the same m-plane. ■

Lines and p-planes in $\mathbb{R}_{m+1}P^n$ are interpreted as $(2m+1)$-planes and $(mp+p+m)$-planes in P^{mn+n+m}, respectively. Adjacent points in $R_{m+1}P^n$ are interpreted as m-planes in P^{mn+n+m} with non-empty intersection and with dimension of the sum $< 2m+1$. Adjacent lines in $\mathbb{R}_{m+1}P^n$ are interpreted as $(2m+1)$-planes in P^{mn+n+m} with dimension of the intersection $> m$.

The manifold of m-planes in the projective space P^n is called the *Grassmann manifold* $Gr^P_{n,m}$. Theorem 2.19 shows that the space $\mathbb{R}_{m+1}P^n$ is homeomorphic to the manifold $Gr^P_{mn+n+m,m}$.

Analogous interpretation for Grassmann manifold in arbitrary P^n can be defined by projective spaces of fractional dimension investigated by I. A. Chakhtauri [Chkh].

Note that Ch. Ehresmann [Ehr] found the cellular decomposition of the Grassmann manifold $\mathbb{F}Gr^P_{n,p}$ of p-planes in the space $\mathbb{F}P^n$ over any division algebra $\mathbb{F}$. For P^n this decomposition corresponds to the cellular polynomial

$$C(t) = \frac{(t^{n+1}-1)(t^n-1)\cdots(t^{n-p+1}-1)}{(t^{p+1}-1)(t^p-1)\cdots(t-1)}. \tag{2.24}$$

For $p=0$ this polynomial has the form (2.19). For $n=3, p=1$ this polynomial has the form

$$C(t) = \frac{(t^4-1)(t^3-1)}{(t^2-1)(t-1)} = (t^2+1)(t^2+t+1) = t^4+t^3+2t^2+t+1. \tag{2.25}$$

If the dimension of the division algebra is m, the cellular polynomial of the Grassmann manifold $\mathbb{F}Gr^P_{n,p}$ of p-planes in $\mathbb{F}P^n$ has the form

$$C(t) = \frac{(t^{(n+1)m}-1)(t^{nm}-1)\cdots(t^{(n-p+1)m}-1)}{(t^{(p+1)m}-1)(t^{pm}-1)\cdots(t^m-1)}. \tag{2.26}$$

As for $\mathbb{F}P^n$, this polynomial coincides with the Poincaré polynomial when there are no differences of powers equal to 1.

Since the Grassmann coordinates (1.175) of p-planes in P^n are related by

$$p^{[i_0 i_1,\dots i_p}p^{i_{p+1}]i_{p+2}\dots i_{2p-1}} = 0, \tag{2.27}$$

the Plücker coordinates (1.174) of lines in P^3 are related by

$$p^{01}p^{23} + p^{12}p^{03} + p^{20}p^{13} = 0. \tag{2.28}$$

Equation (2.28) can be regarded as the equation of a hyperquadric with planar generators in P^5. This is the so-called *Plücker interpretation* of $Gr^P_{3,1}$.

So, the manifold $Gr^P_{3,1}$ and the lines $\mathbb{R}_2P^1$ and $\mathbb{H}'P^1$, which are homeomorphic to this manifold, are homeomorphic to this hyperquadric. Therefore, the cellular decomposition of $\mathbb{R}_2P^1 = \mathbb{H}'P^1$ has the form

$$\mathbb{R}_2P^1 = \mathbb{H}'P^1 = E^4 \cup E^3 \cup 2E^2 \cup E^1 \cup E^0. \tag{2.29}$$

Note that when interpreting the line $\mathbb{R}_2 P^1 = \mathbb{H}' P^1$ by the hyperquadric (2.28) in P^5, adjacent points on this line are interpreted as points on the hyperquadric that are joined by rectilinear generators of it.

The interpretations of $\mathbb{C}E^n$ and $\mathbb{C}'E^n$ in E^{2n}, which are established in Theorem 2.6, imply

Theorem 2.20. *The spaces $\mathbb{C}P^n$ and $\mathbb{C}'P^n$ admit interpretations in the space P^{2n+1} as lines of an elliptic linear congruence determined by the transformations (2.6), and as lines of a hyperbolic linear congruence determined by the transformation (2.7), respectively.*

The interpretations of $\mathbb{H}E^n$ and $\mathbb{H}'E^n$, which are established in Theorem 2.11, imply

Theorem 2.21. *The spaces $\mathbb{H}E^n$ and $\mathbb{H}'E^n$ admit interpretations in the space $\mathbb{C}P^{2n+1}$ as lines of an elliptic Hermitian linear congruence determined by the transformations (2.13), and as lines of a hyperbolic Hermitian linear congruence determined by the transformation (2.12), respectively.*

Quaternionic projective geometry was investigated by J. Bilo [Bi].

§2.3. Affine and Projective Transformations

2.3.1. Affine Transformations. In **0.8.8** we have defined the group of *affine transformations* in the space E^n. The group of affine transformations in the space $\mathbb{A}E^n$ can be defined as the group of bijections in this space that preserve straight lines.

Theorem 2.22. *An affine transformation in the space $\mathbb{A}E^n$ in affine coordinates has the form*

$$'x^i = U^i_j f(x^j) + a^i, \tag{2.30}$$

where $x \to f(x)$ is a continuous ring automorphism in the algebra $\mathbb{A}$. Formula (2.30) can also be written in vector form:

$$x' = Uf(x) + a. \tag{2.31}$$

PROOF. We split the proof into steps.

1) An affine transformation in $\mathbb{A}E^n$ preserves m-planes. For $m = 1$ this assertion coincides with the definition of an affine transformation. Suppose this assertion is correct for all ℓ-planes with $\ell < m$. Consider two $(m-1)$-planes, with as intersection an $(m-2)$-plane. By formula (0.17), which is valid for $(m-1)$-planes in $\mathbb{A}E^n$ without common directions, the sum of these $(m-1)$-planes is an m-plane. Let X be an arbitrary point in this m-plane outside the given $(m-1)$-planes. Then there is a line through X which meets both these $(m-1)$-planes. According to the assumption, these $(m-1)$-planes and its intersection are transformed into two $(m-1)$-planes with as intersection an $(m-2)$-plane, and the line through X is transformed into a line which meets both new $(m-1)$-planes. Therefore, X is transformed into a point X' in the m-plane which is the sum of these new $(m-1)$-planes. This m-plane is the image of the sum of the given $(m-1)$-planes.

2) An affine transformation maps parallel lines to parallel lines. Any two parallel lines belong to a common plane and the assertion follows from the first assertion for $m = 2$ and from the fact that affine transformations are bijections.

3) An affine transformation maps vectors to vectors. Each vector a determines a translation $y = x + a$, and vice versa. If the affine transformation maps the points with position vectors x and y to points with position vectors x' and y', by 2) these points are related by $y' = x' + a$.

4) An affine transformation maps the sum of vectors to the sum of the images of the vectors. Suppose the affine transformation maps vectors a and b to vectors a' and b'. Since $a = \overrightarrow{OA}$, $b = \overrightarrow{OB}$ and $a + b = \overrightarrow{OC}$, the line OC is the diagonal of the parallelogram $OACB$, which is mapped to an analogous parallelogram $O'A'C'B'$.

5) An affine transformation maps the product of a vector by a scalar λ in $\mathbb{A}$ to the product of the image of the vector by the element $f(\lambda)$ in the same algebra, where $\lambda \to f(\lambda)$ is a ring automorphism in this algebra.

By the definition of affine transformation, it maps two collinear vectors a and $a\lambda$ to two collinear vectors a' and $a'\lambda'$. The element λ' does not depend on a since if $a = \overrightarrow{OA}, b = \overrightarrow{OB}, a\lambda = \overrightarrow{OC}, b\lambda = \overrightarrow{OD}$, the line CD is parallel to the line AB and, by 2), in the corresponding figure $O'A'C'D'B'$ determined by a' and b' and the element λ' the lines $A'B'$ and $C'D'$ are also parallel. Therefore, the element λ' can be denoted by $f(\lambda)$.

An affine transformation maps the vector $a(\lambda + \mu) = a\lambda + a\mu$ to the vector $a'f(\lambda + \mu) = a'f(\lambda) + a'f(\mu) = a'[f(\lambda) + f(\mu)]$ and the vector $a(\lambda\mu) = (a\lambda)\mu$ to the vector $a'f(\lambda\mu) = (a'\lambda)f(\mu) = a'f(\lambda)f(\mu)$. Therefore, $f(\lambda + \mu) = f(\lambda) + f(\mu)$ and $f(\lambda\mu) = f(\lambda)f(\mu)$. Since an affine transformation is continuous, this ring automorphism must also be continuous.

6) The assertion of the theorem. By 5) an affine transformation maps a vector a to the vector $Uf(a)$ where U is a linear operator in $\mathbb{A}L^n$ and $f(x)$ is the vector whose coordinates $f(x^i)$ are obtained by a continuous ring automorphism $f(x)$ from the coordinates x^i of a vector x.

Therefore, if an affine transformation maps a point O to a point O', it maps any point X with position vector x to the point X' with position vector $\overrightarrow{OO'} + Uf(x)$. Denoting the vector $\overrightarrow{OO'}$ by a, we obtain formulas (2.31) and (2.30). ■

For $a = 0$ the transformations (2.31), that is,

$$x' = Uf(x), \tag{2.32}$$

are called *affine rotations* in $\mathbb{A}E^n$. For $U = I$ and $f(x) = x$, the transformations (2.31), that is, $x' = x + a$, are called *translations*.

Theorem 2.23. *An affine transformation in the real space E^n has the form*

$$x' = Ux + a. \tag{2.33}$$

An affine transformation in the complex or split complex space $\mathbb{C}E^n$ or $\mathbb{C}'E^n$ has the form (2.33), or

$$x' = U\bar{x} + a. \tag{2.34}$$

An affine transformation in the quaternionic, split quaternionic or matrix space $\mathbb{H}E^n, \mathbb{H}'E^n$, *or* $\mathbb{R}_m E^n$ *has the form*

$$x' = Uxb + a. \tag{2.35}$$

In the definition of affine transformation in the spaces $E^n, \mathbb{H}E^n, \mathbb{H}'E^n$, *and* $\mathbb{R}_m E^n$ *the requirement of continuity is superfluous.*

PROOF. The assertions of this theorem follow from Theorems 1.3, 1.4, 1.15, and 1.16, according to which the only ring automorphism in the field $\mathbb{R}$ is the identity, the continuous ring automorphisms of $\mathbb{C}$ and $\mathbb{C}'$ are $f(x) = x$ and $f(x) = \bar{x}$, and all ring automorphisms of $\mathbb{H}, \mathbb{H}'$, and $\mathbb{R}_m$ are inner, thus can be written in the form $f(x) = b^{-1}xb$ (the matrix U in (2.35) is equal to the product of the matrix U in (2.31) by b^{-1}). ■

2.3.2. Simple Ratio of Three Points. If X, Y, Z are three points on a single line in $\mathbb{A}E^n$ and have position vectors x, y, z, the vectors $z-y$ and $y-x$ are collinear and $z - y = (y - x)v$, where v is an element in the algebra $\mathbb{A}$. This element is called the *simple ratio* of the points X, Y, Z and is denoted by

$$v = \overline{XY, Z} = (y - x)^{-1}(z - y). \tag{2.36}$$

Theorem 2.24. *Under an affine transformation (2.31) the simple ratio (2.35) changes according to the rule* $v \to f(v)$.

PROOF. Since the affine transformation (2.31) maps the position vector x of a point X to the position vector $x' = Uf(x) + a$, it maps the differences $y - x$ and $z-y$ to $y'-x' = Uf(y-x)$ and $z'-y' = Uf(z-y) = Uf((y-x)v) = Uf(y-x)f(v)$, and the ratio $v' = \overline{X'Z', Y'}$ is equal to $f(v)$. ■

Theorem 2.25. *An affine transformation in the real space* E^n *leaves invariant the simple ratios* v.

An affine transformation in the complex or split complex space $\mathbb{C}E^n$ *or* $\mathbb{C}'E^n$ *either leaves invariant the simple ratios* v *or maps them to* $\bar{v}$.

An affine transformation in the quaternionic, split quaternionic or matrix space $\mathbb{H}E^n, \mathbb{H}'E^n$ *or* $\mathbb{R}_m E^n$ *maps the simple ratios* v *to* $v' = a^{-1}va$.

The assertions of this theorem follow from Theorems 1.3, 1.4, 1.15, and 1.16.

2.3.3. Affine lines. An affine line $\mathbb{A}E^1$ can be identified with the algebra $\mathbb{A}$ itself. An affine transformation (2.31) in such a line has the form

$$x' = af(x) + b. \tag{2.37}$$

In particular, an affine transformation (2.33) in the line E^1, an affine transformation (2.33) or (2.34) in the lines $\mathbb{C}E^1$ or $\mathbb{C}'E^1$, and an affine transformation (2.35) in the lines $\mathbb{H}E^1$ or $\mathbb{H}'E^1$ can be expressed by the formula $x' = ax + b$, (1.9), or the first formula in (1.47), respectively.

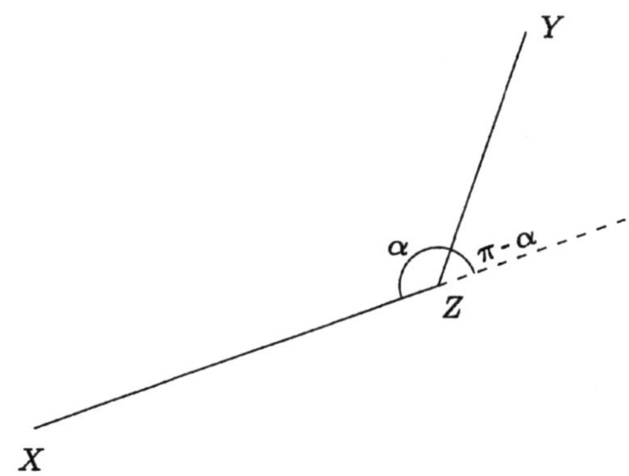

FIG. 2.9

Theorem 2.26. *The affine lines $E^1, \mathbb{C}E^1, \mathbb{C}'E^1, \mathbb{H}E^1$, and $\mathbb{H}'E^1$ admit real interpretations as the line R^1, the planes R^2 and R^2_1, or the 4-spaces R^4 and R^4_2, respectively.*

The affine transformations in these lines are interpreted as similitudes in these real line, planes, and 4-spaces.

PROOF. If we regard the absolute values or moduli $|y - x|$ of differences $y - x$ of elements in the algebras $\mathbb{R}, \mathbb{C}, \mathbb{C}', \mathbb{H}$, and $\mathbb{H}'$ as distances between these elements, we obtain the metrics or pseudometrics of the line R^1, the planes R^2 and R^2_1 and the 4-spaces R^4 and R^4_2 in these algebras. The transformations $x' = ax + b$ in $\mathbb{R}$ become similitudes in R^1, the transformations (1.9) in $\mathbb{C}$ and $\mathbb{C}'$ become similitudes in R^2 and R^2_1 (see **1.1.1** and **1.1.2**), the transformations (1.47) in $\mathbb{H}$ and $\mathbb{H}'$ become similitudes in R^4 and R^4_2 (see **1.2.1** and **1.2.3**). ■

For the algebras $\mathbb{C}$ and $\mathbb{C}'$ the simple ratio $v = \overline{XY, Z}$ is a complex or split complex number whose modulus is equal to the quotient of the lengths of segments XZ and ZY and the argument is equal to $\pi - \alpha$, where α is the angle between these segments (Fig. 2.9).

For the algebra $\mathbb{H}$ the simple ratio v can be reduced by the transformation $'v = a^{-1}va$ to the form $v_0 + iv_1$ with the same geometric meaning as for $\mathbb{C}$. For the algebra $\mathbb{H}'$ the simple ratio v can be reduced by the transformation $'v = a^{-1}va$ to the form $v_0 + iv_1$, $v_0 + ev_1$ or $v_0 + v_1(i + e)$, and in the first two cases the geometric meaning of v is the same as for $\mathbb{C}$ or $\mathbb{C}'$.

2.3.4. Collineations. In **0.8.9** we have defined the group of *collineations* in the space P^n. Like the group of affine transformations in $\mathbb{A}E^n$, the group of collineations in the space $\mathbb{A}P^n$ can be defined as the group of bijections in $\mathbb{A}P^n$ that preserve straight lines.

Theorem 2.27. *A collineation in the space $\mathbb{A}P^n$ has in projective coordinates the form*

$$'x^i = U^i_j f(x^j). \tag{2.38}$$

Formula (2.38) can also be written in the vector form (2.32).

PROOF. Since points and lines in $\mathbb{A}P^n$ are interpreted as lines and planes through a point in $\mathbb{A}E^{n+1}$, this theorem follows from Theorem 2.21 for $\mathbb{A}E^{n+1}$. ■

Theorem 2.28. *A collineation in the real space P^n has the form $x' = Ux$.*

A collineation in the complex or split complex space $\mathbb{C}P^n$ or $\mathbb{C}'P^n$ has the form $x' = Ux$ or $x' = U\bar{x}$.

A collineation in the quaternionic, split quaternionic, or matrix space $\mathbb{H}P^n, \mathbb{H}'P^n$, or $\mathbb{R}_m P^n$ has the form $x' = Uxb$.

In the definitions of collineation in $P^n, \mathbb{H}P^n, \mathbb{H}'P^n$, or $\mathbb{R}_m P^n$ the requirement of continuity is superfluous.

This theorem follows from Theorem 2.23.

2.3.5. Correlations. In **0.8.9** we have defined *correlations* in P^n. Correlations in the space $\mathbb{A}P^n$ can be defined as maps of points in $\mathbb{A}P^n$ to hyperplanes in $\mathbb{A}P^n$ and of lines to pencils of hyperplanes. Since the hyperplanes in $\mathbb{A}P^n$ form an interpretation of $\mathbb{A}P^n$ under which lines in $\mathbb{A}P^n$ are interpreted as pencils of hyperplanes, the correlations in $\mathbb{A}P^n$ can be expressed by formulas similar to (2.30) and (2.31), but the projective coordinates $'x^i$ of points and the vectors representing these points must be replaced by the tangential coordinates u_i of hyperplanes and the covectors u representing them. When $\mathbb{A}$ is noncommutative, the coordinates u_i are determined by a multiplication $u_i \to \lambda u_i$ and the expressions $U^i_j f(x^j)$ must be replaced by expressions $\varphi(x^j)U_{ij}$ where $x \to \varphi(x)$ is a continuous ring antiautomorphism in $\mathbb{A}$ $(\varphi(x + y) = \varphi(x) + \varphi(y), \varphi(xy) = \varphi(y)\varphi(x))$. Therefore, the following theorem holds:

Theorem 2.29. *Correlations in the space $\mathbb{A}P^n$ in projective and tangential coordinates have the form*

$$'u_i = \varphi(x^j)U_{ij}, \tag{2.39}$$

where $x \to \varphi(x)$ is a continuous ring antiautomorphism of the algebra $\mathbb{A}$. The formula (2.39) can be written also in the vector form

$$u' = \varphi(x)U. \tag{2.40}$$

Theorem 2.30. *A correlation in the real space P^n has the form $u' = xU$. A correlation in the complex or split complex space $\mathbb{C}P^n$ or $\mathbb{C}'P^n$ has the form $u' = xU$ or $u' = \bar{x}U$. A correlation in the quaternionic or split quaternionic spaces $\mathbb{H}P^n$ or $\mathbb{H}'P^n$ has the form*

$$u' = b\bar{x}U. \tag{2.41}$$

A correlation in the space $\mathbb{R}_m P^n$ has the form

$$u' = bx^{\top}U; \tag{2.42}$$

in the definitions of correlation in $P^n, \mathbb{H}P^n, \mathbb{H}'P^n$, and $\mathbb{R}_m E^n$ the requirement of continuity is superfluous.

PROOF. This theorem follows from Theorem 2.28. Formulas (2.40) and (2.41) follow from the fact that an antiautomorphism in $\mathbb{H}$ or $\mathbb{H}'$ has the form $\varphi(x) = b\bar{x}b^{-1}$, while an antiautomorphism in $\mathbb{R}_m$ has the form $\varphi(x) = bx^{\top}b^{-1}$. ■

Often, in the formulas $u = xU$, $u = \bar{x}U$ and (2.41) the vector x is also written as an $((n+1) \times 1)$-matrix $x^{\top}$.

The product of two correlations in $\mathbb{A}P^n$ is a collineation, therefore the collineations and correlations in $\mathbb{A}P^n$ form a group, called the group of *projective transformations* in this space.

2.3.6. Cross Ratio of Four Points. Let X, Y, Z, T be four points on a single line in $\mathbb{A}P^n$ and represented by vectors x, y, z, t in $\mathbb{A}L^{n+1}$. Since these four points are coplanar, the vectors z and t are linear combinations of the vectors x and y: $z = x\alpha + y\beta, t = x\gamma + y\delta$. The *cross ratio* of X, Y, Z, T is the following element in the algebra $\mathbb{A}$:

$$w = \overline{XY, ZT} = \delta\gamma^{-1}\alpha\beta^{-1}. \tag{2.43}$$

Note that if z is multiplied by λ, the elements α and β are replaced by $\alpha\lambda$ and $\beta\lambda$ and the element w is invariant; if t is multiplied by λ, the elements γ and δ are replaced by $\gamma\lambda$ and $\delta\lambda$ and the element w is invariant; if x is multiplied by λ, the elements α and γ are replaced by $\lambda^{-1}\alpha$ and $\lambda^{-1}\gamma$ and the element w is invariant; finally, if y is multiplied by λ, the elements β and δ are replaced by $\lambda^{-1}\beta$ and $\lambda^{-1}\delta$ and the element w is replaced by $\lambda^{-1}w\lambda$. Therefore, for a commutative algebra $\mathbb{A}$ the element (2.43) does not depend on scaling of the vectors x, y, z, t, while for a noncommutative algebra $\mathbb{A}$ under such a scaling the element w has to be replaced by $\lambda^{-1}w\lambda$.

Theorem 2.31. *If the four points X, Y, Z, T are regarded as lying in the space $\mathbb{A}E^n$, the cross ratio $w = \overline{XY, ZT}$ can be expressed in terms of the position vectors of these points by the formula*

$$w = \overline{XY, ZT} = \lambda^{-1}(t-x)(y-t)^{-1}(y-z)(z-x)^{-1}\lambda. \tag{2.44}$$

PROOF. If the points X, Y, Z, T are regarded as points in a hyperplane in $\mathbb{A}E^{n+1}$ and vectors x, y, z, t are taken as their position vectors $\overrightarrow{OX}, \overrightarrow{OY}, \overrightarrow{OZ}, \overrightarrow{OT}$, the simple ratios α, β, γ, and δ are equal to $(y-x)^{-1}(y-z), (x-y)^{-1}(z-x), (y-x)^{-1}(y-t), (x-y)^{-1}(t-x)$, respectively. Therefore, we obtain

$$w = \overline{XY, ZT} = (x-y)^{-1}(t-x)(y-t)^{-1}(y-x)(y-x)^{-1}(y-z)(z-x)(x-y)$$

which implies (2.44). ■

Theorem 2.32. *A collineation (2.38) in $\mathbb{A}P^n$ transforms the cross ratio (2.43) according to the rule $w \to f(w)$.*

PROOF. Since the collineation (2.38) maps the vectors x representing points X to vectors $x' = Uf(x)$, it maps linear combinations $z = x\alpha + y\beta$ and $t = x\gamma + y\delta$ to linear combinations $z' = x'f(\alpha) + y'f(\beta)$ and $t' = x'f(\gamma) + y'f(\delta)$. Therefore, the cross ratio $'w$ is equal to $f(\delta)[f(\gamma)]^{-1}f(\alpha)[f(\beta)]^{-1} = f(w)$. ■

Theorem 2.33. *A collineation in the real space P^n leaves invariant the cross ratio w.*

A collineation in the complex or split complex space $\mathbb{C}P^n$ or $\mathbb{C}'P^n$ either leaves invariant the cross ratio v or transforms it to $\overline{v}$.

A collineation in the quaternionic, split quaternionic, or matrix space $\mathbb{H}P^n, \mathbb{H}'P^n$, or $\mathbb{R}_m P^n$ transforms the cross ratio w to $'w = a^{-1}wa$.

The assertions of this theorem follow from Theorems 1.3, 1.4, 1.15, and 1.16.

2.3.7. Projective Lines. The projective line $\mathbb{A}P^1$ can be represented as the extended algebra $\mathbb{A}$. A collineation (2.38) in this line has the form $'x^0 = df(x^0) + cf(x^1)$, $'x^1 = b(x^0) + af(x^1)$. Therefore, $x' = {}'x^1({}'x^0)^{-1}$ can be written as

$$x' = [af(x) + b][cf(x) + d]^{-1}. \tag{2.45}$$

In particular, a collineation $x' = Ux$ in P^1, $\mathbb{C}P^1$ or $\mathbb{C}'P^1$, or $x' = Uxb$ in $\mathbb{H}P^1$ or $\mathbb{H}'P^1$, can be expressed by

$$x' = (ax + b)(cx + d)^{-1}. \tag{2.46}$$

For a collineation $x' = U\bar{x}$ in $\mathbb{C}P^1$ or $\mathbb{C}'P^1$ we have

$$x' = (a\bar{x} + b)(c\bar{x} + d)^{-1}. \tag{2.47}$$

Since the role of hyperplanes in $\mathbb{A}P^1$ is played by points, in such a line there are also correlations (2.39). The correlations $x' = xU$ and (2.41) in $P^1, \mathbb{C}P^1$, or $\mathbb{C}'P^1$ coincide with the collineations (2.46); the correlations $x' = \bar{x}U$ and (2.41) in $\mathbb{C}P^1, \mathbb{C}'P^1, \mathbb{H}P^1$, or $\mathbb{H}'P^1$ have the form (2.47).

Theorem 2.34. *The projective lines $\mathbb{C}P^1, \mathbb{C}'P^1, \mathbb{H}P^1$, or $\mathbb{H}'P^1$ admit real interpretations as the conformal or pseudoconformal planes C^2 and C_1^2, or the 4-spaces C^4 or C_2^4, respectively.*

The collineations and correlations in these lines are represented as conformal transformations in these real planes and 4-spaces.

PROOF. In **2.2.2** we have seen that the lines $\mathbb{C}P^1$, $\mathbb{C}'P^1$, $\mathbb{H}P^1$, $\mathbb{H}'P^1$ are homeomorphic to the planes C^2, C_1^2 and 4-spaces C^4, C_2^4, respectively. Because the fractional linear transformations (2.46) and (2.47) are products of the transformations (1.9) in $\mathbb{C}$ and $\mathbb{C}'$, and of the transformations (1.47) in $\mathbb{H}$ and $\mathbb{H}'$, which, as we have seen in **1.1.1, 1.1.2, 1.2.1**, and **1.2.3**, are represented as similitudes in R^2, R_1^2, R^4, and R_2^4, by transformations $x' = r^2/\bar{x}$, which are represented as inversions in circles and hyperspheres in these planes and 4-spaces, the similitudes and inversions in these planes and 4-spaces generate the conformal transformations in C^2, C_1^2, C^4, and C_2^4. ■

For the algebras $\mathbb{C}$ and $\mathbb{C}'$ the cross ratio $w = \overline{XY, ZT}$ is a complex or split complex number which is invariant under the transformations (2.46) and is complexly conjugated under the transformations (2.47). Therefore, if the cross ratio is written in the form $w_0 + iw_1$ or $w_0 + ew_1$, the real numbers w_0 and w_1 are invariants of the four points in the spaces $\mathbb{C}P^n$ and $\mathbb{C}'P^n$ under collineations in these spaces.

For the algebra $\mathbb{H}$ the cross ratio w can be reduced to the form $w_0 + iw_1$ using the transformation $w' = a^{-1}wa$. For the algebra $\mathbb{H}'$ the cross ratio w can be can be reduced to the form $w_0 + iw_1$, $w_0 + ew_1$, or $w_0 + w_1(i + e)$ using the transformation $w' = a^{-1}wa$. Therefore, the real numbers w_0 and w_1 are invariants of the four points of the spaces $\mathbb{H}P^n$ and $\mathbb{H}'P^n$ under collineations in these spaces.

§2.4. Lines, m-Planes, and Hyperplanes

2.4.1. Matrix Coordinates of m-Planes. m-Planes in an affine space $\mathbb{A}E^n$ are determined by the same equation (0.15) as m-planes in a real space E^n. This equation can be written in affine coordinates in the form

$$x^i = x_0^i + a_a^i t^a \quad (i = 1, 2, \ldots, n,\ a = 1, 2, \ldots, m). \tag{2.48}$$

Let A be the $(m \times n)$-matrix (a_a^i) and let t be a vector $\{t^a\}$ in the linear space or free module $\mathbb{A}L^m$. We can write the equation (2.48) in the vector form

$$x' = x_0 + At. \tag{2.49}$$

Since m-planes in $\mathbb{A}E^n$ are intersections of $n - m$ hyperplanes, m-planes can also be defined by a system of $n - m$ equations

$$u_i^u x^i + v^u = 0 \quad (i = 1, 2, \ldots, n,\ u = m + 1,\ m + 2, \ldots, n). \tag{2.50}$$

Let U be the $(n \times (n - m))$-matrix (u_i^u), and let v be a vector $\{v^u\}$ in the linear space or free module $\mathbb{A}L^{n-m}$. We can write the equation (2.50) in the vector form

$$Ux + v = 0. \tag{2.51}$$

The $(m \times n)$-matrix A is called the *matrix coordinate* of the m-plane in $\mathbb{A}E^n$; the $(n \times (n - m))$-matrix U is called the *tangential matrix coordinate* of this m-plane.

Analogously, the equations $x = a_a t^a$ and $u^u x = 0$ of an m-plane in $\mathbb{A}P^n$ ($a = 0, 1, \ldots, m,\ u = m + 1, \ldots, n$) can be written in projective coordinates as

$$x^i = a_a^i t^a, \quad u_i^u x^i = 0 \quad (i = 0, 1, \ldots, n).$$

Let A be the $((m + 1) \times (n + 1))$-matrix (a_a^i), U the $((n + 1) \times (n - m))$-matrix (u_i^u), and t the vector $\{t^a\}$ in the linear space or free module $\mathbb{A}L^{m+1}$. We can rewrite these equations in the vector form

$$x = At, \quad Ux = 0. \tag{2.52}$$

The $((m + 1) \times (n + 1))$-matrix A is called the *projective matrix coordinate* of the m-plane in $\mathbb{A}P^n$ and the $((n + 1) \times (n - m))$-matrix U is called the *tangential matrix coordinate* of this m-plane.

The matrix A is an one-column matrix if $m = 1$ in $\mathbb{A}E^n$ or $m = 0$ in $\mathbb{A}P^n$. In this case A can be regarded as vector a; t is a vector in $\mathbb{A}L^1$ and can be regarded as a scalar. Therefore, in this case equation (2.49) has the form $x = x_0 + at$ and is the equation of a line, and the first equation in (2.52) has the form $x = at$ and determines a point.

The matrix U is one-row matrix if $m = n - 1$. In this case the matrix U can be regarded as a covector u; v is a vector in $\mathbb{A}L^1$ and can be regarded as a scalar. In this case equation (2.51) has the form $ux + v = 0$ and is the equation of a hyperplane in $\mathbb{A}E^n$, and the second equation in (2.52) has the form $ux = 0$ and is the equation of a hyperplane in $\mathbb{A}P^n$.

Theorem 2.35. *The projective matrix coordinates A of an m-plane in $\mathbb{A}P^n$ are determined up to multiplication $A \to AK$ by an $((m+1)\times(m+1))$-matrix K.*

The tangential matrix coordinates U of an m-plane in $\mathbb{A}P^n$ are determined up to multiplication $U \to LU$ by an $((n-m)\times(n-m))$-matrix L.

PROOF. The columns in the $((m+1)\times(n+1))$-matrix A can be regarded as vectors a_a representing points A_a whose sum is this m-plane. If we replace these points by other points B_a, the vectors b_b representing these points are linear combinations, $b_b = a_a k_b^a$, the coordinates of these vectors are equal to $b_b^i = a_a^i k_b^a$, and this matrix equality can be written as $B = AK$.

The rows in the $((n+1)\times(n-m))$-matrix U can be regarded as covectors u^u determining hyperplanes α^u whose intersection is this m-plane. If we replace these hyperplanes by hyperplanes β^v, the covectors v^v determining these hyperplanes are linear combinations, $v^v = \ell_u^v u^u$, the coordinates of these covectors are equal to $v_i^v = \ell_u^v u_i^u$, and this matrix equality can be written as $V = LU$.■

Theorem 2.36. *A collineation in the space $\mathbb{A}P^n$ has, in projective matrix coordinates X and tangential matrix coordinates U, the form*

$$X' = AX, \quad U' = UA^{-1}. \tag{2.53}$$

In these coordinates a correlation in $\mathbb{A}P^n$ has the form

$$U' = (AX)^\top = X^\top A^\top, \quad X' = (UA^{-1})^\top = (A^{-1})^\top U^\top, \tag{2.54}$$

where A is an $((n+1)\times(n+1))$-matrix.

PROOF. The assertions of this theorem follow from the fact that the entries in the columns of the matrices X and X' are the projective coordinates of points and the entries in the rows of the matrices U and U' are the tangential coordinates of hyperplanes.

In the cases of points ($m = 0$) and hyperplanes ($m = n-1$) the matrices K and L are (1×1)-matrices and can be viewed as scalars, up to multiplication by which the projective coordinates a^i of points and the tangential coordinates u_i of hyperplanes are determined. ■

2.4.2. Intersection of Two m-Planes. Consider two m-planes p, with projective and tangential matrix coordinates A and U,and q, with projective and tangential matrix coordinates B and V.

Theorem 2.37. *If the algebra $\mathbb{A}$ is commutative, a necessary and sufficient condition for the dimension of the intersection of the m-planes p and q to be equal to d is that the ranks of the matrices UB and VA are equal to $m-d$.*

A necessary and sufficient condition for the m-planes p and q to coincide is that the matrices UA and VB are equal to the zero matrix.

A necessary and sufficient condition for the m-planes p and q to have empty intersection is that the matrices UA and VB are nonsingular.

PROOF. If the m-planes p and q have a common point represented by a vector x, according to formulas (2.53) and the analogous formulas for the m-plane q, this vector can be written in the form $x = At = Bu$, and satisfies the conditions $Ux = Vx = 0$. Therefore, there are relations $(UB)u = 0$ and $(VA)t = 0$. If the dimension of the intersection of p and q is equal to d, the vectors x satisfying these conditions form a linear space or free module $\mathbb{A}L^d$, and, therefore, similar spaces or modules $\mathbb{A}L^d$ are formed by the vectors t and u. The ranks of the matrices UB and VA are equal to $m - d$; vice versa, if the ranks of these matrices are equal to $m - d$, the vectors t, u, and x form $\mathbb{A}L^d$. If p and q coincide, $d = m$, while if these m-planes have no common points, that is, their intersection is empty, then $d = -1$ and the matrices UB and VA are nonsingular. ■

This theorem is easily generalized to the case of noncommutative algebras $\mathbb{A}$, since such algebras are isomorphic to subalgebras of an algebra $\mathbb{B}_r$ of $(r \times r)$-matrices over a commutative algebra $\mathbb{B}$. Therefore, the matrices A, B, U, V can be represented by matrices whose entries are elements in $\mathbb{B}$. If the sizes of these matrices are equal to the products of the sizes of the matrices A, B, U, V by r, the condition for equality of ranks in the commutative case must be replaced by the condition that the ranks of the matrices representing UA and VB are equal to $(m - d)r$.

2.4.3. Projection onto an m-Plane in the Direction of an $(\mathbf{n - m - 1})$-Plane. A point Y in an m-plane p is said to be the projection onto p of a point X in the direction of an $(n - m - 1)$-plane q which has no points in common with p if the line XY intersects q (Fig. 2.10).

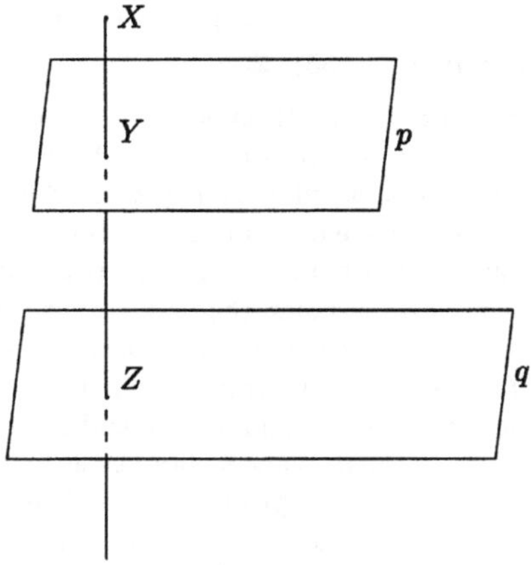

FIG. 2.10

Theorem 2.38. *If the points X and Y are represented by vectors x and y, the projective matrix coordinate of p is an $((m + 1) \times (n + 1))$-matrix A and the tangential matrix coordinate of q is an $((n + 1) \times (m + 1))$-matrix U. Then the vector y has the form*

$$y = A(AU)^{-1}Ux. \tag{2.55}$$

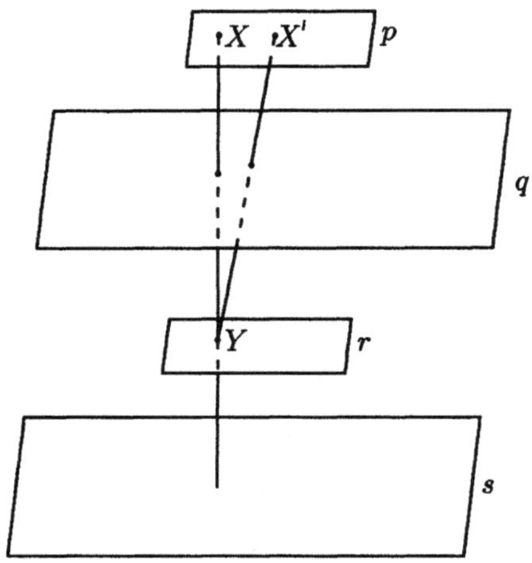

FIG. 2.11

PROOF. Since p and q have no points in common, the matrix AU is nonsingular. Let the point of intersection of the line XY with q be the point Z, which is represented by the vector z. Then the vectors x, y, z are linearly dependent and can be chosen so that $x = y + z$. Since Y is a point in p, $y = At$; since z is a point in q, $Uz = 0$. Therefore, $Uz = U(x - y) = Ux - (UA)t = 0$, and $t = (UA)^{-1}Ux$. Since $y = At$, we obtain formula (2.55). ■

2.4.4. m-Pairs and Their Cross Ratios. A pair consisting of an m-plane and an $(n-m-1)$-plane is called an *m-pair*. An m-pair whose m-plane and $(n-m-1)$-plane have no common points is called a *nondegenerated m-pair*; an m-pair whose m-plane and $(n-m-1)$-plane have common points is called a *degenerated m-pair*. For $m = 0$ an m-pair consists of a point and a hyperplane and is called a *0-pair*.

Consider two nondegenerated m-pairs, the first m-pair consisting of an m-plane p and an $(n-m-1)$-plane q, the second m-pair consisting of an m-plane r and an $(n-m-1)$-plane s. The projective matrix coordinates of p and r are A and B, respectively, the tangential matrix coordinates of q and s are U and V, respectively. Consider the transformation in the m-plane p obtained as follows: we project a point X in p onto r in the direction of s and project the resulting point Y in r onto p in the direction of q. Then the projection X' of Y (Fig. 2.11) is represented by the vector x' obtained by applying twice formula (2.55). Since $y = B(VB)^{-1}Vx$ and $x' = A(UA)^{-1}Uy$, we obtain

$$x' = A(UA)^{-1}(UB)(VB)^{-1}Vx. \tag{2.56}$$

Since $x = At$ and $x' = At'$, formula (2.56) implies

$$t' = (UA)^{-1}(UB)(VB)^{-1}(AV)t. \tag{2.57}$$

The $((m+1) \times (m+1))$-matrix

$$W = (UA)^{-1}(UB)(VB)^{-1}(VA) \tag{2.58}$$

is called the *cross ratio of the two m-pairs.*

For two 0-pairs, consisting of the points X and Y, represented by vectors x and y, and two hyperplanes π and ρ, represented by covectors u and v, we obtain the *cross ratio of two 0-pairs*:

$$w = (ux)^{-1}(uy)(vy)^{-1}(vx). \tag{2.59}$$

Theorem 2.39. *The cross ratio (2.59) of two 0-pairs is equal to the cross ratio $\overline{XY, ZT}$ of the points X and Y in these 0-pairs and the points Z and T of intersection of the line XY with the hyperplanes π and ρ in these 0-pairs.*

PROOF. Let X, Y, Z, T be represented by vectors x, y, z, t and let the equations of π and ρ be $ux = 0$ and $vx = 0$. If $z = x\alpha + y\beta$, $t = x\gamma + y\delta$, then since Z is a point in π and T is a point in ρ, we obtain $uz = ux\alpha + uy\beta = 0$ and $vt = vx\gamma + vy\delta = 0$. Therefore, $(ux)^{-1}(uy) = -\alpha\beta^{-1}, (vy)^{-1}(vx) = -\delta\gamma^{-1}$, and w has the form (2.59). ■

Theorem 2.40. *For $m \leq n - m - 1$ there are in general $m + 1$ lines which meet both m-planes and both $(n - m - 1)$-planes of two m-pairs. The cross ratios of the points of intersection of these lines with the m-planes and $(n-m-1)$-planes of the m-pairs are equal to the eigenvalues of the $((m+1) \times (m+1))$-matrix which is the cross ratio of these m-pairs.*

PROOF. In general, the matrix (2.58) has $m + 1$ eigenvectors, which determine $m + 1$ fixed points of the linear transformation (2.57). Each of these fixed points is the point of intersection of a line which meets both m-planes and both $(n - m - 1)$-planes of two m-pairs with these m-planes and $(n - m - 1)$-planes.

We call these $m + 1$ lines, *transversals* of the two m-pairs.

The points of intersection of transversals with m-planes and $(n-m-1)$-planes of two m-pairs can be regarded as 0-pairs in $\mathbb{A}P^1$ and the cross ratios of these 0-pairs coincide with the eigenvalues of the matrix (2.58).

If two eigenvalues are conjugate imaginary, the corresponding transversals are conjugate imaginary lines.

If there are k equal eigenvalues, the two m-pairs have infinitely many transversals, which are rectilinear generators of a Segrean $\Sigma_{1,k-1}$ in a $(2k - 1)$-plane (see **2.1.8**).

If $k = m + 1$, all transversals of the two m-pairs are rectilinear generators of a Segrean $\Sigma_{1,m}$ in a $(2m + 1)$-plane. ■

2.4.5. Affine Matrix Coordinates. Let A_0 be the $((m+1) \times (m+1))$-matrix consisting of the upper $m + 1$ rows of the projective matrix coordinate A of an m-plane in $\mathbb{A}P^n$ and let A_1 be the $((m + 1) \times (n - m))$-matrix consisting of the lower $n - m$ rows of the matrix A. Then the $((m + 1) \times (n - m))$-matrix

$$\overline{A} = A_1 A_0^{-1} \tag{2.60}$$

is said to be the *affine matrix coordinate* of the given m-plane. Not all m-planes in $\mathbb{A}P^n$ have affine matrix coordinates; a condition for the existence of these coordinates is nonsingularity of the matrix A_0. This condition is equivalent to the condition of that the m-plane and the $(n - m - 1)$-plane $a^0 = a^1 = \cdots = a^m = 0$

have empty intersection: the first $m+1$ columns in the tangential matrix coordinate U of this $(n-m-1)$-plane form the identity $((m+1)\times(m+1))$-matrix I, the remaining $n-m$ columns in the matrix U consist of zeros, therefore, the product UA is equal to A_0 and a necessary condition for the intersection of the m-plane and the $(n-m-1)$-plane to be a d-plane, by Theorem 2.37, is that the rank of the matrix UA be $m-d$.

For $m=0$ the affine matrix coordinate consists of the column with as entries the affine coordinates of a point in $\mathbb{A}P^n$.

Theorem 2.41. *The affine matrix coordinate of an m-plane in $\mathbb{A}P^n$ does not depend on the points determining this m-plane.*

PROOF. In Theorem 2.35 we have seen that the projective matrix coordinates A of an m-plane in $\mathbb{A}P^n$ are determined up to multiplication $A \to AK$ by an $((m+1)\times(m+1))$-matrix $K=(k_a^b)$ whose entries are the coefficients of linear combinations $b_a = a_b k_a^b$, where a_a and b_a are vectors representing the points A_a and B_a determining the m-plane. These linear combinations can be written in coordinate form: $b_a^i = a_b^i k_a^b$. Therefore, if the matrices A_0 and A_1 participating in formula (2.60) are (a_a^c) and (a_a^u), the analogous matrices $B_0=(b_a^c)$ and $B_1=(b_a^c)$ are related to A_0 and A_1 by $B_0 = A_0K$ and $B_1 = A_1K$, and the product $B_1B_0^{-1}$ is equal to $A_1KK^{-1}A_0^{-1} = A_1A_0^{-1} = \overline{A}$. ■

Theorem 2.42. *The real dimension $\dim \mathbb{A}Gr_{n,m}^P$ of the Grassmann manifold $\mathbb{A}Gr_{n,m}^P$ of m-planes in $\mathbb{A}P^n$ over an r-algebra $\mathbb{A}$ is equal to*

$$\dim \mathbb{A}Gr_{n,m}^P = r(m+1)(n-m). \tag{2.61}$$

PROOF. The number of entries in a matrix $\overline{A}$ is equal to $(m+1)(n-m)$, therefore the real dimension of the set of these matrices is equal to the product of this number by the real dimension r of the algebra $\mathbb{A}$. Since the dimension of the set of m-planes in $\mathbb{A}P^n$ that have no affine matrix coordinates is less than $\dim \mathbb{A}Gr_{n,m}^P$, this dimension is equal to the real dimension of the set of matrices $\overline{A}$, that is, $r(m+1)(n-m)$. ■

Theorem 2.43. *The affine matrix coordinate $\overline{A}$ of an m-plane in $\mathbb{A}P^n$ is related to the tangential matrix coordinate U of the same m-plane by*

$$\overline{A} = -U_1^{-1}U_0, \tag{2.62}$$

where U_1 is the $((n-m)\times(n-m))$-matrix consisting of the utmost right $n-m$ columns in the tangential matrix coordinate U and U_0 is the $((m+1)\times(n-m))$-matrix consisting of the utmost left $m+1$ columns in U.

PROOF. In Theorem 2.37 we have seen that the condition of coincidence of an m-plane p with projective matrix coordinate A with an m-plane q with tangential matrix coordinate U has the form $UA=0$. This condition can be written in the form $U_0A_0 + U_1A_1 = 0$. Therefore, we obtain $U_1^{-1}(U_0A_0+U_1A_1)A^{-1} = 0$ and, according to (2.60), formula (2.62). ■

Note that the matrices A_1 and U_0 are rectangular $((m+1)\times(n-m))$-matrices, like $\overline{A}$, and that A_0 and U_1 are a square $((m+1)\times(m+1))$-matrix and $((n-m)\times(n-m))$-matrix, respectively.

Theorem 2.44. *In the affine matrix coordinates $\overline{X}$ of m-planes the collineations in the space $\mathbb{A}P^n$ are written as*

$$\overline{X}' = (A_0^1 + A_1^1\overline{X})(A_0^0 + A_1^0\overline{X})^{-1}. \tag{2.63}$$

For $n = 2m + 1$, the correlations in $\mathbb{A}P^n$ have the form

$$\overline{X}' = (-A_{10}^{\top} - \overline{X}^{\top} A_{11}^{\top})^{-1}(A_{00}^{\top} + \overline{X}A_{01}^{\top}). \tag{2.64}$$

PROOF. The formulas (2.63) and (2.64) can be obtained from the first formulas in (2.54) by replacing the matrices X, X', and U by matrices X_0, X_1, X_0', X_1', and U_0, U_1 and by replacing the matrices A in these formulas by the matrices

$$A = \begin{bmatrix} A_0^0 & A_1^0 \\ A_0^1 & A_1^1 \end{bmatrix} \quad \text{and} \quad A = \begin{bmatrix} A_{00} & A_{01} \\ A_{10} & A_{11} \end{bmatrix}.$$

The first formulas in (2.54) then take the form

$$X_0' = A_0^0 X_0 + A_1^0 X_1, \quad X_1' = A_0^1 X_0 + A_1^1 X_1$$

and

$$U_0 = X_0^{\top} A_{00}^{\top} + X_1^{\top} A_{01}^{\top}, \quad U_1 = X_0^{\top} A_{10}^{\top} + X_1^{\top} A_{11}^{\top},$$

so that formulas (2.63) and (2.64) are obtained. ■

Theorem 2.44 shows that the projective transformations in $\mathbb{A}P^n$ in affine matrix coordinates $\overline{X}$ of m-planes can be expressed by fractional linear transformations of these coordinates.

2.4.6. Duality Principle. Since points in $\mathbb{A}P^n$ are characterized by $n + 1$ coordinates x^i, elements in the algebra $\mathbb{A}$, up to multiplication $x^i \to x^i k$ by an element k in $\mathbb{A}$ and hyperplanes in $\mathbb{A}P^n$ are characterized by $n + 1$ tangential coordinates u_i from $\mathbb{A}$ up to multiplication $u_i \to \ell u_i$ by an element ℓ in $\mathbb{A}$, and the condition $u_i x^i = 0$ of incidence of a point with a hyperplane is symmetric with respect to x^i and u_i, in $\mathbb{A}P^n$ there is a *duality principle*: for each theorem of this geometry there is another theorem obtained from the first one by replacing the words "point" by "hyperplane" and vice versa and by replacing the expressions "a point in a hyperplane" by "a hyperplane through a point" and vice versa. Such pairs of theorems are called *mutually dual* theorems, and figures or configurations obtained one from another by such a replacement are also called *mutually dual* figures or configurations.

Such replacement can be realized by a correlation in $\mathbb{A}P^n$.

By the duality principle in $\mathbb{A}P^n$, a line corresponds to a pencil of hyperplanes through an $(n - 2)$-plane and an m-plane corresponds to a bundle of hyperplanes through an $(n - m - l)$-plane. Therefore, by this principle lines in $\mathbb{A}P^n$ correspond to $(n - 2)$-planes and m-planes in $\mathbb{A}P^n$ correspond to $(n - m - l)$-planes in this space.

Therefore, the projective space $\mathbb{A}P^n$ admits an interpretation as the manifold $\mathbb{A}Gr^P_{n,n-1}$ of hyperplanes in this space.

§2.5. Hyperquadrics

2.5.1. Hyperquadrics in Affine Spaces. A *hyperquadric*, or a *hypersurface of second order*, in the space $\mathbb{A}E^n$ over a commutative algebra $\mathbb{A}$ is a hypersurface

$$a_{ij}x^i x^j + 2b_i x^i + c = 0. \tag{2.65}$$

Since $a_{ij}x^i x^j$ is a quadratic form, the matrix (a_{ij}) is symmetric, that is, $a_{ij} = a_{ji}$. Equation (2.65) can be written in vector form:

$$xAx + 2bx + c = 0, \tag{2.66}$$

where A is a linear operator which transforms a vector x to the covector Ax, xAx is the convolution of this covector with the vector x, and b is a covector whose convolution with x is bx.

The condition $a_{ij} = a_{ji}$ is equivalent to the condition $A = A^\top$. Therefore, for any x and y, $xAy = yAx$.

To find the mutual location of the hyperquadric (2.66) and a line $x = x_0 + \ell t$ we replace x in (2.66) by $x_0 + \ell t$, to obtain

$$\begin{aligned}(x_0 + \ell t)A(x_0 + \ell t) + 2b(x_0 + \ell t) + c = \\ (\ell A \ell)t^2 + 2(Ax_0 + b)\ell t + x_0 A x_0 + 2bx_0 + c = 0.\end{aligned} \tag{2.67}$$

The equation (2.67) is quadratic if $\ell A \ell \neq 0$ and is linear if $\ell A \ell = 0$. The directions determined by the vectors ℓ satisfying the last condition are called *asymptotic directions*.

Lines in E^n with non-asymptotic directions meet the hyperquadric in two real, conjugate complex or coincident points. In $\mathbb{C}E^n$ lines with non-asymptotic directions meet the hyperquadric in two distinct or coincident points.

For lines with asymptotic directions equation (2.67) can be written as $Pt+Q = 0$. If $P \neq 0$ the line meets the hyperquadric in one point, if $P = 0, Q \neq 0$ the line and the hyperquadric do not meet, if $P = Q = 0$ the line lies on the hyperquadric, that is, is a *rectilinear generator* of it.

2.5.2. Hermitian Hyperquadrics in Affine Spaces. A *Hermitian hyperquadric* in the space $\mathbb{A}E^n$ over an algebra $\mathbb{A}$ with involution $x \to \overline{x}$ is a variety

$$\overline{x}^j a_{ij} x^i + \overline{b}_i x^i + \overline{x}^i b_i + c = 0. \tag{2.68}$$

Since $\overline{x}^j a_{ij} x^i$ is a Hermitian quadratic form, the matrix (a_{ij}) is Hermitian symmetric, that is, $a_{ij} = \overline{a}_{ji}$. Equation (2.68) can be written in vector form:

$$\overline{x}Ax + \overline{b}x + \overline{x}b + c = 0, \tag{2.69}$$

where A is a linear operator which transforms a vector x to the covector Ax, $\overline{x}Ax$ is the convolution of this covector with the vector $\overline{x}$, and $\overline{b}$ and b are covectors whose convolutions with x and $\overline{x}$ are $\overline{b}x$ and $\overline{x}b$. The condition $a_{ij} = \overline{a}_{ji}$ is equivalent to the condition $A = \overline{A}^\top$. Therefore, for any x and y, $\overline{x}Ay = \overline{y}Ax$.

To find the mutual location of the Hermitian hyperquadric (2.69) with a line $x = x_0 + \ell t$ we replace x in (2.69) by $x_0 + \ell t$, to obtain

$$\begin{aligned}(\overline{x}_0 + \overline{t}\,\overline{l})A(x_0 + lt) + \overline{b}(x_0 + lt) + (\overline{x}_0 + \overline{t}\,\overline{l})b + c = \\ \overline{t}(\overline{l}\,A\,l)t + (\overline{x}_0 A + \overline{b})lt + \overline{t}\overline{l}(Ax_0 + b) + \overline{x}_0 A x_0 + \overline{b}x_0 + \overline{x}_0 b + c = 0.\end{aligned} \tag{2.70}$$

Theorem 2.45. *For the spaces $\mathbb{C}E^n, \mathbb{C}'E^n, \mathbb{H}E^n$, and $\mathbb{H}'E^n$ equation (2.70) is the equation of a circle in the planes R^2 and R^2_1 (for $\mathbb{C}$, $\mathbb{C}'$) and of a hypersphere in the spaces R^4 and R^4_2 (for $\mathbb{H}$, $\mathbb{H}'$) if $\bar{\ell}A\ell \neq 0$ and of a line in the planes R^2 and R^2_1 (for $\mathbb{C}$, $\mathbb{C}'$) and of a hyperplane in the spaces R^4 and R^4_2 (for $\mathbb{H}$, $\mathbb{H}'$) if $\bar{\ell}A\ell = 0$.*

PROOF. If $\bar{\ell}A\ell \neq 0$ equation (2.70) can be written in the form $P\bar{t}t+\overline{Q}t+\bar{t}Q+R = 0$, where P and R are real numbers. But the equations of circles in R^2 and R^2_1 and the equations of hyperspheres in R^4 and R^4_2 have the form $|t - t_0| = r$ or $(\bar{t} - \bar{t}_0)(t - t_0) = \bar{t}t - \bar{t}_0 t - \bar{t}t_0 + \bar{t}_0 t = r^2$, and if we multiply both parts of this equation by P we obtain the above mentioned equation, with $Q = -Pt_0$ and $R = P(\bar{t}_0 t_0 - r^2)$.

If $\bar{\ell}A\ell = 0$ equation (2.70) can be written in the form $\overline{Q}t + \bar{t}Q + R = 0$, and is the equation of a line in E^2 or of a hyperplane in E^4. ■

The directions determined by vectors ℓ satisfying the condition $\bar{\ell}A\ell = 0$ are also called *asymptotic directions.*

2.5.3. Center of a Hyperquadric. A point of symmetry of a (Hermitian) hyperquadric in $\mathbb{A}E^n$, that is, a point such that reflection with respect to this point preserves this hyperquadric, is called a *center* of this hyperquadric. For hyperquadrics in E^n such a point is the middle point of the segments on all lines through this point between the points of intersection of these lines with the hyperquadric. For Hermitian hyperquadrics in $\mathbb{C}E^n, \mathbb{C}'E^n, \mathbb{H}E^n$, or $\mathbb{H}'E^n$ such a point is the center of a circle or a hypersphere cut out on all lines through this point by the Hermitian hyperquadric, in accordance with Theorem 2.45.

Theorem 2.46. *A point X_0 with position vector*

$$x_0 = -A^{-1}b \tag{2.71}$$

is a center of the hyperquadric (2.66) in E^n and a center of a Hermitian hyperquadric (2.69) in $\mathbb{C}E^n, \mathbb{C}'E^n, \mathbb{H}E^n$ or $\mathbb{H}'E^n$.

PROOF. If the point X_0 is a center of a hyperquadric in E^n, the values t_1 and t_2 of the points of intersection of any line $x = x_0 + \ell t$ through this point are related by $t_1 = -t_2$. Therefore, $t_1 + t_2 = 0$ and in (2.67) the coefficient $2(Ax_0 + b)\ell$ at t is equal to 0.

If X_0 is a center of a Hermitian hyperquadric in $\mathbb{C}E^n, \mathbb{C}'E^n, \mathbb{H}E^n$, or $\mathbb{H}'E^n$, the values of t_0 for centers of the mentioned circles and hyperspheres are equal to 0 and in (2.70) $Q = \bar{\ell}(Ax_0 + b) = 0$.

Since ℓ is arbitrary, in all cases the vector $Ax_0 + b$ is the zero vector. ■

Hyperquadrics and Hermitian hyperquadrics with unique centers are called *central.*

If the center of a central (Hermitian) hyperquadric is the origin of coordinates, $x_0 = 0$ and, therefore, $b = 0$ in equations (2.66) and (2.69).

2.5.4. Conjugate Diameters. Lines through the center of a central hyperquadric are called *diameters* of this hyperquadric, m-planes through the center of a central hyperquadric are called *diametral m-planes* of this hyperquadric.

Theorem 2.47. *The middle points of the segments of parallel lines between the points of intersection with a hyperquadric in E^n, and the centers of the circles or hyperspheres in the parallel lines cut out from them by a Hermitian hyperquadric in accordance with Theorem 2.45, are located in hyperplanes which are diametral hyperplanes of these hyperquadrics.*

PROOF. If the point X_0 is the middle of such a segment on the line $x = x_0 + \ell t$ through X_0, the values t_1 and t_2 for the points of intersection of this line with the hyperquadric (2.66) are related by $t_1 + t_2 = 0$ and the coefficient $2(Ax_0 + b)\ell$ at t in (2.67) is equal to 0.

If X_0 is the center of a circle or hypersphere on the line $x = x_0 + \ell t$ through X_0, cut out by a Hermitian hyperquadric in $\mathbb{C}E^n, \mathbb{C}'E^n, \mathbb{H}E^n$, or $\mathbb{H}'E^n$, the values t_0 for the centers of these circles and hyperspheres are equal to 0 and in (2.70) $Q = \overline{\ell}(Ax_0+b) = 0$. In this case the vector ℓ is not arbitrary and the position vectors x_0 of all points X_0 satisfy the condition $\overline{x}_0 A\ell + \overline{\ell}b = 0$, which is the equation of a hyperplane. Since this hyperplane passes through the center of the hyperquadric, it is a diametral hyperplane of this hyperquadric. ■

The equation of a diametral hyperplane for Hermitian hyperquadric can be written in the form

$$(\overline{\ell A})x + \overline{\ell}b = \overline{x}A\ell + \overline{\ell}b = 0 \tag{2.72}$$

and, therefore, the equation for a hyperquadric in E^n is analogous.

Equation (2.72) shows that the covector u of this hyperplane is related to the direction vector ℓ of the lines $x = x_0 + \ell t$ by

$$u = A\ell. \tag{2.73}$$

Formula (2.73) reveals the geometric sense of the linear operator A for the hyperquadrics (2.66) and (2.69).

A line $x = x_0 + \ell t$ through the center of a central hyperquadric and diameters of this hyperquadric lying in the hyperplane (2.72) are called *conjugate diameters.* If the diameter $x = x_0 + \ell' t$ is conjugate with the diameter $x = x_0 + \ell t$, all points of the diameter $x = x_0 + \ell' t$ are points in the hyperplane (2.72), that is, $(A\overline{\ell})(x_0 + \ell' t) + \overline{\ell}b = 0$ and, therefore, by (2.71)

$$\ell' A\ell = 0 \text{ or } \ell' A\overline{\ell} = 0. \tag{2.74}$$

Since the operator A of a (Hermitian) hyperquadric is (Hermitian) symmetric, (2.74) is equivalent to

$$\ell A\ell' = 0 \text{ or } \ell A\overline{\ell}' = 0, \tag{2.75}$$

that is, conjugacy of diameters is reciprocal.

Theorem 2.48. *A central hyperquadric in E^n can be written as*

$$\sum_i (x^i)^2 + c = 0 \tag{2.76}$$

or

$$-\sum_a (x^a)^2 + \sum_u (x^u)^2 + c = 0. \tag{2.77}$$

A central Hermitian hyperquadric in $\mathbb{C}E^n$ *or* $\mathbb{H}E^n$ *can be written as*

$$\sum_i \overline{x}^i x^i + c = 0 \tag{2.78}$$

or

$$-\sum_a \overline{x}^a x^a + \sum_u \overline{x}^u x^u + c = 0. \tag{2.79}$$

A central Hermitian hyperquadric in $\mathbb{C}' E^n$ *or* $\mathbb{H}' E^n$ *can be reduced to the form (2.78).*

<u>PROOF</u>. If the basis vectors e_i are directed along n mutually conjugate diameters of a central hyperquadric in E^n or of a central Hermitian hyperquadric in $\mathbb{C}E^n$, $\mathbb{C}' E^n$, $\mathbb{H}E^n$, or $\mathbb{H}' E^n$, all coefficients a_{ij} with $i \neq j$ in equations (2.65) and (2.68) of these hyperquadrics are equal to 0. If, moreover, the origin of coordinates coincides with the center of the (Hermitian) hyperquadric, the equations (2.65) and (2.68) have the form

$$\sum_i \lambda_i (x^i)^2 + c = 0 \tag{2.80}$$

and

$$\sum_i \lambda_i \overline{x}^i x^i + c = 0. \tag{2.81}$$

This follows from the fact that the coefficients a_{ij} in equations (2.65) and (2.68) are equal to the convolutions $e_i A e_j$ and $\overline{e}_i A e_j$, respectively, and from the equality $b = 0$ if the origin of coordinates coincides with the center of the (Hermitian) hyperquadric. The coefficients $\lambda_i = a_{ii}$ of all equations (2.81) are real and can be positive or negative.

Formulas (2.76) and (2.77) can be obtained from (2.80), and formulas (2.78) and (2.79) can be obtained from (2.81) by multiplication of the vectors e_i by the corresponding real multipliers.

If the equation of the Hermitian hyperquadric in $\mathbb{C}' E^n$ or $\mathbb{H}' E^n$ has the form (2.79), it can be reduced to the form (2.78) by multiplication of the vectors e_a by e. ■

2.5.5. Classification of Hyperquadrics and Hermitian Hyperquadrics. Central hyperquadrics in E^n that can be brought to the form (2.76), respectively (2.77), are called *ellipsoids*, respectively *hyperboloids*.

Central Hermitian hyperquadrics in $\mathbb{C}E^n$ or $\mathbb{H}E^n$ that can be brought to the form (2.78), respectively (2.79), are called *Hermitian ellipsoids*, respectively *Hermitian hyperboloids*.

If, in equations (2.77) and (2.79), $a = 1, 2, \ldots, \ell$ and $c < 0$ or $a = 1, 2, \ldots, \ell + 1$ and $c > 0$, the (Hermitian) hyperboloids are called *hyperboloids of index* ℓ.

The central Hermitian hyperquadrics in $\mathbb{C}'E^n$ and $\mathbb{H}'E^n$, which can always be brought to the form (2.78), are called *Hermitian ellipsoids.*

If a hyperquadric is not central not all coefficients $\lambda_i = a_{ii}$ are unequal to zero and if $\lambda_i = a_{ii} \neq 0$, the coefficient b_i cannot be reduced to 0, but c still can be reduced to 0. The hyperquadrics (2.76) in E^n and (2.77) in $\mathbb{C}E^n$ and $\mathbb{H}E^n$ whose equations can be brought to the form

$$\sum_{i=1}^{n-1} \lambda_i (x^i)^2 + 2b_n x^n = 0, \tag{2.82}$$

or

$$\sum_{i=1}^{n-1} \lambda_i \overline{x}^i x^i + \overline{b}_n x^n + \overline{x}^n b_n = 0, \tag{2.83}$$

are called *paraboloids*, respectively *Hermitian paraboloids.* These paraboloids are called *elliptic* if all coefficients λ_i have the same sign and *hyperbolic* otherwise.

Straight lines in E^n, $\mathbb{C}E^n$, and $\mathbb{H}E^n$ with the basis vector e_n for the (Hermitian) paraboloids (2.82) and (2.83) as direction vector are called *diameters* of these paraboloids.

Hyperquadrics whose equations can be brought to the form $xAx = 0$ or $\overline{x}Ax = 0$ are called *hypercones.* If the forms $xAx = 0$ and $\overline{x}Ax = 0$ are nondegenerated the hypercone has a *point apex* at the origin of coordinates, if these equations can be reduced to the forms $\sum_{i=1}^{m} (x^i)^2 = 0$ or $\sum_{i=1}^{m} \overline{x}^i x^i = 0$ a hypercone has an $(n-m)$-*planar*, for $m = n-1$, *rectilinear, apex.* For $m = 1$ a hypercone is a couple of hyperplanes, for $m = 0$ it is a double hyperplane.

Hyperquadrics whose equations can be brought to a form involving not all coordinates are called *hypercylinders.*

Theorem 2.49. *Ellipsoids in E^n and Hermitian ellipsoids in $\mathbb{C}E^n$ or $\mathbb{H}E^n$ are homeomorphic to hyperspheres in R^n, R^{2n} or R^{4n}, respectively.*

Hyperboloids of index ℓ in E^n and Hermitian hyperboloids of index ℓ in $\mathbb{C}E^n$ or $\mathbb{H}E^n$ are homeomorphic to the topological product of R^ℓ and a hypersphere in $R^{n-\ell}$, to the topological product of $R^{2\ell}$ and a hypersphere in $R^{2n-2\ell}$, or to the topological product of $R^{4\ell}$ and a hypersphere in $R^{4n-4\ell}$, respectively.

Hermitian ellipsoids in $\mathbb{C}'E^n$ or $\mathbb{H}'E^n$ are homeomorphic to the topological product of R^n and a hypersphere in R^n, or to the topological product of R^{2n} and a hypersphere in R^{2n}, respectively.

PROOF. If we introduce in the space E^n the metric of the space R^n, ellipsoids and hyperboloids of index ℓ in E^n become ellipsoids and hyperboloids of index ℓ in R^n, which have equations (2.80) with $c < 0$ in orthogonal coordinates. For ellipsoids all λ_i are positive and the transformation $'x^i = \sqrt{\lambda_i}x^i$ transforms them to the hyperspheres (2.76). For hyperboloids of index ℓ all λ_a $(a = 1, 2, \ldots, \ell)$ are negative and all λ_u $(u = \ell+1, \ldots, n)$ are positive. The transformation $'x^a = \sqrt{|\lambda_a|}x^a$, $'x^u = \sqrt{\lambda_u}x^u$ transforms them to the hyperboloids (2.77), and the transformation $'x^a = x^a$, $'x^u = x^u[(x^{\ell+1})^2 + \cdots + (x^n)^2]$ establishes a homeomorphism between a hyperboloid (2.77) and the hypercylinder $(x^{\ell+1})^2 + \cdots + (x^n)^2 = 1$,

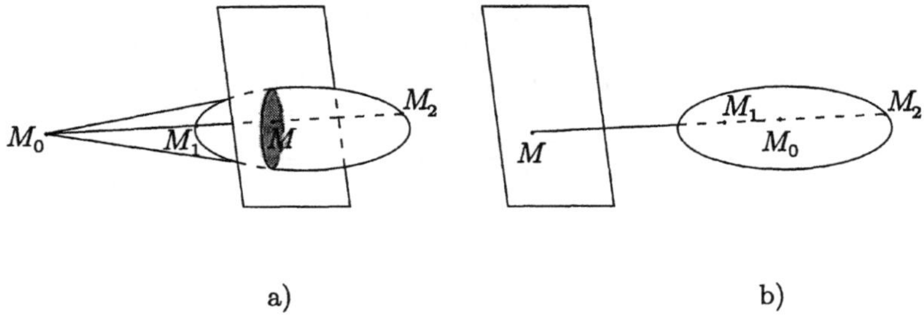

FIG. 2.12

which is the topological product of the hypersphere in the $(n-\ell)$-plane $x^a = 0$, that is, the space $R^{n-\ell}$, and its ℓ-planar generators, which are parallel and homeomorphic to the ℓ-plane $x^u = 0$, that is, the space R^ℓ.

The assertions of the theorem concerning Hermitian ellipsoids and hyperboloids in $\mathbb{C}E^n, \mathbb{C}'E^n, \mathbb{H}E^n$ and $\mathbb{H}'E^n$ follow from the fact that in the real interpretations of these spaces in the spaces E^{2n} and E^{4n}, where the coordinates of points are real coordinates of elements x^i in the algebras $\mathbb{C}, \mathbb{C}'$, $\mathbb{H}$, and $\mathbb{H}'$, Hermitian ellipsoids in these spaces are interpreted by ellipsoids and hyperboloids of index n in E^{2n} and by ellipsoids and hyperboloids of index $2n$ in E^{4n}, and Hermitian hyperboloids of index ℓ in $\mathbb{C}E^n$ or $\mathbb{H}E^n$ are interpreted as hyperboloids of index 2ℓ in E^{2n} and by hyperboloids of index 4ℓ in E^{4n}. We can now use the assertions of this theorem on ellipsoids and hyperboloids in E^n, which have already been proved. ■

2.5.6. Poles and Polar Hyperplanes. Let X_0 be a point in E^n, and let X_1 and X_2 be the points of intersection of a line $x = x_0 + \ell t$ through X_0 with a hyperquadric. The point X that is the fourth harmonic for these three points, that is, such that $\overline{X_0X, X_1X_2} = -1$, is called the *polar conjugate point* of X_0 with respect to the hyperquadric.

If X_0 is a point in $\mathbb{C}E^n, \mathbb{C}'E^n, \mathbb{H}E^n$, or $\mathbb{H}'E^n$, and X is the point in the line $x = x_0 + \ell t$ through X_0 which corresponds to X_0 under inversion in the circle or 3-sphere cut out by the Hermitian hyperquadric from this line, in accordance with Theorem 2.44, the point X is called the *polar conjugate point* of X_0 with respect to the Hermitian hyperquadric.

Theorem 2.50. *All points polar conjugate with a point X_0 in E^n, respectively $\mathbb{C}E^n$, $\mathbb{C}'E^n$, $\mathbb{H}E^n$, or $\mathbb{H}'E^n$, with respect to a hyperquadric (2.66), respectively with respect to a Hermitian hyperquadric (2.69), in these spaces belong to the hyperplane*

$$x_0Ax + b(x_0 + x) + c = 0, \tag{2.84}$$

respectively,

$$\overline{x}_0Ax + \overline{b}x + \overline{x}_0b + c = 0. \tag{2.85}$$

Fig. 2.12 represents a pole and its polar hyperplane in E^n.

PROOF. For a hyperquadric in E^n we denote the values of the parameter t corresponding to the points X_0, X_1, X_2, and X by $t_0 = 0$, t_1, t_2, and t, respectively. By (2.44) we can write the condition $\overline{X_0X, X_1X_2} = -1$ in the form

$$\frac{t_2 - t_0}{t_2 - t} \cdot \frac{t_1 - t_0}{t_1 - t} = \frac{t_2}{t_2 - t} \cdot \frac{t_1}{t_1 - t} = -1$$

or

$$\frac{t_2}{t_2 - t} + \frac{t_1}{t_1 - t} = 0.$$

Hence

$$t = \frac{2t_1t_2}{t_1 + t_2}. \tag{2.86}$$

Since t_1 and t_2 are the roots of the quadratic equation (2.67), their sum and product are equal to

$$t_1 + t_2 = -\frac{2(Ax_0 + b)\ell}{\ell A \ell}, \quad t_1t_2 = \frac{x_0Ax_0 + 2bx_0 + c}{\ell A \ell}. \tag{2.87}$$

Therefore,

$$t = -\frac{x_0Ax_0 + 2bx_0 + c}{(Ax_0 + b)\ell}$$

and

$$x = x_0 + \ell t = x_0 - \frac{x_0Ax_0 + 2bx_0 + c}{(Ax_0 + b)\ell}\ell.$$

The vector x satisfies the condition

$$(x - x_0) \cdot (Ax_0 + b)(x - x_0) + (x_0Ax_0 + 2bx_0 + c)(x - x_0) = 0.$$

The vector on the left hand side of this equality is equal to the product of the vector $x - x_0 = \ell t$ by a number equal to 0, this number being

$$(Ax_0 + b)(x - x_0) + x_0Ax_0 + 2bx_0 + c = 0,$$

that is, equation (2.84).

For a Hermitian hyperquadric in $\mathbb{C}E^n, \mathbb{C}'E^n, \mathbb{H}E^n$, or $\mathbb{H}'E^n$ the equation of the circle or 3-sphere cut out by the hyperquadric on the line $x = x_0 + \ell t$ is, in accordance with Theorem 2.44, (2.70). If we write this equation in the form

$$R(|t - t_c|^2 - r^2) = R((\bar{t} - \bar{t}_c) - r^2) = R(\bar{t}t - \bar{t}t_c - \bar{t}_c t + \bar{t}_c t_c - r^2) = 0,$$

we find that

$$R = \bar{\ell}A\ell, \quad Rt_c = -\bar{\ell}(Ax_0 + b), \quad R(\bar{t}_c t_c - r^2) = \bar{x}_0Ax_0 + \bar{b}x_0 + \bar{x}_0 b + c. \tag{2.88}$$

Therefore,

$$t_c = -\frac{\overline{\ell}(Ax_0+b)}{\overline{\ell}A\ell}, \quad \overline{t}_c t_c - r^2 = \frac{\overline{x}_0 A x_0 + \overline{b}x_0 + \overline{x}_0 b + c}{\ell A \ell}. \tag{2.89}$$

Note that in the real case we can consider $t_c = \frac{t_1+t_2}{2}$ and $r = \frac{t_1-t_2}{2}$ and formulas (2.88) can be written in the form (2.87).

An inversion in the circle or 3-sphere $(\overline{t}-\overline{t}_c)(t-t_c) = r^2$ has the form $t - t_c = r^2(\overline{t}_0 - \overline{t}_c)^{-1} = r^2\overline{t}_c^{-1}$. Therefore,

$$t = t_c - r^2\overline{t}_c^{-1} = (t_c\overline{t}_c - r^2)\overline{t}_c^{-1} = (\overline{x}_0 A x_0 + \overline{b}x_0 + \overline{x}_0 b + c)(\overline{\ell}(Ax_0+b))^{-1}$$

and

$$x = x_0 + \ell t = x_0 + \ell(\overline{x}_0 A x_0 + \overline{b}x_0 + \overline{x}_0 b + c)(\overline{\ell}(Ax_0+b))^{-1}.$$

The vector x satisfies the condition

$$(x - x_0)\cdot(\overline{x} - \overline{x}_0)(Ax_0+b) + (x-x_0)\cdot(\overline{x}_0 A x_0 + \overline{b}x_0 + \overline{x}_0 b + c) = 0.$$

The vector on the left hand side of this equality is equal to the product of the vector $x - x_0 = \ell t$ by an element equal to 0, this element being

$$(\overline{x} - \overline{x}_0)(Ax_0+b) + \overline{x}_0 A x_0 + \overline{b}x_0 + \overline{x}_0 b + c = 0,$$

that is, equation (2.85). ■

The hyperplanes (2.84) and (2.85) are called the *polar hyperplanes* of the point X_0, and the point X_0 is called a *pole* of the hyperplanes (2.84) and (2.85) with respect to the hyperquadrics (2.66) and (2.69).

Note that equation (2.84) is the equation of the *tangent hyperplane* to the hyperquadric (2.66) at the point X_0: in this case $t_1 = t_2 = t_1 + t_2 = t_1t_2 = 0$ and the numerators of both expressions (2.87) are equal to 0. This implies equation (2.84). Therefore, the tangent hyperplanes at the points on a hyperquadric in E^n can be regarded as the polar hyperplanes of the points on these hyperquadrics. In $\mathbb{C}E^n, \mathbb{C}'E^n, \mathbb{H}E^n$ and $\mathbb{H}'E^n$ the hyperplane (2.85) for the point X_0 on a Hermitian hyperquadric (2.69) is also called the *tangent hyperplane* to the Hermitian hyperquadric at X_0.

A transformation in E^n connected with an ellipsoid, hyperboloid or paraboloid which maps each point x in this space to the point x', which is intersection of polar hyperplane of x with the diameter passing through x, is called *inversion in an ellipsoid, hyperboloid or paraboloid.* These inversiuons are particular cases of *birational transformations.* For conics these inversions were considered already by Apollonius (see [Ro4, pp. 114-116]).

2.5.7. Rectilinear and m-Planar Generators. Lines, or m-planes, all points of which belong to a hyperquadric or a Hermitian hyperquadric are called *rectilinear*, respectively *m-planar, generators* of this hyperquadric.

Theorem 2.51. *Ellipsoids in E^n, and Hermitian ellipsoids in $\mathbb{C}E^n$ or $\mathbb{H}E^n$, have no rectilinear and m-planar generators.*

Hyperboloids in E^n, and Hermitian hyperboloids in $\mathbb{C}E^n$ or HE^n, of index 1 do have rectilinear generators but have no m-planar generators.

Hyperboloids in E^n, and Hermitian hyperboloids in $\mathbb{C}E^n$ or $\mathbb{H}E^n$, of index ℓ do have both rectilinear and m-planar generators for $m = 2, 3, \dots, \ell$; the maximal dimension of an m-planar generator of a hyperquadric or Hermitian hyperquadric is equal to $\frac{n-1}{2}$ for n odd and $\frac{n}{2} - 1$ for n even.

PROOF. The assertion on ellipsoids in E^n and Hermitian ellipsoids in $\mathbb{C}E^n$ and $\mathbb{H}E^n$ follows from the fact that, as we have seen in the proof of Theorem 2.49, such an ellipsoid can be obtained by an affine transformation from a hypersphere in R^n, R^{2n}, or R^{4n}, respectively, and from the fact that hyperspheres in Euclidean spaces do not have rectilinear and m-planar generators.

The assertion on hyperboloids in E^n follows from the fact that the equations of such a hyperboloid can be reduced to the form (2.77), which can be written as

$$\sum_a (x^a + x^{k+a})(-x^a + x^{k+a}) + \sum_u (x^u)^2 + c = 0.$$

If $c < 0$, $c = -r^2$, $k = \ell$, the $\ell + n - 2\ell = n - \ell$ linear equations $\pm x^a + x^{\ell+a} = x^u = \pm r + x^n = 0$ $(a = 1, 2, \dots, \ell, u = 2\ell + 1, \dots, n - 1)$ are equations of ℓ-planes, for $\ell = 1$ of lines, which are ℓ-planar generators of this hyperquadric. If $c > 0$, $c = r^2$, $k = \ell + 1$, the $\ell + n - 2\ell - 1 + 1 = n - l$ linear equations $\pm x^a + x^{\ell+1+a} = \pm x^{\ell+1} + r = x^u = 0$ $(a = 1, 2, \dots, \ell,\ u = 2\ell + 2, \dots, n)$ are also equations of ℓ-planes, for $\ell = 1$ of lines, which are ℓ-planar generators of this hyperquadric. For $\ell = 1$ m-planar generators are called rectilinear generators. The assertions on Hermitian hyperboloids in $\mathbb{C}E^n$ and $\mathbb{H}E^n$ follow from the fact that the equation of such a hyperboloid can be reduced to the form (2.77), which can be written as

$$\frac{1}{2}\sum_a (\overline{x}^a + \overline{x}^{k+a})(-x^a + x^{k+a}) + \frac{1}{2}\sum_a (-\overline{x}^a + \overline{x}^{k+a})(x^a + x^{k+a}) + \sum_u \overline{x}^u x^u + c = 0.$$

The remainder of the proof is analogous to the proof for the case of hyperboloids in E^n.

The assertions on Hermitian ellipsoids in $\mathbb{C}'E^n$ and $\mathbb{H}'E^n$ can be proved analogously to the assertions on Hermitian hyperboloids in $\mathbb{C}E^n$ and $\mathbb{H}E^n$. ■

The intersection of all hyperplanes tangent to the points on an m-planar generator is an $(n - m - \ell)$-plane, and can be regarded as the polar $(n - m - \ell)$-plane of this m-plane. Since the dimension m of m-planar generators satisfies the condition $m \leq n - m - 1$, or $n \geq 2m + 1$, the maximal possible value of m is $\frac{n-1}{2}$ for n odd $\frac{n-2}{2}$ for n even.

Let $Q^E_{n,m}, \mathbb{C}\overline{Q}^E_{n,m}, \mathbb{C}'\overline{Q}^E_{n,m}, \mathbb{H}\overline{Q}^E_{n,m}$, and $\mathbb{H}'\overline{Q}^E_{n,m}$ be the manifolds of m-planar generators of a hyperquadric in E^n and of Hermitian hyperquarics in $\mathbb{C}E^n$, $\mathbb{C}'E^n$, $\mathbb{H}E^n$, and $\mathbb{H}'E^n$, respectively.

Theorem 2.52. *The real dimensions* $\dim Q^E_{n,m}$, $\dim \mathbb{C}\overline{Q}^E_{n,m}$, $\dim \mathbb{C}'\overline{Q}^E_{n,m}$, $\dim \mathbb{H}\overline{Q}^E_{n,m}$, *and* $\dim \mathbb{H}'\overline{Q}^E_{n,m}$ *of* $Q^E_{n,m}$, $\mathbb{C}\overline{Q}^E_{n,m}$, $\mathbb{C}'\overline{Q}^E_{n,m}$, $\mathbb{H}\overline{Q}^E_{n,m}$, *and* $\mathbb{H}'\overline{Q}^E_{n,m}$ *are equal to*

$$\dim Q^E_{n,m} = \frac{(2n-3m-2)(m+1)}{2}, \tag{2.90}$$

$$\dim \mathbb{C}\overline{Q}^E_{n,m} = \dim \mathbb{C}'\overline{Q}^E_{n,m} = \frac{(4n-6m-2)(m+1)}{2}, \tag{2.91}$$

$$\dim \mathbb{H}\overline{Q}^E_{n,m} = \dim \mathbb{H}'\overline{Q}^E_{n,m} = \frac{(8n-12m-2)(m+1)}{2}. \tag{2.92}$$

PROOF. Let r be the real dimension of $\mathbb{R}, \mathbb{C}, \mathbb{C}', \mathbb{H}$, or $\mathbb{H}'$ ($r = 1, 2, 2, 4, 4$, respectively). Since any two points X_1 and X_2 on a (Hermitian) hyperquadric lying on a single rectilinear or m-planar generator of this hyperquadric are polar conjugate, X_2 lies on the tangent hyperplane to the hyperquadric at X_1. Therefore, each m-planar generator of a hyperquadric is determined by a point X_0 on this hyperquadric, by a point X_1 in the intersection of this hyperquadric with its tangent hyperplane at X_0, by a point X_2 in the intersection of this hyperquadric with its tangent hyperplane at X_0 and X_1,..., by a point X_m in the intersection of this hyperquadric with its tangent hyperplane at $X_0, X_1, \dots, X_{m-1}$. Since X_0 is an arbitrary point on the hyperquadric, it depends on real parameters whose number is equal to the real dimension of the hyperquadric, that is, $rn-1$. Since X_1 is an arbitrary point in the intersection of the hyperquadric with a hyperplane, it depends on real parameters whose number is equal to the real dimension of the hyperquadric in $(n-1)$-space, that is $r(n-1)-1 = rn-r-1$. Analogously, the point X_2 depends on $r(n-2)-1 = rn-2r-1$ real parameters,..., the point X_m depends on $r(n-m)-1 = rn-rm-1$ real parameters. So all points $X_0, X_1, \dots, X_m$ depend on

$$\begin{aligned}(m+1)(rn-1)+r(1+2+\dots+m) &= (m+1)(rn-1) + r\frac{m(m+1)}{2} \\ &= \frac{(2rn-rm-2)(m+1)}{2}\end{aligned}$$

real parameters.

The dimension of the manifold of m-planar generators of a hyperquadric in n-space is equal to this number minus the number of real parameters of the points $X_0, X_1, \dots, X_m$ in the m-plane. Since each of these $m+1$ points depends on rm real parameters, this number is equal to $rm(m+1)$ and the dimension of the manifold of m-planar generators of a hyperquadric in n-space is equal to

$$\frac{(2rn-rm-2)(m+1)}{2} - rm(m+1) = \frac{(2rn-3rm-2)(m+1)}{2}.$$

For $r = 1, 2, 4$ this number coincides with the numbers (2.90), (2.91), and (2.92), respectively. ■

Note that all manifolds $Q^E_{n,m}, \mathbb{C}\overline{Q}^E_{n,m}, \mathbb{C}'\overline{Q}^E_{n,m}, \mathbb{H}\overline{Q}^E_{n,m}$, and $\mathbb{H}'\overline{Q}^E_{n,m}$ are connected, except for $Q^E_{2m+1,m}$, which consists of two connected families.

In this case the following is true (see [Brt, p. 143]): if m is odd, two m-planar generators of one family have no common point or contain a common p-planar generator with p odd; two m-planar generators of different families have one common point or contain a common p-planar generator with p even. If m is even two m-planar generators of one family have one common point or contain a common p-planar generator with p even; two m-planar generators of different families have no common point or contain a common p-planar generator with p odd.

2.5.8. Hyperquadrics and Hermitian Hyperquadrics in Projective Spaces. *A hyperquadric*, or *hypersurface of second order*, in a space $\mathbb{A}P^n$ over a commutative algebra $\mathbb{A}$ is a hypersurface

$$a_{ij}x^i x^j = 0. \tag{2.93}$$

Since $a_{ij}x^i x^j$ is a quadratic form, the matrix (a_{ij}) is symmetric, that is $a_{ij} = a_{ji}$. The equation (2.93), like the equation (2.65) in $\mathbb{A}E^n$, can be written in the vector form

$$xAx = 0, \tag{2.94}$$

where A is a linear operator mapping the vector x to the covector Ax and xAx is the convolution of this covector with x. The condition $a_{ij} = a_{ji}$ is equivalent to the conditions $A = A^{\top}$ and $xAy = yAx$.

To find the mutual location of the hyperquadric (2.94) and a line $x = x_0 + yt$, we replace in (2.94) x by $x_0 + ly$, to obtain

$$(x_0 + yt)A(x_0 + yt) = (yAy)t^2 + 2Ax_0yt + x_0Ax_0 = 0. \tag{2.95}$$

The equation (2.95) is quadratic if $yAy \neq 0$ and linear if $yAy = 0$, that is, if the point Y represented by the vector y is a point on the hyperquadric, the solutions of this equation determine common points of the hyperquadric and a line. A line in P^n meets the hyperquadric in two real, conjugate complex or coincident points. A line in $\mathbb{C}P^n$ meets the hyperquadric in two distinct or coincident points.

A Hermitian hyperquadric in $\mathbb{A}P^n$, with $\mathbb{A}$ an algebra with a involution $x \to \overline{x}$, is a manifold

$$\overline{x}^j a_{ij} x^i = 0. \tag{2.96}$$

Since $\overline{x}^j a_{ij} x^i$ is a Hermitian quadratic form, the matrix (a_{ij}) is Hermitian symmetric, that is, $a_{ij} = \overline{a}_{ji}$. The equation (2.96), like the equation (2.68) in $\mathbb{A}E^n$, can be written in the vector form

$$\overline{x}Ax = 0, \tag{2.97}$$

where A is a linear operator mapping the vector x to the covector Ax and $\overline{x}Ax$ is the convolution of this covector with $\overline{x}$. The condition $a_{ij} = \overline{a}_{ji}$ is equivalent to the conditions $A = \overline{A}^{\top}$ and $\overline{x}Ay = \overline{y}Ax$.

To find the mutual location of the Hermitian hyperquadric (2.97) and a line $x = x_0 + yt$, we replace in (2.97) x by $x_0 + yt$, to obtain

$$(\overline{x}_0 + \overline{t}\,\overline{y})A(x_0 + yt) = (\overline{y}Ay)\overline{t}t + (\overline{x}_0 Ay)t + \overline{t}(\overline{y}Ax_0) + \overline{x}_0 Ax_0 = 0. \qquad (2.98)$$

Theorem 2.45 implies that equation (2.98) is the equation of a circle in the planes C^2 and C_1^2 in the case of the spaces $\mathbb{C}P^n$, $\mathbb{C}'P^n$, while in the case of $\mathbb{H}P^n$ and $\mathbb{H}'P^n$ it is the equation of a hypersphere in the spaces C^4 and C_2^4.

For hyperquadrics in P^n and for Hermitian hyperquadrics in $\mathbb{C}P^n, \mathbb{C}'P^n, \mathbb{H}P^n$, and $\mathbb{H}P^n$ we can define *polar conjugate points* as for Hermitian hyperquadrics in affine spaces over the same algebras.

From Theorem 2.50 it follows

Theorem 2.53. *All points polar conjugate with a point X_0 in P^n, respectively in $\mathbb{C}P^n$, $\mathbb{C}'P^n$, $\mathbb{H}P^n$, or $\mathbb{H}'P^n$, with respect to the hyperquadric (2.94), respectively the Hermitian hyperquadric (2.97), in this space belong to the hyperplane*

$$x_0 Ax = 0, \qquad (2.99)$$

respectively

$$\overline{x}_0 Ax = 0. \qquad (2.100)$$

The hyperplanes (2.99) and (2.100) are called *polar hyperplanes* of the point X_0, and X_0 itself is called a *pole* of these hyperplanes with respect to the hyperquadrics (2.94) and (2.97).

Note that, as in E^n, equation (2.99) is the equation of the *tangent hyperplane* to the hyperquadric (2.94). Therefore, the tangent hyperplane at a point on a hyperquadric in P^n can be regarded as the polar hyperplane of a point on the hyperquadric.

Points E_i $(i = 0, 1, \ldots, n)$, represented by vectors e_i for which $e_i A e_j = 0$ if $i \neq j$ for a hyperquadric in P^n and $\overline{e}_i A e_j = 0$ if $i \neq j$ for a Hermitian hyperquadric in $\mathbb{C}P^n, \mathbb{C}'P^n, \mathbb{H}P^n$, or $\mathbb{H}'P^n$, are said to be the vertices of the *autopolar simplex* of the hyperquadric. If these points are basis points of a system of projective coordinates, the equations (2.93) and (2.96) have the forms

$$\sum_i \lambda_i (x^i)^2 = 0 \qquad (2.101)$$

and

$$\sum_i \lambda_i \overline{x}^i x^i = 0. \qquad (2.102)$$

If, in the spaces $P^n, \mathbb{C}P^n$, or $\mathbb{H}P^n$ the coefficients λ_i in the equations (2.101) and (2.102) are all positive or all negative, the corresponding hyperquadrics are imaginary. If there are ℓ negative and $n+1-\ell$ positive λ_i $(\ell \leq n+1-\ell)$, the corresponding hyperquadrics are called *real hyperquadrics of index ℓ*; hyperquadrics of index 1 are called *oval hyperquadrics*.

In $\mathbb{C}'P^n$ and $\mathbb{H}'P^n$ the Hermitian hyperquadrics for which all coefficients λ_i in the equations (2.102) are unequal to 0, are real.

In the spaces $P^n, \mathbb{C}P^n, \mathbb{C}'P^n, \mathbb{H}P^n$, and $\mathbb{H}'P^n$, if one of the coefficients λ_i is equal to 0, the hyperquadric is called a *hypercone with a point apex*; if k of the λ_i are equal to 0, the hyperquadric is called a *hypercone with a* $(k+1)$*-planar apex.*

If we view the spaces $P^n, \mathbb{C}P^n$, and $\mathbb{H}P^n$ as being obtained from the spaces $E^n, \mathbb{C}E^n$, and $\mathbb{H}E^n$ by the addition of a hyperplane at infinity, then the center of a central hyperquadric is the pole of the hyperplane at infinity and the diametral hyperplanes are the polar hyperplanes of the points in the hyperplane at infinity. In this case, ellipsoids in $E^n, \mathbb{C}E^n$, and $\mathbb{H}E^n$ can be regarded as hyperquadrics in $P^n, \mathbb{C}P^n$, and $\mathbb{H}P^n$ which have no common points with the hyperplane at infinity. Hyperboloids can be regarded as hyperquadrics that are intersected by this hyperplane.

Paraboloids can be regarded as hyperquadrics for which the hyperplane at infinity is a tangent hyperplane, the point of tangency of the paraboloid with this hyperplane, which is the common point of all diameters of a paraboloid, can be regarded as the center of the paraboloid.

Hypercylinders can be regarded as hypercones with apices in the hyperplane at infinity.

Note that hyperboloids of index ℓ in $E^n, \mathbb{C}E^n$, and $\mathbb{H}E^n$ become hyperquadrics of index $\ell+1$ in $P^n, \mathbb{C}P^n$, and $\mathbb{H}P^n$.

Theorem 2.54. *An oval hyperquadric in P^n, respectively an oval Hermitian hyperquadric in $\mathbb{C}P^n$ or $\mathbb{H}P^n$, is homeomorphic to a hypersphere in R^n, respectively in R^{2n} or R^{4n}.*

A hyperquadric of index ℓ in P^n, respectively a Hermitian hyperquadric of index ℓ in $\mathbb{C}P^n$ or $\mathbb{H}P^n$, is homeomorphic to the fibration with the space $P^{\ell-1}$ as base and the hyperspheres in $R^{n-\ell+1}$ as fibers, respectively to the fibration with the space $\mathbb{C}P^{\ell-1}$ as base and the hyperspheres in $R^{2n-2\ell+2}$ as fibers, or to the fibration with the space $\mathbb{H}P^{\ell-1}$ as base and the hyperspheres in $R^{4n-4\ell+4}$ as fibers.

A Hermitian hyperquadric in $\mathbb{C}'P^n$ or $\mathbb{H}'P^n$ is homeomorphic to the topological product of P^n and P^{n-1}, respectively to the topological product of $\mathbb{C}P^n$ and P^{2n-1}.

PROOF. This theorem is the projective analog of Theorem 2.49. The assertions on a real hyperquadric (2.101) of index ℓ with equation $-\sum_a (x^a)^2 + \sum_u (x^u)^2 = 0$ and on Hermitian hyperquadrics in $\mathbb{C}P^n$ and $\mathbb{H}P^n$ of index ℓ with equations $-\sum_a \overline{x}^a x^a + \sum_u \overline{x}^u x^u = 0$ can be proved by considering of $(\ell-1)$-planes $x^u = 0$ and $(n-\ell)$-planes $x^a = 0$ in $P^n, \mathbb{C}P^n$, and $\mathbb{H}P^n$, respectively, which have no common points with the hyperquadrics, and of $(n-\ell+1)$-planes in these spaces which are sums of an $(n-\ell)$-plane and of points in a $(\ell-1)$-plane. These $(n-\ell+1)$-planes cut out from the hyperquadric oval real, respectively oval complex and quaternionic $(n-l)$-quadrics which are fibers of a trivial fibration of the hyperquadric whose base is the space $P^{\ell-1}$, respectively $\mathbb{C}P^{\ell-1}$ or $\mathbb{H}P^{\ell-1}$. These fibers are homeomorphic to a hypersphere in $R^{n-\ell+1}$, respectively $R^{2n-2\ell+2}$ or $R^{4n-4\ell+4}$.

This reasoning is not valid for Hermitian hyperquadrics in $\mathbb{C}'P^n$ and $\mathbb{H}'P^n$ because in $\mathbb{C}'P^n$ and $\mathbb{H}'P^n$ the $(\ell-1)$-planes $x^u = 0$ and $(n-\ell)$-planes $x^a = 0$ have common points with Hermitian hyperquadrics.

Points in $\mathbb{C}'P^n$ and $\mathbb{H}'P^n$ can be determined by projective coordinates $x^i = x^i_+ e_+ + x^i_- e_-$, where for $\mathbb{C}'P^n$ x^i_+ and x^i_- are real and e_+ and e_- are split complex numbers (1.11) and for $\mathbb{H}'P^n$ $x^i_+ = X^{2i}_+ + iX^{2i+1}_+$, $x^i_- = X^{2i}_- + iX^{2i+1}_-$, where i is split quaternionic basis element and e_+ and e_- also have the form (1.11) where, however, e is also split quaternionic basis element. Therefore the equation of a Hermitian hyperquadric $\sum_i \bar{x}^i x^i = 0$ in $\mathbb{C}' P^n$ can be written as $\sum_i x_+^i x_-^i = 0$ and in $\mathbb{H}' P^n$ it can be written as $\sum_i (\bar{x}_+^i e_+ x_-^i + \bar{x}_-^i e_- x_+^i) \sum_i (X_+^{2i} X_-^{2i} + X_+^{2i+1} X_-^{2i+1}) = 0$. Therefore coordinates x_+^i of points in P^n and $\mathbb{C}P^n$, respectively, and coordinates x_-^i of these points can be regarded as projective coordinates of points in P^{n-1} and P^{2n-1}, respectively. The subspaces P^n and $\mathbb{C}P^n$ in $\mathbb{C}' P^n$ and $\mathbb{H}' P^n$ have no common points with Hermitian hyperquadrics in these spaces and therefore these hyperquadrics are homeomorphic to topological products of P^n and P^{n-1} and of $\mathbb{C}P^n$ and P^{2n-1}, respectively.■

Theorem 2.55. *A hyperquadric in $\mathbb{C}P^n$ is homeomorphic to the Grassmann manifold $\overline{Gr}^P_{n,1}$ of oriented lines in P^n.*

An oval hyperquadric, respectively a hyperquadric of index ℓ, in $\mathbb{C}' P^n$ is homeomorphic to the topological product of two oval hyperquadrics, respectively, two hyperquadrics of index ℓ, in P^n.

PROOF. The first assertion follows from the fact that a hyperquadric in $\mathbb{C}P^n$ can be viewed as an imaginary hyperquadric in P^n, and each real line in P^n meets this hyperquadric in two imaginary conjugate points. We obtain a bijection between the Grassmann manifold $Gr^P_{n,1}$ of non-oriented lines in P^n and the manifold of pairs of imaginary conjugate points on an imaginary hyperquadric in P^n, and also a bijection between the manifold $\overline{Gr}^P_{n,1}$, which covers $Gr^P_{n,1}$ twice, and an oval hyperquadric in P^n, if we put into correspondence two imaginary conjugate points on an imaginary hyperquadric in P^n and the two directions of the line in P^n corresponding to this pair of points. In particular, an oval conic in $\mathbb{C}P^2$ is homeomorphic to $\mathbb{C}P^1$ and to the sphere in R^3, and the manifolds $Gr^P_{2,1}$ and $\overline{Gr}^P_{2,1}$ coincide with P^2 and the sphere in R^3, respectively.

The second assertion follows from Theorem 2.18. ■

Note that all hyperquadrics in P^n, as well as all Hermitian hyperquadrics in $\mathbb{C}P^n, \mathbb{C}' P^n, \mathbb{H}P^n$, and $\mathbb{H}' P^n$, are compact manifolds.

Theorem 2.51 implies that oval hyperquadrics in P^n, and oval Hermitian hyperquadrics in $\mathbb{C}P^n$ and $\mathbb{H}P^n$, do not have rectilinear and m-planar generators. Hyperquadrics in P^n, and Hermitian hyperquadrics in $\mathbb{C}P^n$ and $\mathbb{H}P^n$, of index 2 do have rectilinear generators and do not have m-planar generators. Hyperquadrics in P^n, and Hermitian hyperquadrics in $\mathbb{C}P^n$ and $\mathbb{H}P^n$, of index ℓ have rectilinear and m-planar generators for $m = 2, 3, \ldots, \ell - 1$. The maximal dimension of an m-planar generator is equal to $\frac{n-1}{2}$ for n odd and $\frac{n}{2} - 1$ for n even.

2.5.9. Application of Matrix Coordinates. Let us consider m-planar generators of hyperquadrics in P^n and of Hermitian hyperquadrics in $\mathbb{C}P^n, \mathbb{C}' P^n$, $\mathbb{H}P^n$ or $\mathbb{H}' P^n$ by means of their matrix coordinates.

Theorem 2.56. *The projective matrix coordinates X of an m-planar generator of a hyperquadric (2.99) in P^n, respectively of an m-planar generator of a Hermitian hyperquadric (2.100) in $\mathbb{C}P^n, \mathbb{C}' P^n, \mathbb{H}P^n$ or $\mathbb{H}' P^n$, satisfy the equation*

$$X^\top A X = 0, \tag{2.103}$$

respectively

$$\bar{X}^\top A X = 0. \tag{2.104}$$

The affine matrix coordinates $\overline{X}$ *of an* m*-planar generator of the same hyperquadric satisfy the equation*

$$\overline{X}^\top \hat{A} \overline{X} + B^\top \overline{X} + \overline{X}^\top B + C = 0, \tag{2.105}$$

respectively

$$\bar{\overline{X}}^\top \hat{A} \overline{X} + \bar{B}^\top \overline{X} + \bar{X}^\top B + C = 0. \tag{2.106}$$

In (2.103) and (2.104) the matrices A *are the same* $((n+1)\times(n+1))$*-matrices as in (2.99) and (2.100). In (2.105) and (2.106) the matrices* $\hat{A}$ *are* $((m+1)\times(m+1))$*-matrices with as entries the first* $m+1$ *rows and columns of the matrices* A*, the matrices* B *are* $((m+1)\times(n-m))$*-matrices with as entries the last* $n-m$ *rows and first* $m+1$ *columns of the matrices* A*, and the matrices* C *are* $((n-m)\times(n-m))$*-matrices with as entries the last* $n-m$ *rows and columns of the matrices* A*. The matrices* $\bar{X}$ *and* $\bar{\overline{X}}$ *in (2.104) and (2.106) are obtained from* X *and* $\overline{X}$ *by transforming the entries of the latter matrices in accordance with the transformation* $x \to \overline{x}$*.*

PROOF. Formulas (2.103) and (2.104) can be obtained from formulas (2.99) and (2.100), applied to the $m+1$ vectors x_a representing points X_a determining the m-planar generator of the hyperquadric.

Formulas (2.105) and (2.106) follow from formulas (2.103) and (2.104). ∎

We denote the manifolds of m-planar generators of a hyperquadric in P^n, respectively of a Hermitian hyperquadric in $\mathbb{C}P^n, \mathbb{C}'P^n, \mathbb{H}P^n$ or $\mathbb{H}'P^n$, by $Q^P_{n,m}$, respectively by $\mathbb{C}\overline{Q}^P_{n,m}$, $\mathbb{C}'\overline{Q}^P_{n,m}$, $\mathbb{H}\overline{Q}^P_{n,m}$ or $\mathbb{H}'\overline{Q}^P_{n,m}$.

Theorem 2.57. *The real dimensions dim* $Q^P_{n,m}$*, dim* $\mathbb{C}\overline{Q}^P_{n,m}$*, dim* $\mathbb{C}'\overline{Q}^P_{n,m}$*, dim* $\mathbb{H}\overline{Q}^P_{n,m}$*, and dim* $\mathbb{H}'\overline{Q}^P_{n,m}$ *of* $Q^P_{n,m}, \mathbb{C}\overline{Q}^P_{n,m}, \mathbb{C}'\overline{Q}^P_{n,m}, \mathbb{H}\overline{Q}^P_{n,m}$*, and* $\mathbb{H}'\overline{Q}^P_{n,m}$ *are equal to dim* $Q^E_{n,m}$*, dim* $\mathbb{C}\overline{Q}^E_{n,m}$*, dim* $\mathbb{C}'\overline{Q}^E_{n,m}$*, dim* $\mathbb{H}\overline{Q}^E_{n,m}$*, and dim* $\mathbb{H}'\overline{Q}^E_{n,m}$*, respectively (see (2.90), (2.91), and (2.92)).*

PROOF. This theorem follows from Theorem 2.52, since the manifold of m-planar generators of a hyperquadric in projective space differs from the analogous manifold in affine space by the manifold of m-planes, whose dimension is smaller than the dimension of the manifold of m-planar generators of the hyperquadric in affine space. ∎

Formulas (2.103) and (2.104) make it possible to give a simpler proof of this theorem. The left hand sides of these formulas are $((m+1)\times(m+1))$-matrices which are symmetric and symmetric Hermitian, respectively. These matrices depend on $R = r\frac{m(m+1)}{2} + m + 1$ real parameters. Therefore, the projective matrix coordinates of an m-planar generator of a hyperquadric are connected by R real conditions, and, since, as we have seen in Theorem 2.42, the dimension $\dim \mathbb{A}Gr^P_{n,m}$ of the

Grassmann manifold $\mathbb{A}Gr^P_{n,m}$ is equal to $r(m+1)(n-m)$, the dimension of the manifold of m-planar generators of a (Hermitian) hyperquadric is equal to

$$r(m+1)(n-m) - R = r(m+1)(n-m) - r\frac{m(m+1)}{2} - m - 1$$
$$= \frac{(2rn - 3rm - 2)(m+1)}{2}.$$

For $r = 1, 2, 4$ this number coincides with the numbers (2.90), (2.91) and (2.92), respectively.

2.5.10. Varieties. Hyperquadrics in E^n and P^n are special cases of varieties, that is, algebraic manifolds, in these spaces. Below we give some important facts from the geometry of varieties of order higher than 2.

Among the *hypercubics*, that is, hypersurfaces of third order $a_{ijk}x^i x^j x^k = 0$ in P^n, there is a smooth cubic surface, S_3, in P^3 with 27 rectilinear generators. The equation of this surface can be reduced to the form $ace - bdf = 0$, where a, b, c, d, e, f are linear polynomials. If we denote the intersection of two planes $a = 0$ and $b = 0$ by ab, then 9 of these 27 rectilinear generators are lines: $ab, ad, af, cb, cd, cf, eb, ed, ef$. If we denote the 3 rectilinear generators which meet the lines ab, cd, ef by $(ab, cd, ef)_i$, $i = 1, 2, 3$, etc., then the other 18 rectilinear generators are $(ab, cd, ef)_i$, $(ad, cf, ef)_i$, $(af, cb, ed)_i$, $(ab, cf, ed)_i$, $(ad, cb, ef)_i$, and $(af, cd, eb)_i$. In **1.4.7** we mentioned the surface S_3 since this surface admits the group of collineations in P^3 isomorphic to the finite group E_6, i.e. the Weyl group of the simple Lie group with the same notation.

Among the hyperquartics, that is, hypersurfaces of fourth order $a_{ijkl}x^i x^j x^k x^l = 0$ in P^n, there is a quartic curve, C_4, in P^2 with 28 double tangents. In **1.4.7** we mentioned the curve C_4 since it admits the group of collineations in P^2 isomorphic to the group E_7, i.e. the Weyl group of the simple Lie group with the same notation.

In **2.1.8** we have defined the *Segrean* $\Sigma_{m,n}$; it is a variety (2.9) in P^{mn+m+n} whose order and dimension are equal to $C^n_{n+m} = C^m_{n+m}$.

In **2.2.6** we have given equations (2.27), relating the Grassmann coordinates of m-planes in P^n which form the Grassmann manifold $Gr^P_{n,m}$. These equations can be regarded as equations of a variety, called the *Grassmannian* and denoted by $\Gamma\rho^P_{n,m}$ in P^N, where $N = C^{m+1}_{n+1} - 1$. Like the dimension of $Gr^P_{n,m}$, its dimension is equal to $(m+1)(n-m)$.

In **1.6.3** we have defined the *Lipschitzian* Ω_n by equations (1.162) and (1.63) in $P^{2^{n-1}-1}$; its dimension is $\frac{n(n-1)}{2}$.

Let us also mention the *Veronesean* Υ_n, with parametric equations

$$z^{ij} = x^i x^j \tag{2.107}$$

in P^N, where $N = \frac{(n+1)(n+2)}{2} - 1$; its dimension is equal to n. The Veronesean results from mapping P^n into P^N; the equation of the Veronesean has the form

$$z^{ij} z^{k\ell} = z^{ik} z^{j\ell}. \tag{2.108}$$

The Veronesean for $n = 2$ was defined by G. Veronese [Ver]. Since coordinates of points in P^N can be regarded as coordinates of a symmetric tensor $T^{ij} = T^{ji}$, these points represent hyperquadrics in P^n and the points on the Veronesean Υ_n represent degenerated hyperquadrics which are double hyperplanes. Therefore, since manifolds of nondegenerated imaginary and real hyperquadrics in P^n form models of symmetric Riemannian and pseudo-Riemannian spaces: imaginary hyperquadrics form a model of a Riemannian symmetric space $V^{(n+1)(n+2)/2-1}$ of the class AI in the classification of Cartan [Car12], real hyperquadrics form models of both Riemannian and pseudo-Riemannian symmetric spaces of the same dimension. Therefore Cartan called the variety in P^N representing all degenerated hyperquadrics containing Υ_n the "absolute" of these symmetric spaces [Car13, p. 909].

Like the Segrean, the Grassmannian, and the Lipschitzian, the Veronesean is an intersection of hyperquadrics.

§2.6. Linear Complexes

2.6.1. Linear Complexes and Congruences of Lines. Besides varieties consisting of points we also consider varieties consisting of lines and m-planes.

In **2.1.7** we have defined *congruences of m-planes* and *linear congruences of lines* in P^{2n-1}. In **2.1.9** we have defined *Hermitian linear congruences of lines* in $\mathbb{C}P^{2n-1}$.

Varieties of lines, respectively of m-planes, in P^n which depend on $\dim Gr^P_{n,1} - 1$, respectively $\dim Gr^P_{n,m} - 1$, parameters are called *complexes* of lines, respectively of m-planes. A complex of lines or m-planes is called *algebraic* if the condition put on its lines or m-planes is an algebraic equation in the Plücker coordinates p^{ij} for lines or the Grassmann coordinates $p^{i_0 i_1 \cdots i_m}$ for m-planes. In particular, a complex is called *linear* if this equation is linear, that is, for lines has the form

$$a_{ij} p^{ij} = 0 \tag{2.109}$$

and for m-planes has the form

$$a_{i_0 i_1 \cdots i_m} p^{i_0 i_1 \cdots i_m} = 0. \tag{2.110}$$

A variety of lines or m-planes in P^n which depends on $n-1$, respectively $n-m$, parameters is called a *congruence* of lines or m-planes. A congruence of lines or m-planes is called *algebraic* if the conditions put on its lines or m-planes are algebraic equations in the Plücker coordinates p^{ij} for lines or the Grassmann coordinates $p^{i_0 i_1 \cdots i_m}$ for m-planes.

In P^3, a system of equations of a congruence of lines consists of two equations of complexes (thus, in P^3 the lines in a congruence are the common or coincident lines in two complexes; this is the reason for naming it "congruence"). In particular, in P^3 a linear complex of lines is determined by an equation (2.109) for $i, j = 0, 1, 2, 3$ and a linear congruence of lines is determined by a system of equations $a_{ij} p^{ij} = b_{ij} p^{ij} = 0$.

Note, that according to Theorem 2.19 the space $\mathbb{H}' P^n$ admits the interpretation as Grassmann manifold $Gr^P_{2n+1,1}$ of lines in P^{2n+1}. The equation of a Hermitian hyperquadric $\sum_i \bar{x}^i x^i = 0$ in this space can be written as $\sum_i (X^{2i} Y^{2i+1} - X^{2i+1} Y^{2i}) = 0$ or $\sum_i p^{2i} p^{2i+1} = 0$ where p^{ij} are the Plücker coordinates of lines in P^{2n+1}, this variety of lines is a linear complex of lines in P^{2n+1}. The last assertion of Theorem 2.54 implies that linear complex of lines in P^{2n+1} is homeomorphic to the topological product of $\mathbb{C}P^n$ and P^{2n-1}. In particular, for $n = 1$, a linear complex of lines in P^3 is homeomorphic to the topological product of $\mathbb{C}P^1$ and P^1, that is to hyperquadric of index 2 in P^4. This hyperquadric canbe regarded as intersection of the hyperquadric Q^4 with equation (2.28) in P^5 with hyperplane and linear congruences of lines in P^3 can be interpreted as oval and ruled quadrics which are intersections of Q^4 with 3-planes.

2.6.2. Null-systems. Since the Plücker coordinates p^{ij} of a line in P^n form a skew symmetric tensor, the coefficients a_{ij} also form a skew symmetric tensor. They determine a correlation

$$u_i = a_{ij}x^j, \tag{2.111}$$

where $a_{ij} = -a_{ji}$. This correlation is called a *null-system.* Note that (2.99) shows that a correlation (2.111) in which $a_{ij} = a_{ji}$ is a polar transformation with respect to the hyperquadric (2.93), mapping any point with projective coordinates x^i to the hyperplane $u_ix^i = 0$, which is the polar hyperplane of this point.

Therefore, there is an analogy between null-systems and polar transformations with respect to hyperquadrics. Unlike hyperquadrics, which exist in P^n for any n, null-systems exist only in the spaces P^{2n-1} with odd dimension. This is because skew symmetric matrices can be nonsingular only in linear spaces L^{2n} with even dimension, while the determinant of a skew symmetric $((2n+1)\times(2n+1))$-matrix is always equal to zero.

Unlike polar transformations with respect to hyperquadrics, which map points to hyperplanes through these points only if the these points are points on the hyperquadric (in these cases the polar hyperplanes are the tangent hyperplanes to a hyperquadric), null-system maps each point in the space onto a hyperplane through this point, since if $a_{ij} = -a_{ji}$ we always have $a_{ij}x^ix^j = 0$.

If a point X in P^n (n odd) with coordinates x^i runs through an m-plane, the hyperplanes in P^n corresponding to these points in a null-system pass through an $(n-m-1)$-plane, which is said to correspond to the m-plane.

Unlike a point, an m-plane not always belongs to the $(n-m-1)$-plane corresponding to it in a null-system. The m-planes belonging to their corresponding $(n-m-1)$-planes are called *null m-planes* of the null-system, for $m = 1$ they are lines, and are thus called *null lines.*

Formula (2.109) shows that the linear complex of lines connected with a null system consists of null lines of this null-system.

2.6.3. Hermitian Null-Systems. In $\mathbb{C}P^n, \mathbb{C}'P^n, \mathbb{H}P^n$, and $\mathbb{H}'P^n$ we define *Hermitian null-systems* as correlations

$$u_i = \overline{x}^j a_{ij} \tag{2.112}$$

for which $a_{ij} = -\overline{a}_{ji}$. Note that (2.100) shows that a correlation (2.112) for which $a_{ij} = \overline{a}_{ji}$ is a polar transformation with respect to the Hermitian hyperquadric (2.97). Unlike null-systems in P^n, Hermitian null-systems in $\mathbb{C}P^n, \mathbb{C}'P^n, \mathbb{H}P^n$, and $\mathbb{H}'P^n$ exist for all n, and not every point belongs to the hyperplane corresponding to it in a Hermitian null-system. Points that do belong to their corresponding hyperplanes are called *null points*. The null lines and null m-planes of a Hermitian null-system are defined as for null-systems in P^n.

Theorem 2.58. *In $\mathbb{C}P^n$ Hermitian null-systems coincide with polar transformations with respect to Hermitian hyperquadrics.*

In $\mathbb{H}P^n$, a Hermitian null-system can be reduced to the form

$$u_i = \overline{x}^i i. \tag{2.113}$$

PROOF. The first assertion follows from the fact that if the matrix $A = (a_{ij})$ is skew Hermitian ($a_{ij} = -\overline{a}_{ji}$), the matrix $B = (b_{ij})$ with entries $b_{ij} = a_{ij}i$ is Hermitian symmetric ($b_{ij} = \overline{b}_{ji}$).

The second assertion can be proved by representing the algebra $\mathbb{H}_{n+1}$ by matrices in the algebra $\mathbb{C}_{2n+2}$ and by the reducing the matrices in this algebra to diagonal form. ■

2.6.4. Null m-Planes of a Null-System Let $N^P_{n,m}$ be the manifold of null m-planes of a null-system in P^n and let $\mathbb{C}\overline{N}^P_{n,m}, \mathbb{C}'\overline{N}^P_{n,m}, \mathbb{H}\overline{N}^P_{n,m}$, or $\mathbb{H}'\overline{N}^P_{n,m}$ be the manifold of null m-planes of a Hermitian null-system in $\mathbb{C}P^n, \mathbb{C}'P^n, \mathbb{H}P^n$, or $\mathbb{H}'P^n$.

As for m-planar generators of a hyperquadric in P^n, and of a Hermitian hyperquadric in $\mathbb{C}P^n, \mathbb{C}'P^n, \mathbb{H}P^n$, $\mathbb{H}'P^n$, the projective matrix coordinates of a null m-plane of a null-system in these spaces satisfy the conditions (2.103) and (2.104), where A is the matrix (a_{ij}) of the null-system.

Analogously to Theorem 2.56 we can prove

Theorem 2.59. *The real dimensions dim $N^P_{n,m}$, dim $\mathbb{C}\overline{N}^P_{n,m}$, dim $\mathbb{C}'\overline{N}^P_{n,m}$, dim $\mathbb{H}\overline{N}^P_{n,m}$, and dim $\mathbb{H}'\overline{N}^P_{n,m}$ of $N^P_{n,m}$, $\mathbb{C}\overline{N}^P_{n,m}$, $\mathbb{C}'\overline{N}^P_{n,m}$, $\mathbb{H}\overline{N}^P_{n,m}$, and $\mathbb{H}\overline{N}^P_{n,m}$ are equal to*

$$\dim N^P_{n,m} = \frac{(2n-3m)(m+1)}{2}, \tag{2.114}$$

$$\dim \mathbb{C}\overline{N}^P_{n,m} = \dim \mathbb{C}'\overline{N}^P_{n,m} = \frac{(4n-6m-2)(m+1)}{2}, \tag{2.115}$$

$$\dim \mathbb{H}\overline{N}^P_{n,m} = \dim \mathbb{H}'\overline{N}^P_{n,m} = \frac{(8n-12m-6)(m+1)}{2}. \tag{2.116}$$

PROOF. In this case the left hand sides of formulas (2.103) and (2.104) are $((m+1)\times(m+1))$-matrices which are skew symmetric and skew Hermitian, respectively. These matrices depend on $S = r\frac{m(m+1)}{2} + (r-1)(m+1)$ real parameters. Therefore, the projective matrix coordinates of a null m-plane of a null-system are related by S real conditions, and, since, as we have seen in Theorem 2.42, the dimension $\dim \mathbb{A}Gr^P_{n,m}$ of the Grassmann manifold $\mathbb{A}Gr^P_{n,m}$ is equal to $r(m+1)(n-m)$, the dimension of the manifold of null m-planes of a null-system is equal to

$$r(m+1)(n-m) - S = r(m+1)(n-m) - r\frac{m(m+1)}{2} - (r-1)(m+1) =$$

$$\frac{(2rn - 3rm - 2r + 2)(m + 1)}{2}.$$

For $r = 1, 2, 4$ this coincides with the numbers (2.114), (2.115), and (2.116), respectively.

§2.7. Projective Configurations

2.7.1. Plane Configurations. A *projective plane configuration* is a set of n points and g lines in the projective plane $\mathbb{A}P^2$ such that each point is incident with i lines and each line is incident which k points. These configurations are characterized by the *De Vries symbols* $(n_i g_k)$ introduced by J. De Vries [DeV1]. The numbers n, g, i, k are related by $ni = gk$. Each projective plane configuration $(n_i g_k)$ has a *dual configuration* $(n_k g_i)$, which corresponds to it under the duality principle in $\mathbb{A}P^2$.

If a configuration coincides with the dual, that is, $n = g$ and $i = k$, it is called an *autodual configuration.* Such configurations are characterized by the simpler *Reye symbols* (n_i) introduced by T. Reye [Re].

The points and lines which participate in the projective theorems of Pappus and Desargues (see **2.2.4**) form plane configurations, called the *configurations of Pappus* and *Desargues*, respectively. These configurations are the autodual configurations (9_3) and (10_3), respectively.

The four vertices of a quadrangle and the six lines of this quadrangle, i.e. the sides and diagonals, form the configuration $(4_3 6_2)$ called the *complete quadrangle.* The dual configuration $(6_2 4_3)$ which consists of the four sides of a quadrilateral and the six points, at each of which one line meets the three other lines, is called the *complete quadrilateral.* Note that the complete quadrilateral $ABCDEF$ with sides ABE, CDE, ACF, and BDF has three diagonals, AD, BC and EF, and that the diagonals AD and BC meet the diagonal EF at points G and H for which $\overline{EF, GH} = -1$, that is, these four points form a harmonic quadruple.

Let us mention the *Hesse configuration* $(9_4 12_3)$, whose points are the 9 points of inflection of a plane cubic in the complex plane $\mathbb{C}P^2$, any three of which lie on a single line in the set of 12 lines [Hel], and the configuration $(378_3 28_{27})$, determined by a plane quartic with 28 double tangents such that each double tangent meets the other 27 double tangents at 27 points. This configuration consists of the 28 double tangents and their points of intersection.

2.7.2. Space Configurations. In $\mathbb{A}P^3$ there are *projective space configurations*, that is, sets of n points, g lines, and p planes such that each point is incident with i lines and j planes, each line is incident which k planes and l points, and each plane is incident with m points and n lines.

These configurations are characterized by the *De Vries symbols* $(n_i^j g_k^l p_m^n)$ introduced by J. De Vries in [DeV2]. If such a configuration is autodual, that is, $n = p$, $i = n$, $j = m$, $k = l$, it is characterized by the *Reye symbol* $(n_i g_k)$. The vertices, (extended) edges, and (extended) faces of a regular polyhedron in R^3 form a space configuration. A tetrahedron determines the autodual configuration $(4_3 6_2)$, a cube and an octahedron determine $(8_3^3 12_2^2 6_4^4)$ and $(6_4^4 12_2^2 8_3^3)$, a dodecahedron and an icosahedron determine $(20_3^3 30_2^2 12_5^5)$ and $(12_5^5 30_2^2 20_3^3)$.

An autodual space configuration $(n_i g_k)$ gives rise to a new autodual space configuration by applying a null-system; this new configuration is $(2n_{i+1} 2g_k)$. Repeated application of this process to the configuration $(4_3 6_2)$ gives the configurations $(8_4 12_2)$, $(16_5 24_2)$,..., $(2^n)_{n+1}$ $(3 \cdot 2^{n-1})_2)$.

Let us mention the autodual *Reye configuration* $(12_6 16_3)$, whose points are the 8 vertices of a cube in R^3, its center of symmetry and the 4 points at infinity of its (extended) edges, whose planes are the 6 (extended) faces of the cube and its 6 diagonal planes through pairs of opposite edges, and whose lines are the 12 (extended) edges and its 4 main diagonals. We also mention the autodual configuration $(135_2 27_{10})$ determined by a smooth cubic surface with 27 rectilinear generators. The lines of this configuration are these 27 rectilinear generators, its points are the points of intersection of these lines.

2.7.3. Regular Configurations. A projective plane (space) configuration in AP^2 ($\mathbb{A}P^3$) is called *regular* if it admits a group of collineations in $\mathbb{A}P^2$ ($\mathbb{A}P^3$). The configurations of Pappus, of a complete quadrangle and of a complete quadrilateral, and also the configuration determined by the quartic with 28 double tangents are regular configurations in P^2, the Hesse configuration is a regular configuration in $\mathbb{C}P^2$.

The configurations determined by regular polyhedra in R^3, the configurations obtained from the configuration determined by a tetrahedron by applying null-systems, the Reye configuration, and the configuration determined by the smooth cubic surface with 27 rectilinear generators are regular configurations in P^3.

The groups of collineations admitted by configurations determined by regular polyhedra, by the smooth cubic surface with 27 rectilinear generators, and by the quartic with 28 double tangents are isomorphic to the following finite groups of Euclidean rotations: $A_3, B_3 = C_3, H_3, E_6$, and E_7, respectively.

Projective configurations in $\mathbb{A}P^n$ are similarly defined. They include the configurations determined by regular polyhedra in R^n. The groups of collineations admitted by these configurations are isomorphic to the finite groups of Euclidean rotations $A_n, B_n = C_n, F_4$, and H_4.

§2.8 Symmetry and Parabolic Figures

2.8.1. Symmetry Figures in Affine Spaces. A geometric figure in a homogeneous space is called a *symmetry figure* if it is determined by an involutive element in the fundamental group of this space; this involutive transformation is called a *symmetry* in this figure. In most cases symmetry figures are sets of points which are fixed under symmetries in these figures.

Thus, symmetry figures in $\mathbb{A}E^n$ are determined by involutive affine transformations.

Theorem 2.60. *Symmetry figures in $E^n, \mathbb{C}E^n$, and $\mathbb{H}E^n$ are points and m-planes $(m = 1, 2, \ldots, n-1)$ with $(n-m)$-directions, and, for $\mathbb{C}E^n$ and $\mathbb{H}E^n$, n-chains and complex n-chains, respectively.*

PROOF. Since a transformation $x' = f(x)$ is involutive if and only if its square $x' = f(f(x))$ is the identity, $x' = x$, and since the square of the affine transformation $x' = Ax + b$ is $x' = A^2x + Ab + b$, the affine transformation $x' = Ax + b$ can be the identity if and only if $b = 0$ and $A^2 = I$. Since the matrix A can satisfy the condition $A^2 = I$ only if A is a diagonal matrix E with diagonal entries ± 1, the involutive affine transformations in E^n can be written as

$$'x^i = -x^i \tag{2.117}$$

or

$$'x^a = -x^a, \quad 'x^u = x^u. \tag{2.118}$$

The transformation (2.117), which can also be written in vector form as $x' = -x$, is a reflection in a *point*.

The transformation (2.118) is a symmetry in an m-plane $x^u = 0$ in the direction of an $(n-m)$-plane $x^a = 0$, which we shall call an $(n-m)$-*direction*.

In $\mathbb{C}E^n$, besides the affine transformations $x' = Ax + b$, there are also affine transformations $x' = A\overline{x} + b$, whose squares are $x' = A\overline{A}x + A\overline{b} + b$. These transformations can be the identity only if $b = 0$ and $A\overline{A} = I$. Since A can satisfy the condition $A\overline{A} = I$ only if A is a diagonal matrix with as entries complex numbers $c_i = e^{i\varphi_i}$, the involutive affine transformation $x' = A\overline{x}$ can be written as $'x^i = c_i\overline{x}^i$. The change of variables $x^{i'} = \sqrt{c_i}x^i$ reduces this transformation to the form $x^{i'} = \overline{x}^{i'}$, so this involutive transformation can be written as

$$'x^i = \overline{x}^i. \tag{2.119}$$

The transformation (2.119), which can also be written in the vector form $x' = -\overline{x}$, is a reflection in the set of points which satisfy the condition $x' = \overline{x}^i$, that is, a set of points with real coordinates. Such a set of points in $\mathbb{C}E^n$, and the sets obtained from such a set by affine transformations, are called n-*chains*.

In $\mathbb{H}E^n$, besides $x' = Ex$, the affine transformations $x' = Ax$ with matrices A satisfying $A^2 = I$ can be reduced to the form

$$'x^i = ix^i i^{-1}. \tag{2.120}$$

Since the expression $i(x_0 + ix_1 + jx_2 + kx_3)i^{-1}$ is equal to $x_0 + ix_1 - jx_2 - x_3$, the transformation (2.120), which can also be written in the vector form $x' = ixi^{-1}$, is a reflection in the set of points which satisfy the condition $x^i = ix^i i^{-1}$, that is, a set of points with complex coordinates. Such a set of points in $\mathbb{H}E^n$, and the sets obtained from such a set by affine transformations, are called *complex n-chains*. ∎

We can similarly define symmetry figures in the spaces $\mathbb{C}'E^n$, $\mathbb{H}'E^n$, $\mathbb{C}^0E^n$, $\mathbb{H}^0E^n$, and $\mathbb{H}'^0E^n$.

Since the space $\mathbb{C}'E^n$ admits an interpretation as the manifold of pairs of points in E^n (see **2.1.9**), an involutive affine transformation $x' = Ax + b$ in $\mathbb{C}'E^n$ can be interpreted as an independent involutive affine transformation of both these points or as an involutive affine transformation of one of these points. An involutive affine transformation $x' = A\overline{x} + b$ in $\mathbb{C}'E^n$ can be interpreted as permutation of these points or as the product of this by an independent involutive affine transformation of both these points or by an involutive affine transformation of one of these points.

2.8.2. Symmetry Figures in Projective Spaces. The symmetry figures in the space $\mathbb{A}P^n$ are determined by involutive collineations of this space.

Theorem 2.61. *The symmetry figures in P^n are m-pairs $\left(m = 0, 1, \ldots, \left[\frac{n-1}{2}\right]\right)$ and, for n odd, elliptic linear congruences of lines.*

The symmetry figures in $\mathbb{C}P^n$ are m-pairs $\left(m = 0, 1, \ldots, \left[\frac{n-1}{2}\right]\right)$ and n-chains, and, for n odd, Hermitian linear congruences of lines.

The symmetry figures in $\mathbb{H}P^n$ are m-pairs $\left(m = 0, 1, \ldots, \left[\frac{n-1}{2}\right]\right)$ and complex n-chains.

PROOF. Since the identity in P^n has the form $x' = x\lambda$, the square of a collineation $x' = Ax$ of P^n is $x' = A^2x$, so, the collineation $x' = A^2x$ can be the identity only if $A^2 = I\lambda$. Since the matrix A can satisfy the condition $A^2 = I\lambda$ only if A is a diagonal matrix of the form $E\sqrt{\lambda}$, with diagonal entries equal to $\pm\sqrt{\lambda}$, or if it can be reduced to imaginary diagonal form with diagonal entries $\pm i\sqrt{\lambda}$. Since the matrix A is real, in the last case the number of diagonal entries $i\sqrt{\lambda}$ is equal to the number of diagonal entries $-i\sqrt{\lambda}$, and A can be reduced to the real form $J\sqrt{\lambda}$, where the matrix J has the form (0.62). Therefore, since the group of collineations in P^n is locally isomorphic to the group SL_{n+1} and their determinants must be equal to ± 1, the involutive collineations in P^n can be reduced to the forms (2.118) and (2.6).

The collineation (2.118) represents a symmetry in the *m-pair* consisting of the m-plane $x^u = 0$ and the $(n-m-1)$-plane $x^a = 0$. This collineation maps a point X to a point X' defined as follows: a unique line passes through X and meets both the m-plane and the $(n-m-1)$-plane of the m-pair at points P and Q, respectively; X' is the point on this line which is the fourth harmonic point for the points P, Q, X, that is, such that the cross-ratio $\overline{XX', PQ} = -1$.

The collineation (2.6), which can exist for odd n only, has no real fixed points and preserves an elliptic linear congruence of lines.

We define n-chains in $\mathbb{C}P^n$, respectively complex n-chains in $\mathbb{H}P^n$, as sets of points in these spaces obtained by collineations from a set of points with real coordinates, respectively with coordinates of form $x_0 + ix_1$.

Analogously to the case of P^n and to Theorem 2.60 we can prove that the symmetry figures in $\mathbb{C}P^n$ are m-pairs, elliptic Hermitian linear congruences of lines and n-chains, and that the symmetry figures in $\mathbb{H}P^n$ are m-pairs and complex n-chains. ■

We can similarly determine the symmetry figures in the spaces $\mathbb{C}'P^n$, $\mathbb{H}'P^n$, $\mathbb{C}^0P^n$, $\mathbb{H}^0P^n$, and $\mathbb{H}'^0P^n$.

Since the space $\mathbb{C}'P^n$ admits an interpretation as the manifold of pairs of points in P^n (see **2.2.6**), an involutive collineation $x' = Ax$ in $\mathbb{C}'P^n$ can be interpreted as two independent involutive collineations of both these points or as an involutive collineation of one of these points. An involutive collineation $x' = A\bar{x}$ of $\mathbb{C}'P^n$ can be interpreted as a permutation of these points or as the product of this by two independent involutive collineations of both these points or by an involutive collineation of one of these points.

Since the space $\mathbb{H}'P^n$, like the space $\mathbb{R}_2P^n$, admits an interpretation by lines in P^{2n+1} (see **2.2.6**), an involutive collineation in $\mathbb{H}'P^n$ can be interpreted as an involutive collineation in P^{2n+1}. In $\mathbb{H}'P^n$ there are, next to complex n-chains, also split complex n-chains, with symmetries which can be reduced to the form $'x^i = ex^ie$.

2.8.3. Cosymmetry Figures in Projective Spaces. In $\mathbb{A}P^n$, next to symmetry figures there are also *cosymmetry figures*, which are determined by involutive correlations in this space.

Theorem 2.62. *The cosymmetry figures in P^n are imaginary and real hyperquadrics of index ℓ $\left(\ell = 1, 2, \ldots, \left[\frac{n+1}{2}\right]\right)$, and, for n odd, linear complexes of lines.*

The cosymmetry figures in $\mathbb{C}P^n$ are hyperquadrics, imaginary and real Hermitian hyperquadrics of index ℓ $\left(\ell = 1, 2, \ldots, \left[\frac{n+1}{2}\right]\right)$, and, for n odd, linear complexes of lines and Hermitian linear complexes.

The cosymmetry figures in $\mathbb{H}P^n$ are imaginary and real Hermitian hyperquadrics of index ℓ $\left(\ell = 1, 2, \ldots, \left[\frac{n+1}{2}\right]\right)$, and Hermitian linear complexes.

PROOF. Since the identity in P^n has the form $x' = x\lambda$, a correlation $u = xA$ in P^n maps the hyperplane represented by the covector u to the point represented by the vector $x' = A^{-1}u$. The square of this correlation is $x' = A^{-1}(xA) = A^{-1}A^{\top}x$, so the correlation $u = xA$ can be involutive only if $A^{-1}A^{\top} = I\lambda$. Since the condition $A^{-1}A^{\top} = I\lambda$ is equivalent to the condition $A^{\top} = A\lambda$, hence the condition $A = A\lambda^2$, and the equalities $\lambda^2 = 1$ and $\lambda = \pm 1$ follow. So, a correlation $u = xA$ in P^n can be involutive only if the matrix A satisfies the condition $A = \pm A^{\top}$, that is, only if this correlation is a *polar transformation with respect to a hyperquadric* or a *null-system*, which is a symmetry in a *linear complex of lines*.

The same condition holds for a correlation $u = xA$ in $\mathbb{C}P^n$ and in $\mathbb{C}'P^n$ to be involutive.

Analogously, a correlation $u = \overline{x}A$ in $\mathbb{C}P^n, \mathbb{C}'P^n, \mathbb{H}P^n$, or $\mathbb{H}'P^n$ maps the hyperplane represented by the covector u to the point represented by the vector $x' = A^{-1}\overline{u}$. The square of this correlation is $x' = A^{-1}(\overline{x\overline{A}}) = A^{-1}\overline{A}^{\top}x$, so the correlation $u = \overline{x}A$ can be the identity only if $A^{-1}\overline{A}^{\top} = I\lambda$. Since the condition $A^{-1}\overline{A}^{\top} = I\lambda$ is equivalent to the condition $A^{\top} = \overline{A}\lambda$, hence follows the condition $A = \overline{\lambda}A\lambda$. In the case of a noncommutative algebra the equality $\lambda = \pm 1$ follows, while in the case of a commutative algebra we have $\overline{\lambda}\lambda = 1$, and so $|\lambda| = 1$. However, the condition $A^{\top} = \overline{A}\lambda$ implies that under multiplication of the matrix A of a correlation by an element κ in the commutative algebra the element λ is multiplied by $\kappa\overline{\kappa}^{-1}$, and $|\kappa\overline{\kappa}^{-1}| = 1$. Therefore, we can always choose κ such that λ is a real number, and then the equality $\overline{\lambda}\lambda = 1$ leads to $\lambda = \pm 1$. Therefore, in all spaces $\mathbb{C}P^n, \mathbb{C}'P^n, \mathbb{H}P^n$, and $\mathbb{H}'P^n$ a correlation $u = \overline{x}A$ can be involutive only if the matrix A satisfies $A = \pm\overline{A}^{\top}$, that is, only if this correlation is a *polar transformation with respect to a Hermitian hyperquadric* or a *Hermitian null-system*, which is a symmetry in a *Hermitian linear complex*. ■

We can similarly determine the cosymmetry figures in the spaces $\mathbb{C}'P^n$, $\mathbb{H}'P^n$, $\mathbb{C}^0P^n$, $\mathbb{H}^0P^n$, and $\mathbb{H}'^0P^n$.

Since the space $\mathbb{C}'P^n$ admits an interpretation by pairs of points in P^n, an involutive correlation $u = xA$ in $\mathbb{C}'P^n$ can be interpreted as two independent involutive correlations of both these points, and a correlation $u = \overline{x}A$ of $\mathbb{C}'P^n$ can be interpreted as the product of the permutation of these points by two independent involutive correlations of both these points.

Since the space $\mathbb{H}'P^n$ admits an interpretation by lines of P^{2n+1}, an involutive correlation in $\mathbb{H}'P^n$ can be interpreted as an involutive correlation in P^{2n+1}.

Note that in $\mathbb{C}^0 P^n$ there are both imaginary and real hyperquadrics as well as Hermitian hyperquadrics of index ℓ $\left(\ell = 1, 2, \ldots, \left[\frac{n+1}{2}\right]\right)$, and, for n odd, both linear and Hermitian linear complexes of lines.

2.8.4. Fundamental and Parabolic Figures in Projective Spaces. In **1.6.4** we have defined fundamental α_i-figures and parabolic $(\alpha_{i_1}, \alpha_{i_2}, \ldots, \alpha_{i_k})$-figures in homogeneous spaces whose fundamental groups are simple Lie groups.

Theorem 2.63. *The fundamental α_i-figures in the space P^n are $(i-1)$-planes (points for $i = 1$, lines for $i = 2$, planes for $i = 3$, hyperplanes for $i = n$). The parabolic $(\alpha_{i_1}, \alpha_{i_2}, \ldots, \alpha_{i_k})$-figures in this space are flags consisting of incident $(i_1 - 1)$-planes, $(i_2 - 1)$-planes $, \ldots, (i_k - 1)$-planes.*

The fundamental α_i-figures and the parabolic $(\alpha_{i_1}, \alpha_{i_2}, \ldots, \alpha_{i_k})$-figures in the space $\mathbb{C}P^n$ are the same as in P^n.

The fundamental α_{2i}-figures in the space $\mathbb{H}P^n$ are real $(i-1)$-planes (points for $i = 1$, lines for $i = 2$, planes for $i = 3$, hyperplanes for $i = n$). The fundamental α_{2i-1}-figures in $\mathbb{H}P^n$ are imaginary $(i - \frac{1}{2})$-planes (halfpoints for $i = 1$, halflines for $i = 2$, halfplanes for $i = 3$, halfhyperplanes for $i = n$, halfspaces for $i = n+1$). The parabolic $(\alpha_{i_1}, \alpha_{i_2}, \ldots, \alpha_{i_k})$-figures in this space are flags consisting of incident $(i_1 - 1)$-planes, $(i_2 - 1)$-planes, $, \ldots, (i_k - 1)$-planes, where $i_1, i_2, \ldots, i_k$ can be integers or half-integers.

PROOF. The first assertion was proved in **1.6.4**, the second is proved analogously. Since the space $\mathbb{H}'P^n$ admits an interpretation as the manifold of lines in P^{2n+1}, the $(i-1)$-planes in P^{2n+1} are interpreted as $(2i-1)$-planes in P^{2n+1}. Therefore, the fundamental α_{2i-1}-figures in $\mathbb{H}'P^n$, which are interpreteted as $(i-\frac{1}{2})$-planes in P^{2n+1}, are called $(i - \frac{1}{2})$-planes. Therefore, the imaginary fundamental α_{2i-1}-figures in $\mathbb{H}P^n$ are also called $(i - \frac{1}{2})$-planes. ■

Note that the Satake diagram for the group of collineations in P^n, which is locally isomorphic to the group SL_{n+1}, coincides with the Dynkin diagram for the simple Lie group A_n (Fig. 1.3, A_n).

The Satake diagram for the group of collineations in $\mathbb{C}P^n$, which is locally isomorphic to the group $\mathbb{C}SL_{n+1}$ regarded as a real Lie group, consists of two identical Dynkin diagrams of the simple Lie group A_n, whose dots corresponding to the simple root vectors α_i are joined by two-sided arrows.

The Satake diagram for the real Lie group of collineations in $\mathbb{H}P^{\frac{n-1}{2}}$, which is locally isomorphic to the group $\mathbb{H}SL_{\frac{n+1}{2}}$, has the form depicted in Fig. 1.9, AII.

Theorem 1.39 implies that if the tangent Lie algebra $\mathbb{G}$ of simple Lie group G has the form $\mathbb{G} = \mathbb{G}_{-1} \oplus \mathbb{G}_0 \oplus +\mathbb{G}_1$ two fundamental figures in the space with fundamental group G with isometric stabilizers determine a symmetry figure. In P^n two fundamental figures with isomorphic stabilizers are m-plane and $n-m-1$-plane, they determine an m-pair, which is a symmetry figure in P^n.

§2.9. Finite Geometries

2.9.1. Finite Affine Geometries. If we replace in the definitions of the spaces $E^n, \mathbb{C}E^n$, and $\mathbb{H}E^n$ the fields $\mathbb{R}$ and $\mathbb{C}$ and the skew field $\mathbb{H}$ by a Galois field $\mathbb{F}_q$, we obtain the *finite affine space* $\mathbb{F}_q E^n$.

Lines, m-planes, hyperplanes, and affine transformations in $\mathbb{F}_q E^n$ are defined analogously to lines, m-planes, hyperplanes, and affine transformations in $E^n, \mathbb{C}E^n$, and $\mathbb{H}E^n$.

Let $\mathbb{F}_q Gr^E_{n,m}$ be the manifold of m-planes in $\mathbb{F}_q E^n$ (lines for $m = 1$, hyperplanes for $m = n-1$). The number of elements in the field $\mathbb{F}_q$ and the cardinal number of points, lines, m-planes, or hyperplanes in $\mathbb{F}_q E^n$, and of elements in the fundamental group of this space, is called the *cardinality* of this field, respectively of this space, of the manifold of lines, m-planes or hyperplanes, and of the group. We denote these cardinalities by card $\mathbb{F}_q$, card $\mathbb{F}_q E^n$, card $\mathbb{F}_q Gr^E_{n,m}$, etc. Note that

$$\text{card } \mathbb{F}_q = q. \tag{2.121}$$

Theorem 2.64. *The cardinality of the space $\mathbb{F}_q E^n$ is equal to*

$$\text{card } \mathbb{F}_q E^n = q^n. \tag{2.122}$$

The cardinality of the manifold $\mathbb{F}_q Gr^E_{n,n-1}$ of hyperplanes of this space is equal to

$$\text{card } \mathbb{F}_q Gr^E_{n,n-1} = q^n + q^{n-1} + \cdots + q^2 + q = \frac{q(q^n - 1)}{q-1}. \tag{2.123}$$

PROOF. Since each point in $\mathbb{F}_q E^n$ is determined by n coordinates x^i and card $\mathbb{F}_q = q$, the cardinality of $\mathbb{F}_q$ is equal to q^n.

Since each hyperplane in $\mathbb{F}_q E^n$ is determined by equation (2.2), the manifold $\mathbb{F}_q Gr^E_{n,n-1}$ consists of $n+1$ submanifolds of hyperplanes whose equations (2.2):

1) have coefficient $u_0 \neq 0$;

2) have coefficients $u_0 = 0$, $u_1 \neq 0$;

3) have coefficients $u_0 = u_1 = 0$, $u_2 \neq 0, \ldots$;

n) have coefficients $u_0 = u_1 = \cdots = u_{n-1} = 0$, $u_n \neq 0$. The equations (2.2) for the hyperplanes of the i-th submanifold can be normalized by the condition $u_{i+1} = 1$; therefore, the cardinality of the 1-st submanifold is q^n, the cardinality of the 2-nd submanifold is $q^{n-1}, \ldots$, the cardinality of the n-th submanifold is q, and the cardinality of the whole manifold $\mathbb{F}_q Gr^E_{n,n-1}$ is equal to (2.123). ∎

Theorem 2.65. *The points and lines in $\mathbb{F}_q E^2$ form the plane configuration $((q^2)_{q+1}(q^2+q)_q)$.*

PROOF. Theorem 2.64 implies that card $\mathbb{F}_q E^2 = q^2$ and card $\mathbb{F}_q Gr^E_{2,1} = q^2 + q$. Each line in $\mathbb{F}_q E^2$ contains card $\mathbb{F}_q E^1 =$ card $\mathbb{F}_q = q$ points, and the number of lines through each point in $\mathbb{F}_q E^2$ is equal to $q+1$. ∎

In particular, for $q = 2, 3$ these configurations are $(4_3 6_2)$ and $(9_4 12_3)$, respectively. Therefore, the affine planes $\mathbb{F}_2 E^2$ and $\mathbb{F}_3 E^2$ admit interpretations as the complete quadrangle and Hesse configurations (see **2.7.1**).

2.9.2. Finite Projective Geometries. If we replace in the definitions of the spaces $P^n, \mathbb{C}P^n$, and $\mathbb{H}P^n$ the fields $\mathbb{R}$ and $\mathbb{C}$ and the skew field $\mathbb{H}$ by a Galois field $\mathbb{F}_q$, we obtain the *finite projective space* $\mathbb{F}_q P^n$.

Lines, m-planes, hyperplanes, collineations, and correlations in $\mathbb{F}_q P^n$ are defined analogously to lines, m-planes, hyperplanes, collineations, and correlations in $P^n, \mathbb{C}P^n$, and $\mathbb{H}P^n$.

Let $\mathbb{F}_q Gr^P_{n,m}$ be the manifold of m-planes (lines for $m = 1$, hyperplanes for $m = n-1$) in $\mathbb{F}_q P^n$. We again have corresponding cardinalities card $\mathbb{F}_q P^n$ and card $\mathbb{F}_q Gr^P_{n,m}$ of $\mathbb{F}_q P^n$ and $\mathbb{F}_q Gr^P_{n,m}$, respectively:

Theorem 2.66. *The cardinality of the space $\mathbb{F}_q P^n$ is equal to*

$$\text{card}\, \mathbb{F}_q P^n = q^n + q^{n-1} + q^{n-2} + \dots + q + 1 = \frac{q^{n+1}-1}{q-1}. \tag{2.124}$$

The cardinality of the manifold $\mathbb{F}_q Gr^P_{n,m}$ is equal to

$$\text{card}\, \mathbb{F}_q Gr^P_{n,m} = \frac{(q^{n+1}-1)(q^n-1)\dots(q^{n-m+1}-1)}{(q^{m+1}-1)(q^m-1)\dots(q-1)}. \tag{2.125}$$

PROOF. Formula (2.124) follows from the cellular decomposition of the space $\mathbb{F}P^n$ over an arbitrary field or division algebra $\mathbb{F}$:

$$\mathbb{F}P^n = \mathbb{F}E^n \cup \mathbb{F}E^{n-1} \cup \mathbb{F}E^{n-2} \cup \dots \cup \mathbb{F}E^1 \cup \mathbb{F}E^0. \tag{2.126}$$

Formula (2.125) follows from the cellular decomposition of the manifold $\mathbb{F}Gr^P_{n,m}$ of m-planes of this space corresponding to the cellular polynomial (2.24). Note that the right hand side of (2.125) can be written as $C(q)$. ∎

Since the manifold $\mathbb{F}_q Gr^E_{n,m}$ can be regarded as the difference of $\mathbb{F}_q Gr^P_{n,m}$ and $\mathbb{F}_q Gr^P_{n-1,m}$, the cardinality of this manifold is equal to

$$\begin{aligned}\text{card}\, \mathbb{F}_q Gr^E_{n,m} =& \frac{(q^{n+1}-1)(q^n-1)\dots(q^{n-m+1}-1)}{(q^{m+1}-1)(q^m-1)\dots(q-1)} \\ &- \frac{(q^n-1)(q^{n-1}-1)\dots(q^{n-m}-1)}{(q^{m+1}-1)(q^m-1)\dots(q-1)}.\end{aligned} \tag{2.127}$$

For $m = n-1$ formula (2.127) becomes formula (2.123).

The geometry of $\mathbb{F}_q E^n$ and $\mathbb{F}_q P^n$ was founded by O. Veblen and W. H. Bussey [VB].

Theorem 2.67. *The points and lines in $\mathbb{F}_q P^2$ form the autodual plane configuration $(q^2+q+1)_{q+1}$.*

The points, lines, and planes in $\mathbb{F}_q P^3$ form the autodual space configuration $((q^3+q^2+q+1)_{q^2+q+1}(q^4+q^3+2q^2+q+1)_{q+1})$.

PROOF. Theorem 2.65 implies that card $\mathbb{F}_q P^2 =$ card $\mathbb{F}_q Gr^P_{2,1} = q^2+q+1$, card $\mathbb{F}_q P^3 = q^3+q^2+q+1$, card $\mathbb{F}_q Gr^P_{3,1} = q^4+q^3+2q^2+q+1$. Each line in $\mathbb{F}_q P^2$ or $\mathbb{F}_q P^3$ contains card $\mathbb{F}_q P^1 = q+1$ points. This is also the number of lines through each point in $\mathbb{F}_q P^2$ and of planes through each line in $\mathbb{F}_q P^3$. Each plane in $\mathbb{F}_q P^3$ contains card $\mathbb{F}_q P^2 = q^2+q+1$ points. This is also the number of planes through each point in $\mathbb{F}_q P^3$. ∎

In particular, for $q = 2, 3$ these autodual plane configurations are (7_3) and (13_4), respectively. The configuration (7_3), which cannot be realized in the real plane P^2, is given in Fig. 1.1 and was used by H. Freudenthal to represent the law of multiplication of basis elements in the algebra $\mathbb{O}$ of octonions (see **1.3.1**). Here, points in $\mathbb{F}_q P^2$ are represented by dots, which also represent the basis elements i, j, k, ℓ, p, q, r in $\mathbb{O}$, and lines are represented by the 3 sides, the 3 altitudes in the triangle, and the circle inscribed in it.

For $q = 2$ and 3 these autodual space configurations are $(15_7 35_3)$ and $(40_{13} 130_4)$, respectively. The configuration $(15_7 35_3)$ can be interpreted as the famous problem about the 15 schoolgirls. In this problem it is required to make a week schedule for the walks of 15 schoolgirls in groups of three such that during the week no schoolgirl is more than once in the same group. Girls are represented by the points in $\mathbb{F}_2 P^3$, their groups of three by the lines.

For the history of this problem see [Ro4, p. 421].

Besides the finite projective spaces $\mathbb{F}_q P^n$ there are other finite projective spaces which have the same incidence structure as P^n and $\mathbb{C}P^n$. For instance (see [Rys]), finite projective planes with $q + 1$ points on each line exist for $q = 2, 3, 4, 5, 7, 8, 9, 11, 13, 16, 17, 19, 23$, and 25, and do not exist for $q = 6, 14, 21$, and 22, while the question on the existence of planes for $q = 10$ and 12 is open (the values of q which determine fields $\mathbb{F}_q$ are powers p^k of a prime integers p).

2.9.3. Projective Spaces over Finite Rings. Finite projective spaces can also be defined over finite rings. Consider the ring $\mathbb{F}_q \oplus \mathbb{F}_q$, whose elements can be written in the form $x = x_0 + ex_1$ with x_0 and x_1 elements in $\mathbb{F}_q$ and $e^2 = +1$. This ring can be called a "split complex extension of $\mathbb{F}_q$".

Consider also the ring whose elements can be written in the form $x = x_0 + \epsilon x_1$ with x_0 and x_1 elements in $\mathbb{F}_q$ and $\epsilon^2 = 0$. This ring can be called a "dual extension of $\mathbb{F}_q$". We denote this ring by $(\mathbb{C}^0)_q$.

The role of complex extension of $\mathbb{F}_q$ is played by the field $\mathbb{F}_{q^2}$.

The projective spaces $(\mathbb{F}_q \oplus \mathbb{F}_q)P^n$ and $(\mathbb{C}^0)_q P^n$ were investigated by N. I. Kharitonova [Kha] and S. B. Kapralova [Kap] (see also [RoKK]). We have:

$$\operatorname{card} \mathbb{F}_{q^2} Gr^P_{n,m} = \frac{(q^{n+1}-1)(q^{n+1}+1)(q^n-1)(q^n+1)\dots(q^{n-m+1}-1)(q^{n-m+1}+1)}{(q^{m+1}-1)(q^{m+1}+1)(q^m-1)(q^m+1)\dots(q-1)(q+1)}, \tag{2.128}$$

$$\operatorname{card} (\mathbb{F}_q \oplus \mathbb{F}_q) Gr^P_{n,m} = \frac{(q^{n+1}-1)^2(q^n-1)^2\dots(q^{n-m+1}-1)^2}{(q^{m+1}-1)^2(q^m-1)^2\dots(q-1)^2}, \tag{2.129}$$

$$\operatorname{card} (\mathbb{C}^0)_q Gr^P_{n,m} = \frac{(q^{n+1}-1)q^{n+1}(q^n-1)q^n\dots(q^{n-m+1}-1)q^{n-m+1}}{(q^{m+1}-1)q^{m+1}(q^m-1)q^m\dots(q-1)q}. \tag{2.130}$$

2.9.4. Hyperquadrics and Hermitian Hyperquadrics in Finite Projective Spaces. In $\mathbb{F}_q P^n$ a hyperquadric can be defined by the same equation (2.93) as in P^n and $\mathbb{C}P^n$. As in these spaces, the equation (2.93) can be reduced to a canonical form: in $\mathbb{F}_q P^{2n}$ to

$$x^0x^1 + x^2x^3 + \cdots + x^{2n-2}x^{2n-1} + (x^{2n})^2 = 0, \tag{2.131}$$

in $\mathbb{F}_q P^{2n+1}$ to

$$x^0x^1 + x^2x^3 + \cdots + x^{2n-2}x^{2n-1} + x^{2n}x^{2n+1} = 0 \tag{2.132}$$

and

$$x^0x^1 + x^2x^3 + \cdots + x^{2n-2}x^{2n-1} + (x^{2n})^2 + \alpha(x^{2n+1})^2 = 0, \tag{2.133}$$

where α is an element in $\mathbb{F}_q$ which is not a square. The hyperquadric (2.132) is called the *hyperbolic hyperquadric* and the hyperquadric (2.133) is called the *elliptic hyperquadric* in $\mathbb{F}_q P^{2n+1}$.

The hyperquadrics in $\mathbb{F}_q P^{2n}$ have rectilinear and m-planar generators for $m = 2, 3, ..., n-1$, the hyperbolic and elliptic hyperquadrics in $\mathbb{F}_q P^{2n+1}$ have rectilinear and m-planar generators for $m = 2, 3, ..., n$ and $m = 2, 3, ..., n-1$, respectively.

In $\mathbb{F}_q P^3$ the elliptic quadrics do not have rectilinear generators and the hyperbolic quadrics do have rectilinear generators. Let $\mathbb{F}_q Q^P_{2n,m}$, $\mathbb{F}_q Qh^P_{2n+1,m}$, and $\mathbb{F}_q Qe^P_{2n+1,m}$ be the varieties of m-planar (rectilinear for $m = 1$) generators of the hyperquadrics (2.131), (2.132), and (2.133) in $\mathbb{F}_q P^{2n}$ and $\mathbb{F}_q P^{2n+1}$, respectively.

Theorem of B. Segre [SeB]. *The cardinalities of the varieties of m-planar generators of hyperquadrics in $\mathbb{F}_q P^{2n}$ and $\mathbb{F}_q P^{2n+1}$ are equal to*

$$\operatorname{card} \mathbb{F}_q Q^P_{2n,m} = \frac{(q^{2n}-1)(q^{2n-1}-1)\dots(q^{2n-2m}-1)}{(q^{m+1}-1)(q^m-1)\dots(q-1)}, \tag{2.134}$$

$$\operatorname{card} \mathbb{F}_q Qh^P_{2n+1,m} = \frac{(q^{n+1}-1)(q^n-1)\dots(q^{n-m+1}-1)(q^n+1)(q^{n-1}+1)\dots(q^{n-m}+1)}{(q^{m+1}-1)(q^m-1)\dots(q-1)}, \tag{2.135}$$

$$\operatorname{card} \mathbb{F}_q Qe^P_{2n+1,m} = \frac{(q^{n+1}+1)(q^n+1)\dots(q^{n-m+1}+1)(q^n-1)(q^{n-1}-1)\dots(q^{n-m}-1)}{(q^{m+1}-1)(q^m-1)\dots(q-1)}. \tag{2.136}$$

In particular, for $m = 0$ these formulas give the cardinalities of the hyperquadrics themselves:

$$\operatorname{card} \mathbb{F}_q Q^P_{2n,0} = \operatorname{card} \mathbb{F}_q P^{2n-1} = \frac{q^{2n}-1}{q-1}, \tag{2.137}$$

$$\operatorname{card} \mathbb{F}_q Qh^P_{2n+1,0} = \frac{(q^{n+1}-1)(q^n+1)}{q-1}, \tag{2.138}$$

$$\text{card } \mathbb{F}_q Qe^P_{2n+1,0} = \frac{(q^{n+1}+1)(q^n-1)}{q-1}. \tag{2.139}$$

For $n = 1$ formulas (2.137), (2.138), and (2.139) show that the cardinality of a conic in $\mathbb{F}_q P^2$ is $q+1$, while the cardinalities of an elliptic or a hyperbolic quadric in $\mathbb{F}_q P^3$ are q^2+1 and $(q+1)^2$, respectively.

Note that the cardinalities (2.137), (2.138), and (2.139) can be written in the form of products

$$\text{card } \mathbb{F}_q Q^P_{2n,0} = \text{card } \mathbb{F}_q P^{n-1}(\text{card } \mathbb{F}_q E^n + 1), \tag{2.140}$$

$$\text{card } \mathbb{F}_q Qh^P_{2n+1,0} = \text{card } \mathbb{F}_q P^n(\text{card } \mathbb{F}_q E^n + 1), \tag{2.141}$$

$$\text{card } \mathbb{F}_q Qe^P_{2n+1,0} = \text{card } \mathbb{F}_q P^{n-1}(\text{card } \mathbb{F}_q E^{n+1} + 1). \tag{2.142}$$

These formulas are finite analogs of Theorem 2.54 for P^n, since according to this theorem a hyperquadric of index n in P^{2n} is homeomorphic to the fibration with the space P^{n-1} as base and the hyperspheres in R^{n+1} as fibers, and a hyperquadric of index $n+1$, respectively of index n, in P^{2n+1} is homeomorphic to the fibration with the space P^n as base and the hyperspheres in R^{n+1} as fibers, respectively to the fibration with the space P^{n-1} as base and the hyperspheres in R^{n+2} as fibers, while a hypersphere in R^{N+1} is homeomorphic to the union of E^N and a point.

The field F_{q^2} has an involutive automorphism $x \to x^q$, which is the analog of the automorphism $x \to \overline{x}$ in $\mathbb{C}$. Therefore, $\mathbb{F}_{q^2} P^n$ contains *Hermitian hyperquadrics* with equations which can be reduced to the form

$$(x_0)^q x^0 + (x^1)^q x^1 + \cdots + (x^n)^q x^n = 0. \tag{2.143}$$

The Hermitian hyperquadric (2.143) has rectilinear and m-planar generators for $m = 2, 3, ..., \frac{n}{2} - 1$ if n is even and for $m = 2, 3, ..., \frac{n-1}{2}$ if n is odd. Let $\mathbb{F}_{q^2}\overline{Q}^P_{n,m}$ be the variety of m-planar (rectilinear for $m = 1$) generators of the Hermitian hyperquadric (2.143) in $\mathbb{F}_{q^2} P^n$.

Theorem 1 of Wan Zhe-Xian [Wan, p. 247]. *The cardinality of the variety of m-planar generators of the Hermitian hyperquadric in $\mathbb{F}_{q^2} P^n$ is equal to*

$$\text{card } \mathbb{F}_{q^2}\overline{Q}^P_{n,m} = \frac{(q^{n+1}-(-1)^{n+1})(q^n-(-1)^n)\dots(q^{n-2m}-(-1)^{n-2m})}{(q^{2m+2}-1)(q^{2m}-1)\dots(q^2-1)}. \tag{2.144}$$

In particular, for $m = 0$ this formula gives the cardinality of the Hermitian hyperquadric itself:

$$\text{card } \mathbb{F}_{q^2}\overline{Q}^P_{n,0} = \frac{(q^{n+1}-(-1)^{n+1})(q^n-(-1)^n)}{q^2-1}. \tag{2.145}$$

This cardinality can be written in the form of a product:

$$\text{card } \mathbb{F}_{q^2}\overline{Q}^P_{n,0} = \text{card } \mathbb{F}_{q^2} P^{\ell-1}(\text{card } E^{2n-2\ell+1} + 1), \tag{2.146}$$

where $\ell = \left[\frac{n}{2}\right]$. This formula is a finite analog of Theorem 2.54 for $\mathbb{C}P^n$.

2.9.5. Null-systems in Finite Projective Spaces. Null-systems in the spaces $\mathbb{F}_q P^n$ can be defined by the same equation (2.111), where $a_{ij} = -a_{ji}$, as in P^n and $\mathbb{C}P^n$. Null lines and null m-planes of a null-system in $\mathbb{F}_q P^{2n+1}$ are also defined as in these spaces. Let $\mathbb{F}_q N^P_{2n+1,m}$ be the variety of null m-planes in $\mathbb{F}_q P^{2n+1}$.

Theorem 2 of Wan Zhe-Xian [Wan, p. 126]. *The cardinality of the variety of null m-planes, for $m = 1$ of null lines, in $\mathbb{F}_q P^{2n+1}$ is equal to*

$$\textit{card}\ \mathbb{F}_q N^P_{2n+1,m} = \frac{(q^{2n+2}-1)(q^{2n}-1)\dots(q^{2n-2m+2}-1)}{(q^{m+1}-1)(q^m-1)\dots(q-1)}. \tag{2.147}$$

For $m = 1$ this formula gives the cardinality of a linear complex of lines:

$$\text{card}\ \mathbb{F}_q N^P_{2n+1,1} = \frac{(q^{2n+2}-1)(q^{2n}-1)}{(q^2-1)(q-1)}. \tag{2.148}$$

This cardinality can be written in the form of a product:

$$\text{card}\ \mathbb{F}_q N^P_{2n+1,1} = \text{card}\ \mathbb{F}_{q^2} P^n \cdot \text{card}\ P^{2n-1}. \tag{2.149}$$

In **6.2.2** we will prove that a linear complex of lines in P^{2n+1} is homeomorphic to a Hermitian hyperquadric in $\mathbb{H}' P^n$. Therefore, formula (2.149) is a finite analog of Theorem 2.54 for $\mathbb{H}' P^n$.

2.9.6. Affine and Projective Transformations in Finite Spaces. Since a field $\mathbb{F}_q$ with $q = p^h$ has the h automorphisms $x \to x^{p^k}$, $k = 0, 1, 2, \dots, h-1$, the affine transformations in the space $\mathbb{F}_q E^n$ have the form

$$x^i = A^i_j (x^j)^{p^k} + b^i, \tag{2.150}$$

the collineations in the space $\mathbb{F}_q P^n$ have the form

$$x^i = A^i_j (x^j)^{p^k}, \tag{2.151}$$

and the correlations in $\mathbb{F}_q P^n$ have the form

$$u_i = A_{ij} (x^j)^{p^k}. \tag{2.152}$$

The matrix (A^i_j) of an affine transformation (2.150) belongs to the group $\mathbb{F}_q GL_n$, the matrix (A^i_j) of a collineation (2.151) belongs to the group $\mathbb{F}_q SL_{n+1}$.

Theorem 2.68. *The cardinalities of the groups $\mathbb{F}_q GL_n$ and $\mathbb{F}_q SL_n$ are equal to*

$$\textit{card}\ \mathbb{F}_q GL_n = q^{n(n-1)/2}(q^n-1)(q^{n-1}-1)\dots(q-1) \tag{2.153}$$

and

$$\textit{card}\ \mathbb{F}_q SL_n = q^{n(n-1)/2}(q^n-1)(q^{n-1}-1)\dots(q-2). \tag{2.154}$$

The cardinality of the group of affine transformations in $\mathbb{F}_q E^n$, respectively of collineations in $\mathbb{F}_q P^n$, is equal to (card $\mathbb{F}_q GL_n$)$\cdot q^n \cdot h$, respectively to (card $\mathbb{F}_q SL_{n+1}$)$\cdot h/(n+1, q-1)$, where h is the number of automorphisms in $\mathbb{F}_q$ and (a, b) is the greatest common divisor of the integers a and b.

PROOF. The number card $\mathbb{F}_q GL_n$ of all nonsingular $(n \times n)$-matrices with entries from $\mathbb{F}_q$ is the product of all possible entries of rows of matrices in $\mathbb{F}_q GL_n$. Since the entries in the first row in such a matrix are arbitrary elements in $\mathbb{F}_q$ not all equal to zero, the number of all possible entries in the first row is $q^n - 1$. Suppose that the entry $A_1^1 \neq 0$. Then the entry A_2^1 in the second row is any element in $\mathbb{F}_q$, while the other entries in the second row are arbitrary elements in $\mathbb{F}_q$ not all equal to zero. So, the number of all possible entries in the second row is $q(q^{n-1} - 1)$. Suppose that the entry $A_2^2 \neq 0$. Then the entries A_3^1 and A_3^2 in the third row are arbitrary elements in $\mathbb{F}_q$, and the other entries in the third row are arbitrary elements in $\mathbb{F}_q$ not all equal to zero. So, the number of all possible entries in the third row is $q^2(q^{n-2} - 1)$. Analogously, we find that the number of all possible entries in the $(i+1)$-th row is $q^i(q^{n-i} - 1)$, and

$$\text{card } \mathbb{F}_q GL_n = (q^n - 1)q(q^{n-1} - 1)q^2(q^{n-2} - 1)\dots q^{n-1}(q - 1), \tag{2.155}$$

that is, we obtain formula (2.153). The number card $\mathbb{F}_q SL_n$ is equal to card $\mathbb{F}_q GL_n/(q-1)$, that is, we obtain formula (2.154).

The cardinality of the group of affine transformations (2.150) in $\mathbb{F}_q E^n$ is equal to the product of the sum of the cardinalities of the groups of matrices (A_j^i) and vectors $b = \{b^i\}$ by the number of automorphisms in the field $\mathbb{F}_q$. Since the group of collineations (2.151) in $\mathbb{F}_q P^n$ for $k = 0$, that is, $'x^i = A_j^i x^j$, is isomorphic to the quotient group of the group $\mathbb{F}_q SL_{n+1}$ by its center, which consists of the matrices $(\alpha\delta_j^i)$, and the number of factors α of such matrices is equal to $(n+1, q-1)$, the cardinality of the group of all collineations of $\mathbb{F}_q P^n$ is equal to (card $\mathbb{F}_q SL_{n+1}$) $\cdot$ $h/(n+1, q-1)$. ■

Note that the group of collineations $'x^i = A_j^i x^j$ in $\mathbb{F}_q P^n$ is a simple finite group of Lie type of class A_n. The cardinality of this group is equal to the number (1.129) for the group A_n since, according to (1.114), the numbers $a_i + 1$, with a_i the exponents of A_n, are equal to $2, 3, \dots, n, n+1$, the number $2N$ of root vectors of this group is $r - n = n(n+2) - n = n(n+1)$, therefore the number N of positive root vectors of this group is $\frac{n(n+1)}{2}$, while $u = (n+1, q-1)$.

CHAPTER III

EUCLIDEAN, PSEUDO-EUCLIDEAN, CONFORMAL AND PSEUDOCONFORMAL GEOMETRIES

§3.1. Euclidean and Pseudo-Euclidean Spaces

3.1.1. Quadratic Euclidean and Pseudo-Euclidean Spaces. In **0.5.3** we have defined the real Euclidean space R^n as the affine space E^n whose vectors satisfy axioms $E.1° - 5°$. Since the inner product x^2 of a vector $x = x^i e_i$ of this space by itself is a positive definite quadratic form $x^2 = e_{ij}x^i x^j$, this space is also called a *quadratic Euclidean space.* In **0.5.3** we have seen that the space R^n also satisfies the axioms $M.1° - 3°$ of a metric space.

In **0.5.6** we have defined the real pseudo-Euclidean space R^n_ℓ as the affine space E^n whose vectors satisfy axioms $E.1° - 4°$ and $E.5'$. Since the inner product xEx of a vector $x = x^i e_i$ in this space by itself is an indefinite quadratic form $xEx = e_{ij}x^i x^j$, this space is also called a *quadratic pseudo-Euclidean space.* In **0.5.6** we have seen that the space R^n_ℓ is a pseudometric space.

Analogously, we can define the quadratic Euclidean and pseudo-Euclidean spaces $\mathbb{A}R^n$ and $\mathbb{A}R^n_\ell$ over a commutative algebra $\mathbb{A}$, as the spaces $\mathbb{A}E^n$ whose vectors satisfy axioms $E.1° - 4°$ and an axiom replacing the axiom $E.5°$ or $E.5'$. In particular, the *complex quadratic Euclidean space* $\mathbb{C}R^n$ is the complex affine space $\mathbb{C}E^n$ whose vectors satisfy axioms $E.1° - 4°$ and the axiom

$E.5''$. There are n mutually orthogonal vectors a_i with nonzero inner square.

Therefore, in this space the quadratic form $x^2 = e_{ij}x^i x^j$ is nondegenerate. Since each nondegenerate quadratic form in the complex linear space $\mathbb{C}L^n$ can be reduced to the form $\sum_i (x^i)^2$, pseudo-Euclidean spaces over the field $\mathbb{C}$ do not exist.

The *split complex quadratic Euclidean* and *pseudo-Euclidean spaces* $\mathbb{C}' R^n$ and $\mathbb{C}' R^n_\ell$ are the split complex affine spae $\mathbb{C}' E^n$ whose vectors satisfy axioms $E.1° - 4°$ and

$E.5'''$. There are n mutually orthogonal vectors a_i with positive inner square; and $E.5'$, respectively.

The algebras $\mathbb{A}$ over which quadratic Euclidean and pseudo-Euclidean spaces can be defined must be commutative, because the transformation law of the coordinates of a symmetric tensor e_{ij} which determines the inner product of vectors xy or xEy in these spaces has the form $a_{i'j'} = a_{ij}A^i_{i'}A^j_{j'}$, and such a law is possible only for commutative algebras. These algebras $\mathbb{A}$ must also be *Frobenius algebras* (see **1.1.12**), since the tensor e_{ij} determines a correlation $u_i = e_{ij}x^j$ in the projective space $\mathbb{A}P^{n-1}$ for the hyperplanes at infinity of the spaces $\mathbb{A}R^n$ and $\mathbb{A}R^n_\ell$, but this correlation can be bijective from $\mathbb{A}P^{n-1}$ onto the manifold of hyperplanes in $\mathbb{A}P^{n-1}$ only if $\mathbb{A}$ is a Frobenius algebra.

The spaces $\mathbb{C}^0 R^n$ and $\mathbb{C}^0 R^n_\ell$ over the algebra $\mathbb{C}^0$ of dual numbers can be defined, like the spaces $\mathbb{C}' R^n$ and $\mathbb{C}' R^n_\ell$, as the space $\mathbb{C}^0 E^n$ whose vectors satisfy axioms $E.1^\circ - 4^\circ$ and $E.5'''$ and $E.5'$, respectively.

In $\mathbb{C}R^n, \mathbb{C}' R^n$, and $\mathbb{C}' R^n_\ell$ the distance $\rho(X,Y)$ between two points X and Y is defined as the root of the inner square $(y-x)^2$ or $(y-x)E(y-x)$ of the difference between the position vectors x and y of X and Y. The distance $\rho(X,Y)$ between two points in $\mathbb{C}R^n, \mathbb{C}' R^n$ or $\mathbb{C}' R^n_\ell$ is a complex number in the first case, a split complex number in the second case, and a split complex number or bicomplex number of the form $x_1 i + x_2 I$ in the third case.

In $\mathbb{C}^0 R^n$ and $\mathbb{C}^0 R^n_\ell$ the distance $\rho(X,Y)$ between two points X and Y is, in general, defined as in $\mathbb{C}R^n, \mathbb{C}' R^n$, and $\mathbb{C}' R^n_\ell$, but if $\rho(X,Y) = 0$ we can define a second distance $\sigma(X,Y)$: if the vectors x and y are written in the form $x_0 + x_1\epsilon$ and $y_0 + y_1\epsilon$, the distance $\sigma(X,Y)$ is defined as the root of the inner square $(y_1 - x_1)^2$ or $(y_1 - x_1)E(y_1 - x_1)$. Since $\mathbb{C}^0$ is a *Hjelmslev algebra* (see **1.1.11**), $\mathbb{C}^0 R^n$ and $\mathbb{C}^0 R^n_\ell$ are Hjelmslev spaces.

3.1.2. Hermitian Euclidean and Pseudo-Euclidean Spaces. For a Frobenius algebra $\mathbb{A}$ with involution $x \to \overline{x}$ we can define the *Hermitian Euclidean* and *Hermitian pseudo-Euclidean spaces* $\mathbb{A}\overline{R}^n$ and $\mathbb{A}\overline{R}^n_\ell$.

The complex and quaternionic Hermitian Euclidean spaces $\mathbb{C}\overline{R}^n$ and $\mathbb{H}\overline{R}^n$ are defined as the affine space $\mathbb{C}E^n$ and $\mathbb{H}E^n$ whose vectors satisfy axioms $E.2^\circ - 5^\circ$ and

$E.1'$. $xy = \overline{yx}$ (Hermitian symmetry).

We denote the inner product of vectors in $\mathbb{A}\overline{R}^n$ by $\overline{x}y$.

The complex and quaternionic Hermitian pseudo-Euclidean spaces $\mathbb{C}\overline{R}^n_\ell$ and $\mathbb{H}\overline{R}^n_\ell$ are defined as the affine spaces $\mathbb{C}E^n$ and $\mathbb{H}E^n$ whose vectors satisfy axioms $E.1'$, $E.2^\circ - 4^\circ$ and $E.5'$.

We denote the inner product of vectors in $\mathbb{A}\overline{R}^n_\ell$ by $\overline{x}Ey$.

These spaces are called Hermitian because the inner squares $\overline{x}x$ and $\overline{x}Ey$ of a vector $x = x^i e_i$ in a Hermitian Euclidean space can be expressed by a positive definite Hermitian form $\overline{x}x = \overline{x}^j e_{ij} x^i$ and in a Hermitian pseudo-Euclidean space by an indefinite Hermitian form $\overline{x}Ex = \overline{x}^j e_{ij} x^i$.

The split complex and split quaternionic Hermitian Euclidean spaces $\mathbb{C}'\overline{R}^n$ and $\mathbb{H}'\overline{R}^n$ are defined as the affine spaces $\mathbb{C}' E^n$ and $\mathbb{H}' E^n$ whose vectors satisfy axioms $E.1', E.2^\circ - 4^\circ$, and $E.5''$. Since each nondegenerate Hermitian form in the split complex or split quaternionic free module $\mathbb{C}' L^n$, respectively, $\mathbb{H}' L^n$, can be reduced to the form $\sum_i \overline{x}^i x^i$, there are no Hermitian pseudo-Euclidean spaces over the algebras $\mathbb{C}'$ and $\mathbb{H}'$.

Both Hermitian Euclidean and Hermitian pseudo-Euclidean spaces can be defined over the algebras $\mathbb{C}^0, \mathbb{H}^0$, and $\mathbb{H}'^0$ of dual numbers, of semiquaternions, and of split semiquaternions, respectively. The spaces $\mathbb{C}^0\overline{R}^n, \mathbb{H}^0\overline{R}^n$, and $\mathbb{H}'^0\overline{R}^n$, are defined like $\mathbb{C}'\overline{R}^n$ and $\mathbb{H}'\overline{R}^n$, as the spaces $\mathbb{C}^0 E^n, \mathbb{H}^0 E^n$, and $\mathbb{H}'^0 E^n$, respectively, whose vectors satisfy axioms $E.1', E.2^\circ - 4^\circ$, and $E.5'''$. The spaces $\mathbb{C}^0\overline{R}^n_\ell$ and $\mathbb{H}^0\overline{R}^n_\ell$ can be defined, like $\mathbb{C}\overline{R}^n_\ell$ and $\mathbb{H}\overline{R}^n_\ell$, as the spaces $\mathbb{C}^0 E^n$ and $\mathbb{H}^0 E^n$, respectively, whose vectors satisfy axioms $E.1', E.2^\circ - 4^\circ$, and $E.5'$.

In $\mathbb{C}\overline{R}^n$, $\mathbb{C}\overline{R}^n_\ell$, $\mathbb{C}'\overline{R}^n$, $\mathbb{H}\overline{R}^n$, $\mathbb{H}\overline{R}^n_\ell$, and $\mathbb{H}'\overline{R}^n$ we define the distance $\rho(X,Y)$ between two points X and Y as the root of the Hermitian inner square $(\overline{y}-\overline{x})(y-x)$

or $(\overline{y}-\overline{x})E(y-x)$ of the difference of the position vectors x and y of the points X and Y. The distances $\rho(X,Y)$ between points in $\mathbb{C}\overline{R}^n$ and $\mathbb{H}\overline{R}^n$ are real numbers, the distances $\rho(X,Y)$ between points in $\mathbb{C}\overline{R}^n_\ell, \mathbb{C}'\overline{R}^n, \mathbb{H}\overline{R}^n_\ell$ or $\mathbb{H}'\overline{R}^n$ are real or pure imaginary numbers.

In $\mathbb{C}^0\overline{R}^n, \mathbb{C}^0\overline{R}^n_\ell, \mathbb{H}^0\overline{R}^n, \mathbb{H}^0\overline{R}^n_\ell, \mathbb{H}'^0\overline{R}^n$, and $\mathbb{H}'^0\overline{R}^n_\ell$ the distance $\rho(X,Y)$ between two points X and Y is defined in the general case as in $\mathbb{C}\overline{R}^n, \mathbb{C}\overline{R}^n_\ell, \mathbb{C}'\overline{R}^n, \mathbb{H}\overline{R}^n, \mathbb{H}\overline{R}^n_\ell$, and $\mathbb{H}'\overline{R}^n$, but if $\rho(X,Y)=0$ for two points X and Y, we define a second distance $\sigma(X,Y)$: if the position vectors x and y in $\mathbb{C}^0\overline{R}^n$ or $\mathbb{C}^0\overline{R}^n_\ell$ are written in the form $x_0+x_1\epsilon$ and $y_0+y_1\epsilon$, the distance $\sigma(X,Y)$ is defined as the root of the inner square $(y_1-x_1)^2$ or $(y_1-x_1)E(y_1-x_1)$. Analogously, if x and y in the semiquaternionic or split semiquaternionic Hermitian Euclidean or pseudo-Euclidean space are written in the same form, the distance $\sigma(X,Y)$ is defined as the root of the Hermitian inner square $(\overline{y}_1-\overline{x}_1)(y_1-x_1)$ or $(\overline{y}_1-\overline{x}_1)E(y_1-x_1)$.

Since $\mathbb{C}^0$ and $\mathbb{H}^0$ are Hjelmslev algebras, $\mathbb{C}^0\overline{R}^n, \mathbb{C}^0\overline{R}^n_\ell, \mathbb{H}^0\overline{R}^n$, and $\mathbb{H}^0\overline{R}^n_\ell$ are Hjelmslev spaces; since $\mathbb{H}'^0$ is not a Hjelmslev algebra, $\mathbb{H}'^0\overline{R}^n$ is not a Hjelmslev space.

Theorem 3.1. *The spaces $\mathbb{C}\overline{R}^n$ and $\mathbb{H}\overline{R}^n$ are isometric to the spaces R^{2n} and R^{4n}, respectively.*

The spaces $\mathbb{C}\overline{R}^n_\ell, \mathbb{C}'\overline{R}^n, \mathbb{H}\overline{R}^n_\ell$, and $\mathbb{H}'\overline{R}^n$ are isometric to the spaces $R^{2n}_{2\ell}$, R^{2n}_n, $R^{4n}_{4\ell}$, and R^{4n}_{2n}, respectively.

PROOF. We reduce the Hermitian inner square $\overline{x}x$ of a vector x in $\mathbb{C}\overline{R}^n$ or $\mathbb{H}\overline{R}^n$ to the form $\sum_i \overline{x}^i x^i$. Putting, for $\mathbb{C}\overline{R}^n$, $x^i = X^{2i-1}+iX^{2i}$ and, for $\mathbb{H}\overline{R}^n$, $x^i = X^{4i-3}+iX^{4i-2}+jX^{4i-1}+kX^{4i}$, the form $\sum_i \overline{x}^i x^i$ can be written as $\sum_I (X^I)^2$, where in the first case $I=1,2,\dots,2n$ and in the second case $I=1,2,\dots,4n$. These forms are equal to the inner square of the vector X in R^{2n} or R^{4n}.

We reduce the inner Hermitian square $\overline{x}Ex$ of a vector x in $\mathbb{C}\overline{R}^n_\ell$ or $\mathbb{H}\overline{R}^n_\ell$ to the form $-\sum_a \overline{x}^a x^a + \sum_u \overline{x}^u x^u$. Using the same notation as for $\mathbb{C}\overline{R}^n$ and $\mathbb{H}\overline{R}^n$, this form can be written as $-\sum_A (X^A)^2 + \sum_U (X^U)^2$, where in the first case $A = 1,2,\dots,2\ell$, $U = 2\ell+1,\dots,2n$ and in the second case $A=1,2,\dots,4\ell$, $U=4\ell+1,\dots,4n$. These forms are equal to the inner square of a vector X in $R^{2n}_{2\ell}$ or $R^{4n}_{4\ell}$.

We reduce the Hermitian inner square $\overline{x}x$ of a vector x in $\mathbb{C}'\overline{R}^n$ or $\mathbb{H}'\overline{R}^n$ to the form $\sum_i \overline{x}^i x^i$. Putting, for $\mathbb{C}'\overline{R}^n$, $x^i = X^{2i-1}+eX^{2i}$ and, for $\mathbb{H}'\overline{R}^n$, $x^i = X^{4i-3}+iX^{4i-2}+eX^{4i-1}+fX^{4i}$, the form $\sum_i \overline{x}^i x^i$ can be written as $\sum_i (X^{2i-1})^2 - \sum_i (X^{2i})^2$ and $\sum_i (X^{4i-3})^2 + \sum_i (X^{4i-2})^2 - \sum_i (X^{4i-1})^2 - \sum_i (X^{4i})^2$. These forms are equal to the inner square of a vector X in R^{2n}_n or R^{4n}_{2n}. ∎

This theorem implies that $\mathbb{C}\overline{R}^n$ and $\mathbb{H}\overline{R}^n$ are metric spaces and that $\mathbb{C}\overline{R}^n_\ell$, $\mathbb{C}'\overline{R}^n$, $\mathbb{H}\overline{R}^n_\ell$, and $\mathbb{H}'\overline{R}^n$ are pseudometric spaces.

The metricity of $\mathbb{C}\overline{R}^n$ and $\mathbb{H}\overline{R}^n$ can also be proved analogously to Theorems 0.1 and 0.2 (see **0.5.3**). The Cauchy inequality (0.18) in these spaces is proved as follows. Since the Hermitian inner square $(\overline{x}-\overline{t}\overline{y})(x-yt)$ of a vector $x-yt$ in $\mathbb{C}\overline{R}^n$ or $\mathbb{H}\overline{R}^n$ is nonnegative, the inequality $\overline{x}x - \overline{x}yt - \overline{t}\overline{y}x + \overline{t}\overline{y}yt \geq 0$ holds for all complex numbers or quaternions t. If $y=0$, the inequality (0.18) is fulfilled. If $y \neq 0$ we take $t = \frac{\overline{y}x}{\overline{y}y}$. In this case the inequality takes the form

$$\overline{x}x - 2\frac{\overline{x}y \cdot \overline{y}x}{\overline{y}y} + \frac{\overline{x}y \cdot \overline{y}x}{\overline{y}y} \geq 0,$$

that is, $\overline{x}x \cdot \overline{y}y \geq \overline{x}y \cdot \overline{y}x$. Denoting $\overline{x}x = |x|^2$, this is equivalent to (0.18). The proof of axiom $M.3$ for $\mathbb{C}\overline{R}^n$ and $\mathbb{H}\overline{R}^n$ is the same as for R^n in Theorem 0.2.

3.1.3. Real Metrics and Pseudometrics in Complex and Split Complex Quadratic Spaces. We can also define a real metric (pseudometric) in complex (split complex) quadratic Euclidean (pseudo-Euclidean) spaces.

Theorem 3.2. *If we define in the space $\mathbb{C}'R^n$ endowed with the split complex distance between points $\omega = \omega_1 + e\omega_2$ a real distance $\widetilde{\omega}$ between the same points by*

$$\widetilde{\omega}^2 = \mathrm{Re}\,(\omega^2), \tag{3.1}$$

then we obtain the metric of the Euclidean space R^{2n}.

If we define in $\mathbb{C}R^n$ endowed with the complex distance between points $\omega = \omega_1 + i\omega_2$ a real distance $\widetilde{\omega}$ between the same points by (3.1), we obtain the pseudometric of the pseudo-Euclidean space R^{2n}_n.

If we define in $\mathbb{C}'R^n_\ell$ endowed with the split complex or bicomplex distance between points $\omega = \omega_1 + e\omega_2$, where ω_1 and ω_2 are real or complex conjugate numbers, a real or pure imaginary distance $\widetilde{\omega}$ between the same points by (3.1), we obtain the pseudometric of the pseudo-Euclidean space $R^{2n}_{2\ell}$.

The assertions of this theorem can be immediately verified.

3.1.4. Bicomplex, Biquaternionic, and Quaterquaternionic Euclidean Spaces. Since the algebra $\mathbb{C} \otimes \mathbb{C}$ is commutative and has three involutions $x \to \overline{x}$, $x \to \tilde{x}$, and $x \to \tilde{\overline{x}}$, the first and second of which are equivalent, we can define the quadratic Euclidean space $(\mathbb{C}\otimes\mathbb{C})R^n$ and two kinds of Hermitian Euclidean spaces $(\mathbb{C}\otimes\mathbb{C})\overline{R}^n$ and $(\mathbb{C}\otimes\mathbb{C})\,\tilde{\overline{R}}^n$ over it.

Since the algebras $\mathbb{C}\otimes\mathbb{H}$ and $\mathbb{H}\otimes\mathbb{H}$ are noncommutative and $\mathbb{C}\otimes\mathbb{H}$ has two involutions $x \to \overline{x}$ and $x \to \tilde{\overline{x}}$, while $\mathbb{H}\otimes\mathbb{H}$ has only the single involution $x \to \tilde{\overline{x}}$, we can define the Hermitian Euclidean spaces $(\mathbb{C}\otimes\mathbb{H})\overline{R}^n$, $(\mathbb{C}\otimes\mathbb{H})\,\tilde{\overline{R}}^n$ and $(\mathbb{H}\otimes\mathbb{H})\,\tilde{\overline{R}}^n$ over these algebras.

The definition of the quadratic space $(\mathbb{C}\otimes\mathbb{C})R^n$ is analogous to the definition of $\mathbb{C}R^n$. The definition of the Hermitian spaces $(\mathbb{C}\otimes\mathbb{C})\overline{R}^n$ and $(\mathbb{C}\otimes\mathbb{C})\,\tilde{\overline{R}}^n$ is analogous to the definition of $\mathbb{C}\overline{R}^n$.

The distance between two points is a bicomplex number in $(\mathbb{C}\otimes\mathbb{C})R^n$, but in Hermitian spaces over tensor products distances between points are not general elements in these algebras. In $(\mathbb{C}\otimes\mathbb{C})\overline{R}^n$, respectively in $(\mathbb{C}\otimes\mathbb{C})\,\tilde{\overline{R}}^n$, the Hermitian inner square α satisfies the condition $\alpha = \overline{\alpha}$, respectively $\alpha = \tilde{\overline{\alpha}}$. These inner squares and distances are real linear combinations of the basis elements 1 and I, respectively of the basis elements 1 and iI, in this algebra. Therefore, the distance between two points in these spaces can be regarded as a complex, respectively a split complex, number.

Analogously we can prove that the distances between points in $(\mathbb{C}\otimes\mathbb{H})\overline{R}^n$, respectively in $(\mathbb{C}\otimes\mathbb{H})\,\tilde{\overline{R}}^n$, are real linear combinations of the basis elements 1 and

I in $\mathbb{C}\otimes\mathbb{H}$, respectively of the basis elements 1 and iI in this algebra. Therefore, in these spaces distances between points can also be regarded as complex, respectively split complex, numbers.

We can similarly prove that the distances between points in $(\mathbb{H}\otimes\mathbb{H})\ \overline{\widetilde{R}}^n$ are real linear combinations of the basis elements $1, iI, jJ$, and kK in $\mathbb{H}\otimes\mathbb{H}$. Therefore, in this space the distance between points can be regarded as an element in the algebra $\mathbb{R}^4$ of 4-multiple numbers (see **1.1.7**).

Theorem 3.3. *If we define in* $(\mathbb{C}\otimes\mathbb{C})\overline{R}^n$, $(\mathbb{C}\otimes\mathbb{C})\ \overline{\widetilde{R}}^n$, $(\mathbb{C}\otimes\mathbb{H})\overline{R}^n$, $(\mathbb{C}\otimes\mathbb{H})\ \overline{\widetilde{R}}^n$, *or* $(\mathbb{H}\otimes\mathbb{H})\ \overline{\widetilde{R}}^n$ *a real distance* $\widetilde{\omega}$ *between points by (3.1), then we obtain the metrics (pseudometrics) of the Euclidean (pseudo-Euclidean) spaces* $R^{4n}_{2n}, R^{4n}, R^{8n}_{4n}, R^{8n}$, *and* R^{16n}.

PROOF. The assertions of this theorem can be proved analogously to the assertions of Theorem 3.2. In the last case we have to note that the algebra $\mathbb{R}^4$ is isomorphic to the algebra of diagonal real (4×4)-matrices and if $\omega = \sum_i \omega_i e_i$ is an element in this algebra, then $\omega^2 = \sum_i \omega_i^2 e_i$ and the role of Re (ω^2) is played by a quarter of the trace of the corresponding scalar matrix, that is, by $\frac{1}{4}\sum_i \omega_i^2$. ∎

3.1.5. Applications of Complex, Split Complex, and Dual Numbers to Real Plane and Spherical Geometry. In **1.1.1** and **1.1.2** we have considered the application of the algebras $\mathbb{C}$ and $\mathbb{C}'$ of complex and split complex numbers to the geometry of the planes R^2 and R^2_1. Now we will consider the application of the algebras $\mathbb{C}$ and $\mathbb{C}'$ to the geometry on a sphere in the space R^3 or R^3_1 and the application of the algebra $\mathbb{C}^0$ of *dual numbers* to the geometry in the planes R^2 and R^2_1.

Theorem 3.4. *There is a bijection from pairs of antipodal points on a sphere in* R^3 *onto the extended algebra* $\mathbb{C}$.

There is a bijection from pairs of antipodal points on a sphere of real, respectively pure imaginary, radius in R^3_1 *onto the extended algebra* $\mathbb{C}'$, *respectively onto the extended algebra* $\mathbb{C}$.

Under these bijections, a rotation in R^3 *or* R^3_1 *is mapped to a transformation*

$$x' = (ax+b)(-\bar{b}x+\bar{a})^{-1} \quad or \quad x' = (a\bar{x}+b)(-\bar{b}\bar{x}+\bar{a})^{-1} \tag{3.2}$$

in $\mathbb{C}$ *and a rotation in* R^3_1 *is mapped to a transformation (3.2) in* $\mathbb{C}'$ *or to*

$$x' = (ax+b)(\bar{b}x+\bar{a})^{-1} \quad or \quad x' = (a\bar{x}+b)(\bar{b}\bar{x}+\bar{a})^{-1}. \tag{3.3}$$

in $\mathbb{C}$.

PROOF. These maps are determined by the same stereographic projections of spheres onto planes as considered in Theorem 2.13 for the case $m = 2$ and in Theorem 2.15 (see **2.2.2**).

For the sphere $(X^1)^2 + (X^2)^2 + (X^3)^2 = 1$ in R^3, projection from the pole S with coordinates $(0,0,1)$ (Fig. 2.3) maps a point on the sphere with coordinates (X^1, X^2, X^3) to a point in the plane $x^3 = 0$ with coordinates $(x^1, x^2, 0)$. Therefore, these coordinates are related by $\frac{X^1}{x^1} = \frac{X^2}{x^2} = \frac{X^3-1}{-1} = k$, that is, $X^1 = kx^1, X^2 =$

$kx^2, X^3 = 1 - k$. Since these coordinates satisfy the equation of the sphere, we obtain $k^2((x^1)^2 + (x^2)^2 + 1) = 2k$, or, since $x = x^1 + ix^2$, $k = \frac{2}{\overline{x}x+1}$ and

$$X^1 + iX^2 = \frac{2x}{\overline{x}x + 1}, \quad X^3 = \frac{\overline{x}x - 1}{\overline{x}x + 1}. \tag{3.4}$$

Formula (3.4) shows that the reflection $'X^i = X^i$ in the center of the sphere corresponds to the transformation $x' = -\frac{1}{\overline{x}}$ in the complex plane, that is, to the inversion with respect to the imaginary circle $\overline{x}x = -1$ in this plane. It is easy to verify that a linear transformation of the coordinates X^i corresponds to a fractional linear transformation (2.46) or (2.47) in the complex plane. Therefore, since a rotation of the sphere is an affine rotation preserving antipodal points on the sphere, it becomes a fractional linear transformation preserving the imaginary circle $\overline{x}x = -1$, that is, a transformation (3.2) in the complex plane.

For the sphere $(X^1)^2 + (X^2)^2 - (X^3)^2 = -1$ in R^3_1, the stereographic projection from the pole S with coordinates $(0, 0, 1)$ also maps a point on the sphere with coordinates (X^1, X^2, X^3) to a point in the plane $x^3 = 0$ with coordinates $(x^1, x^2, 0)$. Like in the case of R^3, these coordinates are related by $\frac{X^1}{x^1} = \frac{X^2}{x^2} = \frac{X^3-1}{-1} = k$, that is, $X^1 = kx^1, X^2 = kx^2, X^3 = 1 - k$. Since these coordinates satisfy the equation of the sphere of radius i, we find that $k^2((x^1)^2 + (x^2)^2 - 1) = -2k$, or, since $x = x^1 + ix^2$, $k = \frac{-2}{\overline{x}x-1}$ and

$$X^1 + iX^2 = \frac{2x}{\overline{x}x - 1}, \quad X^3 = \frac{\overline{x}x + 1}{\overline{x}x - 1}. \tag{3.5}$$

Formula (3.5) shows that the reflection $'X^i = -X^i$ in the center of the sphere corresponds to the transformation $x' = \frac{1}{\overline{x}}$ in the complex plane, that is, the inversion with respect to the real circle $\overline{x}x = 1$ in this plane. Since a linear transformation of the coordinates X^i corresponds to a fractional linear transformation (2.46) or (2.47) in the complex plane and a rotation of the sphere is an affine rotation preserving antipodal points on the sphere, it becomes a fractional linear transformation preserving the real circle $\overline{x}x = 1$, that is, the transformation (3.3) in the complex plane.

For the sphere $(X^1)^2 - (X^2)^2 + (X^3)^2 = 1$ in R^3_1, stereographic projection from the pole S with coordinates $(0, 0, 1)$ (Fig. 2.6) maps a point on the sphere with coordinates (X^1, X^2, X^3) to a point in the plane $x^3 = 0$ with coordinates $(x^1, x^2, 0)$. Therefore, these coordinates are also related by $\frac{X^1}{x^1} = \frac{X^2}{x^2} = \frac{X^3-1}{-1} = k$, that is, $X^1 = kx^1, X^2 = kx^2, X^3 = 1 - k$. Since these coordinates satisfy the equation of the sphere of radius 1, we find that $k^2((x^1)^2 - (x^2)^2 + 1) = 2k$, or, since $x = x^1 + ex^2$, $k = \frac{2}{\overline{x}x+1}$ and (3.4). Formula (3.4) shows that the reflection $'X^i = -X^i$ in the center of the sphere corresponds to the transformation $x' = -\frac{1}{\overline{x}}$ in the split complex plane, that is, inversion with respect to the circle $\overline{x}x = -1$ in this plane. It is easy to verify that a linear transformation of the coordinates X^i corresponds to a fractional linear transformation (2.46) or (2.47) in the split complex plane. Therefore, since a rotation of the sphere is an affine rotation preserving antipodal points of the sphere, it becomes a fractional linear transformation preserving the circle $\overline{x}x = -1$, that is, a transformation (3.2) in the split complex plane. ■

Theorem 3.5. *There are bijections from the manifolds of oriented lines in the planes R^2 and R_1^2 onto the extended algebra $\mathbb{C}^0$. If an oriented line not parallel to the x^1-axis meets this axis at a point with abscissa s and the angle between the positive direction of this line and the positive direction of the x^1-axis is φ, this line is mapped to the dual number*

$$x = \tan\frac{\varphi}{2}(1+\epsilon s). \tag{3.6}$$

The motions in R^2 and R_1^2 are mapped to the transformations (3.2) and (3.3), respectively.

PROOF. If a line not parallel to the x^1-axis tends to a line parallel to this axis, equi-directed with it and with distance t from this axis, the dual number (3.5) tends to the number $x = \frac{t}{2}\epsilon$. Therefore, a line parallel to the x^1-axis at distance t from this axis and equi-directed with it is mapped to the dual number $x = \frac{t}{2}\epsilon$.

If we replace the direction of the line mapped to the number (3.6) by the opposite direction, the number (3.6) will be replaced by

$$x' = \tan\frac{\varphi+\pi}{2}(1+\epsilon s) = -\cot\frac{\varphi}{2}(1+\epsilon s) = -1/\tan\frac{\varphi}{2}(1-\epsilon s) = -\frac{1}{\overline{x}}. \tag{3.7}$$

(Note that $(1+\epsilon s)(1-\epsilon s) = 1$.) Therefore, lines parallel to the x^1-axis at distance t from this axis and with direction opposite to this axis are mapped to "ideal dual numbers", which play the role of the numbers $-\frac{1}{\overline{x}}$ for the dual numbers $x = b\epsilon$.

Under translation along the x^1-axis over a distance t, the line mapped to the dual number (3.6) is transformed into the line mapped to the dual number $x' = \tan\frac{\varphi}{2}(1+\epsilon(s+t)) = (1+\epsilon t)x$. Under translation along the x^2-axis over a distance u, the line mapped to the dual number (3.6) is transformed into the line mapped to the dual number $x' = \tan\frac{\varphi}{2}(1+\epsilon(s+u\cot\varphi)) = \frac{2x+\epsilon u}{\epsilon ux+2}$. Under rotation about O over an angle ψ, the line mapped to the dual number (3.6) is transformed into the line mapped to the dual number $x' = \frac{x+\tan(\psi/2)}{-x\tan(\psi/2)+1}$. Reflection in the x^1-axis becomes the transformation $x' = \overline{x}$ in the dual plane.

Therefore, an arbitrary motion in the plane R^2 becomes a transformation (3.2) in the dual plane. Such a transformation maps a pair of points x, $-\frac{1}{\overline{x}}$ in the dual plane to a pair x', $-\frac{1}{\overline{x}'}$. The points in this pair can be obtained from each other by inversion with respect to the locus $\overline{x}x = -1$ in this plane. This locus is a pair of imaginary isotropic lines.

As we have seen in **0.5.6** in R_1^2 the angle φ is equal to $i\psi$, $\frac{\pi}{2} \pm i\psi$, or $\pi - i\psi$. In the first case formula (3.6) can be written as

$$x = s + \epsilon\tanh\frac{\psi}{2}. \tag{3.8}$$

As for R^2 we can prove that a motion in R_1^2 becomes a transformation (3.3) in the dual plane. Such a transformation maps a pair of points x, $\frac{1}{\overline{x}}$ in the dual plane to a pair x', $\frac{1}{\overline{x}'}$. The points in this pairs can be obtained from each other by inversion with respect to the locus $\overline{x}x = 1$ in this plane. This locus is a pair of real isotropic lines. ■

Note that since each pair of antipodal points on the sphere in R^3 corresponds to a great circle of this sphere with these points as poles, and since each pair of antipodal points on a sphere of real or pure imaginary radius in R^3_1 corresponds to a great circle of the conjugate sphere, that is, the sphere with the same center but with pure imaginary, respectively real, radius, with these points as poles (in this case the axis joining the points is orthogonal to the plane of the circle), in Theorem 3.4 we can replace pairs of antipodal points by oriented great circles on the sphere or on the conjugate sphere.

§3.2. Motions and Similitudes

3.2.1. Motions. In **0.8.9** we have defined a *motion* in the real space R^n or R^n_ℓ as a bijection in the space which preserves distances between points. We have seen that the motions in R^n or R^n_ℓ form a subgroup of the group of affine transformations in the space E^n.

In Theorem 3.1 we have proved that the Hermitian Euclidean and pseudo-Euclidean spaces over $\mathbb{C}, \mathbb{C}', \mathbb{H}$, and $\mathbb{H}'$ are isometric to real Euclidean or pseudo-Euclidean spaces. Now, if we retain the definitions of the groups of motions in these Hermitian spaces, these groups will be isomorphic to the groups of motions in the corresponding real spaces. Therefore, the group of *motions* in a quadratic or Hermitian, Euclidean or pseudo-Euclidean, space over an algebra $\mathbb{A}$ is defined as the subgroup of the group of affine transformations in $\mathbb{A}E^n$ which preserve distances between points.

Therefore, the groups of motions in the spaces $\mathbb{C}R^n, \mathbb{C}'R^n, \mathbb{C}'R^n_\ell, \mathbb{C}^0R^n$, and $\mathbb{C}^0R^n_\ell$ can be written in the form (2.33), where U is an orthogonal or pseudoorthogonal matrix in $\mathbb{C}O_n, \mathbb{C}'O_n, \mathbb{C}'O_{n,\ell}, \mathbb{C}^0O_n$, or $\mathbb{C}^0O_{n,\ell}$, respectively.

The groups of motions in the spaces $\mathbb{C}\overline{R}^n, \mathbb{C}\overline{R}^n_\ell, \mathbb{C}'\overline{R}^n, \mathbb{C}^0\overline{R}^n$, and $\mathbb{C}^0\overline{R}^n_\ell$ can be written in the form (2.33) or (2.34), where U is a unitary or pseudounitary matrix in $\mathbb{C}U_n, \mathbb{C}U_{n,\ell}, \mathbb{C}'U_n, \mathbb{C}^0U_n$, or $\mathbb{C}^0U_{n,\ell}$, respectively.

The groups of motions in the spaces $\mathbb{H}\overline{R}^n, \mathbb{H}\overline{R}^n_\ell, \mathbb{H}'\overline{R}^n, \mathbb{H}^0\overline{R}^n$, $\mathbb{H}'^0\overline{R}^n$, and $\mathbb{H}'^0_\ell$ can be written in the form (2.35), where U is a unitary or pseudounitary matrix in $\mathbb{H}U_n, \mathbb{H}U_{n,\ell}, \mathbb{H}'U_n, \mathbb{H}^0U_n, \mathbb{H}^0U_{n,\ell}$, or $\mathbb{H}'^0U_n$, respectively.

Motions preserving a point are called *rotations*. An arbitrary motion is the product of a rotation and a translation. Therefore, the dimension of a group of motions considered above is equal to the sum of the dimensions of the group of rotations and the group of translations.

Theorem 3.6. *The dimensions of the groups of motions in the spaces $\mathbb{C}R^n$, $\mathbb{C}'R^n$, and $\mathbb{C}'R^n_\ell$ are equal to $n(n+1)$.*

The dimensions of the groups of motions in the spaces $\mathbb{C}\overline{R}^n, \mathbb{C}\overline{R}^n_\ell$, and $\mathbb{C}'\overline{R}^n$ are equal to $n(n+2)$.

The dimensions of the groups of motions in the spaces $\mathbb{H}\overline{R}^n, \mathbb{H}\overline{R}^n_\ell$, and $\mathbb{H}'\overline{R}^n$ are equal to $(n+1)(2n+3)$.

PROOF. A rotation in a complex or split complex space can be reduced to the form $x' = Ux$ or $x' = U\overline{x}$, and so the dimensions of the groups of rotations in these spaces are equal to the dimensions of the groups of matrices U.

For the groups of rotations in $\mathbb{C}R^n, \mathbb{C}'R^n$, and $\mathbb{C}'R^n_\ell$ these dimensions are equal to $2 \cdot \frac{n(n-1)}{2} = n(n-1)$.

For the groups of rotations in $\mathbb{C}\overline{R}^n$, $\mathbb{C}\overline{R}^n_\ell$, and $\mathbb{C}'\overline{R}^n$ these dimensions are equal to $2 \cdot \frac{n(n-1)}{2} + n = n(n-1) + n = n^2$.

The dimensions of the groups of translations $x' = x + a$ are equal to the real dimension of the spaces themselves, that is, $2n$.

Therefore, the dimensions of the groups of motions in $\mathbb{C}R^n$, $\mathbb{C}'R^n$, and $\mathbb{C}'_\ell$ are equal to $n(n-1) + 2n = n(n+1)$.

Analogously, the dimensions of the groups of motions in $\mathbb{C}\overline{R}^n, \mathbb{C}\overline{R}^n_\ell$, and $\mathbb{C}'\overline{R}^n$ are equal to $n^2 + 2n = n(n+2)$.

A rotation in a quaternionic or split quaternionic space can be reduced to the form $x' = Uxb$ where $|b| = 1$, and so the dimensions of the groups of rotations in these spaces are equal to the dimensions of the groups of matrices U plus 3.

The dimensions of the groups of matrices U for the groups of rotations in $\mathbb{H}\overline{R}^n, \mathbb{H}\overline{R}^n_\ell$, and $\mathbb{H}'\overline{R}^n$ are equal to $4 \cdot \frac{n(n-1)}{2} + 3n = 2n(n-1) + 3n = 2n^2 + n$, the dimensions of the groups of rotations in these spaces are equal to $2n^2 + n + 3$.

The dimensions of the groups of translations in these spaces are equal to $4n$.

Therefore, the dimensions of the groups of motions in $\mathbb{H}\overline{R}^n, \mathbb{H}\overline{R}^n_\ell$, and $\mathbb{H}'\overline{R}^n$ are equal to $2n^2 + n + 3 + 4n = 2n^2 + 5n + 3 = (n+1)(2n+3)$. ■

3.2.2. Similitudes. In **0.8.9** we have defined a *similitude* in the real space R^n or R^n_ℓ as the product of a motion and a homothety. This definition is valid for all Euclidean and pseudo-Euclidean spaces.

In the space $\mathbb{A}E^n$ a homothety is an affine transformation that can be reduced to the form $x' = xt$, with t an arbitrary invertible element in $\mathbb{A}$, since the inverse homothety for $x' = xt$ is $x' = xt^{-1}$ and, for a noncommutative algebra, the transformation $x \to t^{-1}xt$ is a ring automorphism in the algebra.

Therefore, the dimensions of the groups of similitudes in complex and split complex, Euclidean and pseudo-Euclidean, spaces are equal to the dimensions of the groups of motions in these spaces plus 2.

For the algebras $\mathbb{H}$ and $\mathbb{H}'$ we put $t = t_0 t_1$, where t_0 is real and $|t_1| = 1$. Since t_1 occurs in the coefficient b, the dimensions of the groups of similitudes in quaternionic and split quaternionic, Euclidean and pseudo-Euclidean, spaces are equal to the dimensions of the groups of motions in these spaces plus 1.

3.2.3. Antimotions and Antisimilitudes. In **1.1.2** and **1.2.3** we have mentioned *antimotions* and *antisimilitudes* in the plane R^2_1 and the space R^4_2. An *antimotion* in R^{2n}_n is a bijection mapping any points with real distance ρ to points with imaginary distance $i\rho$, and mapping any two points with pure imaginary distance $i\rho$ to points with real distance ρ.

An antimotion in $\mathbb{C}'R^{2n}_n$ is an affine transformation mapping any two points with split complex distance ρ to points with bicomplex distance $i\rho$, and mapping any two points with bicomplex distance $i\rho$ to points with split complex distance ρ.

An antimotion in $\mathbb{C}\overline{R}^{2n}_n, \mathbb{C}'\overline{R}^n, \mathbb{H}\overline{R}^{2n}_n$, or $\mathbb{H}'\overline{R}^n$ is an affine transformation in the space mapping any two points with real distance ρ to points with pure imaginary distance $i\rho$, and mapping any two points with pure imaginary distance $i\rho$ to points with real distance ρ.

An antimotion in the space R^{2n}_n or $\mathbb{C}'R^{2n}_n$, with inner squares of vectors $xEx = -\sum_i (x^{2i-1})^2 + \sum_i (x^{2i})^2$, or in the space $\mathbb{C}\overline{R}^{2n}_n$ or $\mathbb{H}\overline{R}^{2n}_n$, with Hermitian inner squares of vectors $\overline{x}Ex = -\sum_i \overline{x}^{2i-1}x^{2i-1} + \sum_i \overline{x}^{2i}x^{2i}$, is the product of a motion in the corresponding space by the transformation (2.7).

An antimotion in the space $\mathbb{C}'\overline{R}^n$ or $\mathbb{H}'\overline{R}^n$ is the product of a motion in the space by the transformation $'x^i = x^i e$.

The product of two antimotions is a motion, therefore, the motions and antimotions in a space form the group of motions and antimotions. The dimension of this group is the same as the dimension of the group of motions in the space.

Since antimotions change the sign of inner squares, the matrices U occurring in the expressions $x' = Ux + a$ describing antimotions in R^{2n}_n satisfy the conditions $(Ux)^2 = xU^\top Ux = -x^2 = -xIx$, that is, $U^\top U = -I$.

The matrices U of antimotions in $\mathbb{C}\overline{R}^{2n}_n$ and $\mathbb{H}\overline{R}^{2n}_n$ satisfy the conditions $\overline{Ux}(Ux) = \overline{x}\overline{U}^\top Ux = -\overline{x}x = -\overline{x}Ix$, that is, $\overline{U}^\top U = -I$.

An *antisimilitude* in one of these spaces is the product of an antimotion and a homothety.

§3.3. Lines, m-Planes and Hyperplanes

3.3.1. Matrix Coordinates of m-Planes. In a quadratic Euclidean space the vectors a_a in equations (0.15) and (2.48) and covectors u^u in equations (2.50), which in Euclidean spaces can be regarded as vectors, are assumed to be unit and to be orthogonal ($a_a a_b = \delta_{ab}$, $u^u u^v = \delta^{uv}$). Therefore, the $(m \times n)$-matrices A in equation (2.49) and $(n \times m)$-matrices U in equations (2.51) satisfy the conditions

$$A^\top A = I, \quad UU^\top = I. \tag{3.9}$$

In a Hermitian Euclidean space the vectors a_a and covectors u^u in (2.48) and (2.50), which can too be regarded as vector, are assumed to be unit and unitary orthogonal ($\overline{a}_a a_b = \delta_{ab}$, $\overline{u}^u u^v = \delta_{uv}$). Therefore the matrices A in (2.49) and U in (2.51) satisfy the conditions

$$\overline{A}^\top A = I, \quad \overline{U}U^\top = I. \tag{3.10}$$

3.3.2. Shortest Distance between Two m-Planes. Let us find the *shortest distance* between two m-planes in the space R^n without common directions.

Theorem 3.7. *The shortest distance δ between two m-planes $x = x_0 + a_a t^a$ and $y = y_0 + b_a u^a$ in R^n without common directions is equal to*

$$\delta = \frac{|[a_1 a_2 \ldots a_m b_1 b_2 \ldots b_m, y_0 - x_0]|}{|[a_1 a_2 \ldots a_m b_1 b_2 \ldots b_m]|}. \tag{3.11}$$

<u>PROOF</u>. Since the m-planes have no common directions, their $2m$ direction vectors a_a, b_a are linearly independent. The shortest distance δ between these m-planes, which is the length of a common perpendicular, is equal to the altitude of the parallelepiped built on the vectors $a_a, b_a, y_0 - x_0$ whose base is the parallelepiped

built on a_a, b_a. Therefore, δ is equal to the quotient of the volume of the first parallelepiped by the volume of the second one, that is, (3.11). ■

In R^n_ℓ and in quadratic and Hermitian Euclidean and pseudo-Euclidean spaces the rules for determining the shortest distance between two m-planes are analogous.

3.3.3. Stationary Angles between Two m-Planes. Besides shortest distance, two m-planes in a quadratic or Hermitian, Euclidean or pseudo-Euclidean, space have other metric invariants, called *stationary angles*. Among these angles there are the minimal and the maximal angles between vectors in these m-planes, as well as all other angles satisfying only the necessary conditions for being an extremal angle between these vectors. These angles are realized in planes orthogonal to both m-planes.

Theorem 3.8. *The squares of the cosines of the stationary angles ω_a between two m-planes $x = x_0 + a_a t^a$ and $y = y_0 + b_a u^a$ in $\mathbb{A}R^n$ or $\mathbb{A}\overline{R}^n$ without common directions are the eigenvalues of the matrix (2.58), where A and B are the projective matrix coordinates of the intersections of the m-planes with the hyperplane at infinity in $\mathbb{A}R^n$ or $\mathbb{A}\overline{R}^n$ and U and V are the tangential matrix coordinates of the intersections with the same hyperplane of the two $(n-m)$-planes totally orthogonal to the m-planes.*

<u>PROOF</u>. Since the planes that are orthogonal to both m-planes in $\mathbb{A}R^n$ or $\mathbb{A}\overline{R}^n$ are also orthogonal to the $(n-m)$-planes totally orthogonal to these m-planes, the lines at infinity of these planes are the transversals of the $(m-1)$-planes and $(n-m-1)$-planes at infinity for the m-planes and $(n-m)$-planes, and the eigenvalues of the $(m \times m)$- matrix (2.58) are the cross ratios of the points of intersection of these transversals with the $(m-1)$-planes and $(n-m-1)$-planes. Let the a-th common orthogonal plane of the m-planes intersect the m-planes in lines with direction vectors e_a and $e'_a = e_a \cos\, \omega_a + e_{m+a} \sin\, \omega_a$, and intersect the $(n-m)$-planes in lines with direction vectors e_{m+a} and $e'_{m+a} = -e_a \sin\, \omega_a + e_{m+a} \cos\, \omega_a$. Then $e_{m+a} = -e_a \cot\, \omega_a + e'_a / \sin\, \omega_a$, $e'_{m+a} = -e_a/\sin\, \omega_a + e'_a \cot\, \omega_a$, that is, $\alpha = -\delta = \cot\, \omega_a$, $\beta = -\gamma = 1/\sin\, \omega_a$ and the cross ratio (2.43) is equal to $\cos^2 \omega_a$.

Let us show that in $\mathbb{A}R^n$ the angles ω_a are stationary values of the angles between vectors in the m-planes. Let $a = a_a t^a$ and $b = b_a u^a$ $(a^2 = b^2 = 1)$ be arbitrary unit vector in the m-planes. The Lagrange function is $U = ab + \frac{\lambda}{2} a^2 + \frac{\mu}{2} b^2$. The partial derivatives $\frac{\partial U}{\partial t^a}$ and $\frac{\partial U}{\partial u^a}$ are equal to $a_a b + \lambda a_a a$ and $a b_a + \mu b b_a$. A necessary condition for a conditional extremum of the function $ab = \cos\, \omega$ under the conditions $a^2 = b^2 = 1$ is that both these partial derivatives vanish. This can be rewritten in the form $a_a b_b u^b + \lambda a_a a_b t^b = 0$, $a_a b_a t^a + \mu b_a b_b u^a = 0$ or, if we introduce vectors $t = \{t^a\}$ and $u = \{u^a\}$ in L^m, in the form $A^\top B u + \lambda A^\top A t = 0$, $B^\top A t + \mu B^\top B u = 0$. Hence $t = -\lambda^{-1}(A^\top A)^{-1}(A^\top B)u$, $u = -\mu^{-1}(B^\top B)^{-1}(B^\top A)t$, and

$$(A^\top A)^{-1}(A^\top B)(B^\top B)^{-1}(B^\top A)t = \lambda\mu t. \tag{3.12}$$

If we multiply the partial derivatives $\frac{\partial U}{\partial t^a}$ and $\frac{\partial U}{\partial u^a}$ by t^a and u^a, respectively, and take their sum, we obtain the equalities $ab + \lambda = 0$ and $ab + \mu = 0$, hence $\lambda\mu = (ab)^2 = \cos^2\, \omega$. It is easy to check that the matrices $A^\top$ and $B^\top$ coincide with the tangential matrix coordinates U and V of the $(n-m)$-planes totally orthogonal to the m-planes. If we write formula (3.12) as $Wt = \lambda\mu t$, the matrix W can be written as

$$W = (A^\top A)^{-1}(A^\top B)(B^\top B)^{-1}(B^\top A). \tag{3.13}$$

The case of $\mathbb{A}\overline{R}^n$ differs from that of $\mathbb{A}R^n$ by the replacement of the matrices $A^\top$ and $B^\top$ by the matrices $\overline{A}^\top$ and $\overline{B}^\top$. Therefore, in this case formula (3.13) must be replaced by

$$W = (\overline{A}^\top A)^{-1}(\overline{A}^\top B)(\overline{B}^\top B)^{-1}(\overline{B}^\top A). \tag{3.14}$$

The eigenvalues of matrices (3.13) and (3.14) determine stationary angles of m-planes in $\mathbb{A}R^n$ and $\mathbb{A}\overline{R}^n$. ■

If the matrix coordinates A and B of the m-planes in $\mathbb{A}R^n$ and $\mathbb{A}\overline{R}^n$ satisfy the conditions (3.9) and (3.10), the matrices (3.13) and (3.14) have the form

$$W = (A^\top B)(B^\top A) \tag{3.15}$$

and

$$W = (\overline{A}^\top B)(\overline{B}^\top A), \tag{3.16}$$

respectively.

In particular, for lines in $R^n, \mathbb{C}R^n, \mathbb{C}\overline{R}^n$, and $\mathbb{H}\overline{R}^n$ $(m = 1)$, when the role of matrices A and B is played by unit direction vectors a and b of the lines, we obtain the formulas $w = (ab)^2$ and $w = \overline{a}b \cdot \overline{b}a = |\overline{a}b|^2$. In this case the number w is equal to $\cos^2 \omega$, where ω is the angle between the two lines with direction vectors a and b.

If all stationary angles ω_a are equal, the m-planes are called *isoclinic*, if all these angles are equal to $\frac{\pi}{2}$, the m-plane are said to be *totally orthogonal*. Isoclinic m-planes have an infinite set of common orthogonal planes.

The cases of the spaces $\mathbb{A}R^n_\ell$ and $\mathbb{A}\overline{R}^n_\ell$ differ from the cases of the spaces $\mathbb{A}R^n$ and $\mathbb{A}\overline{R}^n$ by replacement of the Euclidean inner products xy and $\overline{x}y$ by the pseudo-Euclidean inner products xEy and $\overline{x}Ey$ and by replacement of the $(m\times m)$-matrices $A^\top A, A^\top B, B^\top B$, and $B^\top A$ by the $(m \times m)$-matrices $A^\top EA$, $A^\top EB$, $B^\top EB$, and $B^\top EA$, or of the $(m\times m)$-matrices $\overline{A}^\top A$, $\overline{A}^\top B$, $\overline{B}^\top B$, and $\overline{B}^\top A$ by $\overline{A}^\top EA$, $\overline{A}^\top EB$, $\overline{B}^\top EB$, and $\overline{B}^\top EA$.

3.3.4. Application of Complex, Split Complex, and Dual Numbers to the Geometry of Real 3-Spaces and 3-Spheres. In **1.6.4** we have defined the Plücker coordinates (1.174) of lines in P^n. These coordinates can also be defined for lines in R^n and R^n_ℓ and also, since P^n can be regarded as a hypersphere in R^{n+1} with identified antipodal points, for great circles of hyperspheres in R^{n+1} and R^{n+1}_ℓ.

Theorem 3.9. *There are bijections from the manifolds of oriented lines in the spaces R^3 and R^3_1 and from the manifolds of oriented great circles on a hypersphere in R^4, R^4_1, or R^4_2 onto a sphere in $\mathbb{C}^0 R^3, \mathbb{C}^0 R^3_1, \mathbb{C}' R^3, \mathbb{C}R^3$, and $\mathbb{C}' R^3_1$, respectively. The motions in R^3 and R^3_1 and the rotations in the hypersphere in R^4, R^4_1 and*

R_2^4 are mapped to rotations of this sphere in $\mathbb{C}^0R^3, \mathbb{C}^0R_1^3, \mathbb{C}'R^3, \mathbb{C}R^3$, and $\mathbb{C}'R_1^3$, respectively.

PROOF. In R^3 we characterize points X by their affine coordinates x^1, x^2, x^3. The projective coordinates are then $x^0 = 1, x^1, x^2, x^3$. Let Y be the point with coordinates 1, $y^1 = x^1 + a^1$, $y^2 = x^2 + a^2$, $y^3 = x^3 + a^3$, where $a = \{a^1, a^2, a^3\}$ is the unit direction vector of the line XY. Then the Plücker coordinates of this line are $p^{01} = a^1$, $p^{02} = a^2$, $p^{03} = a^3$, $p^{12} = x^1a^2 - x^2a^1$, $p^{23} = x^2a^3 - x^3a^2$, $p^{31} = x^3a^1 - x^1a^3$, that is, these coordinates coincide with the coordinates of the vectors a and $x \times a$. Let us characterize the line XY by the dual coordinates

$$P^1 = p^{01} + \epsilon p^{23}, \quad P^2 = p^{02} + \epsilon p^{31}, \quad P^3 = p^{03} + \epsilon p^{12}. \tag{3.17}$$

Then, according to (2.28),

$$(P^1)^2 + (P^2)^2 + (P^3)^2 = 1 + 2\epsilon(p^{01}p^{23} + p^{02}p^{31} + p^{03}p^{12}) = 1, \tag{3.18}$$

that is, these coordinates are the coordinates of a point on a sphere of radius 1 in $\mathbb{C}^0R^3$. If we replace a by $-a$, we obtain the antipodal point on this sphere with coordinates $-P^1, -P^2, -P^3$. For two lines XY and ZT with unit direction vectors a and b we obtain

$$P^1Q^1 + P^2Q^2 + P^3Q^3 = ab + \epsilon([ayb] + [bxa]) = ab + \epsilon([a(y-x)b]).$$

But, by Theorem 3.7, the product $[a(y-x)b]$, that is, the volume of the parallelepiped built on the vectors a, b, and $y - x$, is equal to the product of the shortest distance δ between the two lines by the area of the base of this parallelepiped, that is, $[ab] = \sin\varphi$, where φ is the angle between the directions of the oriented lines. By (1.23) we obtain

$$P^1Q^1 + P^2Q^2 + P^3Q^3 = \cos\varphi - \epsilon\delta\sin\varphi = \cos(\varphi + \epsilon\delta). \tag{3.19}$$

Therefore, the spherical distance between the points P and Q on the dual sphere can be expressed in terms of the metric invariants of the corresponding lines in R^3 and a motion in R^3 is mapped to a rotation of this sphere.

We can similarly consider a bijection from the manifold of oriented lines with unit and imaginary unit direction vectors ($|a| = 1$ and i) in R_1^3 onto the spheres of radius 1 and i in $\mathbb{C}^0R_1^3$.

For R^4 we characterize the oriented great circles on the sphere by orthogonal unit vectors x and y (that is, $x^2 = y^2 = 1$, $xy = 0$), whose coordinates can be regarded as the projective coordinates of points X and Y in P^3. Let us characterize the oriented great circles on this sphere, like the lines XY in P^3, also by split complex coordinates:

$$P^1 = p^{01} + ep^{23}, \quad P^2 = p^{02} + ep^{31}, \quad P^3 = p^{03} + ep^{12}. \tag{3.20}$$

Then, in according with the relations $(p^{01})^2 + (p^{02})^2 + (p^{03})^2 + (p^{12})^2 + (p^{23})^2 + (p^{31})^2 = x^2 \cdot y^2 - (xy)^2 = 1$ and (2.28),

$$(P^1)^2 + (P^2)^2 + (P^3)^2 = (p^{01})^2 + (p^{02})^2 + (p^{03})^2 + \\ +(p^{12})^2 + (p^{23})^2 + (p^{31})^2 + 2e(p^{01}p^{23} + p^{02}p^{31} + p^{03}p^{12}) = 1, \tag{3.21}$$

that is, these coordinates are the coordinates of the points on a sphere of radius 1 in $\mathbb{C}'R^3$. If we replace the vectors x and y by y and x, we obtain the antipodal point on this sphere with coordinates $-P^1, -P^2, -P^3$. Let the points X and Z on two great circles XY and ZT be the points of intersection of these circles with the common orthogonal plane of the planes of these circles containing the stationary angle ω_1, and let the points Y and T on these circles be the points of intersection of these circles with the common orthogonal plane of the planes of these circles containing the stationary angle ω_2. Let $x = e_0$, $y = e_1$, $z = e_0 \cos\,\omega_1 + e_2 \sin\,\omega_1$, $t = e_1 \cos\,\omega_2 + e_3 \sin\,\omega_2$ be the position vectors of these points. Then $x^0 = 1$, $x^1 = x^2 = x^3 = 0$, $y^1 = 1, y^0 = y^2 = y^3 = 0$, and $p^{01} = 1, p^{02} = p^{03} = p^{12} = p^{31} = p^{23} = 0$, $z^0 = \cos\,\omega_1, z^1 = 0$, $z^2 = \sin\,\omega_1$, $z^3 = 0$, $t^0 = 0$, $t^1 = \cos\,\omega_2$, $t^2 = 0$, $t^3 = \sin\,\omega_2$, and $q^{01} = \cos\,\omega_1 \cos\,\omega_2$, $q^{02} = 0, q^{03} = \cos\,\omega_1\,\sin\,\omega_2$, $q^{12} = -\sin\,\omega_1 \cos\,\omega_2$, $q^{31} = 0$, $q^{23} = \sin\,\omega_1 \sin\,\omega_2$. Hence

$$P^1Q^1 + P^2Q^2 + P^3Q^3 = \cos\,\omega_1 \cos\,\omega_2 + e \sin\,\omega_1 \sin\,\omega_2 = \cos(\omega_1 + e\omega_2). \tag{3.22}$$

Therefore, the spherical distance between two points P and Q on the split complex sphere can be expressed in terms of the metric invariants of the corresponding great circles on a hypersphere in R^4, and a rotation in R^4 is mapped to a rotation of this split complex sphere.

We can similarly define bijections from the manifolds of oriented great circles on the hyperspheres in R^4_1 and R^4_2 onto the spheres $\mathbb{C}R^3$ and $\mathbb{C}'R^3_1$. ■

3.3.5. Symmetry Figures in Euclidean and Pseudo-Euclidean Spaces. Since motions in $\mathbb{A}R^n$, $\mathbb{A}R^n_\ell$, $\mathbb{A}\overline{R}^n$, and $\mathbb{A}\overline{R}^n_\ell$ are particular cases of affine transformations in $\mathbb{A}E^n$, symmetry figures in $\mathbb{A}R^n$, $\mathbb{A}R^n_\ell$, $\mathbb{A}\overline{R}^n$, and $\mathbb{A}\overline{R}^n_\ell$ are particular cases of symmetry figures in $\mathbb{A}E^n$. Since reflections in points in $\mathbb{A}E^n$ are motions in $\mathbb{A}R^n$, $\mathbb{A}R^n_\ell$, $\mathbb{A}\overline{R}^n$, and $\mathbb{A}\overline{R}^n_\ell$, points in these spaces are also symmetry figures.

Since a reflection in an m-plane in $\mathbb{A}E^n$ $(m = 1, 2, \ldots, n-1)$ with an $(n-m)$-direction is a motion in $\mathbb{A}R^n$, $\mathbb{A}R^n_\ell$, $\mathbb{A}\overline{R}^n$, or $\mathbb{A}\overline{R}^n_\ell$ only if the $(n-m)$-direction is orthogonal to the m-plane, we see that lines, m-planes $(m = 2, 3, \ldots, n-2)$, and hyperplanes in $\mathbb{A}R^n$, $\mathbb{A}R^n_\ell$, $\mathbb{A}\overline{R}^n$, and $\mathbb{A}\overline{R}^n_\ell$ are also symmetry figures in these spaces.

Since a reflection in an n-chain in $\mathbb{C}R^n$, $\mathbb{C}'R^n$, $\mathbb{C}'R^n_\ell$, $\mathbb{C}\overline{R}^n$, $\mathbb{C}\overline{R}^n_\ell$ or $\mathbb{C}'\overline{R}^n$ is a motion in this space only if this n-chain consists of points with real coordinates or is obtained from this n-chain by a motion in this space, these n-chains are symmetry figures in these spaces. Such n-chains are called *normal n-chains* in these spaces.

Since a reflection in a complex or split complex n-chain in $\mathbb{H}\overline{R}^n$, $\mathbb{H}\overline{R}^n_\ell$, or $\mathbb{H}'\overline{R}^n$ is a motion in this space only if this (split) complex n-chain consists of points with coordinates of the form $x^i_0 + ix^i_1$ or $x_0 + ex^i_1$ or is obtained from this (split) complex n-chain by a motion in this space, these complex and split complex n-chains are symmetry figures in these spaces. Such complex and split complex n-chains are colled *normal complex* and *split complex n-chains* in these spaces.

Theorem 3.10. *The symmetry figures in* $\overset{n}{R}$, $\mathbb{C}\overset{n}{R}$, $\mathbb{C}\overline{R}^n$, *and* $\mathbb{H}\overline{R}^n$ *are the points, lines, m-planes* $(m = 1, 2, \ldots, n-2)$, *and hyperplanes. For* $\mathbb{C}\overline{R}^n$ *and* $\mathbb{H}\overline{R}^n$ *we have to add normal n-chains and normal complex n-chains, respectively.*

PROOF. The assertions of this theorem follow from Theorem 2.58. A reflection in a point of the spaces can be reduced to the form (2.117), a reflection in an m-plane can be reduced to the form (2.118). A reflection in $\mathbb{C}\overline{R}^n$ in a *normal n-chain* can be reduced to the form (2.119) (note that in $\mathbb{C}R^n$ this transformation maps points with distance ω to points with distance $\overline{\omega}$ and is not a motion). A reflection in $\mathbb{H}\overline{R}^n$ in a *normal complex n-chain* can be reduced to the form (2.120). ■

We can similarly determine the symmetry figures in the spaces $\mathbb{C}\overline{R}^n_\ell$, $\mathbb{C}'\overset{n}{R}$, $\mathbb{C}'\overset{n}{R}_\ell$, $\mathbb{C}'\overline{R}^n$, $\mathbb{C}^0\overset{n}{R}$, $\mathbb{C}^0\overset{n}{R}_\ell$, $\mathbb{C}^0\overline{R}^n$, $\mathbb{C}^0\overline{R}^n_\ell$, $\mathbb{H}\overline{R}^n_\ell$, $\mathbb{H}'\overline{R}^n$, $\mathbb{H}^0\overline{R}^n$, $\mathbb{H}^0\overline{R}^n_\ell$, and $\mathbb{H}'^0\overline{R}^n$.

3.3.6. Holomorphy Angle, Holomorphic and Antiholomorphic Real 2-Directions. Since the spaces $\mathbb{C}\overline{R}^n$ and $\mathbb{H}\overline{R}^n$ are isometric to the real spaces R^{2n} and R^{4n}, we can consider two orthogonal unit vectors a and b in real planes in these spaces, represented by vectors A and B in R^{2n} or R^{4n} (the vectors A and B satisfy the conditions $A^2 = B^2 = 1$, $AB = 0$) and consider the Hermitian inner product $\overline{a}b$ of these vectors. In $\mathbb{C}\overline{R}^n$ the coordinates of the vector a are $a^i = A^{2i-1} + iA^{2i}$, the coordinates of the vector A are A^{2i-1}, A^{2i}; the coordinates of the vectors b and B can be similarly written.

Therefore, the Hermitian inner product $\overline{a}b$ is equal to

$$\begin{gathered}\overline{a}b = \sum_i (A^{2i-1} - iA^{2i})(B^{2i-1} + iB^{2i}) = \\ \sum_i (A^{2i-1}B^{2i-1} + A^{2i}B^{2i}) + i\sum_i (A^{2i-1}B^{2i} - A^{2i}B^{2i-1}) = \\ AB + iAJB,\end{gathered} \tag{3.23}$$

where J is a $(2n \times 2n)$-matrix which can be reduced to the form (0.62). Since $AB = 0$ and J is an orthogonal matrix, the vector JB is a unit vector ($|JB| = 1$) and the inner product AJB can be regarded as the cosine of the angle α between A and JB. The equality (3.23) can be written as

$$\overline{a}b = i\,\cos\,\alpha. \tag{3.24}$$

Since the Hermitian inner product $\overline{a}'b'$ of two vectors $a' = a\cos\,\varphi + b\sin\,\varphi$ and $b' = -a\sin\,\varphi + b\cos\,\varphi$ is equal to $\overline{a}b$, the angle α does not depend on a and b but depends only on the real plane containing these vectors. This angle is called the *holomorphy angle* of the real plane or of the 2-direction in $\mathbb{C}\overline{R}^n$.

In $\mathbb{H}\overline{R}^n$ the coordinates of the vector a are $a^i = A^{4i-3} + iA^{4i-2} + jA^{4i-1} + kA^{4i}$, the coordinates of the vector A are $A^{4i-3}, A^{4i-2}, A^{4i-1}, A^{4i}$, and the coordinates of the vectors b and B can be written down similarly. In this case, we obtain instead of (3.24) the formula

$$\overline{a}b = u\,\cos\alpha, \tag{3.25}$$

where u is a unit vector quaternion ($\overline{u} = -u$, $|u| = 1$). The vector u, like the angle α, depends only on the real plane determined by a and b, and, therefore, it is called the *holomorphy vector* of this plane.

Theorem 3.11. *The holomorphy angle α of a real plane in $\mathbb{C}\overline{R}^n$ or $\mathbb{H}\overline{R}^n$ is equal to the angle between two lines in this space whose direction vectors are orthogonal in this plane.*

PROOF. Formulas (3.24) and (3.25) imply that $\cos\alpha = |\overline{a}b|$. However, in **3.3.3** we have seen that $|\overline{a}b|$ is equal to the cosine of the angle between the lines with unit direction vectors a and b. ■

The real planes (or 2-directions) in $\mathbb{C}\overline{R}^n$ and $\mathbb{H}\overline{R}^n$ with $\alpha = 0$ are called *holomorphic* planes; the real planes with $\alpha = \frac{\pi}{2}$ are called *antiholomorphic* planes.

Theorem 3.12. *The set of holomorphic real planes in $\mathbb{C}\overline{R}^n$ coincides with the set of complex lines in this space.*

The holomorphic real planes in $\mathbb{H}\overline{R}^n$ are the real planes that are contained in the quaternionic lines in this space.

The antiholomorphic real planes in $\mathbb{C}\overline{R}^n$ or $\mathbb{H}\overline{R}^n$ are the real planes that are contained in the normal real n-chains in these spaces.

PROOF. The first two assertions follow from the fact that if vectors a and b are the direction vectors in the same complex or quaternionic line in $\mathbb{C}\overline{R}^n$ or $\mathbb{H}\overline{R}^n$, the lines determined by these vectors coincide and therefore $\alpha = 0$.

The third assertion follows from the fact that a normal real n-chain in $\mathbb{C}\overline{R}^n$ or $\mathbb{H}\overline{R}^n$ is isometric to R^n and the Hermitian inner product $\overline{a}b$ of two vectors a and b in this n-chain coincides with the inner product of these vectors in R^n. Therefore, since a and b are orthogonal, the product $\overline{a}b$ is also equal to zero. Hence $\cos\alpha = 0$ and $\alpha = \frac{\pi}{2}$. ■

Theorem 3.13. *The angle $\mathcal{A}$ between two real lines in $\mathbb{C}\overline{R}^n$ or $\mathbb{H}\overline{R}^n$, the angle $\mathfrak{A}$ between the two complex or quaternionic lines determined by the directions of these real lines, and the holomorphy angle α of the real plane through these real lines are related by*

$$\sin\mathfrak{A} = \sin\mathcal{A}\sin\alpha. \tag{3.26}$$

PROOF. Consider two non-orthogonal unit vectors a and b in real planes in $\mathbb{C}\overline{R}^n$ or $\mathbb{H}\overline{R}^n$, represented by vectors A and B in R^{2n} or R^{4n}. Then $AB = \cos\mathcal{A}$ and $|\overline{a}b| = \cos\mathfrak{A}$.

Let c be a unit vector in this real plane orthogonal to a. Then $b = a\cos\mathcal{A} + c\sin\mathcal{A}$, and we have $\overline{a}b = \cos\mathcal{A} + u\cos\alpha\sin\mathcal{A}$ and $\cos^2\mathfrak{A} = \cos^2\mathcal{A} + \cos^2\alpha\sin^2\mathfrak{A}$ and $\sin^2\mathfrak{A} = \sin^2\mathcal{A} - \cos^2\alpha\sin^2\mathcal{A} = \sin^2\mathcal{A}\sin^2\alpha$. In $\mathbb{C}\overline{R}^n$ we have to replace u by i. In both cases we obtain the same formula. ■

§3.4. Polyhedra

3.4.1. Real Polyhedra. In **1.4.6** we have mentioned regular polyhedra in R^n. In this section we consider these polyhedra in more detail. Each hyperplane in R^n, with equation $ux + v = 0$, determines two ***half-spaces***, $ux + v \geq 0$ and $ux + v \leq 0$; the hyperplane $ux + v = 0$ itself is called the ***boundary*** of these half-spaces. An ***n-polyhedron*** (or ***n-polytope***) in R^n is the closed intersection of certain half-spaces.

The boundary of an n-polyhedron consists of ***faces***, which are $(n-1)$-polyhedra. The intersection of two adjacent faces is an $(n-2)$-face and can contain m-faces $(m = 0, 1, \ldots, n-3)$. 0-faces are called ***vertices***, 1-faces are called ***edges*** of the polyhedron.

2-polyhedra are also called ***polygons***; their faces are called ***sides***.

The m-polyhedron whose boundary consists of parts of m pairs of parallel hyperplanes is called an ***m-parallelepiped***, for $m = 2$ a ***parallelogram***.

The m-parallelepiped built on m linearly independent vectors a_a is the set of points with position vectors of the form (0.15), where all parameters t_a $(a = 1, 2, \ldots, m)$ satisfy the conditions $0 \leq t^a \leq 1$. The vertices of a parallelepiped are the points for which which all t^a are equal to 0 or 1.

An m-parallelepiped has $2m$ faces and 2^m vertices. The skew product $[a_1 a_2 \ldots a_m]$ is called the ***oriented volume*** of this m-parallelepiped (we have already met these volumes in formula (3.11)).

A set of points with position vectors of the same form (0.15) but where all parameters t^a satisfy the conditions $0 \leq t^a \leq 1$ and $\sum_a t^a \leq 1$ is called an ***m-simplex***, for $m = 2$ a ***triangle***, for $m = 3$ a ***tetrahedron***.

An m-simplex has $m+1$ faces and $m+1$ vertices. The number $\frac{1}{m!}[a_1 a_2 \ldots a_m]$ is called the ***oriented volume*** of this m-simplex.

3.4.2. Regular Polyhedra. A ***regular polygon*** $\{p\}$ is a polygon in R^2 with p vertices and p sides such that all sides and all angles at the vertices are equal.

A regular 3-polyhedron $\{p, q\}$ is a polyhedron in R^3 all faces of which are regular polygons $\{p\}$ and all ***vertex polygons***, with the middle points of the edges issueing from a vertex of the polyhedron as vertices, are regular polygons $\{q\}$. It is well known that there are regular polygons $\{p\}$ for all $p \geq 3$ and five regular 3-polyhedra: $\{3,3\}$ (tetrahedron), $\{4,3\}$ (cube), $\{3,4\}$ (octahedron), $\{5,3\}$ (dodecahedron), and $\{3,5\}$ (icosahedron).

A regular n-polyhedron $\{p_1 p_2 \ldots p_{n-1}\}$ is an n-polyhedron all faces of which are regular $(n-1)$-polyhedra $\{p_1 p_2 \ldots p_{n-2}\}$ and all ***vertex (n − 1)-polyhedra*** with as vertices the middle points of the edges issueing from a vertex of the n-polyhedron are regular $(n-1)$-polyhedra $\{p_2 p_3 \ldots p_{n-1}\}$.

Theorem of L. Schläfli ([Scl, pp. 212–226; Cox2, pp. 126–128]). *There are six regular 4-polyhedra:* $\{3,3,3\}$ *(4-simplex),* $\{4,3,3\}$ *(4-cube),* $\{3,3,4\}$ *(4-cocube),* $\{3,4,3\}$ *(bicube),* $\{5,3,3\}$ *(120-hedron), and* $\{3,3,5\}$ *(600-hedron).*

For $n > 4$ there are only three regular n-polyhedra: $\{3,3,\ldots,3,3\}$ *(n-simplex),* $\{4,3,\ldots,3,3\}$ *(n-cube), and* $\{3,3,\ldots,3,4\}$ *(n-cocube).*

The classification of the regular polyhedra in R^n is based on the condition that the n-simplex with as vertices a vertex A_0 of an n-polyhedron, a middle point A_1 of the edge through A_0, a center A_2 of symmetry of its 2-face through $A_0 A_1$, etc, till the center A_n of symmetry of the n-polyhedron, has nonzero volume. This condition is equivalent to the inequality

$$\begin{vmatrix} 1 & \cos\frac{\pi}{p_1} & 0 & \dots & 0 & 0 \\ \cos\frac{\pi}{p_1} & 1 & \cos\frac{\pi}{p_2} & \dots & 0 & 0 \\ 0 & \cos\frac{\pi}{p_2} & 1 & \dots & 0 & 0 \\ \vdots & \vdots & \vdots & \ddots & \vdots & \vdots \\ 0 & 0 & 0 & \dots & 1 & \cos\frac{\pi}{p_{n-1}} \\ 0 & 0 & 0 & \dots & \cos\frac{\pi}{p_{n-1}} & 1 \end{vmatrix} > 0. \qquad (3.27)$$

Note that a regular n-simplex, like any n-simplex, has $n+1$ faces and $n+1$ vertices.

A n-cube, like any parallelepiped, has $2n$ faces and 2^n vertices.

A n-cocube, which corresponds to the n-cube under the duality principle in P^n, has 2^n faces and $2n$ vertices (for $n = 3$ the cocube is the octahedron).

A bicube has 24 faces and 24 vertices (the bicube was described in **1.4.7**; its vertices can be represented by the 24 quaternions (1.112).

A 120-hedron has 120 faces and 600 vertices, and the 600-hedron, which corresponds to a 120-hedron under the duality principle in P^4 has 600 faces and 120 vertices.

In **1.4.7** we have noted that the group of symmetries of a regular n-simplex is the finite group A_n of rotations in R^n, the group of symmetries of an n-cube or an n-cocube is the finite group $B_n = C_n$ of rotations in R^n.

The group of symmetries of a regular r-gon is the finite group $G_2(r)$ of rotations in R^2, the group of symmetries of a dodecahedron or an icosahedron is the finite group H_3 of rotations in R^3, the group of symmetries of a bicube is the finite group F_4 in R^4, and the group of symmetries of the 120-hedron or 600-hedron is the finite group H_4 of rotations in R^4.

3.4.3. Regular Honeycombs. The limit case for regular n-polyhedra are *regular* $(n-1)$*-honeycombs*, that is, decompositions of R^{n-1} into congruent regular $(n-1)$-polyhedra with congruent vertex $(n-1)$-polyhedra.

Regular $(n-1)$-honeycombs can be regarded as regular n-polyhedra with infinitely many faces and vertices and with as center of symmetry a point at infinity. Therefore, the regular honeycombs are denoted by the same symbols $\{p_1 p_2, \dots, p_{n-1}\}$ as n-polyhedra.

In the case of regular honeycombs the determinant (3.27) is equal to zero and this makes it possible to classify regular honeycombs.

Theorem of Schläfli-Coxeter ([Scl, p. 224, Cox1, pp. 126–128]).
There are three kinds of regular 2*-honeycombs:* $\{3,6\}, \{4,4\}$, *and* $\{6,3\}$*;*
one kind of regular 3*-honeycombs:* $\{4,3,4\}$*;*
three kinds of regular 4*-honeycombs:* $\{3,3,4,3\}, \{3,4,3,3\}$, *and* $\{4,3,3,4\}$*;*
and one kind of regular n*-honeycombs for* $n > 4$*:* $\{4,3,\dots,3,4\}$.

The faces of the regular 2-honeycombs $\{3,6\}, \{4,4\}$, and $\{6,3\}$ are equilateral triangles, squares, and regular hexagons, respectively.

The faces of the regular 3-honeycomb $\{4,3,4\}$ are cubes.

The faces of the regular 4-honeycombs $\{3,3,4,3\}$, $\{3,4,3,3\}$, and $\{4,3,3,4\}$ are 4-cocubes, bicubes, and 4-cubes, respectively.

The faces of the regular n-honeycomb $\{4,3,\dots,3,4\}$ are n-cubes.

There is also the theory of regular polyhedra in $\mathbb{C}\overline{R}^n$ built by H. S. M. Coxeter [Cox9].

§3.5. Hyperquadrics

3.5.1. Principal Axes of Hyperquadrics. In **2.5.4** we have considered conjugate diameters of hyperquadrics and of Hermitian hyperquadrics in the affine space $\mathbb{A}E^n$.

If we introduce in this space the metric or pseudometric of $\mathbb{A}R^n$, $\mathbb{A}R^n_\ell$, $\mathbb{A}\overline{R}^n$, or $\mathbb{A}\overline{R}^n_\ell$, a diameter of a (Hermitian) hyperquadric orthogonal to the hyperplane conjugate with it is called a *principal axis* of this (Hermitian) hyperquadric.

Theorem 3.14. *The direction vectors of the principal axes of a hyperquadric (2.66) or Hermitian hyperquadric (2.69) in the Euclidean space $\mathbb{A}R^n$ or $\mathbb{A}\overline{R}^n$ are the eigenvectors of the matrix A in (2.66), respectively (2.69).*

PROOF. In **2.5.4** we have seen that the hyperplane conjugate with the diameter of the hyperquadric (2.66) or Hermitian hyperquadric (2.69), with direction vector ℓ, is the hyperplane $ux + v = 0$, where the covector u is equal to $A\ell$. In $\mathbb{A}R^n$ and $\mathbb{A}\overline{R}^n$ the covector u can be regarded as a vector which is orthogonal or unitary orthogonal to this hyperplane, and the convolution ux can be regarded as the inner or Hermitian inner product of two vectors. Therefore, the direction vector ℓ of a principal axis must satisfy the condition

$$A\ell = \lambda\ell, \tag{3.28}$$

that is, ℓ is an eigenvector of the matrix A. ■

In the pseudo-Euclidean spaces $\mathbb{A}R^n_\ell$ and $\mathbb{A}\overline{R}^n_\ell$ the covector u can also be regarded as a vector which is orthogonal or unitary orthogonal to the hyperplane $ux+v=0$ and the convolution ux can be regarded as the inner or Hermitian inner product of two vectors. Therefore, if the inner or Hermitian inner products of vectors in these spaces is xEy or $\overline{x}Ey$, the direction vector ℓ of a principal axis in these spaces must satisfy the condition

$$EA\ell = \lambda\ell, \tag{3.29}$$

that is, ℓ is an eigenvector of the matrix EA.

Theorem 3.15. *A central hyperquadric in R^n or R^n_ℓ can be reduced to the form (2.76) or (2.77).*

A central Hermitian hyperquadric in $\mathbb{C}\overline{R}^n, \mathbb{H}\overline{R}^n, \mathbb{C}\overline{R}^n_\ell$, or $\mathbb{H}\overline{R}^n_\ell$ can be reduced to the form (2.78) or (2.79).

A central Hermitian hyperquadric in $\mathbb{C}'\overline{R}^n$ or $\mathbb{H}'\overline{R}^n$ can be reduced to the form (2.78).

PROOF. A central hyperquadric or central Hermitian hyperquadric can be reduced to these forms by taking the principal axes of the hyperquadric as coordinate axes. ■

3.5.2. Invariants of Equations of Hyperquadrics. If a motion $x = Ux' + a$ in R^n is applied to a hyperquadric (2.66), this hyperquadric becomes the hyperquadric

$$x'A'x' + 2b'x' + c' = 0, \tag{3.30}$$

where

$$A' = U^\top AU, \quad b' = U^\top(Aa + b), \quad c' = aAa + 2ba + c. \tag{3.31}$$

If a motion $x = Ux' + a$ in $\mathbb{C}\overline{R}^n$ is applied to a Hermitian hyperquadric (2.69), this hyperquadric becomes the hyperquadric

$$\overline{x}'A'x' + \overline{b}'x' + \overline{x}'b' + c' = 0, \tag{3.32}$$

where

$$A' = \overline{U}^\top AU, \quad b' = \overline{U}^\top(Aa + b), \quad c' = \overline{a}Aa + \overline{b}a + \overline{a}b + c. \tag{3.33}$$

Theorem 3.16. *The metric invariants of the equations of hyperquadrics in R^n under a rotation in this space are the coefficients $\delta, \delta_1, \ldots, \delta_{n-1}$ of the polynomial*

$$\det(A - \lambda I) = \delta - \delta_1\lambda + \delta_2\lambda^2 - \cdots + (-1)^{n-1}\delta_{n-1}\lambda^{n-1} + (-1)^n c\lambda^n \tag{3.34}$$

and the coefficients $\Delta, \Delta_1, \ldots, \Delta_{n-1}$ of the polynomial

$$\det\begin{bmatrix} A - \lambda I & b \\ b & c \end{bmatrix} = \Delta - \Delta_1\lambda + \Delta_2\lambda^2 - \cdots + (-1)^{n-1}\Delta_{n-1}\lambda^{n-1} + (-1)^n c\lambda^n. \tag{3.35}$$

Under a translation these invariants are the coefficients a_{ij} and the determinant Δ.

Under an arbitrary motion in R^n these invariants are the coefficients δ, $\delta_1, \ldots,$ δ_{n-1} of the polynomial (3.34) and the determinant Δ.

The metric invariants of equations of Hermitian hyperquadrics in $\mathbb{C}\overline{R}^n$ under a rotation are the coefficients $\delta, \delta_1, \ldots, \delta_{n-1}$ of the polynomial (3.34) and the coefficients $\Delta, \Delta_1, \ldots, \Delta_{n-1}$ of the polynomial which is obtained from (3.35) by replacing the matrix $\begin{bmatrix} A - \lambda I & b \\ b & c \end{bmatrix}$ by $\begin{bmatrix} A - \lambda I & b \\ \overline{b} & c \end{bmatrix}$.

Under a translation these invariants are the coefficients a_{ij} and the determinant Δ.

Under an arbitrary motion in $\mathbb{C}\overline{R}^n$ these invariants are the coefficients δ, $\delta_1, \ldots,$ δ_{n-1} of (3.34) and the determinant Δ.

PROOF. For a rotation $x = Ux'$, that is, if $a = 0$, formulas (3.31) give $A' = U^\top AU$, $b' = U^\top b$, $c' = c$. Therefore, since $A' - \lambda I = U^\top(A - \lambda I)U$ and det $U = \pm 1$, det $(A' - \lambda I) =$ det $(A - \lambda I)$, and since

$$\begin{bmatrix} A' - \lambda I & b \\ b & c \end{bmatrix} = \begin{bmatrix} U^{-1} & 0 \\ 0 & 1 \end{bmatrix} \begin{bmatrix} A - \lambda I & b \\ b & c \end{bmatrix} \begin{bmatrix} U & 0 \\ 0 & 1 \end{bmatrix}$$

we obtain $\det \begin{bmatrix} A' - \lambda I & b \\ b & c \end{bmatrix} = \det \begin{bmatrix} A - \lambda I & b \\ b & c \end{bmatrix}$ and all coefficients of the polynomials (3.34) and (3.35) are invariant under a rotation.

For a translation $x =' x + a$ $A' = A$, $b' = Aa + b$, $c' = aAa + 2ba + c$, all entries of the matrix A are invariant; since $\Delta = \det \begin{bmatrix} A & b \\ b & c \end{bmatrix} = \det \begin{bmatrix} A & Aa + b \\ b & ba + c \end{bmatrix}$ and by adding to the entries of the last row of this matrix the products of the entries of the upper row by the vector a, we obtain the determinant Δ'. Therefore, Δ' is equal to Δ, that is, this determinant is also invariant.

The third assertion of the theorem follows from the first and second ones.

The proof of the second part of the theorem is similar to the proof of the first part. ■

A similar theorem is valid also for Hermitian hyperquadrics in $\overline{\mathbb{H}R}^n$, but instead of determinants of real and complex matrices here we must consider semideterminants of quaternionic matrices.

The coefficients of the equations (2.76) and (2.78) of hyperquadrics in Euclidean spaces can be expressed in terms of the metric invariants of these hyperquadrics.

§3.6. Hyperspheres

3.6.1. Hyperspheres and Hermitian Hyperspheres. *Hyperspheres* in quadratic Euclidean or pseudo-Euclidean spaces and *Hermitian hyperspheres* in Hermitian Euclidean or pseudo-Euclidean spaces are sets of points in these spaces whose distance from a fixed point is constant. The fixed point is called the *center*, the constant distance is called the *radius* of the hypersphere. The equation of a hypersphere with as center the point with position vector x_0 and with radius r in a quadratic space has the form (0.21), in a Hermitian space this equation has the form

$$(\overline{x} - \overline{x}_0)(x - x_0) = r^2. \tag{3.36}$$

Theorem 3.17. *Hyperspheres in R^n and $\mathbb{C}R^n$ are particular cases of hyperquadrics in these spaces. Hermitian hyperspheres in $\overline{\mathbb{C}R}^n$ and $\overline{\mathbb{H}R}^n$ are particular cases of Hermitian hyperquadrics in these spaces.*

The matrix A occuring in the equation of such a hyperquadric has the form λI, where λ is a real number for R^n, $\overline{\mathbb{C}R}^n$, and $\overline{\mathbb{H}R}^n$, and a complex number for $\mathbb{C}R^n$.

PROOF. Equation (0.21) can be written in the form $\lambda(x^2 - 2x_0x + x_0^2 - r^2) = 0$, where λ is real for R^n and complex $\mathbb{C}R^n$. The equations (3.36) can be written in the form $\lambda(\overline{x}x - \overline{x}_0x - \overline{x}x_0 + \overline{x}_0x_0 - r^2) = 0$, where λ is real. ■

We can similarly prove that hyperspheres in R^n_ℓ are particular cases of hyperquadrics in these spaces and that Hermitian hyperspheres in $\overline{\mathbb{C}R}^n_\ell$ and $\overline{\mathbb{H}R}^n_\ell$ are particular cases of Hermitian hyperquadrics in these spaces. In these cases the matrix A in the equation of a hyperquadric has the form λE with λ real.

Note that a Hermitian hypersphere in $\mathbb{C}\overline{R}^n$, $\mathbb{C}\overline{R}^n_\ell$, $\mathbb{H}\overline{R}^n$, or $\mathbb{H}\overline{R}^n_\ell$ is isometric to a hypersphere in R^{2n}, $R^{2n}_{2\ell}$, R^{4n}, or $R^{4n}_{4\ell}$, respectively.

3.6.2. Angle between Hyperspheres. The angle between two hyperspheres in R^n is defined as the angle between the tangent hyperplanes to these hyperspheres at a point of their intersection. This angle is the same for all intersection points of hyperspheres.

Theorem 3.18. *The angle φ between two hyperspheres in R^n with centers at points X_1 and X_2 with position vectors x_1 and x_2 and radii r_1 and r_2 is determined by*

$$\cos\varphi = \left|\frac{r_1^2 + r_2^2 - (x_2 - x_1)^2}{2r_1r_2}\right|. \tag{3.37}$$

PROOF. If X is a common point of the two hyperspheres, consider the triangle X_1X_2X. The angle $\angle X_1XX_2$ of this triangle is equal to φ (Fig. 3.1).

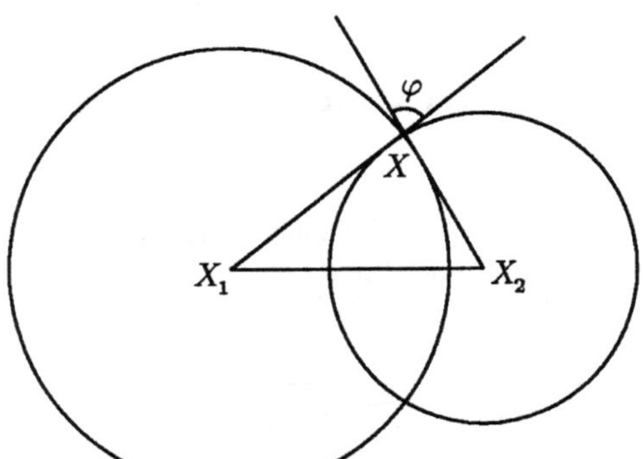

FIG. 3.1

Since $X_1X = r_1$, $X_2X = r_2$, and $X_1X_2 = |x_2 - x_1|$, the cosine law in R^n gives

$$|x_2 - x_1|^2 = r_1^2 + r_2^2 \mp 2r_1r_2 \cos\varphi, \tag{3.38}$$

and hence we obtain (3.37). ■

If the hyperspheres are tangent, $|x_2 - x_1| = |r_1 \pm r_2|$, hence $\cos\varphi = 1$ and $\varphi = 0$.

If the hyperspheres have no real common points, $\cos\varphi > 1$, that is, $\cos\varphi$ can be regarded as $\cosh\psi = \cos(i\psi)$ and therefore we can call the imaginary quantity $i\psi$ the *imaginary angle* between the hyperspheres.

If the hyperspheres are concentric, $x_1 = x_2$ and $\cos\varphi = |\frac{r_1^2+r_2^2}{2r_1r_2}|$. Hence, if $r_1 > r_2$ we have

$$i\sin\varphi = \frac{r_1^2 - r_2^2}{2r_1r_2},\ e^{i\varphi} = \cos\varphi + i\sin\varphi = \frac{2r_1^2}{2r_1r_2} = \frac{r_1}{r_2} \quad \text{and } \varphi = i\ln\frac{r_1}{r_2}.$$

3.6.3. Spherical Trigonometry. Since on spheres in R^3 the role of straight lines is played by their great circles, that is, sections of spheres by diametral planes, a *spherical triangle* on a sphere is defined as the part of the sphere bounded by arcs of great circles.

Theorem 3.19. *The angles A, B, C at the vertices A, B, C of a spherical triangle ABC on a sphere of radius r in R^3 and its sides a, b, c which are opposite to the vertices A, B, C, respectively, are related by the following laws of spherical trigonometry in R^3:*

$$\cos\frac{a}{r} = \cos\frac{b}{r}\cos\frac{c}{r} + \sin\frac{b}{r}\sin\frac{c}{r}\cos A \tag{3.39}$$

(cosine law),

$$\frac{\sin A}{\sin\frac{a}{r}} = \frac{\sin B}{\sin\frac{b}{r}} = \frac{\sin C}{\sin\frac{c}{r}} \tag{3.40}$$

(sine law), and

$$\cos A = -\cos B\,\cos C + \sin B\,\sin C\cos\frac{a}{r} \tag{3.41}$$

(dual cosine law).

PROOF. Let O be the center of the sphere and let the position vectors of the vertices A, B, C with common initial point at O be x, y, z (Fig. 3.2), respectively. Then the sides a, b, c of the triangle ABC are determined by

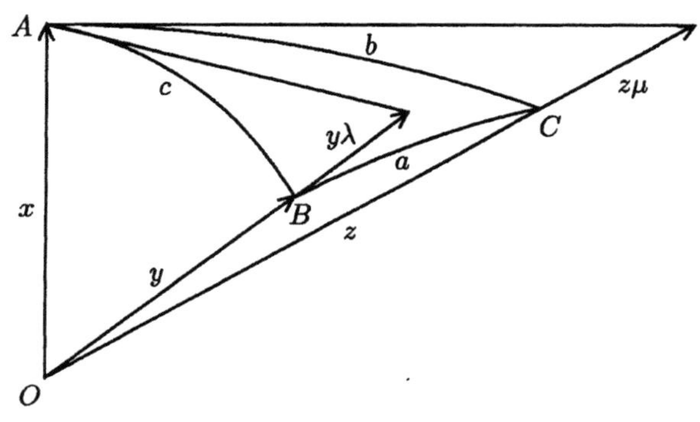

FIG. 3.2

$$\cos\frac{a}{r} = \frac{yz}{r^2}, \quad \cos\frac{b}{r} = \frac{zx}{r^2}, \quad \cos\frac{c}{r} = \frac{xy}{r^2}. \tag{3.42}$$

Let the points of intersection of the tangents to the arcs AB and AC with the lines OB and OC be B' and C', and let the position vectors of these points be $y\lambda$ and $z\mu$. Then the angle A is determined by

$$\cos A = \frac{(y\lambda - x)(z\mu - x)}{|y\lambda - x|\,|z\mu - x|}. \tag{3.43}$$

Since $y\lambda - x$ and $z\mu - x$ are vectors orthogonal to x, that is, $xy\lambda - x^2 = xz\mu - x^2 = 0$, and by (3.42) we have

$$\lambda = \frac{r^2}{xy} = \frac{1}{\cos\frac{c}{r}}, \quad \mu = \frac{r^2}{xz} = \frac{1}{\cos\frac{b}{r}}. \tag{3.44}$$

Substitution of (3.42) and (3.44) into (3.43) gives formula (3.39).

Formulas (3.40) and (3.41) can be obtained from (3.39) by analytical transformations. ■

Formula (3.41) can also be obtained from formula (3.39) as follows. With each point on the sphere the *polar great circle* is connected, and all points on this circle have distance $\frac{\pi}{2}r$ from this point. The point itself is called the *pole* of this great circle. Therefore, with each vertex A, B, C of the spherical triangle ABC a polar great circle is connected. These three great circles form the *polar triangle* $A'B'C'$ for the triangle ABC (Fig. 3.3).

Since great circles AB and AC cut out the arc Ar from the side of the polar triangle opposite to the vertex A', the whole side a' of the polar triangle is equal to $Ar + 2(\frac{\pi}{2} - a)r = (\pi - A)r$. Analogously, $b' = (\pi - B)r$ and $c' = (\pi - C)r$. Since the polar triangle for $A'B'C'$ is ABC, we also obtain $a = (\pi - A')r$, $b = (\pi - B')r$, and $c = (\pi - C')r$. If we express by formula (3.39) a', b', c', and A' in terms of A, B, C, and a, we obtain formula (3.41).

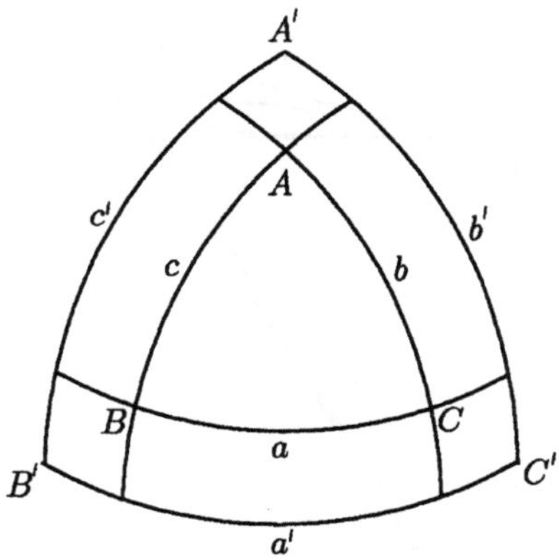

FIG. 3.3

Theorem 3.20. *The angles A, B, C at the vertices A, B, C of a spherical triangle ABC on a sphere of radius iq in the space R^3_1 and the sides a, b, c of this* *triangle which are opposite to the vertices A, B, C, respectively, are related by the following laws of spherical trigonometry in R^3_1:*

$$\cosh\frac{a}{q} = \cosh\frac{b}{q}\cosh\frac{c}{qr} - \sinh\frac{b}{q}\sinh\frac{c}{qr}\cos A \tag{3.45}$$

(cosine law),

$$\frac{\sin A}{\sinh\frac{a}{q}} = \frac{\sin B}{\sinh\frac{b}{r}} = \frac{\sin C}{\sinh\frac{c}{q}} \tag{3.46}$$

(sine law), and

$$\cos A = -\cos B\ \cos C + \sin B\ \sin C\cosh\frac{a}{q} \tag{3.47}$$

(dual cosine law).

PROOF. Let O be the center of the sphere and let the position vectors of the vertices A, B, C with common initial points at O be x, y, z, respectively. Then the sides a, b, c of the triangle ABC are determined by

$$\cosh\frac{a}{q} = \frac{yz}{r^2}, \quad \cosh\frac{b}{q} = \frac{zx}{r^2}, \quad \cosh\frac{c}{q} = \frac{xy}{r^2}. \tag{3.48}$$

Let the points of intersection of the tangents to the arcs AB and AC with the lines OB and OC be B' and C', and let the position vectors of these points be $b\lambda$ and $c\mu$ (Fig. 3.4).

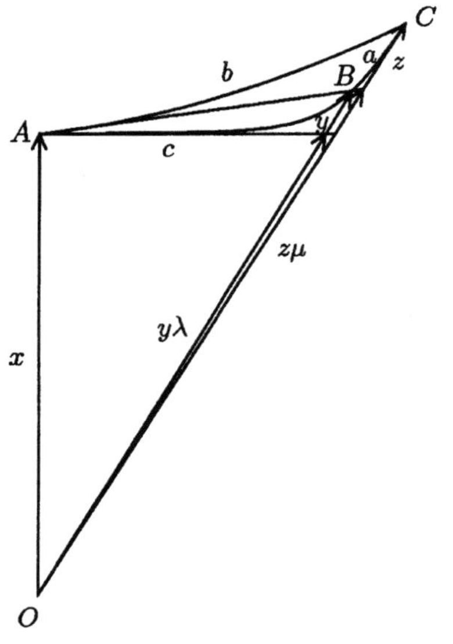

FIG. 3.4

Then the angle A is determined by (3.43). Since $y\lambda - x$ and $z\mu - x$ are vectors orthogonal to the vector x, that is, $xy\lambda - x^2 = xz\mu - x^2 = 0$, by (3.48) we have

$$\lambda = \frac{r^2}{xy} = \frac{1}{\cosh \frac{c}{qr}}, \quad \mu = \frac{r^2}{xz} = \frac{1}{\cosh \frac{b}{qr}}. \tag{3.49}$$

Substitution of (3.48) and (3.49) into (3.43) gives formula (3.45).

Formulas (3.46) and (3.47) can be obtained from (3.45) by analytical transformations. ■

Since the trigonometric functions $\cos x$ and $\sin x$ are related with the hyperbolic functions $\cosh x$ and $\sinh x$ by

$$\cos(ix) = \cos x, \quad \sin(ix) = i \sinh x, \tag{3.50}$$

formulas (3.45), (3.46), and (3.47) coincide with formulas (3.39), (3.40), and (3.41) for $r = iq$. For small values of a, b, c, or for large values of r and q, these formulas are close to the formulas of plane trigonometry:

$$a^2 = b^2 + c^2 \mp 2\, bc \cos A, \tag{3.51}$$

$$\frac{\sin A}{a} = \frac{\sin B}{b} = \frac{\sin C}{c}, \tag{3.52}$$

$$\cos A = -\cos B \cos C + \sin B \sin C. \tag{3.53}$$

Formula (3.53) is equivalent to the equality $A + B + C = \pi$. Note that, since $\cos x < 1$ and $\cosh x > 1$, formulas (3.41) and (3.47) are equivalent to the inequalities $A + B + C > \pi$ and $A + B + C < \pi$, respectively.

3.6.4. Volumes of Spherical Figures. The *geographical coordinates* on a hypersphere of radius r in R^n are the curvilinear coordinates $u^1, u^2, \dots, u^{n-1}$ related with the coordinates $x^1, x^2, \dots, x^n$ of points by the formulas

$$\left.\begin{aligned} x^1 &= r \cos u^1 \cos u^2 \dots \cos u^{n-1}, \\ x^2 &= r \sin u^1 \cos u^2 \dots \cos u^{n-1}, \\ x^3 &= r \sin u^2 \dots \cos u^{n-1}, \\ &\dots\dots\dots\dots\dots\dots\dots\dots \\ x^{n-1} &= r \sin u^{n-2} \cos u^{n-1}, \\ x^n &= r \sin u^{n-1}. \end{aligned}\right\} \tag{3.54}$$

The angle u^1 satisfies $0 \le u^1 < 2\pi$; the angles $u^2, u^3, \dots, u^{n-1}$ satisfy $-\frac{\pi}{2} \le u^i < \frac{\pi}{2}$.

For $n = 3$ the coordinates u^1 and u^2 are the geographical *longitude* and *latitude*, respectively.

The coordinate lines for u^1 are called *meridians* of the hypersphere, that is, great circles through its *poles* (with coordinates $\pm r, 0, \dots, 0$). The coordinate lines for

$u^2, u^3, \dots, u^{n-1}$ are circles, which are in general small circles (for $n = 3$ these circles are *parallels* of the sphere).

If r is variable in (3.54), we obtain *spherical coordinates* in R^n, a multidimensional generalization of polar coordinates in the plane.

The volume element dS of any *cubeable figure* on a hypersphere of radius r in R^n is

$$dS = r^{n-1} \cos u^2 \cos^2 u^3 \dots \cos^{n-2} u^{n-1} du^1 du^2 \dots du^{n-1}. \tag{3.55}$$

The volume element dV of any cubeable figure in R^n in spherical coordinates is $dV = dSdr$.

Theorem 3.21 (JACOBI [Jac], SCHLÄFLI [Scl, p. 229]). *The volume S_n of a hypersphere of radius r in R^n is equal to*

$$\begin{aligned} S_{2m} &= \frac{2^m \pi^m}{(2m-2)!!} r^{2m-1} \quad (\textit{for } n = 2m), \\ S_{2m+1} &= \frac{2^{m+1} \pi^m}{(2m-1)!!} r^{2m} \quad (\textit{for } n = 2m+1), \end{aligned} \tag{3.56}$$

where $(2m-2)!!$, respectively $(2m-1)!!$, denotes the product of all even, respectively odd, numbers from 2 to $2m-2$, respectively from 1 to $2m-1$.

PROOF. These formulas can be obtained by integrating the differential (3.55) for all values of coordinates $u^1, u^2, \dots, u^{n-1}$. ■

Theorem 3.22 (SCHLÄFLI [Scl, p. 229]). *The volume V_n of a ball of radius r in R^n is equal to*

$$\begin{aligned} V_{2m} &= \frac{2^m \pi^m}{(2m)!!} r^{2m}, \quad (\textit{for } n = 2m), \\ V_{2m+1} &= \frac{2^{m+1} \pi^m}{(2m+1)!!} r^{2m+1} \quad (\textit{for } n = 2m+1). \end{aligned} \tag{3.57}$$

PROOF. These formulas can be obtained by integrating the differential $S_n dr$ from 0 to r. ■

Geographical coordinates can also be introduced on hyperspheres in R^n_ℓ. In particular, on a hypersphere of radius iq in R^n_1 these coordinates are the curvilinear coordinates $u^1, u^2, \dots, u^{n-1}$ related with the coordinates $x^1, x^2, \dots, x^n$ of points by the formulas

$$\left.\begin{aligned} x^1 &= q \cos u^1 \cos u^2 \dots \sinh u^{n-1}, \\ x^2 &= q \sin u^1 \cos u^2 \dots \sinh u^{n-1}, \\ x^3 &= q \sin u^2 \dots \sinh u^{n-1}, \\ &\dots\dots\dots\dots\dots\dots\dots\dots \\ x^{n-1} &= q \sin u^{n-2} \sinh u^{n-1}, \\ x^n &= q \cosh u^{n-1}. \end{aligned}\right\} \tag{3.58}$$

The volume element dS of any cubeable figure on a hypersphere of radius iq in R_1^n is

$$dS = q^{n-1} \cos u^2 \cos^2 u^3 \dots \sinh^{n-2} u^{n-1} du^1 du^2 \dots du^{n-1}. \tag{3.59}$$

3.6.5. Areas of Spherical Triangles and Volumes of Spherical Simplexes. A figure on a hypersphere of radius r in R^n or on a hypersphere of radius iq in R_1^n bounded by n great $(n-2)$-spheres is called a *spherical simplex*; for $n = 3$ it is a *spherical triangle*. Integration of the differentials (3.55) and (3.59) over spherical simplexes gives the volumes of these spherical simplexes (for $n = 3$ the areas of spherical triangles on a sphere of radius r in R^3 or on a sphere of radius iq in R_1^3).

For $p < m-1$, each p-face of a spherical m-simplex in R^n or R_1^n has $(m-1)$-faces adjacent to this p-face. The number of these $(m-1)$-faces is equal to the number of p-faces contained in one $(m-1)$-face, that is, C_m^{p+1} multiplied by the number of $(m-1)$-faces, that is, $m+1$, and divided by the number of p-faces of the m-simplex, that is, C_{p+1}^{m+1}. This number is equal to $m-p$. Therefore, the $(m-1)$-faces of an m-simplex adjacent to a p-face form an *$(m-p)$-hedral angle*, for $p = m-2$ a *dihedral angle.*

The part of the hypersphere bounded by the great $(m-1)$-spheres containing the $(m-1)$-faces of the m-simplex adjacent to a p-face of this simplex consists of two congruent spherical $(m-p)$-polyhedra with congruent $(m-p)$-hedral angles at the given p-face. Each of these $(m-p)$-polyhedra is called an $(m-p)$-hedral biangle. For $p = 2$ this figure is a dihedral biangle bounded by two great $(m-1)$-spheres, for $m = 2$ it is a biangle on a sphere bounded by two great circles.

We measure the $(m-p)$-hedral angle at a p-face of a spherical m-simplex by the quotient of the $(m-p)$-hedral biangle to the volume of the whole $(m-p)$-sphere. We call this number the "angle at the p-face" and, for a p-face $A_{i_1}A_{i_2}\dots A_{i_p}$ denote it by $\alpha_{i_1 i_2 \dots i_p}$. In particular, for a triangle ABC the areas of the parts of the sphere bounded by the great circles containing the sides of this triangle are equal to $4Ar^2, 4Br^2$, and $4Cr^2$ (since the area of the dihedral biangle bounded by the two semicircles containing the sides AB and AC of this triangle is to the area of the whole sphere as the angle A to 2π, the area of this biangle is $2Ar^2$ and the part of the sphere bounded by the great circles containing the sides AB and AC is equal to $4Ar^2$). The dihedral angles $\alpha_1, \alpha_2, \alpha_3$ are equal to $\frac{A}{2\pi}, \frac{B}{2\pi}, \frac{C}{2\pi}$, respectively.

Theorem of Schläfli [Scl, p. 238]. *The alternated sum of the $(m-p)$-hedral angles of a spherical m-simplex in R^n,*

$$W_m = \sum_i \alpha_i - \sum_{ij} \alpha_{ij} + \sum_{ijk} \alpha_{ijk} - \dots + (-1)^m \sum_{i_1 i_2 \dots i_{m-1}} \alpha_{i_1 i_2 \dots i_{m-1}} \tag{3.60}$$

is related with the quotient I_m of the volume of this simplex to the volume of the whole great m-sphere by the formulas

$$W_m = 2I_m + \frac{m-1}{2}, \qquad \textit{for } m \textit{ even} \tag{3.61}$$

and

$$W_m = -\frac{m-1}{2}, \qquad \textit{for } m \textit{ odd.} \tag{3.62}$$

Formula (3.61) implies the following expression for the volume Σ_p of a spherical p-simplex in R^n for $p = 2m$:

$$\Sigma_{2m} = S_{2m+1} \cdot I_{2m} = \frac{2^{m+1}\pi^m}{(2m-1)!!} r^{2m} \cdot \frac{2W_{2m} - 2m + 1}{4}. \tag{3.63}$$

For $m = 1$ formula (3.63) becomes $\Sigma_2 = 4\pi r^2 \cdot \frac{2W_2 - 1}{4}$. Since in this case $W_2 = \alpha_1 + \alpha_2 + \alpha_3 = \frac{A+B+C}{2\pi}$, we obtain

$$\Sigma_2 = r^2(A + B + C - \pi). \tag{3.64}$$

Formula (3.64) can also be obtained immediately. Consider all six biangles bounded by the great circles containing the sides of the spherical triangle ABC. Three of these biangles cover the triangle ABC and three other biangles cover the antipodal triangle $A'B'C'$ (Fig. 3.5). We have seen that the areas of the first three of these biangles are equal to $2Ar^2, 2Br^2$, and $2Cr^2$. The second three of these biangles are equal to the same quantities. All six biangles cover the whole sphere, but the triangles ABC and $A'B'C'$ are covered three times and the surface of the sphere outside these triangles is covered only once.

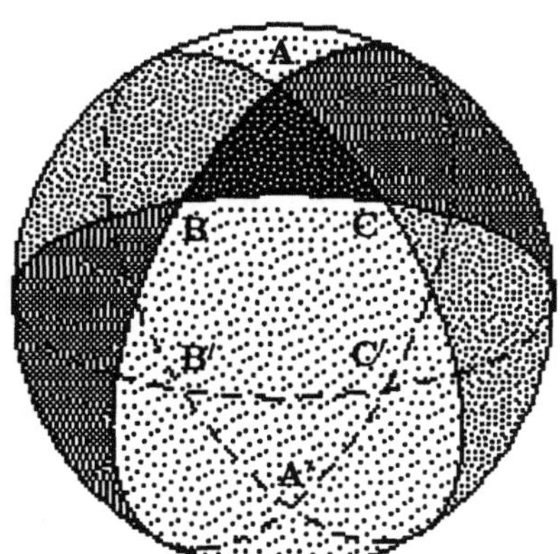

FIG. 3.5

Therefore, the sum of the areas of all six biangles is equal to $4Ar^2 + 4Br^2 + 4Cr^2 = 4\pi r^2 + 4\Sigma_2$. This implies formula (3.64).

We can also define "angles $\alpha_{i_1 i_2 \ldots i_p}$ at p-faces" of spherical simplexes on a hypersphere of radius iq in R^n_1.

Theorem of H. Kneser [Kne]. *The alternated sum (3.60) of the $(m-p)$-hedral angles of spherical m-simplex on a hypersphere of radius iq in R^n_1 is, for m even related by (3.61) with the quotient I_m of the volume of this simplex to the* *number obtained from the volume of the m-surface of the whole great m-sphere of the hypersphere of radius r in R^n by the replacing of r by iq. For m odd W_m is equal to (3.62).*

Formula (3.61) implies the following expression for the volume Σ_p of a spherical p-simplex in R_1^n for $p = 2m$:

$$\Sigma_{2m} = (-1)^m \frac{2^{m+1}\pi^m}{(2m-1)!!} q^{2m} \cdot I_{2m} = (-1)^m \frac{2^{m+1}\pi^m}{(2m-1)!!} qr^{2m} \cdot \frac{2W_{2m} - 2m + 1}{4}. \tag{3.65}$$

For $m = 1$ formula (3.65) becomes $\Sigma_2 = 4\pi q^2 \cdot \frac{1-2W_2}{4}$. Since in this case $W_2 = \alpha_1 + \alpha_2 + \alpha_3 = \frac{A+B+C}{2\pi}$, we obtain

$$\Sigma_2 = q^2(\pi - A - B - C). \tag{3.66}$$

The expressions

$$\epsilon = A + B + C - \pi \quad \text{and} \quad \delta = \pi - A - B - C \tag{3.67}$$

are called the *angular excess* and the *angular defect* of a spherical triangle. The formulas (3.64) and (3.66) can be obtained also as follows. The area $S(\mathfrak{F})$ of a plane figure $\mathfrak{F}$ in R^2 is a real number satisfying the following axioms:

$Ar.1°$ If $\mathfrak{F}$ is transformed by a motion into a figure $\mathfrak{F}'$, then $S(\mathfrak{F}') = S(\mathfrak{F})$ (*axiom of invariance*),

$Ar.2°$ If $\mathfrak{F}$ is the union of figures $\mathfrak{F}_1$ and $\mathfrak{F}_2$, then $S(\mathfrak{F}) = S(\mathfrak{F}_1)+S(\mathfrak{F}_2)$ (*axiom of additivity*),

$Ar.3°$ If $\mathfrak{F}$ contains a triangle, then $S(\mathfrak{F}) > 0$ (*axiom of positivity*),

$Ar.4°$ If $\mathfrak{F}$ is a square with side lengths one, then $S(\mathfrak{F}) = 1$ (*axiom of normalization*).

The area $S(\mathfrak{F})$ of a spherical figure $\mathfrak{F}$ on a sphere in R^n or on a sphere of pure imaginary radius in R_1^n is a real number which satisfies axioms $Ar.1° - 3°$ and the axiom

$Ar.4'$ If $\mathfrak{F}$ is a small figure, $S(\mathfrak{F})$ can be taken equal to the Euclidean area of this figure.

The quantities (3.67) satisfy the axioms $Ar.1° - 3°$ since the angles of triangles are invariant under rotation of the sphere and positive, and if the triangle ABC is the union of triangles ABD and BDC and if the angular excesses and defects of these triangles are $\epsilon, \epsilon_1, \epsilon_2$ and $\delta, \delta_1, \delta_2$, respectively, then $\epsilon = \epsilon_1 + \epsilon_2$ and $\delta = \delta_1 + \delta_2$ (Fig. 3.6).

Therefore the area Σ_2 of a spherical triangle is proportional to its angular excess or defect: $\Sigma_2 = k\epsilon$ and $\Sigma_2 = k\delta$.

The constants k can be found from the condition $Ar.4'$. Consider a small rectangular isosceles triangle with side length a (Fig. 3.7a,b).

Then the Euclidean area of this triangle is $\frac{a^2}{2}$. By formulas (3.41) and (3.47), $\cos A = \sin A \cos \frac{a}{r}$ or $\cos A = \sin A \cosh \frac{a}{q}$, that is, $\cot A = \cos \frac{a}{r}$ or $\cot A = \cosh \frac{a}{q}$, respectively. For small triangles we can replace $\cos \frac{a}{r}$ and $\cosh \frac{a}{q}$ by $1 - \frac{a^2}{2r^2}$ and

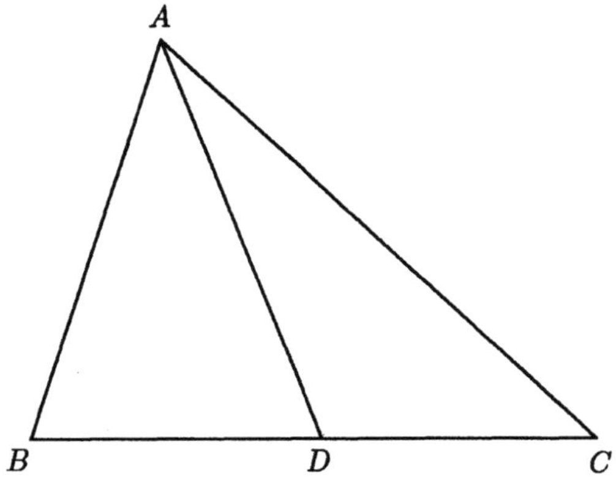

FIG. 3.6

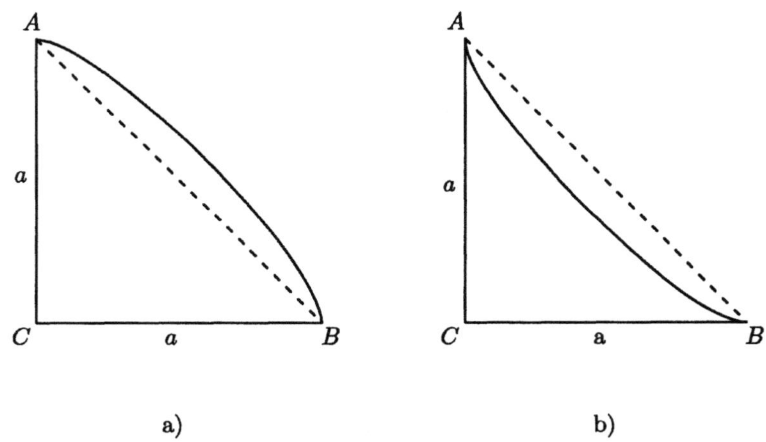

FIG. 3.7

$1+\frac{a^2}{2q^2}$, respectively. Since the angles A are close to $\frac{\pi}{4}$,

$$\cot A = \cot\frac{\pi}{4} - \frac{1}{\sin^2\frac{\pi}{4}}\left(A-\frac{\pi}{4}\right) = 1-2\left(A-\frac{\pi}{4}\right).$$

In R^n we have $A > \frac{\pi}{4}$ and $A - \frac{\pi}{4} = \frac{\epsilon}{2}$. Therefore, from $\cot A = \cos\frac{a}{r}$ we obtain $\frac{a^2}{2r^2} = \epsilon$, that is, $\Sigma_2 = r^2\epsilon$. In R_1^n we have $A < \frac{\pi}{4}$ and $A - \frac{\pi}{4} = -\frac{\delta}{2}$. Therefore, from $\cot A = \cosh\frac{a}{q}$ we obtain $\frac{a^2}{2q^2} = \delta$, that is, $\Sigma_2 = q^2\delta$.

§3.7. Sliding Vectors

3.7.1. Sliding Vectors and Systems of Them. In **2.1.1** we have defined vectors in linear spaces and free modules. These vectors are called *free vectors* since vectors which are equal or equivalent to a given vector can be applied at any point of the affine space. Many problems in Mechanics and Physics require *sliding vectors*, that is, oriented segments on lines which can be equivalent only under translation along the direction they have. In particular, forces and angular velocities can be described by sliding vectors.

Two vectors a and $-a$ on parallel lines are called a *couple of vectors*. The distance between these lines is called the *arm* of the couple. The product of the modulus of a vector of a couple by the arm is called the *moment* of the couple.

Two systems of sliding vectors are called *equivalent* if these systems can be obtained from each other by the following operations:

*Eq.*1° Translation of each vector along its line.

*Eq.*2° Addition of vectors with a common initial point in accordance with the parallelogram rule (the sum of two vectors $\overrightarrow{AB}$ and $\overrightarrow{AC}$ is the diagonal $\overrightarrow{AD}$ of the parallelogram $ABDC$).

*Eq.*3° Decomposition of a vector into a sum of vectors in accordance with the same rule.

*Eq.*4° Addition of a zero vector with initial point at any point.

Like in classical Rational Mechanics, the following theorems can be proved:

Theorem M.1. *A system of two sliding vectors a_1 and a_2 with parallel lines and with initial points O_1 and O_2 is equivalent to the system consisting of the vector $a = a_1 + a_2$ on the line parallel to the lines of these vectors and with initial point O on the line O_1O_2. If $a_1 = a\lambda_1, a_2 = a\lambda_2$, the simple ratio $\overline{O_1O, O_2}$ is equal to $-\lambda_1 : (\lambda_1 + \lambda_2)$.*

Theorem M.2. *A couple of vectors is equivalent to any other couple obtained from the first couple by a rotation in its plane or by translation in its plane or in a parallel plane.*

Theorem M.3. *A couple of vectors is equivalent to any other couple obtained from the first couple by multiplication of the vectors of the couple by a positive number λ and division of the arm of the couple by the same λ.*

3.7.2. Main Vector and Main Moment of a System of Sliding Vectors. For each system of sliding vectors a_i with initial points O_i with position vectors x_i we can make up the vector $a = \sum_i a_i$ and the skew symmetric linear operator $M = \sum_i x_i \wedge a_i$. The vector a is called the *main vector* of the system, the linear operator M is called the *main moment* of the system.

A common initial point O of the position vectors x_i is called a *point of reference* of the main moment. If M is a main moment with point of reference O and M' is a main moment of the same system with point of reference O' and $\overrightarrow{OO'} = x$, then $M' = \sum_i (x_i - x) \wedge a_i = M - x \wedge a$.

Theorem 3.23. *The possible points of references of a main moment M of a system of sliding vectors for which the main vector of the system is an eigenvector of M are located on a line.*

<u>Proof</u>. If $a = \sum_i a_i$ is an eigenvector of $M' = M - x \wedge a$, then $M'a = Ma - x \cdot a^2 + a \cdot xa = \lambda a$. Therefore, $Ma - x \cdot a^2 = \mu a$. For $\mu = 0$ the vector x is

equal to $\frac{Ma}{a^2}$, and since under translation of the point of reference in the direction of a the main moment becomes $M' = M - a \wedge a = M$, that is, does not change, the locus of the points of references with this property is the line $x = \frac{Ma}{a^2} + at$. ■

This line is called the *principal axis* of the system of sliding vectors.

The point of this axis corresponding to $\mu = 0$ is called the *center* of the system.

For an arbitrary point O each system of sliding vectors a_i can be transformed into an equivalent system consisting of vectors a_i with common initial point O and of couples consisting of vectors a_i of the system and of vectors $-a_i$ at O. These couples of vectors can be transformed into equivalent couples consisting of vectors $-a_i'$ at O and of vectors a_i' tangent to the hypersphere with center at O and with radius 1. If on this hypersphere a point X with position vector x is given, all couples of vectors which contain vectors tangent to great circles which pass through X can be rotated so that all these vectors will be applied at X. We will denote the sum of these vectors by a_x and call it the "result of reduction of the system to the point X".

Theorem 3.24. *The main moment M of a system of sliding vectors in R^n transforms the position vector x of the point X on the unit hypersphere to the vector opposite to the result of reduction of the system to the point X,*

$$a_x = -Mx. \tag{3.68}$$

PROOF. Let a sliding vector a_i in our system be at an initial point X_i with position vector x_i on the unit hypersphere. Then the great circle tangent to this vector is cut out from the hypersphere by the diametral plane determined by x_i and a_i. Rotation of this great circle over the angle φ_i transforms these vectors into the vectors

$$a_i' = |a_i| \left(\frac{a_i}{|a_i|} \cos\varphi_i + x_i \sin\varphi_i \right) \text{ and } x_i' = -\frac{a_i}{|a_i|} \sin\varphi_i + x_i \cos\varphi_i. \tag{3.69}$$

Let φ_i be such that x_i is orthogonal to x. Then

$$x_i' x = -\frac{a_i x}{|a_i|} \sin\varphi_i + x x_i \cos\varphi_i = 0,$$

hence $\tan\varphi_i = \frac{x_i x}{a_i x} \cdot |a_i|$. If we substitute the corresponding values of $\sin\varphi_i$ and $\cos\varphi_i$ into (3.69), we obtain

$$a_i' = \frac{a_i x \cdot a + a_i^2 \cdot x_i x \cdot x_i}{\sqrt{(x_i x)^2 a_i^2 + (a_i x)^2}}, \quad x_i' = \frac{-x_i x \cdot a + a_i x \cdot x_i}{\sqrt{(x_i x)^2 a_i^2 + (a_i x)^2}}.$$

The component of the vector a_i' tangent to the great circle passing through X, that is, the orthogonal projection of a_i' onto the direction of the vector x, is equal to

$$a_i' x \cdot x = \frac{(a_i x)^2 + a_i^2 (x_i x)^2}{\sqrt{(x_i x)^2 a_i^2 + (a_i x)^2}} = \sqrt{(x_i x)^2 a_i^2 + (a_i x)^2}.$$

A rotation of the great circle tangent to the last vector over a right angle transforms this vector into the vector a_{ix} with initial point X. It has the same modulus as the vector $a_i^{\prime}x \cdot x$ and the direction of the vector $-x_i^{\prime}$. Therefore, $x_{ix} = a_i x \cdot x_i^{\prime} = -x_i \cdot a^i x + a_i \cdot x_i x = a_i \wedge x_i x$, and $a_x = \sum_i a_{ix}$ is equal to $-Mx$. ∎

3.7.3. Canonical Forms of Systems of Sliding Vectors. In Theorem 2.15 we have seen that hyperquadrics in R^n can be reduced to the canonical form (2.76), which is related to the canonical form of the symmetric linear operator $A = A^{\top}$, that is, the form $A = \sum_i \lambda_i e_i \cdot e_i$. Analogously, the canonical forms of systems of sliding vectors in R^n are related to the canonical form of the skew symmetric linear operator $M = -M^{\top}$, that is, the form

$$M = \sum_i \mu_i e_{2i-1} \wedge e_{2i}. \tag{3.70}$$

The vectors e_{2i-1} and e_{2i} determine invariant planes of M. The eigenvalues of M are equal to $i\mu_i$.

Theorem 3.25. *Each system of sliding vectors in R^n can be reduced to the following canonical form: the sliding vector a equal to the main vector of the system on its principal axis and couples of sliding vectors a_i and $-a_i$ on invariant planes of the main moment M of the system, the moments of these couples being equal to the moduli $|\mu_i|$ of the eigenvalues of the linear operator M.*

PROOF. Replace any vector a_i of the system by an equivalent system consisting of the vector a_i applied at the center of the system and of a couple consisting of the given vector and the vector $-a_i$ at the center of the system. Replace the vectors at the center by their sum, that is, the main vector a of the system directed along the principal axis of the system. Reduce all couples to points with position vectors x_i in invariant planes of the main moment M of the system, to obtain a unique couple in each principal plane which has the form of a summand of the expression (3.70) of M. ∎

Systems of sliding vectors in canonical form are called *screws*. Particular cases of screws are the *dynamical screws* and the *kinematical twists* of classical Mechanics.

Theorem 3.26. *The screws in R^n, with as sum of screws consisting of moments M_1 and M_2 and vectors a_1 and a_2 the screw consisting of the moment $M_1 + M_2$ and the vector $a_1 + a_2$ and with as commutator of these screws the screw consisting of the moment $M_1M_2 - M_2M_1$ and the vector $M_1a_2 - M_2a_1$, form the Lie algebra of the group of motions in R^n.*

PROOF. In **0.8.4** we have seen that the skew symmetric linear operators in R^n form the Lie algebra of the group of rotations in R^n. We can similarly prove that the pairs consisting of a skew symmetric linear operator M and a vector a with addition and commutation as above form the Lie algebra of the group of motions in R^n. ∎

3.7.4. Sliding Vectors in Pseudo-Euclidean Spaces. The theory of sliding vectors in the spaces R^n_l is constructed similarly. As in Theorem 3.25, we can prove that in R^n_l the canonical systems consist of the vector equal to the main vector of the system directed along the principal axis of the system and of pairs of vectors on totally orthogonal invariant planes of the main moment. The main moment ME in R^n_ℓ is the product $(\sum_i x_i \wedge a_i)E$ of the skew symmetric linear operator $M = \sum_i x_i \wedge a_i$ by the operator E determining the inner product of vectors in this space. In general, the eigenvalues μ_i of the operator ME in R^n_ℓ are not pure imaginary, as in R^n but can also be real. The pure imaginary and real eigenvalues of ME correspond to invariant planes of ME, which are R^2 and R^2_1, respectively. For R^n_1 there is only one real eigenvalue of ME and all other eigenvalues are pure imaginary. There are also limit cases, when the invariant planes of ME are isotropic. The screws, consisting of a main moment and a main vector of these systems, form the Lie algebra of the group of motions in R^n_ℓ.

3.7.5. Spherical Sliding Vectors in Euclidean Spaces. Oriented arcs of great circles on a hypersphere in R^n that are equivalent only under rotation along their circles are called *spherical sliding vectors.* The modulus of the projection of a spherical vector from the center of the hypersphere to the tangent hyperplane to the hypersphere at the initial point of this vector is called the *modulus* of this spherical vector.

Two systems of spherical sliding vectors on a hypersphere in R^n are called *equivalent* if these systems can be obtained from another by the operations listed in *Eq.*$1° - 4°$, where lines are replaced by great circles, and addition of vectors with a common initial point is replaced by addition of spherical vectors with common initial point A as follows: project these vectors from the center of the hypersphere to the tangent hyperplane to the hypersphere at A. This gives vectors $\overrightarrow{AB}, \overrightarrow{AC}$. Their sum $\overrightarrow{AD}$ according to the parallelogram rule is the projection of the sum of the spherical vectors. This rule is equivalent to the following rule: the sum $\breve{c}$ of two spherical sliding vectors $\breve{a}$ and $\breve{b}$ which are represented by arcs OA and OB is represented by the arc OC, where C is the point of intersection of the arcs OA' and OB' joining O with the points A' and B' of intersection of the circles OA and OB with the polar great circle for O (Fig. 3.8).

With each system of spherical sliding vectors $\breve{a}_i$ in R^n with initial points O_i with position vectors x_i on a hypersphere we can relate the skew symmetric linear operator $M = \sum_i x_i \wedge a_i$. This linear operator is called the *main moment* of the system (a_i is the projection of $\breve{a}_i$).

Note that a spherical sliding vector is equivalent to the couple of vectors consisting of the projection a of this vector from the center of the hypersphere to the tangent hyperplane to the hypersphere at the initial point of the vector and of vector $-a$ with initial point at the center of the hypersphere.

Theorem 3.27. *A system of spherical sliding vectors on the hypersphere of radius r in R^n can be reduced to the canonical form consisting of spherical sliding vectors $\breve{a}_i$ on totally orthogonal invariant planes of the main moment M of the system with moduli equal to the moduli $|\mu_i|$ of the eigenvalues of the linear operator M divided by r.*

PROOF. The first assertion follows from Theorem 3.25. The last assertion follows from the fact that the moduli $|\mu_i|$ are equal to the moments of the couples of sliding vectors in R^n equivalent to the spherical sliding vectors of the canonical

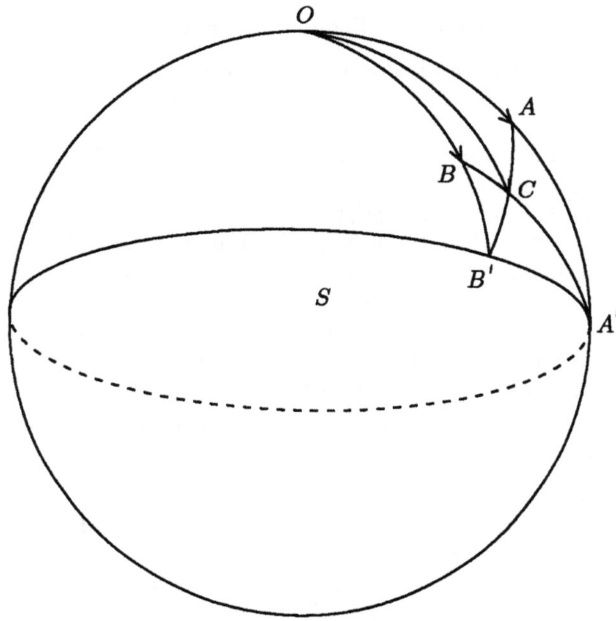

FIG. 3.8

system, and these moments are equal to the product of the moduli of these vectors by the radius r of the hypersphere. ■

The moments of the systems of spherical sliding vectors in R^n form the Lie algebra of the group of rotations in R^n.

3.7.6. Spherical Sliding Vectors in Pseudo-Euclidean Spaces. We can similarly construct the theory of spherical sliding vectors on hyperspheres in R^n_ℓ. With each system of spherical sliding vectors $\breve{a}_i$ in R^n_ℓ with initial points O_i with position vectors x_i on the hypersphere we can associate the linear operator $M = \sum_i x_i \wedge a_i$ and the linear operator ME called the *main moment* of the system. As in Theorem 3.27 we can prove that the canonical systems of sliding vectors on hyperspheres in R^n_ℓ consist of spherical sliding vectors $\breve{a}_i$ in totally orthogonal invariant planes of the main moment. For a hypersphere of real radius r, the vectors $\breve{a}_i$ of a canonical system which belong to an invariant plane R^2 are sliding vectors on this hypersphere and the vectors $\breve{a}_i$ of a canonical system which belong to an invariant plane R^2_1 are sliding vectors on the hypersphere with the same center and with pure imaginary radius ir. For a hypersphere of pure imaginary radius iq, the vectors $\breve{a}_i$ of a canonical system which belong to an invariant plane R^2_1 are sliding vectors on this hypersphere and the vectors $\breve{a}_i$ of canonical system which are in invariant planes R^2 are sliding vectors on the hypersphere with the same center and with real radius q.

The moments of systems of spherical sliding vectors in R^n_ℓ form the Lie algebra of the group of rotations in R^n_ℓ.

§3.8. Conformal and Pseudoconformal Spaces

3.8.1 Conformal Space and Conformal Transformations. In **0.5.7** we have defined conformal space C^n as the space R^n supplemented by one point at infinity.

A conformal transformation in R^n is a transformation $x' = f(x)$ in this space determined by a differentiable vector function of vector argument which preserves angles between curves. Since a differentiable vector function has, for each x, a derivative $A = \frac{df}{dx}$ (see **0.6.6**) which is a linear operator, for functions which determine a conformal transformation this operator A must be the operator of a similitude, that is, be the product kU of an orthogonal operator U by a real number k. The similitudes $x' = kUx + a$ are the simplest conformal transformations. Inversions in hyperspheres (see **0.5.7**) also are conformal transformations. Inversion in a hypersphere in R^n is a particular case of inversion in an ellipsoid in E^n (see **2.5.6**).

Theorem 3.28 (LIOUVILLE [Lio]). *For $n \geq 3$ a conformal transformation in R^n maps a hypersphere to a hypersphere or a hyperplane.*

PROOF. Since conformal transformations preserve angles between curves, they also preserve angles between hypersurfaces. But each hypersurface $x = x(u^1, u^2, \ldots, u^{n-1})$ with normal vector $n = n(u^1, u^2, \ldots, u^{n-1})$ can be included into a family of parallel hypersurfaces $x = x(u^1, u^2, \ldots, u^{n-1}) + n(u^1, u^2, \ldots, u^{n-1})t$. This family can be adjoined to the $n-1$ families of surfaces consisting of the normal lines to the first hypersurface at the points of its $n-1$ curvature lines. Under a conformal transformation this n-orthogonal system is transformed into an other n-orthogonal system and, since the hypersurfaces of an n-orthogonal system are intersected by surfaces of this system in curvature lines, the conformal transformation preserves the curvature lines of the hypersurfaces. Since hyperplanes and hyperspheres are the only hypersurfaces all lines on which are curvature lines, a conformal transformation maps hyperplanes and hyperspheres to hyperplanes and hyperspheres. Therefore, hyperplanes and hyperspheres can be included into a n-orthogonal system of hypersurfaces in infinity many ways, so the image of the hypersphere under a conformal transformation is again a hypersphere or a hyperplane.

This proof is valid only for $n \neq 2$ since a family of concentric circles or parallel straight lines can be included into a biorthogonal system only in one way. ■

Since for $n \geq 3$ a hypersphere in R^n is mapped to a hypersphere or hyperplane by a conformal transformation, m-spheres, which are intersections of these hypersurfaces, are mapped by a conformal transformation to m-spheres or m-planes. In particular, for $m = 1$, circles are mapped by a conformal transformation to circles or straight lines.

Since inversion in a hypersphere maps the center of the hypersphere to the point at infinity, in the space C^n obtained from R^n by the addition of this point at infinity, inversion in a hypersphere is a bijection. In C^n hyperplanes can be regarded as hyperspheres that pass through the point at infinity.

In C^2 conformal transformations mapping circles to circles or straight lines are called *circular transformations.*

Theorem 3.29. *A conformal transformation in C^n for $n > 2$ or circular transformation in C^2 is the product of inversions in hyperspheres, respectively circles.*

PROOF. If a conformal transformation preserves the point at infinity, it maps lines to lines and, therefore, it is an affine transformation in E^n. Since it also preserves angles between curves, it is a similitude, that is, the product of a motion by a homothety. But each motion in R^n is a product of reflections in hyperplanes, and a homothety is the product of inversions in two concentric hyperspheres. If a conformal transformation T does not preserve the point in infinity, it maps this point to a point X_0. Therefore, the product TJ of T and the inversion J in a hypersphere with center at X_0 is the product of inversions in hyperspheres or reflections in hyperplanes. Since a reflection in a hyperplane can be regarded as an inversion in a hypersphere, $T = (TJ)J$ is the product of inversions. ■

3.8.2. Projective Interpretation of Conformal Space. In the proof of Theorem 2.13 (see **2.2.2**) we have seen that the space C^m is homeomorphic to a hypersphere Σ^m in R^{m+1}. This homeomorphism is given by stereographic projection of the hypersphere $\sum_i (x^i)^2 = 1$ in R^{m+1} from its pole S with coordinates $0, 0, \dots, 0, 1$ onto the hyperplane $x^{m+1} = 0$.

Theorem 3.30. *The group of conformal transformations in C^n is isomorphic to the group of collineations in P^{n+1} that preserve an oval hyperquadric in P^{n+1}.*

PROOF. Consider the analogous stereographic projection of the hypersphere $\sum_i (x^i)^2 = 1$ in R^{n+1} from its pole S with coordinates $0, 0, \dots, 0, 1$ onto the hyperplane $x^{n+1} = 0$. This hyperplane is the space R^n, and this hyperplane supplemented by the point at infinity, corresponding to the pole S of the hypersphere, can be regarded as C^n. Since images of hyperspheres in C^n on this hypersphere are cut out from it by hyperplanes in R^{n+1} (in particular, images of hyperplanes in R^n on this hypersphere are cut out from it by hyperplanes in R^{n+1} passing through S), the transformations in C^n preserving hyperspheres are imaged in R^{n+1} by transformations preserving hyperplanes in R^{n+1}. The most general such transformations are collineations in the space P^{n+1} obtained from R^{n+1} by supplementing the latter with a hyperplane at infinity. ■

We considered the analogous stereographic projection for $n = 2$ in **3.1.5**.

If we denote the coordinates of a point X on the hypersphere by X^i $(i = 1, 2, \dots, n, n+1)$ and the coordinates of its stereographic projection x from the pole S onto the hyperplane $x^{n+1} = 1$ by x^i, then we obtain the following formulas, which are similar to formulas (3.4):

$$X^i = \frac{2x^i}{\sum\limits_i (x^i)^2 + 1}, \quad X^{n+1} = \frac{\sum\limits_i (x^i)^2 - 1}{\sum\limits_i (x^i)^2 + 1}. \tag{3.71}$$

Therefore, the projective coordinates s^i $(i = 0, 1, \dots, n, n+1)$ of a point X on the hypersphere are equal to

$$s^0 = \frac{1}{2}\left(\sum_i (x^i)^2 + 1\right), \quad s^i = x^i, \quad s^{n+1} = \frac{1}{2}\left(\sum_i (x^i)^2 - 1\right). \tag{3.72}$$

These coordinates satisfy the equation

$$-(s^0)^2 + \sum_i (s^i)^2 + (s^{n+1})^2 = 0. \tag{3.73}$$

3.8.3. (n + 2)-Spherical Coordinates of Hyperspheres. Coordinates s^i that do not satisfy the equation (3.73) are the projective coordinates of a point in P^{n+1} which is not on the hyperquadric (3.73). If the polar hyperplane of such a point intersects this hyperquadric, it cuts out from this hyperquadric an $(n-1)$-quadric that can be regarded as a real hypersphere in C^n. If the polar hyperplane of such a point does not intersect this hyperquadric, the hyperplane has an imaginary intersection with the hyperquadric. Such an imaginary $(n-1)$-quadric can be regarded as an imaginary hypersphere in C^n.

Since $n+2$ basis points in P^{n+1} are thus the images of $n+2$ hyperspheres, $n+1$ real hyperspheres and one imaginary hyperphere, these coordinates are called $(n+2)$-*spherical*. For $n = 2$ they are called *tetracyclic*, for $n = 3$ *pentaspherical*. The $(n+2)$-spherical coordinates of a real hypersphere in C^n can be normalized by the condition

$$-(s^0)^2 + \sum_i (s^i)^2 + (s^{n+1})^2 = 1. \tag{3.74}$$

The $(n+2)$-spherical coordinates of an imaginary hypersphere in C^n can be normalized by the equation obtained from (3.72) by replacing 1 by -1.

Theorem 3.31. *The angle φ between two hyperspheres with normalized $(n+2)$-spherical coordinates s^i and t^i can be expressed by*

$$\cos\varphi = |-s^0t^0 + \sum_i s^it^i + s^{n+1}t^{n+1}|. \tag{3.75}$$

PROOF. Consider the intersection of the hypersphere $\sum_i (X^i)^2 = 1$ in R^{n+1} by the hyperplane $u_0 + u_iX^i + u_{n+1}X^{n+1} = 0$. If we replace in this equation X^i and X^{n+1} by their values (3.71), we obtain

$$\begin{aligned} u_0\left(\sum_i (x^i)^2 + 1\right) + 2u_ix^i + u_{n+1}\left(\sum_i (x^i)^2 - 1\right) = \\ = (u_0 + u_{n+1})\sum_i (x^i)^2 + 2u_ix^i + (u_0 - u_{n+1}) = 0. \end{aligned} \tag{3.76}$$

This is the equation of a hypersphere if $u_0 + u_{n+1} \neq 0$ and the equation of a hyperplane if $u_0 + u_{n+1} = 0$. In the first case, comparing (3.75) with the equation $x^2 - 2x_0x + x_0^2 - r^2 = 0$, we find that

$$u_0 = \frac{\lambda}{2}\left(1 + \sum_i (x_0^i)^2 - r^2\right), \; u_i = \lambda x_0^i, \; u_{n+1} = \frac{\lambda}{2}\left(1 - \sum_i (x_0^i)^2 + r^2\right).$$

If these coordinates are normalized by the condition $-u_0^2 + \sum_i (u_i)^2 + (u_{n+1})^2 = 1$, which is equivalent to (3.74), we must take $\lambda = \frac{1}{r}$. Therefore,

$$u_0 = \frac{1}{2r}\left(1 + \sum_i (x_0^i)^2 - r^2\right), \; u_i = \frac{x_0^i}{r}, \; u_{n+1} = \frac{1}{2r}\left(1 - \sum_i (x_0^i)^2 + r^2\right). \quad (3.77)$$

The coordinates s^i of the pole of the hyperplane $u_0 + u_i X^i + u_{n+1} X^{n+1} = 0$ with respect to the hyperquadric (3.73) are related with the coefficients u_i by $s^0 = -u_0$, $s^i = u_i$, $u_{n+1} = s^{n+1}$. Therefore, for two hyperspheres with centers having position vectors x_1 and x_2 and with radii r_1 and r_2 the expression $|-s^0 t^0 + \sum_i s^i t^i + s^{n+1} t^{n+1}|$ is equal to the right hand side of (3.37), that is, $\cos\varphi$.

In the second case, comparing (3.76) with the equation $a_i x^i + a_{n+1} = 0$ of a hyperplane ($\sum_i a_i^2 = 1$ for $i = 1, 2, \ldots, n$), we obtain $u_0 = -u_{n+1} = -\frac{\lambda}{2} a_{n+1}$, $u_i = \frac{\lambda}{2} a_i$, and $\lambda = 2$, and the cosine of the angle between the hypersphere and the hyperplane, and also between two hyperplanes, is equal to the right hand side of (3.75). ■

3.8.4. Conformal Configurations. Similar to projective configurations in P^2, in C^2 we have *conformal configurations*, where the role of lines in P^2 is played by circles in C^2. Like projective configurations in P^2, conformal configurations in C^2 are denote by the Reye and the De Vries symbols (n_i) and $(n_i g_k)$.

The simplest conformal configurations can be obtained by means of polyhedra inscribed in a sphere, which can be regarded as the plane C^2: these configurations are formed by the vertices of the polyhedra and by the circles that are the intersections of the planes of the faces of the polyhedra with the sphere. Therefore, for the tetrahedron we have the corresponding conformal configuration (4_3); for the cube, octahedron, dodecahedron, and icosahedron we have the corresponding conformal configurations $(8_3 6_4)$, $(6_4 8_3)$, $(20_3 12_5)$, and $(12_5 20_3)$, respectively.

The configuration (4_3) can be thought of as consisting of the three sides of a triangle and the circle circumscribed around this triangle; the points of this configurations are the vertices of the triangle and the point at infinity.

The configuration $(8_3 6_4)$ can be obtained as follows:

Miquel's First Theorem [Miq1]. *If four points A, B, C, D are on one circle $ABCD$ and if through A and B, B and C, C and D, and D and A we draw circles (Fig. 3.9), then the second points A', B', C', and D' of intersection of these drawn circles also are on one circle.*

The same configuration can be thought of as consisting of three circles meeting at one point and three lines such that two lines join a point on one circle with the points of intersection of this circle with the two other circles while the last line joins the points of intersection of the first two lines with the two other circles and passes through the second point of intersections of these circles.

The autodual configuration (8_4) can be obtained as follows:

Miquel's Second Theorem [Miq2]. *The circumscribed circles around the four triangles ADC, CBF, AEF, and BDF of a complete quadrilateral $ADEFBC$ (Fig. 3.10) meet at one point.*

This configuration is formed by the four sides ADE, ACF, BCD, and BFE of the complete quadrilateral $ADEFBC$, the four circles ADC, CBF, AEF, and

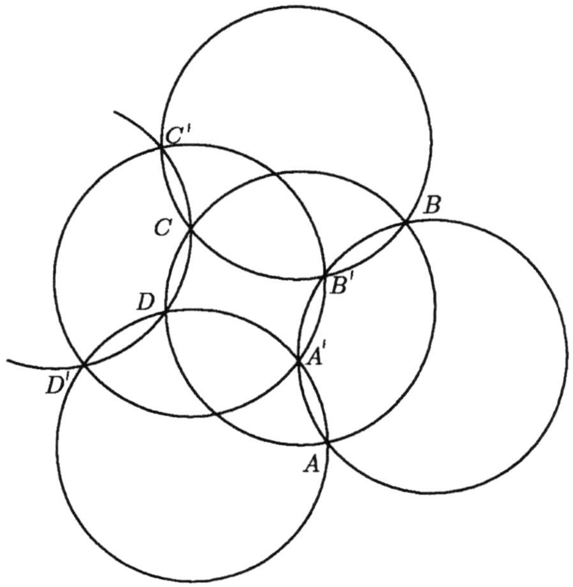

FIG. 3.9

BDF, the six vertices of this quadrilateral, the point of intersection I and the point at infinity.

A series of conformal configurations can be obtained by means of the following theorems of W. K. Clifford, who came to them by developing the ideas of A. Miquel:

" (1) Given three lines, circle may be drawn through their intersections...

(1') Given four lines, the four circles so determined meet in a point...

(2) Given five lines, the five points so found lie on a circle...

(2') Given six lines, the six circles so determined meet in a point...

I shall now show that the series is interminable that is that $2n$ lines determine $2n$ circles all meeting in a point and that for $2n+1$ lines the $2n+1$ points lie on the same circle " [Cli1, pp. 51–52].

If these configurations undergo a circular transformation, the planar analog of a space conformal transformation, mapping the point at infinity to a point S and mapping lines to circles through S, we obtain the series of conformal configurations first found by H. S. M. Coxeter:

"Clifford's First Theorem. *Let $\sigma_1, \sigma_2, \sigma_3, \sigma_4$ be four circles of general position through a point S. Let S_{ij} be the second intersection of the circles σ_i and σ_j. Let σ_{ijk} denote the circle $S_{ij}S_{ik}S_{jk}$. Then the four circles $\sigma_{234}, \sigma_{134}, \sigma_{124}, \sigma_{123}$ all pass through one point S_{1234}.*

Clifford's Second Theorem. *Let σ_5 be a fifth circle through S. Then the five points $S_{2345}, S_{1345}, S_{1245}, S_{1234}$ all lie on one circle σ_{12345}.*

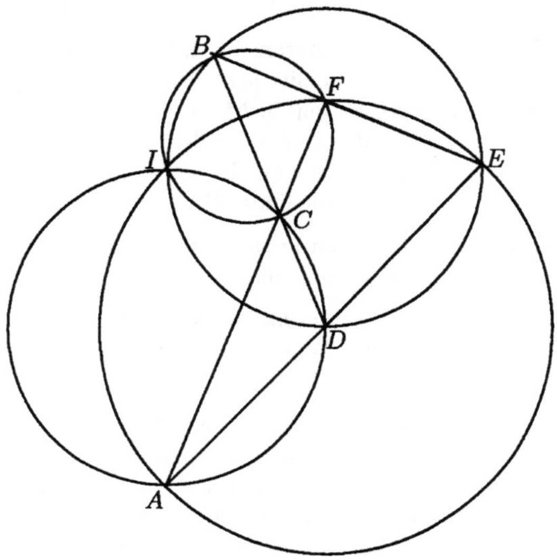

FIG. 3.10

Clifford's Third Theorem. *The six circles* σ_{23456}, σ_{13456}, σ_{12456}, σ_{12356}, σ_{12346}, σ_{12345} *all pass through one point* S_{123456}.

And so on!" [Cox1, p. 262].

Note that the configuration in Clifford's theorem (1) is the above mentioned conformal configuration (4_3), the configuration of his Theorem (1') coincides with the configuration (8_4) of Miquel's second theorem, the configurations of the Clifford's Theorems (2) and (2') are conformal configurations (16_5) and (32_6); in general Clifford's configurations are conformal configurations of the form $((2^n)_{n+1})$. Like the configurations (4_3) in Clifford's theorem (1) can be obtained from a tetrahedron inscribed in a sphere, the conformal configurations $((2^n)_{n+1})$ can analogously be obtained from the projective configurations $((2^n)_{n+1}(3 \cdot 2^{n-1})_2)$ obtained from a tetrahedron by means of null-systems (see **2.7.2**).

A conformal configuration in C^n is called *regular* if it admits a group of conformal transformations in this space (a group of circular transformations for the plane C^2). In particular, a conformal configuration in C^n which is determined by a regular projective configuration in P^{n+1} is a regular conformal configuration.

3.8.5. Pseudoconformal Spaces. In **0.5.8** we have defined the pseudoconformal space C^n_ℓ as the space R^n_ℓ supplemented by one point at infinity and by ideal points. Conformal transformations in R^n_ℓ are defined as for R^n. The simplest conformal transformations in R^n_ℓ are similitudes, and, for $n = 2\ell$, antisimilitudes. An inversion in a hypersphere in R^n_ℓ is also a conformal transformation.

Inversion in a hypersphere in R^n_ℓ is a particular case of inversion in a hyperboloid in E^n (see **2.5.6**).

Theorem 3.32. *A conformal transformation in R^n_ℓ preserves isotropic lines.*

PROOF. Since a conformal transformation is defined by a differentiable function $x' = f(x)$, for which $dx' = Adx$, the equality $|dx| = 0$ implies the equality $|d's| = 0$.

As in R^n, for conformal transformations in R^n_ℓ Liouville's Theorem, analogous to Theorem 3.28, is valid. In this case its proof is simpler than for R^n since the spheres, m-spheres and hyperspheres in R^n_ℓ are the quadrics, m-quadrics and hyperquadrics whose rectilinear generators are isotropic lines. ■

Analogously to Theorem 3.29 we can prove that a conformal transformation in C^n_ℓ is the product of inversions in hyperspheres.

Theorem 3.33. *The group of conformal transformations in C^n_ℓ is isomorphic to the group of collineations in P^{n+1} that preserve a hyperquadric of index $\ell+1$ in P^{n+1}.*

PROOF. Consider the stereographic projection of the hypersphere $-\sum_a (x^a)^2 + \sum_u (x^u)^2 = 1$ in R^{n+1}_ℓ from its pole S with coordinates $0, 0, \dots, 0, 1$ onto the hyperplane $x^{n+1} = 0$. This hyperplane is R^n_ℓ, and by supplementing it with a point at infinity, corresponding to the pole S of the hypersphere, and ideal points, corresponding to points on isotropic lines through S, it can be regarded as C^n_ℓ. Images of hyperspheres in C^n_ℓ on this hypersphere are cut out from it by hyperplanes in R^{n+1}_ℓ. In particular, since images of hyperplanes in R^n_ℓ on this hypersphere are cut from it by hyperplanes in R^{n+1}_ℓ passing through S, the transformations in C^n_ℓ preserving hyperspheres are imaged in R^{n+1}_ℓ by transformations preserving hyperplanes in R^{n+1}_ℓ. The most general such transformations are collineations in P^{n+1} obtained from R^{n+1}_ℓ by supplementing it with a hyperplane at infinity. ■

In **3.1.5** we have considered the analogous stereographic projection for $n = 2$.

If we denote the coordinates of a point X on the hypersphere by X^i $(i = 1, 2, \dots, n, n+1)$ and the coordinates of its stereographic projection x from the pole S onto the hyperplane $x^{n+1} = 1$ by x^i, then we obtain the following formulas, which are similar to formulas (3.4) and (3.71):

$$X^i = \frac{2x^i}{-\sum\limits_a (x^a)^2 + \sum\limits_u (x^u)^2 + 1}, \quad X^{n+1} = \frac{-\sum\limits_a (x^a)^2 + \sum\limits_u (x^u)^2 - 1}{-\sum\limits_a (x^a)^2 + \sum\limits_u (x^u)^2 + 1}. \tag{3.78}$$

Therefore the projective coordinates s^i $(i = 0, 1, \dots, n, n+1)$ of a point X on the hypersphere are equal to

$$\begin{aligned} s^0 &= \frac{1}{2}\left(-\sum_a (x^a)^2 + \sum_u (x^u)^2 + 1\right), \\ s^i &= x^i, \\ s^{n+1} &= \frac{1}{2}\left(-\sum_a (x^a)^2 + \sum_u (x^u)^2 - 1\right). \end{aligned} \tag{3.79}$$

These coordinates satisfy the equation

$$-(s^0)^2 - \sum_a (s^a)^2 + \sum_u (s^u)^2 + (s^{n+1})^2 = 0, \tag{3.80}$$

which represents a hyperquadric in P^{n+1} of index $\ell + 1$.

The coordinates s^i not satisfying the equation (3.80) are projective coordinates of a point in P^{n+1} not on the hyperquadric (3.80). The polar hyperplane of such a point cuts out from this hyperquadric the image of a hypersphere in C^n_ℓ. The points in the two domains in P^{n+1} determined by this hyperquadric give rise to hyperspheres in C^n_ℓ with real and pure imaginary radius. The $(n+2)$-spherical coordinates of a hypersphere in C^n_ℓ with real or pure imaginary radius can be normalized by equation (3.80), with 0 replaced by 1 or -1, respectively.

In C^n_ℓ, like Theorem 3.33 in C^n we can prove that the angle φ between two hyperspheres with real radii and normalized $(n+2)$-spheric coordinates s^i and t^i is given by the formula

$$\cos\varphi = |-s^0t^0 - \sum_a s^at^a + \sum_u s^ut^u + s^{n+1}t^{n+1}|. \tag{3.81}$$

3.8.6. Geometry of m-Spheres. Since hyperspheres in C^n and C^n_ℓ can be represented by points and their polar hyperplanes with respect to the hyperquadrics (3.73) and (3.80) in P^{n+1}, the m-spheres in C^n and C^n_ℓ can be represented by $(m+1)$-planes and their polar $(n-m-1)$-planes with respect to the same hyperquadrics.

m-Spheres are spheres for $m = 2$, circles for $m = 1$, and pairs of points for $m = 0$. In P^{n+1} they are represented by 3-planes and their polar $(n-3)$-planes, by planes and their polar $(n-2)$-planes, and by lines and their polar $(n-1)$-planes, respectively.

Theorem 3.34. *The symmetry figures in C^n and C^n_ℓ are m-spheres; hyperspheres for $m = n-1$, spheres for $m = 2$, circles for $m = 1$, pairs of points for $m = 0$, and for $n = 2\ell$ pairs of isotropic ℓ-planes.*

<u>PROOF</u>. The groups of conformal transformations in C^n and C^n_ℓ are isomorphic to the subgroups of the group of collineations in P^{n+1} whose symmetries preserve the respective hyperquadrics. These symmetry figures in P^{n+1} are $(m+1)$-pairs consisting of an $(m+1)$-plane and its polar $(n-m-1)$-plane with respect to these hyperquadrics and, for $n = 2\ell$, ℓ-pairs consisting of two ℓ-planar generators of the hyperquadric (3.80) representing isotropic ℓ-planes in C^n_ℓ.

The symmetries in these symmetry figures are represented by symmetries in the corresponding symmetry figures in P^{n+1}. Therefore, these symmetries can be reduced to the form (2.118) in coordinates s^i, hence a conformal symmetry in a hypersphere is an inversion in this hypersphere and a conformal symmetry in an m-sphere is the product of inversions in the orthogonal hyperspheres through this m-sphere. In particular, such a symmetry is the product of the reflections in the $(m+1)$-plane of this m-sphere by the inversion in the hypersphere whose great m-sphere is this m-sphere.

For $m = 0$ this symmetry is the product of the reflection in the line joining the pair of points by the inversion in the hypersphere whose antipodal points are the points of this pair. These symmetries are called *inversions in m-spheres, in circles* for $m = 1$, *in pairs of points* for $m = 0$. ■

Theorem 3.35. *The conformal invariants of two m-spheres in C^n or C^n_ℓ are the eigenvalues of the matrices (2.58), where A and B are the projective matrix coordinates of the $(m+1)$-planes representing the m-spheres in C^n and C^n_ℓ and U and V are the tangential matrix coordinates of the polar $(n-m-1)$-planes of these $(m+1)$-planes with respect to the hyperquadrics (3.73), respectively (3.80).*

PROOF. These invariants coincide with the invariants of the m-planes and their polar $(n-m-1)$-planes in P^{n+1} with respect to the hyperquadrics (3.73) and (3.80) representing these m-spheres. Like in **3.3.3**, for the stationary angles between the m-planes in R^n we can prove that the eigenvalues of the matrices (2.58) are equal to the squares $\cos^2\varphi_a$ of the cosines of the angles φ_a which are stationary angles of the hyperspheres passing through the m-spheres. ■

The hyperspheres represented by the points of intersection of transversals of the $(m+1)$-planes representing the m-spheres and their polar $(n-m-1)$-planes with respect to the hyperquadrics (3.73) and (3.80) with these $(m+1)$-planes and $(n-m-1)$-planes are geometric covariants of the two m-spheres. The role of these hyperspheres is analogous to the role of common perpendiculars and planes of stationary angles of two m-planes in R^n and R^n_ℓ.

3.8.7. Application of Complex Numbers, Quaternions, Split Complex Numbers, and Split Quaternions to the Geometry of Conformal and Pseudoconformal Planes and 4-Spaces. In **1.1.1**, **1.1.2**, **1.2.1**, and **1.2.3** we have seen that the planes R^2 and R^2_1 and the spaces R^4 and R^4_2 can be interpreted as the algebras $\mathbb{C}$, $\mathbb{C}'$, $\mathbb{H}$, and $\mathbb{H}'$, respectively. Since the similitudes in these planes and 4-spaces can be written as (1.9) and (1.47), and inversions in circles in $\mathbb{C}$ and $\mathbb{C}'$ and in 3-spheres in $\mathbb{H}$ and $\mathbb{H}'$ with equations $\overline{x}x = r^2$ can be written as transformations $x' = r^2\overline{x}^{-1}$ in these algebras, circular transformations in C^2 and C^2_1 and conformal transformations in C^4 and C^4_2, being products of similitudes and inversions, can be expressed by the fractional linear transformations (2.46) and (2.47).

In **2.2.2** we have seen that the projective lines $\mathbb{C}P^1, \mathbb{H}P^1, \mathbb{C}'P^1$, and $\mathbb{H}'P^1$ are homeomorphic to C^2, C^4, C^2_1, and C^4_2, respectively. The fractional linear transformations (2.46) and (2.47) in $\mathbb{C}$, $\mathbb{C}'$, $\mathbb{H}$, and $\mathbb{H}'$ can be regarded as projective transformations in these projective lines.

Note that inversions in general circles or 3-spheres $A\overline{x}x + \overline{B}x + \overline{x}B + C = 0$ (A and C real numbers) in these planes and 4-spaces have the form

$$x' = (A\overline{x} + \overline{B})^{-1}(-B\overline{x} - C). \tag{3.82}$$

Inversions with respect to pairs of points in these planes and 4-spaces can be reduced to the form $x' = x^{-1}$.

Theorem 3.36. *The conformal invariant of two pairs of points in C^n and of two general pairs of points in C^n_ℓ is the cross ratio of four complex or split complex numbers.*

PROOF. Through any four points in C^n and any four points in C^n_ℓ in general position there passes a plane C^2 or C^2_1. The invariant of four points in C^2 and of four points in C^2_1 in general position, regarded as the extended complex or split complex plane, is the cross ratio (2.44) of the corresponding four complex or split complex numbers. ■

In particular, for C^4 and C^4_2, regarded as the extended 4-spaces of quaternions and split quaternions, in formula (2.44) the element λ can always be chosen so that the right hand side of (2.44) is a real linear combination of the basis elements 1 and i or 1 and e. Therefore this expression can be regarded as a complex or split complex number.

Each two pairs of points in C^2, regarded as the extended complex plane, can be reduced by a fractional linear transformation to the pairs of complex numbers $1, -1$ and $a, -a$ which have cross ratio $w = (-a-1)(-1+a)^{-1}(-1-a)(a-1)^{-1} = (1+a)^2/(a-1)^2$. If $a = |a|e^{i\varphi_1} = e^{\varphi_0 + i\varphi_1}$, this cross ratio is equal to $(\cos\frac{\varphi_0+i\varphi_1}{2}/\sin\frac{\varphi_0+i\varphi_1}{2})^2 = \cot^2\frac{\varphi_0+i\varphi_1}{2}$. This invariant is equivalent to two real invariants of these two pairs of points: to the angle φ_1 between the lines through these pairs of points and to the quotient $|a| = e^{\varphi_0}$ of the radii of the two concentric circles which intersect these lines at the pairs of points. Note that by formula (3.37) (see **3.6.2**) the imaginary angle between two concentric circles with radii 1 and $|a|$ is equal to $i \ln |a| = i\varphi_0$.

§3.9. Finite Geometries

3.9.1. Finite Conformal and Pseudoconformal Spaces. In **2.9.1** and **2.9.2** we have defined the finite affine and projective spaces $\mathbb{F}_q E^n$ and $\mathbb{F}_q P^n$, and in **2.9.4** we have defined hyperquadrics in $\mathbb{F}_q P^n$. Since the real spaces C^n and C^n_ℓ admit interpretations as hyperquadrics in P^{n+1}, we can define the finite conformal and pseudoconformal n-spaces over the field $\mathbb{F}_q$ as hyperquadrics in $\mathbb{F}_q P^{n+1}$. Therefore, the hyperquadrics (2.131) and (2.133) in $\mathbb{F}_q P^{2n}$ and $\mathbb{F}_q P^{2n+1}$ are called *finite conformal spaces* and are denoted by $\mathbb{F}_q C^{2n-1}$ and $\mathbb{F}_q C^{2n}$, respectively; the hyperquadric (2.132) in $\mathbb{F}_q P^{2n+1}$ is called the *finite pseudoconformal space* and is denoted by $\mathbb{F}_q C^{2n}_1$. Sections of these hyperquadrics by planes and $(m+1)$-planes are called *circles* and *m-spheres* in these conformal and pseudoconformal spaces, respectively.

Formulas (2.137), (2.138), and (2.139) (see **2.9.4**) imply the following formulas for the cardinalities of the finite conformal and pseudoconformal spaces $\mathbb{F}_q C^n$ and $\mathbb{F}_q C^n_1$.

Theorem 3.37. *The cardinalities of the finite conformal and pseudoconformal spaces are equal to*

$$\operatorname{card} \mathbb{F}_q C^{2n-1} = \operatorname{card} \mathbb{F}_q P^{2n-1} = \frac{q^{2n}-1}{q-1}, \tag{3.83}$$

$$\operatorname{card} \mathbb{F}_q C^{2n} = \frac{(q^{n+1}+1)(q^n-1)}{q-1}, \tag{3.84}$$

$$\operatorname{card} \mathbb{F}_q C^{2n}_1 = \frac{(q^{n+1}-1)(q^n+1)}{q-1}. \tag{3.85}$$

In particular, for $n = 1, 2$ we obtain

$$\operatorname{card} \mathbb{F}_q C^1 = \operatorname{card} \mathbb{F}_q P^1 = q+1 \tag{3.86}$$

and

$$\operatorname{card} \mathbb{F}_q C^2 = q^2 + 1, \qquad \operatorname{card} \mathbb{F}_q C_1^2 = (q+1)^2. \tag{3.87}$$

The conformal transformations in $\mathbb{F}_q C^{2n-1}$ and $\mathbb{F}_q C^{2n-2}$ form the finite groups of Lie type $\mathbb{F}_q B_n = \mathbb{F}_q SO_{2n+1}$ and $\mathbb{F}_q D_n = \mathbb{F}_q SO_{2n}$, which contain the simple invariant subgroups $\mathbb{F}_q S\Omega_{2n+1}$ and $\mathbb{F}_q S\Omega_{2n}$.

The conformal transformations in $\mathbb{F}_q C_1^{2n-2}$ form the 2-*twisted finite group of Lie type* $\mathbb{F}_q D_n^{(2)} = \mathbb{F}_q SO_{2n}^{(2)}$, which contains the simple invariant subgroup $\mathbb{F}_q S\Omega_{2n}^{(2)}$.

Many theorems for C^2, in particular Miquel's first theorem, are also valid for $\mathbb{F}_q C^2$.

3.9.2. Steiner Triple Systems. Finite affine, projective and conformal spaces are particular cases of *Steiner triple systems.* The Steiner triple system $S(v, k, t)$ is a set with incidence structure consisting of v elements, in which each subset, called a *block*, contains k elements, and which has the property that each t elements belong to one unique block.

The number b of blocks of this system is equal to $b = C_v^t / C_k^t$.

The planes $\mathbb{F}_q E^2$, $\mathbb{F}_q P^2$, and $\mathbb{F}_q C^2$ are the Steiner triple systems $S(q^2, q, 2)$, $S(q^2+q+1,\, q+1, 2)$, and $S(q^2+1,\, q+1, 3)$, respectively. In the first and second cases the blocks are called *lines*, in the third case the blocks are called *circles*.

The number b in these systems is equal to q^2+q, q^2+q+1, and q^3+q, respectively.

The plane $\mathbb{F}_q C_1^2$ which can be regarded as $(\mathbb{F}_q \oplus \mathbb{F}_q)P^1$, is not a Steiner triple system.

The notion of Steiner triple system was introduced by E. Witt [Wit2] on the basis of the note [Ste] by J. Steiner.

The Steiner triple systems which admit a group of transformations are called *regular Steiner triple systems.* The Steiner triple systems which are finite affine, projective or conformal planes are regular; their groups of transformations are the groups of affine, projective and conformal transformations in these planes.

The transition from a Steiner triple system $S(v, k, t)$ to the Steiner triple system $S(v+1,\, k+1,\, t+1)$ is called *one-element extension* of $S(v, k, t)$. The plane $\mathbb{F}_q C^2$ is obtained from $\mathbb{F}_q E^2$ by such an extension.

Not all Steiner triple systems admit such an extension.

3.9.3. Mathieu Planes. The planes $\mathbb{F}_3 E^2$ and $\mathbb{F}_4 P^2$, which are the Steiner triple systems $S(9,3,2)$ and $S(21,5,2)$, admit one-element extensions. Such extensions for $\mathbb{F}_3 E^2$ are $\mathbb{F}_3 C^2 = S(10,4,3)$, $S(11,5,4)$, and $S(12,6,5)$. Such extensions for $\mathbb{F}_4 P^2 = S(21,5,3)$ are $S(22,6,3)$, $S(23,7,4)$, and $S(24,8,5)$. All these Steiner triple systems are regular and their groups of transformations are simple finite groups, that are not of Lie type (see **1.4.17**), these groups were discovered by Émile Mathieu [Mat1-2]. They are called *Mathieu groups,* and are denoted by $M_{11}, M_{12}, M_{22}, M_{23}$, and M_{24}, respectively (the index in the notation of such group is equal to the number v of the corresponding Steiner triple system).

These Steiner triple systems were investigated by E. Witt [Wit1], H. Lüneburg [Lün] and V. V. Afanas'yev [Af1–3].

Lüneburg and Afanas'yev have proved that these Steiner triple systems can be regarded as finite planes, which can be called *Mathieu planes.* The cardinality of each such plane is equal to the number v of the Steiner triple system.

The blocks in Mathieu planes are determined by 3, 4, and 5 elements and, therefore, are called *circles, spheres*, and *hyperspheres*, respectively.

The geometry of these planes is close to the geometry of conformal planes. In particular, in these planes theorems similar to Miquel's first theorem for $\mathbb{F}_q C^2$ are valid.

§3.10. Applications to Physics

3.10.1. Space-time of Special Relativity. 3-Space with extra time coordinate can be regarded as 4-space with spatial orthogonal coordinates x^1, x^2, x^3 and temporal coordinate t. In the classical Mechanics of Galilei and Newton, the transition from one inertial (that is, moving along a straight line with constant velocity) system of coordinates x^1, x^2, x^3, t to another inertial system $x^{1'}, x^{2'} x^{3'}, t'$ is expressed by formulas the

$$x^{i'} = A_i^{i'} x^i + v^{i'} t + a^{i'}, \ t' = t + a, \tag{3.88}$$

where $(A_i^{i'})$ is an orthogonal (3×3)-matrix and $v^{i'}$ are the coordinates of the vector of relative velocity of the two inertial systems of coordinates. These formulas imply that if the velocity of light in one inertial system is c, the velocity of light in the other inertial system moving in the same direction with relative velocity v is equal to $c + v$ or $c - v$.

However, the famous Michelson-Morley experiment showed that the velocity of light is the same in all inertial systems of coordinates. Therefore, the formulas (3.88) are valid only for velocities that are small in comparison with the velocity of light.

For a more exact description of the relative motion of two inertial systems for these systems the condition

$$\sum_i \left(\frac{x_2^i - x_1^i}{t_2 - t_1} \right)^2 = \sum_i \left(\frac{x_2^{i'} - x_1^{i'}}{t'_2 - t'_1} \right)^2 = c^2,$$

that is,

$$\sum_i (x_2^i - x_1^i)^2 - c^2(t_2 - t_1) = \sum_i (x_2^{i'} - x_1^{i'})^2 - c^2(t'_2 - t'_1), \tag{3.89}$$

must hold. This condition, which is the basic principle of the *Special Relativity of Albert Einstein* [Ei1], requires that formulas (3.88) be replaced by a more general transformation of the four coordinates $x^1, x^2 x^3$ and $x^4 = ct$:

$$x^{i'} = A_i^{i'} x^i + a^{i'}, \tag{3.90}$$

where $(A_i^{i'})$ is a pseudoorthogonal (4×4)-matrix of index 1. Since this transformation is a transformation of orthogonal coordinates in the pseudo-Euclidean space R_1^4, we see that *the space-time of the Special Relativity of Einstein can be taken to be the pseudo-Euclidean space* R_1^4.

The points X in this space represent *events*, and if the distance $\rho(X_1, X_2)$ is real, this distance is equal to the spatial distance between these events in any inertial system of coordinates in which these events are simultaneous. If this distance is a pure imaginary number $i\tau$, the number τ is equal the temporal distance between these events in any inertial system of coordinates in which these events have the same space coordinates; if this distance is 0, these events can be joined by a light ray.

The general transformations (3.90) are called *Poincaré transformations*; the transformations (3.90) in which $a^{i'} = 0$ are called *Lorentz transformations.*

The Lorentz transformations in which $x^{2'} = x^2, x^{3'} = x^3$ have the form

$$x^{1'} = x^1 \cosh \psi - x^4 \sinh \psi, \quad x^{4'} = -x^1 \sinh \psi + x^4 \cosh \psi. \tag{3.91}$$

Put

$$\tanh \psi = \frac{v}{c}. \tag{3.92}$$

Then

$$\cosh \psi = \frac{1}{\sqrt{1 - \frac{v^2}{c^2}}}, \quad \sinh \psi = \frac{\frac{v}{c}}{\sqrt{1 - \frac{v^2}{c^2}}}. \tag{3.93}$$

Therefore, if we replace x^1 by x and x^4 by ct, we obtain in (3.91)

$$x' = \frac{x - vt}{\sqrt{1 - \frac{v^2}{c^2}}}, \quad t' = \frac{at - \frac{v}{c}x}{\sqrt{1 - \frac{v^2}{c^2}}}. \tag{3.94}$$

3.10.2. Addition of Velocities in Special Relativity. Formula (3.92) shows that the velocity of a material particle in an inertial coordinate system is determined by the angle $\varphi = i\psi$ between the tangent to the line described by the particle in space-time and the temporal axis, that is, the axis x^4, of this coordinate system.

Therefore, the velocity of one inertial system of coordinates with respect to another such system is determined by the angle $\varphi = i\psi$ between the temporal axes of these systems. Since the angle between two temporal axes in the space-time of Special Relativity, that is, between two lines of pure imaginary length in R_1^4, is proportional to the spherical distance between the two points of intersection of these lines with the hypersphere of pure imaginary radius in this space, this velocity can be represented by the arc of the great circle of this hypersphere through these points.

Therefore, if we consider three inertial coordinate systems, (I), (II), and (III), and the velocities of the system (III) with respect to the systems (I) and (II) are v and v_2, respectively, and the velocity of the system (II) with respect to the system (I) is v_1, these three velocities are represented by the three sides of a spherical triangle ABC on the hypersphere of radius i in R_1^4 whose vertices are the points of intersection of this hypersphere with the temporal axes of the three inertial coordinate systems.

Denote the sides of this triangle corresponding to v, v_1, and v_2 by ψ, ψ_1, and ψ_2, respectively. Since the angle between the sides ψ and ψ_1 is equal to the complement $\pi - \alpha$ of the angle α between the corresponding velocities v and v_1, by the cosine law (3.48) for the hypersphere of radius iq in R_1^n we obtain

$$\cosh\psi = \cosh\psi_1 \cosh\psi_2 - \sinh\psi_1 \sinh\psi_2 \cos(\pi - \alpha)$$

$$= \cosh\psi_1 \cosh\psi_2 + \sinh\psi_1 \sinh\psi_2 \cos\alpha.$$

According to (3.93) and analogous relations for the velocities v_1 and v_2, this formula can be written as

$$\cosh\psi = \frac{1}{\sqrt{1-\frac{v^2}{c^2}}} = \frac{1+\frac{v_1v_2}{c_2}\cos\alpha}{\sqrt{1-\frac{v^1}{c^2}}\sqrt{1-\frac{v^2}{c_2}}}.$$

Hence we obtain Einstein's formula for the addition of velocities:

$$\frac{v}{c} = \frac{\sqrt{\frac{v_1^2}{c^2}+\frac{v_2^2}{c_2}+2\frac{v_1v_2}{c^2}\cos\alpha-\left(\frac{v_1v_2}{c_2}\sin\alpha\right)^2}}{1+\frac{v_1v_2}{c^2}\cos\alpha}. \tag{3.95}$$

The equivalence of this formula with the cosine law (3.45) was established by A. Sommerfeld [Smf].

3.10.3. Electromagnetic Field. In Special Relativity the electric and the magnetic fields are two manifestations of a single phenomenon, called the *electromagnetic field.* The coordinates E^1, E^2, E^3 of the strength vector E of the electric field and the coordinates H^1, H^2, H^3 of the strength vector H of the magnetic field determine the coordinates of the skew symmetric tensor $F^{ij} = -F^{ji}$ of the electromagnetic field. Under Poincaré (and Lorentz) transformations the coordinates E^1, E^2, E^3 and H^1, H^2, H^3 can be expressed in terms of each other. The coordinates F^{ij} have the form

$$\begin{aligned} H^1 &= F^{32} = -F^{23}, \\ H^2 &= F^{13} = -F^{31}, \\ H^3 &= F^{21} = -F^{12}, \\ cE^i &= F^{4i} = -F^{i4} (i = 1, 2, 3). \end{aligned} \tag{3.96}$$

The skew symmetric tensor F^{ij} gives rise to a skew symmetric linear operator $F = -F^\top$. In **3.7.4** and **3.7.6** we have seen that the product FE of the skew symmetric linear operator F in R^n_ℓ by the operator E determined by the inner product of vectors in this space can be regarded as the main moment of a system of sliding vectors with zero main vector, or as the main moment of a system of sliding spherical vectors on a hypersphere in this space.

This interpretation was proposed by O. M. Uzdenov [RoU].

In R^4_1 these systems can be reduced to canonical form, consisting of two pairs of vectors on two totally orthogonal planes or of two sliding spherical vectors, one on a hypersphere of imaginary radius iq and the other on the hypersphere with the same center and real radius q on two totally orthogonal planes in R^4_1. If the arms of these pairs of vectors are equal to i and 1, or if the radii of the hyperspheres are equal to i and 1 (that is, $q = 1$), the moduli of these vectors are equal to the eigenvalues μ_1 and $i\mu_2$ of the linear operator FE. These eigenvalues are the roots of the equation

$$\begin{bmatrix} -\lambda & F^{12} & F^{13} & -F^{14} \\ -F^{12} & -\lambda & F^{23} & -F^{24} \\ -F^{13} & -F^{23} & -\lambda & -F^{34} \\ -F^{14} & -F^{23} & -F^{34} & \lambda \end{bmatrix} = -\lambda^4 + I\lambda^2 + J^2 = 0 \tag{3.97}$$

and have the form

$$\lambda_{1,2,3,4} = \pm\sqrt{I \pm \sqrt{I^2 + 4J^2}} \tag{3.98}$$

where

$$I = (F^{14})^2 + (F^{24})^2 + (F^{34})^2 - (F^{23})^2 - (F^{13})^2 - (F^{12})^2 = c^2E^2 - H^2,$$
$$J = F^{14}F^{23} - F^{24}F^{13} + F^{34}F^{12} = cEH.$$

Formula (3.98) shows that the roots λ_1 and $\lambda_2 = -\lambda_1$ are real and the roots λ_3 and $\lambda_4 = -\lambda_3$ are pure imaginary. The moments of two pairs of vectors of the canonical systems determined by the linear operator FE are equal to $|\lambda_1|$ and $i|\lambda_3|$. One of these pairs of vectors lies in a Euclidean plane, the second pair of vectors lies in a pseudo-Euclidean plane.

The last plane contains two real isotropic lines through the given point of the electromagnetic field, the first plane contains two imaginary isotropic lines through this point. The real and imaginary isotropic vectors with directions along these isotropic lines form the so-called *light quadruple* or *light tetrad* of the electromagnetic field.

In the case of a plane electromagnetic wave, the vectors E and H are orthogonal and $|cE| = |H|$, that is, $I = J = 0$. Therefore, in this case both vectors of the canonical system are isotropic. Since the condition $J = 0$ coincides with the equation (2.28), in this case the operator F has the form $x \wedge y$ and determines a plane. This plane is isotropic and passes through the given point of the electromagnetic field; in this case both isotropic vectors of the canonical system lie on this plane.

The invariants $|\lambda_1|$ and $i|\lambda_3|$ are equivalent to the invariants I and J of the electromagnetic field.

Since the invariants $|\lambda_1|$ and $|\lambda_3|$ are equal or proportional to the lengths of the vectors of the canonical system determined by the operator F, they can be regarded as angles in R^4_1. Therefore *the electromagnetic field is invariant under conformal transformations in R^4_1, that is, under the transformations in the fundamental group of the pseudoconformal space C^4_1.*

CHAPTER IV

ELLIPTIC, HYPERBOLIC, PSEUDOELLIPTIC, AND PSEUDOHYPERBOLIC GEOMETRIES

§4.1. Elliptic, Hyperbolic, Pseudoelliptic, and Pseudohyperbolic Spaces

4.1.1 Quadratic Elliptic, Hyperbolic, Pseudoelliptic, and Pseudohyperbolic Spaces. In **0.8.8** we have seen that the groups of rotations of hyperspheres in the real Euclidean and pseudo-Euclidean spaces R^{n+1} and R_ℓ^{n+1} are imprimitive and that the classes of primitivity of these hyperspheres are pairs of their antipodal points. Therefore, if we identify pairs of antipodal points of these hyperspheres we obtain homogeneous spaces with primitive groups of transformations.

The space obtained by this identification from the hypersphere in R^{n+1} is called a *real elliptic space* and is denoted by S^n.

The space obtained by this identification from the hypersphere of imaginary radius in R_1^{n+1} is called a *real hyperbolic space* and is denoted by H^n.

The space obtained by this identification from the hypersphere of real radius in R_ℓ^{n+1} is called a *real pseudoelliptic space* of index ℓ and is denoted by S_ℓ^n.

The space obtained by this identification from the hypersphere of imaginary radius in $R_{\ell+1}^{n+1}$ is called a *real pseudohyperbolic space* of index ℓ and is denoted by H_ℓ^n.

The radius of the hypersphere is called the *radius of curvature* of S^n, H^n, S_ℓ^n, or H_ℓ^n. The *curvature* of the space obtained from a hypersphere with real radius r is the positive real number $\frac{1}{r^2}$. The *curvature* of the space obtained from a hypersphere with imaginary radius qi is the negative real number $-\frac{1}{q^2}$.

The spaces S^n and H^n are Riemannian manifolds V^n of constant positive, respectively negative, curvature. The spaces S_ℓ^n and H_ℓ^n are pseudo-Riemannian manifolds V_ℓ^n of constant positive, respectively negative, curvature.

The space S^n can also be defined as projective space P^n endowed with a metric. The spaces H^n, S_ℓ^n, and H_ℓ^n can also be regarded as domains in P^n endowed with a metric or pseudometric. The metrics in S^n and H^n are determined by the metrics of the hyperspheres in R^{n+1}, respectively R_1^{n+1}. The pseudometrics in S_ℓ^n and H_ℓ^n are determined by the pseudometrics of the hyperspheres in R_ℓ^{n+1}, respectively $R_{\ell+1}^{n+1}$.

The position vector of a point on a hypersphere is called a *vector representing the point* in S^n, H^n, S_ℓ^n, or H_ℓ^n.

In the space S^n with radius of curvature r the distance between two points X and Y, represented by vectors x and y, is defined by the formula

$$\cos\frac{\delta}{r} = \frac{xy}{r^2}. \tag{4.1}$$

In the space S_ℓ^n with the same radius of curvature the distance between X and Y is defined by the formula obtained from (4.1) by replacing the inner product xy by xEy.

In the spaces H^n and H_ℓ^n with radius of curvature iq the distance between X and Y is defined by the analogous formula

$$\cosh\frac{\delta}{q} = -\frac{xEy}{q^2}. \tag{4.2}$$

Theorem 4.1. *The elliptic space S^n can be defined as the projective space P^n in which the distance δ between two points X and Y is defined by*

$$\cos^2\frac{\delta}{r} = \overline{XY, \alpha\beta}, \tag{4.3}$$

where $\overline{XY, \alpha\beta}$ is the cross ratio of these points and their polar hyperplanes with respect to the imaginary hyperquadric $x^2 = 0$.

PROOF. If x and y are not the position vectors of two points on a hypersphere of radius r in R^{n+1} but are two arbitrary vectors in E^{n+1} representing two points in P^n, formula (4.1) can be written as

$$\cos^2\frac{\delta}{r} = \frac{xy \cdot yx}{x^2 \cdot y^2}. \tag{4.4}$$

Since the equation $x^2 = xIx = 0$ is the equation of an imaginary hyperquadric in P^n, if the vectors x and y represent the points X and Y in this space, the same vectors can be regarded also as covectors representing the hyperplanes polar to these points with respect to this imaginary hyperquadric. The right hand side of (4.4) can be regarded as the cross ratio $\overline{XY, \alpha\beta}$ of these points and the polar hyperplanes of these points with respect to this imaginary hyperquadric. ■

The imaginary hyperquadric $x^2 = 0$ is called the *absolute hyperquadric* of the space S^n.

Theorem 4.2. *The hyperbolic space H^n can be defined as the interior domain of the real oval hyperquadric $xEx = 0$ in P^n, where the distance δ between two points X and Y is defined by*

$$\cosh^2\frac{\delta}{q} = \overline{XY, \alpha\beta}, \tag{4.5}$$

where $\overline{XY, \alpha\beta}$ is the cross ratio of these points and their polar hyperplanes with respect to the oval hyperquadric $xEx = 0$.

PROOF. If x and y are not the position vectors of two points on a hypersphere of radius qi in R_1^{n+1} but are two arbitrary vectors in E^{n+1} representing two points in P^n, formula (4.2) can be written as

$$\cosh^2 \frac{\delta}{q} = \frac{xEy \cdot yEx}{xEx \cdot yEy}. \tag{4.6}$$

Since the equation $xEx = 0$ is the equation of an oval hyperquadric in P^n, if the vectors x and y represent points X and Y in this space, the vectors Ex and Ey can be regarded as covectors representing the hyperplanes polar to these points with respect to this oval hyperquadric. The right hand side of (4.6) can be regarded as the cross ratio $\overline{XY,\alpha\beta}$ of these points and their polar hyperplanes with respect to the oval hyperquadric $xEx = 0$. Since the rectilinear generators of the isotropic hypercone $xEx = 0$ in R_1^{n+1} with inner product xEy of vectors correspond to the points on the oval hyperquadric with the same equation, and since the lines of imaginary length in R_1^{n+1}, that is, the lines with direction vectors x satisfying the condition $xEx < 0$, correspond to the points in the interior domain of this hyperquadric, the space H^n can be represented as this domain. ■

The oval hyperquadric $xEx = 0$ is called the ***absolute hyperquadric*** of H^n.

Theorem 4.3. *The pseudoelliptic space S_ℓ^n and the pseudohyperbolic space H_ℓ^n can be defined as one of the two domains into which the hyperquadric $xEx = 0$ divides the projective space P^n, where the distance δ between two points X and Y is defined by (4.3), or (4.5), in the case of real radius of curvature r, respectively imaginary radius of curvature qi. Here $\overline{XY,\alpha\beta}$ is the cross ratio of these points and their polar hyperplanes with respect to the hyperquadric $xEx = 0$.*

<u>Proof</u>. If x and y are not the position vectors of two points X and Y of a hypersphere in R_ℓ^{n+1} but are two arbitrary vectors in E^{n+1} representing two points in P^n, the distance δ between the corresponding points in S_ℓ^n of radius of curvature r can be defined by the formula whose left hand side coincides with the left hand side of (4.4) and whose right hand side has the form of the right side of (4.6); the distance δ between the corresponding points in $H_{\ell-1}^n$ of radius of curvature iq can be defined by formula (4.6). In both cases E is the linear operator determining the inner product xEy of vectors in R_ℓ^{n+1}. Since the equation $xEx = 0$ is the equation of a real hyperquadric in P^n, if the vectors x and y represent points X and Y in S_ℓ^n and $H_{\ell-1}^n$, the vectors Ex and Ey can be regarded as covectors representing the hyperplanes polar to these points with respect to this hyperquadric. Therefore, the right hand side of (4.6) can be regarded as the cross ratio $\overline{XY,\alpha\beta}$ of these points and their polar hyperplanes with respect to the hyperquadric $xEx = 0$. Since the rectilinear generators of the isotropic hypercone $xEx = 0$ in R_ℓ^{n+1} correspond to the points of a hyperquadric with the same equation, and since the lines of real or imaginary length in R_ℓ^{n+1}, that is, the lines with direction vectors x satisfying the conditions $xEx > 0$, respectively $xEx < 0$, correspond to the points in the two domains into which the hyperquadric $xEx = 0$ divides the space P^n, the spaces $S_{\ell-1}^n$ and H_ℓ^n can be represented as these domains. ■

The hyperquadric $xEx = 0$ is called the ***absolute hyperquadric*** of the pseudoelliptic space S_ℓ^n and of the pseudohyperbolic space $H_{\ell-1}^n$. The hyperbolic space H^n also has the same absolute hyperquadric as the pseudoelliptic space S_1^n.

The space S_ℓ^n with radius of curvature r together with the space obtained from $H_{\ell-1}^n$, for $\ell = 1$ from H^n, with radius of curvature ir by replacing all real distances

δ by $i\delta$ and all pure imaginary distances $i\delta$ by δ is called the *extended pseudoelliptic space* S^n_ℓ. This space can also be defined as the space P^n endowed with the pseudometric determined by (4.6). Analogously we can define the *extended hyperbolic space* H^n and the *extended pseudohyperbolic space* H^n_ℓ.

The analogous quadratic elliptic and extended hyperbolic, pseudoelliptic, and pseudohyperbolic spaces $\mathbb{A}S^n, \mathbb{A}H^n, \mathbb{A}S^n_\ell$, and $\mathbb{A}H^n_\ell$ can be defined over all commutative Frobenius algebras $\mathbb{A}$. These spaces are the spaces $\mathbb{A}P^n$ in which an absolute hyperquadric is given and the distances are determined by (4.3) and (4.5) or (4.4) and (4.6).

Since the equation of a nondegenerate hyperquadric in $\mathbb{C}P^n$ can be reduced to the form $\sum_i (x^i)^2 = 0$ over the field $\mathbb{C}$, quadratic hyperbolic, pseudoelliptic and pseudohyperbolic spaces do not exist in $\mathbb{C}P^n$ and only the quadratic elliptic spaces $\mathbb{C}S^n$ are possible. On the contrary, there exist all elliptic, hyperbolic, pseudoelliptic, and pseudohyperbolic spaces over the algebras $\mathbb{C}'$ and $\mathbb{C}^0$; over $\mathbb{C}^0$ these spaces are Hjelmslev spaces.

The tangent spaces for the spaces $\mathbb{A}S^n$ and $\mathbb{A}H^n$ are the Euclidean spaces $\mathbb{A}R^n$, the tangent spaces for the spaces $\mathbb{A}S^n_\ell$ and $\mathbb{A}H^n_\ell$ are the pseudo-Euclidean spaces $\mathbb{A}R^n_\ell$. Therefore, the tangent spaces for the spaces $\mathbb{C}S^n, \mathbb{C}'S^n$, and $\mathbb{C}'S^n_\ell$ are the spaces $\mathbb{C}R^n$, $\mathbb{C}'R^n$, and $\mathbb{C}'R^n_\ell$, respectively, and if we introduce into $\mathbb{C}R^n, \mathbb{C}'R^n$, and $\mathbb{C}'R^n_\ell$ the real metrics and pseudometrics of the spaces R^{2n}_n, R^{2n}, and $R^{2n}_{2\ell}$, defined in Theorem 3.2 (see **3.1.3**), we thereby introduce into the spaces $\mathbb{C}S^n, \mathbb{C}'S^n$, and $\mathbb{C}'S^n_\ell$ the real metrics and pseudometrics of the Riemannian and pseudo-Riemannian manifolds V^{2n}_n, V^{2n}, and $V^{2n}_{2\ell}$, whose curvature is not constant.

The geometry in H^3 was founded by C. F. Gauss [Gau]. N. I. Lobachevsky [Lob] and J.Bolyai [Boi]. The geometry in S^3 was founded by A. Cayley [Cay3] and W. K. Clifford [Cli2]. The term *non-Euclidean geometry*, for the geometry in H^3, was introduced by Gauss [Gau], this term was extended to the geometry in S^3 by F.Klein in [Kl1], in which also the terms *elliptic space* and *hyperbolic space* were introduced. The geometry in S^n and H^n was founded by B. Riemann [Rie2] as the geometry in V^n of constant curvature. The terms *pseudoelliptic space* and *pseudohyperbolic space* were introduced by the author [Ro5] (see also [Car10] and [Fed1]). The notations of these spaces were introduced by J. A. Wolf [Wol]. For the history of the geometry of the real non-Euclidean spaces see [Ro4, pp. 206-236].

4.1.2. Hermitian Elliptic, Hyperbolic, Pseudoelliptic, and Pseudohyperbolic Spaces. The groups of rotations of Hermitian hyperspheres in the complex and split complex Hermitian Euclidean and pseudo-Euclidean spaces $\mathbb{C}\overline{R}^{n+1}$ and $\mathbb{C}\overline{R}^{n+1}_\ell$ also are imprimitive groups, the classes of primitivity of these hyperspheres are their intersections with the lines through the centers of these hyperspheres. In the real interpretations of these spaces in R^{2n} and $R^{2n}_{2\ell}$ these intersections are interpreted as great circles on hyperspheres in R^{2n} and $R^{2n}_{2\ell}$. If we regard these classes of primitivity as points in a new space, we obtain a homogeneous space with primitive group of transformations.

The space so obtained from the Hermitian hypersphere in $\mathbb{C}\overline{R}^{n+1}$ is called a *complex Hermitian elliptic space* and is denoted by $\mathbb{C}\overline{S}^n$. The space so obtained from the Hermitian hypersphere of imaginary radius in $\mathbb{C}\overline{R}^{n+1}_1$ is called a *complex Hermitian hyperbolic space* and is denoted by $\mathbb{C}\overline{H}^n$. The space so obtained from the Hermitian hypersphere of real radius in $\mathbb{C}\overline{R}_{\ell}^{n+1}$ is called a *complex Hermitian pseudoelliptic space of index* ℓ and is denoted by $\mathbb{C}\overline{S}_{\ell}^{n}$. The space so obtained from the Hermitian hypersphere of imaginary radius in $\mathbb{C}\overline{R}_{\ell+1}^{n+1}$ is called a *complex Hermitian pseudohyperbolic space of index* ℓ and is denoted by $\mathbb{C}\overline{H}_{\ell}^{n}$.

The radius of the Hermitian hypersphere is called the *radius of curvature* of $\mathbb{C}\overline{S}^{n}$, $\mathbb{C}\overline{H}^{n}$, $\mathbb{C}\overline{S}_{\ell}^{n}$, and $\mathbb{C}\overline{H}_{\ell}^{n}$. The *curvature* of the space obtained from a Hermitian hypersphere with real radius r is the positive real number $\frac{1}{r^2}$. The *curvature* of the space obtained from a Hermitian hypersphere with imaginary radius qi is the negative real number $-\frac{1}{q^2}$.

The space $\mathbb{C}\overline{S}^{n}$ can also be defined as projective space $\mathbb{C}P^n$ endowed with a metric, the spaces $\mathbb{C}\overline{H}^{n}, \mathbb{C}\overline{S}_{\ell}^{n}$, and $\mathbb{C}\overline{H}_{\ell}^{n}$ can be regarded as domains in $\mathbb{C}P^n$ endowed with a metric or pseudometric.

The position vectors of points on the hyperspheres are called *vectors representing points* in $\mathbb{C}\overline{S}^{n}, \mathbb{C}\overline{H}^{n}, \mathbb{C}\overline{S}_{\ell}^{n}$, and $\mathbb{C}\overline{H}_{\ell}^{n}$.

In the space $\mathbb{C}\overline{S}^{n}$ with radius of curvature r the distance δ between two points X and Y, represented by vectors x and y, is defined by the formula

$$\cos^2\frac{\delta}{r} = \frac{\overline{x}y \cdot \overline{y}x}{\overline{x}x \cdot \overline{y}y}. \tag{4.7}$$

In the space $\mathbb{C}\overline{S}_{\ell}^{n}$ with the same radius of curvature the distance δ between X and Y is defined by the formula obtained from (4.7) by replacing the inner product $\overline{x}y$ by $\overline{x}Ey$.

In the spaces $\mathbb{C}\overline{H}^{n}$ and $\mathbb{C}\overline{H}_{\ell}^{n}$ with radius of curvature $r = iq$ the distance δ between X and Y is defined by the analogous formula

$$\cosh^2\frac{\delta}{q} = \frac{\overline{x}Ey \cdot \overline{y}Ex}{\overline{x}Ex \cdot \overline{y}Ey}. \tag{4.8}$$

Theorem 4.4. *The complex Hermitian elliptic space* $\mathbb{C}\overline{S}^{n}$ *can be defined as the projective space* $\mathbb{C}P^n$ *in which the distance* δ *between two points* X *and* Y *is defined by (4.3), where* $\overline{XY,\alpha\beta}$ *is the cross ratio of these points and the polar hyperplanes of these points with respect to the imaginary Hermitian hyperquadric* $\overline{x}x = 0$.

Proof. If x and y are not the position vectors of two points of a Hermitian hypersphere of radius r in $\mathbb{C}\overline{R}^{n+1}$ but are two arbitrary vectors in $\mathbb{C}E^{n+1}$ representing two points in $\mathbb{C}P^n$, formula (4.7) can be written as (4.3). Since $\overline{x}x = \overline{x}Ix = 0$ is the equation of an imaginary Hermitian hyperquadric in $\mathbb{C}P^n$, if x and y represent points X and Y in this space, the complex conjugate vectors $\overline{x}$ and $\overline{y}$ can be regarded as covectors representing the hyperplanes polar to these points with respect to this imaginary Hermitian hyperquadric. The right hand side of (4.7) can be regarded as the cross ratio $\overline{XY,\alpha\beta}$ of these points and the polar hyperplanes of these points with respect to the imaginary Hermitian hyperquadric $\overline{x}x = 0$. ■

The imaginary Hermitian hyperquadric $\overline{x}x = 0$ is called the *absolute Hermitian hyperquadric* of the space $\mathbb{C}\overline{S}^{n}$.

Theorem 4.5. *The complex Hermitian hyperbolic space $\overline{\mathbb{C}H}^n$ can be defined as the interior domain of the oval Hermitian hyperquadric $\bar{x}Ex = 0$ in $\mathbb{C}P^n$, in which the distance δ between two points X and Y is determined by (4.5), where $\overline{XY, \alpha\beta}$ is the cross ratio of these points and their polar hyperplanes with respect to the oval Hermitian hyperquadric $\bar{x}Ex = 0$.*

PROOF. If x and y are not the position vectors of two points on a Hermitian hypersphere of radius qi in $\overline{\mathbb{C}R}_1^{n+1}$ but are two arbitrary vectors of $\mathbb{C}E^{n+1}$ representing two points in $\mathbb{C}P^n$, formula (4.8) can be written as (4.5). Since $\bar{x}Ex = 0$ is the equation of an oval Hermitian hyperquadric in $\mathbb{C}P^n$, if the vectors x and y represent points X and Y in this space, the vectors $E\bar{x}$ and $E\bar{y}$ can be regarded as covectors representing the hyperplanes polar to these points with respect to this oval Hermitian hyperquadric, and the right hand side of (4.8) can be regarded as the cross ratio $\overline{XY, \alpha\beta}$ of these points and their polar hyperplanes with respect to the Hermitian hyperquadric $\bar{x}Ex = 0$. Since the rectilinear generators of the isotropic Hermitian hypercone $\bar{x}Ex = 0$ in $\overline{\mathbb{C}R}_1^{n+1}$ with inner product $\bar{x}Ey$ correspond to the points on the oval Hermitian hyperquadric with the same equation, and since the lines in $\overline{\mathbb{C}R}_1^{n+1}$ with direction vectors x satisfying the condition $\bar{x}Ex < 0$ correspond to the points in the interior domain of this oval Hermitian hyperquadric, the space $\overline{\mathbb{C}H}^n$ can be represented as this domain. ■

The oval Hermitian hyperquadric $\bar{x}Ex = 0$ is called the *absolute Hermitian hyperquadric* of $\overline{\mathbb{C}H}^n$.

Theorem 4.6. *The complex Hermitian pseudoelliptic space $\overline{\mathbb{C}S}_\ell^n$ and the complex Hermitian pseudohyperbolic space $\overline{\mathbb{C}H}_\ell^n$ can be defined as one of the two domains into which the hyperquadric $\bar{x}Ex = 0$ divides the projective space $\mathbb{C}P^n$, where the distance δ between points X and Y is determined by (4.3) or (4.5) in the case of real radius of curvature r, respectively imaginary radius of curvature qi. Here, $\overline{XY, \alpha\beta}$ is the cross ratio of these points and their polar hyperplanes with respect to the Hermitian hyperquadric $\bar{x}Ex = 0$.*

PROOF. If x and y are not the position vectors of two points of a hypersphere in $\overline{\mathbb{C}R}_\ell^{n+1}$ but are two vectors of $\mathbb{C}E^{n+1}$ representing two points in $\mathbb{C}P^n$, the distance δ between the corresponding points in the space $\overline{\mathbb{C}S}_\ell^n$ of radius of curvature r, respectively on $\overline{\mathbb{C}H}_{\ell-1}^n$, can be determined by the formula whose left hand side coincides with the left hand side of (4.7) and right hand side has the form of the right hand side of (4.8), respectively by formula (4.8). In both cases E is the linear operator determining the inner product $\bar{x}Ey$ of vectors in $\overline{\mathbb{C}R}_\ell^{n+1}$. Since the equation $\bar{x}Ex = 0$ is the equation of a real Hermitian hyperquadric in $\mathbb{C}P^n$, if the vectors x and y represent points X and Y of this space, the vectors $E\bar{x}$ and $E\bar{y}$ can be regarded as covectors representing the hyperplanes polar to these points with respect to this Hermitian hyperquadric, the right hand side of (4.8) can be regarded as the cross ratios $\overline{XY, \alpha\beta}$ of these points and their polar hyperplanes with respect to the Hermitian hyperquadric $\bar{x}Ex = 0$. Since the rectilinear generators of the Hermitian hypercone $\bar{x}Ex = 0$ in $\overline{\mathbb{C}R}_\ell^{n+1}$ correspond to the points on the Hermitian hyperquadric with the same equation, and since the lines in $\overline{\mathbb{C}R}_\ell^{n+1}$ with direction vectors x satisfying the conditions $\bar{x}Ex > 0$, respectively $\bar{x}Ex < 0$, correspond to the points in the two domains into which the Hermitian hyperquadric $xEx = 0$ divides $\mathbb{C}P^n$, the spaces $\mathbb{C}\overline{S}^n_\ell$ and $\mathbb{C}\overline{H}^n_\ell$ can be interpreted as these domains. ■

The Hermitian hyperquadric $\bar{x}Ex = 0$ is called the *absolute Hermitian hyperquadric* of the complex Hermitian pseudoelliptic space $\mathbb{C}\overline{S}^n_\ell$ or of the complex Hermitian pseudohyperbolic space $\mathbb{C}\overline{H}^n_{\ell-1}$. The space $\mathbb{C}\overline{H}^n$ has the same absolute as $\mathbb{C}\overline{S}^n_1$.

The space $\mathbb{C}\overline{S}^n_\ell$ with radius of curvature r, as well as the space obtained from $\mathbb{C}\overline{H}^n_{\ell-1}$, for $\ell = 1$ from $\mathbb{C}\overline{H}^n$, with radius of curvature ir by replacing all real distances δ by $i\delta$ and all pure imaginary distances $i\delta$ by δ, is called the *extended complex Hermitian pseudoelliptic space* $\overline{\mathbb{C}S}^n_\ell$. This space can also be defined as the space $\mathbb{C}P^n$ with pseudometric determined by the formula obtained from (4.7) by replacing the Hermitian inner product $\bar{x}y$ by $\bar{x}Ey$.

We can analogously define the *extended complex Hermitian hyperbolic space* $\mathbb{C}\overline{H}^n$ and *extended complex Hermitian pseudohyperbolic space* $\mathbb{C}\overline{H}^n_\ell$.

The analogous Hermitian elliptic and extended hyperbolic, pseudoelliptic, and pseudohyperbolic spaces $\mathbb{A}\overline{S}^n, \mathbb{A}\overline{H}^n, \mathbb{A}\overline{S}^n_\ell$, and $\mathbb{A}\overline{H}^n_\ell$ can be defined over all Frobenius algebras $\mathbb{A}$ with involution. These spaces are the spaces $\mathbb{A}P^n$ in which an absolute Hermitian hyperquadric is given and the distances are determined by (4.3) and (4.5) or (4.7) and (4.8).

Since the equation of a nondegenerate Hermitian hyperquadric in $\mathbb{C}'P^n$ or $\mathbb{H}'P^n$ can be reduced to the form $\sum_i \bar{x}^i x^i = 0$, Hermitian hyperbolic, pseudoelliptic and pseudohyperbolic spaces do not exist over the algebras $\mathbb{C}'$ and $\mathbb{H}'$ and only the Hermitian elliptic spaces $\mathbb{C}'\overline{S}^n$ and $\mathbb{H}'\overline{S}^n$ are possible. On the contrary, all Hermitian elliptic, hyperbolic, pseudoelliptic, and pseudohyperbolic spaces exist over the algebras $\mathbb{C}$, $\mathbb{H}$, $\mathbb{C}^0$, $\mathbb{H}^0$, and $\mathbb{H}'^0$; over $\mathbb{C}^0$ and $\mathbb{H}^0$ these spaces are Hjelmslev spaces.

The geometry in $\mathbb{C}\overline{S}^3$ and $\mathbb{C}\overline{H}^3$ was founded by G.G. Fubini [Fu2] and E. Study [Stu2], see also the books by J. L. Coolidge [Cool] and É. Cartan [Car3]. The spaces $\mathbb{C}\overline{S}^n, \mathbb{C}\overline{H}^n, \mathbb{H}\overline{S}^n$ and $\mathbb{H}\overline{H}^n$ were first considered by Cartan [Car12-13] as symmetric Riemannian spaces V^{2n} and V^{4n} of positive or negative curvature. For $\mathbb{C}\overline{S}^n$, $\mathbb{C}\overline{H}^n$, $\mathbb{H}\overline{S}^n$, $\mathbb{H}\overline{H}^n$, $\mathbb{C}'\overline{S}^n$, and $\mathbb{H}'\overline{S}^n$ see [Ro1, pp. 620-654]; for $\mathbb{C}\overline{H}^n$ see also [Gol]; for $\mathbb{C}\overline{H}^n$ and $\mathbb{H}\overline{H}^n$ see also [Mos]. For the history of the geometry in Hermitian non-Euclidean spaces see [Ro4, pp. 361-363]. Note that fibering of hyperspheres in R^{2n+2} and R^{4n+4} determined by definition of the spaces $\mathbb{C}\overline{S}^n$ and $\mathbb{H}\overline{S}^n$ is the fibering of Theorem 2.54.

4.1.3. Bicomplex, Biquaternionic, and Quaterquaternionic Elliptic and Hyperbolic Spaces. Since the algebra $\mathbb{C} \otimes \mathbb{C}$ is commutative and has the three involutions $x \to \bar{x}, x \to \tilde{x}$, and $x \to \tilde{\bar{x}}$, over this algebra we can define the quadratic elliptic and hyperbolic spaces $(\mathbb{C}\otimes\mathbb{C})S^n$ and $(\mathbb{C}\otimes\mathbb{C})H^n$, and the Hermitian elliptic and hyperbolic spaces $(\mathbb{C}\otimes\mathbb{C})\overline{S}^n, (\mathbb{C}\otimes\mathbb{C})\overline{H}^n, (\mathbb{C}\otimes\mathbb{C})\ \tilde{\bar{S}}^n$, and $(\mathbb{C}\otimes\mathbb{C})\ \tilde{\bar{H}}^n$.

Since the algebras $\mathbb{C}\otimes\mathbb{H}$ and $\mathbb{H}\otimes\mathbb{H}$ are not commutative, $\mathbb{C}\otimes\mathbb{H}$ has the two involutions $x \to \tilde{x}$, and $x \to \tilde{\bar{x}}$, and $\mathbb{H}\otimes\mathbb{H}$ has only the involution $x \to \tilde{\bar{x}}$, over these algebras we can define the Hermitian elliptic and hyperbolic space $(\mathbb{C}\otimes\mathbb{H})\overline{S}^n, (\mathbb{C}\otimes\mathbb{H})\overline{H}^n, (\mathbb{C}\otimes\mathbb{H})\ \tilde{\bar{S}}^n, (\mathbb{C}\otimes\mathbb{H})\ \tilde{\bar{H}}^n, (\mathbb{H}\otimes\mathbb{H})\ \tilde{\bar{S}}^n$, and $(\mathbb{H}\otimes\mathbb{H})\ \tilde{\bar{H}}^n$.

The definitions of these spaces are analogous to the definitions of the spaces

$\mathbb{C}S^n$, $\mathbb{C}'H^n$, $\overline{\mathbb{C}S}^n$, and $\overline{\mathbb{C}}H^n$ by formulas (4.3) and (4.5). Since the cross ratio w of two 0-pairs in $\mathbb{A}P^n$, over a noncommutative algebra $\mathbb{A}$, is determined up to the transformation $w \to pwp^{-1}$, in the spaces over the algebras $\mathbb{C} \otimes \mathbb{H}$ and $\mathbb{H} \otimes \mathbb{H}$ the cross ratio w in (4.3) and (4.5) must be replaced by an expression pwp^{-1} where p is an element in the same algebra which is a real linear combination of elements 1 and iI or $1, iI, jJ$, and kK. Therefore, like for Euclidean spaces over the same three algebras (see **3.1.4**), the distances between points in $(\mathbb{C}\otimes\mathbb{C})S^n$ and $(\mathbb{C}\otimes\mathbb{C})H^n$ are bicomplex numbers, the distances between points in $(\mathbb{C} \otimes \mathbb{C})\overline{S}^n, (\mathbb{C} \otimes \mathbb{C})\overline{H}^n, (\mathbb{C} \otimes \mathbb{C})\,\overset{\approx}{S}{}^n, (\mathbb{C} \otimes \mathbb{C})\,\overset{\approx}{H}{}^n, (\mathbb{C} \otimes \mathbb{H})\overline{S}^n, (\mathbb{C} \otimes \mathbb{H})\overline{H}^n, (\mathbb{C} \otimes \mathbb{H})\,\overset{\approx}{S}{}^n$, and $(\mathbb{C} \otimes \mathbb{H})\,\overset{\approx}{H}{}^n$ are split complex numbers, and in $(\mathbb{H} \otimes \mathbb{H})\,\overset{\approx}{S}{}^n$ and $(\mathbb{H} \otimes \mathbb{H})\,\overset{\approx}{H}{}^n$ they are elements in the algebra $\mathbb{R}^4 = \mathbb{R} \oplus \mathbb{R} \oplus \mathbb{R} \oplus \mathbb{R}$.

Theorem 4.7. *The spaces* $(\mathbb{C} \otimes \mathbb{C})\,\overset{\approx}{S}{}^n$ *and* $(\mathbb{C} \otimes \mathbb{C})\,\overset{\approx}{H}{}^n$ *admit interpretations by pairs of spaces* $\overline{\mathbb{C}S}^n$, *respectively* $\overline{\mathbb{C}H}^n$. *If the distance* δ *between two points in* $(\mathbb{C}\otimes\mathbb{C})\,\overset{\approx}{S}{}^n$ *or* $(\mathbb{C}\otimes\mathbb{C})\,\overset{\approx}{H}{}^n$ *is a split complex number* $\delta_+e_+ + \delta_-e_-$, *then the distance between the corresponding points in* $\overline{\mathbb{C}S}^n$ *and* $\overline{\mathbb{C}H}^n$ *is* δ_+, *respectively* δ_-.

PROOF. Since the algebra $\mathbb{C} \otimes \mathbb{C}$ is isomorphic to the tensor product $\mathbb{C} \otimes \mathbb{C}'$ and to the direct sum $\mathbb{C} \oplus \mathbb{C}$, similarly to Theorem 2.18 (see **2.2.6**) we can prove that the space $(\mathbb{C} \otimes \mathbb{C})P^n$ admits an interpretation in $\mathbb{C}P^n$ such that any point x in $(\mathbb{C} \otimes \mathbb{C})P^n$ with coordinates $x^i = x^i_+e_+ + x^i_-e_-$ is interpreted as two points x_+ and x_- with complex coordinates x^i_+ and x^i_-. The points x_+ and x_- can also be regarded as points in two spaces $\mathbb{C}P^n$. If in $(\mathbb{C} \otimes \mathbb{C})P^n$ we introduce the distances δ between points in $(\mathbb{C}\otimes\mathbb{C})\,\overset{\approx}{S}{}^n$ and $(\mathbb{C}\otimes\mathbb{C})\,\overset{\approx}{H}{}^n$, these two spaces $\mathbb{C}P^n$ become the space $\overline{\mathbb{C}S}^n$, respectively $\overline{\mathbb{C}H}^n$. The last assertion is easy to verify. ■

The spaces $(\mathbb{C} \otimes \mathbb{C})\,\overset{\approx}{S}{}^n, (\mathbb{C} \otimes \mathbb{H})\,\overset{\approx}{S}{}^n$, and $(\mathbb{H} \otimes \mathbb{H})\,\overset{\approx}{S}{}^n$ were first considered by N. T. Abbasov [Ab1-2] and L. V. Rumyantseva [Ru2].

4.1.4. Complex, Split Complex, Quaternionic and Split Quaternionic Hermitian Lines. Consider the Hermitian elliptic and hyperbolic lines $\overline{\mathbb{C}S}^1$, $\overline{\mathbb{H}S}^1$, $\overline{\mathbb{C}H}^1$, $\overline{\mathbb{H}H}^1$, $\mathbb{C}'\overline{S}^1$, and $\mathbb{H}'\overline{S}^1$.

Theorem 4.8. *The Hermitian elliptic line* $\overline{\mathbb{C}S}^1$, *respectively* $\overline{\mathbb{H}S}^1$, *of curvature* $\frac{1}{r^2}$ *is isometric to a sphere of radius* $\frac{r}{2}$ *in* R^3, *respectively to a hypersphere of radius* $\frac{r}{2}$ *in* R^5.

The Hermitian hyperbolic line $\overline{\mathbb{C}H}^1$, *respectively* $\overline{\mathbb{H}H}^1$, *of curvature* $-\frac{1}{q^2}$ *is isometric to a sphere of radius* $\frac{qi}{2}$ *in* R^3_1, *respectively to a hypersphere of radius* $\frac{qi}{2}$ *in* R^5_1.

The Hermitian elliptic line $\mathbb{C}'\overline{S}^1$, *respectively* $\mathbb{H}'\overline{S}^1$, *of curvature* $\frac{1}{r^2}$ *is isometric to a sphere of radius* $\frac{r}{2}$ *in* R^3_1, *respectively to a hypersphere of radius* $\frac{r}{2}$ *in* R^5_2.

PROOF. In Theorem 2.13 we have seen that the lines $\mathbb{C}P^1$ and $\mathbb{H}P^1$ are homeomorphic to a sphere in R^3 and to a hypersphere in R^5, respectively. We have also considered the stereographic projection of the sphere in R^3 or the hyperspere in R^5 onto a plane R^2 or space R^4 on which these assertions are based in the proof of Theorem 3.33. Formula (3.78) in this proof (stereographic projection of the sphere and hypersphere of radius 1 when R^2 and R^4 are the complex plane and quaternionic 4-space) can be written as

$$X^0 = \frac{\bar{x}x - 1}{\bar{x}x + 1}, \quad X^i = \frac{2x_i}{\bar{x}x + 1}, \tag{4.9}$$

where for the complex plane $2x_1 = x + \bar{x}$, $2x_2 = i(x - \bar{x})$ and for the quaternionic 4-space $2x_1 = x + \bar{x}$, $2x_2 = i\bar{x} - xi$, $2x_3 = j\bar{x} - xj$, $2x_4 = k\bar{x} - xk$. If the angle between the position vectors of the points X and Y on the sphere or hypersphere is equal to ω, its cosine $\cos\omega$ is equal to $X^0Y^0 + \sum_i X^iY^i$ and

$$\cos^2\frac{\omega}{2} = \frac{1+\cos\omega}{2} = \frac{(\bar{x}y+1)(\bar{y}x+1)}{(\bar{x}x+1)(\bar{y}y+1)}. \tag{4.10}$$

If we write the right hand side of (4.10) using the projective coordinates x^0, x^1 of points on the lines $\mathbb{C}P^1$ or $\mathbb{H}R^1$ related with complex numbers or quaternions x by $x = x^1(x^0)^{-1}$, that is, write $(\bar{x}^0)^{-1}\bar{x}^1y^1(y^0)^{-1}$ and similarly write $\bar{y}x, \bar{x}x$ and $\bar{y}y$, then the right hand side of (4.10) coincides with the right hand side of (4.7) for $n = 1$. Therefore, $\frac{\omega}{2} = \frac{\delta}{r}$, and, since ω is a central angle of the sphere or hypersphere and δ is the length of the arc of a great circle of this sphere or hypersphere corresponding to this angle, the radius of this sphere or hypersphere is equal to $\frac{\delta}{\omega} = \frac{r}{2}$.

The second and third assertions of the theorem are proved similarly by means of stereographic projection of spheres of radius i and 1 in R_1^3 onto planes R^2 and R_1^2 of this space and of hyperspheres of radius i and 1 in R_1^5 and R_2^5 onto hyperplanes R^4 and R_2^4 of these spaces, respectively. ■

§4.2. Motions

4.2.1. Motions in Elliptic, Hyperbolic, Pseudoelliptic, and Pseudohyperbolic Spaces. The *motions* in the quadratic and Hermitian elliptic, hyperbolic, pseudoelliptic, and pseudohyperbolic spaces $\mathbb{A}S^n$, $\mathbb{A}H^n$, $\mathbb{A}S_\ell^n$, $\mathbb{A}H_\ell^n$, $\mathbb{A}\overline{S}^n$, $\mathbb{A}\overline{H}^n$, $\mathbb{A}\overline{S}_\ell^n$, $\mathbb{A}\overline{H}_\ell^n$ are determined by rotations and, for $\mathbb{A}S_\ell^n$ and $\mathbb{A}\overline{S}_\ell^n$ if $n+1 = 2\ell$, and for $\mathbb{A}H_\ell^n$ and $\mathbb{A}\overline{H}_\ell^n$ if $n+1 = 2\ell-2$, by antirotations in the spaces $\mathbb{A}R^{n+1}, \mathbb{A}R_\ell^{n+1}, \mathbb{A}\overline{R}^{n+1}$, and $\mathbb{A}\overline{R}_\ell^{n+1}$. The motions in these spaces can be defined as the collineations in $\mathbb{A}P^n$ preserving absolute hyperquadrics of the quadratic or Hermitian spaces. Therefore, the motions in these spaces can be written in the form (2.38) where the matrix U satisfies the condition of preserving the absolute hyperquadrics.

Therefore, a motion in $S^n, \mathbb{C}S^n$, or $\mathbb{C}'S^n$ can be written as $x' = Ux$, where the matrices U satisfy the condition $U^\top U = I$, and the motions of $H^n, S_\ell^n, H_\ell^n, \mathbb{C}'H^n$, $\mathbb{C}'S_\ell^n$, and $\mathbb{C}'H_\ell^n$ can be written as $x' = Ux$, where the matrices U satisfy the conditions $U^\top EU = E$ or $U^\top EU = -E$.

Analogously, the motions in $\mathbb{C}\overline{S}^n$ and $\mathbb{C}'\overline{S}^n$ can be written as $x' = Ux$ and $x' = U\bar{x}$ where the matrices U satisfy the conditions $\det U = 1$ and $\overline{U}^\top U = I$, the motions in $\mathbb{C}'\overline{H}^n, \mathbb{C}\overline{S}_\ell^n$, and $\mathbb{C}'\overline{H}_\ell^n$ can be written as $x' = Ux$ and $x' = U\bar{x}$, where the matrices U satisfy the conditions $\det U = 1$ and $\overline{U}^\top EU = E$ or $\overline{U}^\top EU = -E$, the motions in $\mathbb{H}\overline{S}^n$ and $\mathbb{H}'\overline{S}^n$ can be written as $x' = Ux$, where the matrices U

satisfy the condition $\overline{U}^{\top} U = I$, and the motions in $\mathbb{H}'\overline{H}^n, \mathbb{H}\overline{S}^n_\ell$, and $\mathbb{H}'\overline{H}^n_\ell$ can be written as $x' = Ux$, where the matrices U satisfy the condition $\overline{U}^{\top} EU = E$ or $\overline{U}^{\top} EU = -E$.

4.2.2. Groups of Motions. The motions in the quadratic and Hermitian elliptic, hyperbolic, pseudoelliptic, and pseudohyperbolic spaces form groups.

From **2.4.1** it follows that the group of motions in S^n is locally isomorphic to the group O_{n+1}, and that the groups of motions in H^n, S^n_ℓ, and H^n_ℓ are locally isomorphic to the groups $O_{n+1,1}, O_{n+1,\ell}$, and $O_{n+1,\ell+1}$, respectively.

Analogously, the group of motions in $\mathbb{C}S^n$ is locally isomorphic to the group $\mathbb{C}O_{n+1}$, the groups of motions in $\mathbb{C}\overline{S}^n$ and $\mathbb{H}\overline{S}^n$ are locally isomorphic to the groups $\mathbb{C}SU_{n+1}$ and $\mathbb{H}U_n$, respectively, and the groups of motions in $\mathbb{C}\overline{H}^n$, $\mathbb{C}\overline{S}^n_\ell$, $\mathbb{C}\overline{H}^n_\ell$ and $\mathbb{H}\overline{H}^n$, $\mathbb{H}\overline{S}^n_\ell$, $\mathbb{H}\overline{H}^n_\ell$ are locally isomorphic to the groups $\mathbb{C}SU_{n+1,1}$, $\mathbb{C}SU_{n+1,\ell}$, $\mathbb{C}SU_{n+1,\ell+1}$ and $\mathbb{H}U_{n-1}$, $\mathbb{H}U_{n,\ell}$, $\mathbb{H}U_{n,\ell+1}$, respectively.

The real interpretations of split complex spaces imply that the groups of motions in $\mathbb{C}'S^n$, $\mathbb{C}'H^n$, $\mathbb{C}'S^n_\ell$, and $\mathbb{C}'H^n_\ell$ are isomorphic to the direct products of pairs of isomorphic groups of motions in S^n, H^n, S^n_ℓ, and H^n_ℓ, respectively.

From **4.1.2** it follows also that the groups of motions in $\mathbb{C}'\overline{S}^n$ and $\mathbb{H}'\overline{S}^n$ are locally isomorphic to the groups $\mathbb{C}'SU_{n+1}$ and $\mathbb{H}'U_n$, respectively.

These local isomorphisms imply

Theorem 4.9. *The groups of motions in S^{2n} and S^{2n-1} are real compact simple Lie groups of the classes B_n, respectively D_n, except for the group of motions in S^3, which is semisimple.*

The groups of motions in H^{2n}, S^{2n}_ℓ, H^{2n}_ℓ and H^{2n-1}, S^{2n-1}_ℓ, H^{2n-1}_ℓ are real noncompact simple Lie groups of the classes B_n, respectively D_n, except for the group of motions in S^3_2, which is semisimple.

The groups of motions in $\mathbb{C}S^{2n}$ and $\mathbb{C}S^{2n-1}$ are complex simple Lie groups of the classes B_n, respectively D_n, except for the group of motions in $\mathbb{C}S^3$, which is semisimple.

The groups of motions in $\mathbb{C}\overline{S}^n$ and $\mathbb{H}\overline{S}^n$ are real compact simple Lie groups of the classes A_n, respectively C_n.

The groups of motions in $\mathbb{C}\overline{H}^n$, $\mathbb{C}\overline{S}^n_\ell$, $\mathbb{C}\overline{H}^n_\ell$ and $\mathbb{H}\overline{H}^n$, $\mathbb{H}\overline{S}^n_\ell$, $\mathbb{H}\overline{H}^n_\ell$ are real noncompact simple Lie groups of the classes A_n, respectively C_n.

The groups of motions in $\mathbb{C}'\overline{S}^n$ and $\mathbb{H}'\overline{S}^n$ are real noncompact simple Lie groups of the classes A_n, respectively C_n.

4.2.3. Isomorphisms of the Groups of Motions. In **1.4.14** we have listed the isomorphisms of the real simple Lie groups A_1 and B_1, B_2 and C_2, A_3 and D_3, and of the real semisimple groups D_2 and the groups $A_1 \otimes A_1$ or $\mathbb{C}A_1$. Since isomorphism of the fundamental groups of two spaces implies a geometric interpretation of one of these spaces in the other, we obtain

Theorem 4.10. *There are geometric interpretations of the line $\mathbb{C}\overline{S}^1$ in the plane S^2, of the lines $\mathbb{C}\overline{S}^1_1$ and $\mathbb{C}'\overline{S}^1$ in the plane S^2_1, of the line $\mathbb{H}\overline{S}^1$ in the space S^4, of the line $\mathbb{H}\overline{S}^1_1$ in the space S^4_1, of the line $\mathbb{H}'\overline{S}^1$ in the space S^4_2, of the space $\mathbb{C}\overline{S}^3$ in the space S^5, of the space $\mathbb{C}\overline{S}^3_2$ in the space S^5_2, of the space $\mathbb{C}'\overline{S}^3$ in the space S^5_3, of the line $\mathbb{H}P^1$ in the space S^5_1, of the space S^3 in the plane $\mathbb{C}'S^2$ and* *in a pair of planes S^2, of the space H^3 in the plane $\mathbb{C}S^2$, and of the space S^3_2 in the plane $\mathbb{C}'S^2_1$ and in the pair of planes S^2_1.*

PROOF. This theorem follows from Theorem 4.9 and the isomorphisms $A0 = B0$ and $AI = BI$ of the groups A_1 and $B_1, B0 = C0, BII = CII$, and $BI = CI$ of the groups B_2 and $C_2, A0 = D0, AIII = DI, AII - DI, AI = DII$ of the groups A_3 and D_3, and $D0 = A0 \otimes A0, DII = \mathbb{C}AI$, and $DI = AI \otimes AI$ (see **1.4.14**), respectively.

The geometric interpretations corresponding to the isomorphisms of the groups A_1 and B_1 and of the groups B_2 and C_2 follow from Theorem 4.8.■

§4.3. Lines, m-Planes and Hyperplanes

4.3.1. Matrix Coordinates of m-Planes. In quadratic and Hermitian elliptic spaces the vectors a_a and u^a in the equations $x = a_a t^a$ and $u^u x = 0$ of an m-plane can be supposed to be orthonormal for $\mathbb{A}S^n$ ($a_a a_b = \delta_{ab}$, $u^u u^v = \delta^{uv}$) or unitary orthonormal for $\mathbb{A}\overline{S}^n$ ($\overline{a}_a a_b = \delta_{ab}$, $\overline{u}^u u^v = \delta^{uv}$). The points A_a represented by the vectors a_a are the vertices of an *autopolar simplex* with respect to the absolute hyperquadric and the hyperplanes $u^u x = 0$ are faces of this simplex. In these cases the matrix coordinates $A = (a^i_a)$ and $U = (u^u_i)$ satisfy the conditions (3.9) and (3.10), respectively.

4.3.2. Stationary Distances between Two m-Planes. Since the spaces $\mathbb{A}S^n$ and $\mathbb{A}\overline{S}^n$ can be regarded as hyperplanes at infinity of the Euclidean spaces $\mathbb{A}R^{n+1}$, respectively $\mathbb{A}\overline{R}^{n+1}$, since m-planes in $\mathbb{A}S^n$ or $\mathbb{A}\overline{S}^n$ can be regarded as m-planes at infinity of $(m+1)$-planes in $\mathbb{A}R^{n+1}$ or $\mathbb{A}\overline{R}^{n+1}$, and since the angles between two lines in $\mathbb{A}R^{n+1}$ or $\mathbb{A}\overline{R}^{n+1}$ are equal to the distances between the two points in $\mathbb{A}S^n$ or $\mathbb{A}\overline{S}^n$ that are the points at infinity of these lines divided by the radius r of curvature of $\mathbb{A}S^n$ or $\mathbb{A}\overline{S}^n$, Theorem 3.8 implies

Theorem 4.11. *The squares of the cosines of the quotients $\frac{\omega_a}{r}$ of the stationary distances ω_a between two m-planes in $\mathbb{A}S^n$ or $\mathbb{A}\overline{S}^n$ which have no common points by the radius r of curvature of the space are the eigenvalues of an $((m+1)\times(m+1))$-matrix (3.13) and (3.14), respectively, where A and B are the matrix coordinates of these m-planes, which are determined by the orthonormal or unitary orthonormal vectors a_a and b_a.*

The stationary distances are realized on common perpendiculars of the m-planes which are transversals of these m-planes and their polar $(n-m-1)$-planes. The eigenvalues of the $((m+1)\times(m+1))$-matrices (3.13) and (3.14) are the cross ratios of the points of intersection of these common perpendiculars with the m-planes and their polar $(n-m-1)$-planes. In the generic case the number of common perpendiculars of two m-planes which have no common points is equal to $m+1$.

If all stationary distances ω_a of two m-planes are equal to a number ω, these m-planes are called *paratactic m-planes.* Such m-planes have an infinite set of common perpendiculars, each point of these m-planes is an end point of a common perpendicular. For paratactic m-planes the matrices (3.13) and (3.14) have the form $(\cos^2\omega)I$. If $\omega = \frac{\pi}{2}r$, the m-planes are *polar* and each line which meets both polar m-planes is a common perpendicular.

The cases of the spaces $\mathbb{A}H^n, \mathbb{A}S^n_\ell, \mathbb{A}H^n_\ell$ and $\mathbb{A}\overline{H}^n, \mathbb{A}\overline{S}^n_\ell, \mathbb{A}\overline{H}^n_\ell$ differ from those of $\mathbb{A}S^n$ and $\mathbb{A}\overline{S}^n$ by replacing the Euclidean inner products xy or $\bar{x}y$ by the pseudo-Euclidean inner products xEy or $\bar{x}Ey$.

In these cases the $((m+1)\times(m+1))$-matrices $A^\top A$, $A^\top B$, $B^\top B$, $B^\top A$ and $\overline{A}^\top A$, $\overline{A}^\top B$, $\overline{B}^\top B$, $\overline{B}^\top A$ must be replaced by the $((m+1)\times(m+1))$-matrices $A^\top EA$, $A^\top EB$, $B^\top EB$, $B^\top EA$ and $\overline{A}^\top EA$, $\overline{A}^\top EB$, $\overline{B}^\top EB$, $\overline{B}^\top EA$.

For detailed proof for m-planes in $\mathbb{C}\overline{S}^n$ and $\mathbb{C}\overline{S}^n S_{m+1}$ see paper [Bec] by S. Berceanu.

4.3.3. Paratactic Congruences. In the spaces S^n and $\mathbb{C}\overline{S}^n$ a very important role is played by *paratactic congruences of lines*, that is, congruences of lines all lines of which are paratactic.

Theorem 4.12. *The lines in S^{2n+1} joining its points with coordinates x^{2i-1}, x^{2i} and (2.6) and the lines in $\mathbb{C}\overline{S}^{2n+1}$ joining its points with coordinates x^{2i-1}, x^{2i} and (2.13) form paratactic congruences.*

PROOF. Make up the matrices (3.13) and (3.14) for the matrix coordinates A and B of the lines $X'X$ and $Y'Y$ in S^{2n+1} and $\mathbb{C}\overline{S}^{2n+1}$, respectively. Here, the matrices A are $(2\times(2n+2))$-matrices consisting of the (2×2)-matrices $\begin{bmatrix} x^{2i} & x^{2i+1} \\ -x^{2i+1} & x^{2i}\end{bmatrix}$ and $\begin{bmatrix} x^{2i} & x^{2i+1} \\ -\bar{x}^{2i+1} & \bar{x}^{2i}\end{bmatrix}$, respectively, and the matrices B are analogous matrices for the coordinates y^{2i} and y^{2i+1}. Therefore, if the vectors X and Y are $\{x^{2i}, x^{2i+1}\}$ and $\{y^{2i}, y^{2i+1}\}$, respectively, then

$$W = \begin{bmatrix} XY & XJY \\ YJX & YX \end{bmatrix} \begin{bmatrix} XY & YJX \\ XJY & YX \end{bmatrix} = \begin{bmatrix} w & 0 \\ 0 & w \end{bmatrix}, \tag{4.11}$$

where J is the $(2n\times 2n)$-matrix (0.62) and $w = (XY)^2 + (XJY)^2$. Since the matrices (4.11) are scalar matrices, the lines XX' and YY' are paratactic, and since through each point X in the space only the line XX' passes, these lines form a paratactic congruence. ■

The shortest distance between the lines XX' and YY' is realized for the points X and Y that satisfy the condition $XJY = 0$, in this case X and Y are end points of a common perpendicular of these lines.

Since paratactic congruences in S^{2n+1} and $\mathbb{C}\overline{S}^{2n+1}$ are preserved by the transformations (2.6) and (2.13), respectively, these congruences are particular cases of linear congruences in P^{2n+1} and of Hermitian linear congruences in $\mathbb{C}P^{2n+1}$. Since the transformations (2.6) and (2.13) are involutive motions in these spaces, paratactic congruences are symmetry figures in S^{2n+1} and $\mathbb{C}\overline{S}^{2n+1}$.

Theorem 4.13 [Ro1,5]. *Paratactic congruences in S^{2n+1} and $\mathbb{C}\overline{S}^{2n+1}$ are isometric to the spaces $\mathbb{C}\overline{S}^n$ and $\mathbb{H}\overline{S}^n$, respectively.*

PROOF. If we replace in (4.11) the real and complex (2×2)-matrices in the matrix coordinates A and B by the complex numbers $x^{2i}+ix^{2i+1}$ and the quaternions $x^{2i}+jx^{2i+1}$ represented by these matrices, we obtain the right hand side of (4.7) for $\bar{x}x = \bar{y}y = 1$. Therefore, if we denote the shortest distance between the lines

XX' and YY' in the spaces S^{2n+1} and $\mathbb{C}\overline{S}^{2n+1}$ of curvature $\frac{1}{r^2}$ by ω, the distance δ is equal to this distance ω. ■

We can analogously, define paratactic congruences in the spaces S^{2n+1}_{n+1} and $\mathbb{C}\overline{S}^{2n+1}_{n+1}$; we can, similarly to Theorems 4.10 and 4.11, prove that the lines in S^{2n+1}_{n+1} joining its points with coordinates x^{2i-1}, x^{2i} and (2.7) and the lines in $\mathbb{C}\overline{S}^{2n+1}_{n+1}$ joining its points with coordinates x^{2i-1}, x^{2i} and (2.12) form paratactic congruences in these spaces, and that these paratactic congruences are isometric to the spaces $\mathbb{C}'\overline{S}^n$ and $\mathbb{H}'\overline{S}^n$, respectively.

Theorem 4.14. *The volumes Vol S^{2n} and Vol S^{2n+1} of the real elliptic spaces S^{2n} and S^{2n+1} with radius of curvature r are equal to*

$$\mathit{Vol}\ S^{2n} = \frac{2^n\pi^n}{(2n-1)!!}r^{2n}, \qquad \mathit{Vol}\ S^{2n+1} = \frac{2^n\pi^{n+1}}{(2n)!!}r^{2n+1} = \frac{\pi^{n+1}}{n!}r^{2n+1}, \tag{4.12}$$

respectively.

The volumes Vol $\mathbb{C}\overline{S}^n$ and Vol $\mathbb{H}\overline{S}^n$ of the complex and quaternionic Hermitian elliptic spaces $\mathbb{C}\overline{S}^n$ and $\mathbb{H}\overline{S}^n$ with radius of curvature r are equal to

$$\mathit{Vol}\ \mathbb{C}\overline{S}^n = \frac{\pi^n}{n!}r^{2n}, \qquad \mathit{Vol}\ \mathbb{H}\overline{S}^n = \frac{\pi^{2n}}{(2n+1)!}r^{4n}, \tag{4.13}$$

respectively.

PROOF. Formulas (4.12) are obtained from (3.56) since the space S^n is obtained from a hypersphere Σ^n in R^{n+1} by identification of its antipodal points and, therefore, Vol S^n is equal to the volume of Σ^n divided by 2.

The first formula in (4.13) is obtained from the second formula in (4.12), since by Theorem 4.13 the space $\mathbb{C}\overline{S}^n$ is isometric to a paratactic congruence of lines in S^{2n+1} and Vol $\mathbb{C}\overline{S}^n$ is equal to Vol S^{2n+1} divided by the length πr of the line in S^n (r is the radius of curvature of both spaces).

The second formula in (4.13) is obtained from the first formula in (4.13), since by Theorem 4.13 the space $\mathbb{H}\overline{S}^n$ is isometric to a paratactic congruence of lines in $\mathbb{C}\overline{S}^{2n+1}$ and Vol $\mathbb{H}\overline{S}^n$ is equal to Vol $\mathbb{C}\overline{S}^{2n+1}$ divided by the area πr^2 of the sphere of radius $\frac{r}{2}$ which is isometric to a line in $\mathbb{C}\overline{S}^{2n+1}$ (here r is also the radius of curvature of both spaces). ■

§4.4. Interpretations of Quadratic and Hermitian Spaces

4.4.1. The Conformal Interpretations. In **4.1.1** we have defined the real spaces S^n, H^n, S^n_ℓ, and H^n_ℓ as hyperspheres in $R^{n+1}, R^{n+1}_1, R^{n+1}_\ell$, and $R^{n+1}_{\ell+1}$ with identified antipodal points, as the projective space P^n with an imaginary absolute hyperquadric or as domains in P^n bounded by real hyperquadrics.

These spaces can also be defined by means of the spaces C^n and C^n_ℓ.

Theorem 4.15. *The spaces S^n and H^n can be defined as the space C^n with points that are in inversion with respect to an imaginary, respectively real, hypersphere identified.*

The spaces S^n_ℓ and H^n_ℓ can be defined as the space C^n_ℓ with points that are in inversion with respect to a hypersphere of imaginary, respectively real, radius identified.

The distance δ between two points X and Y in S^n or S^n_ℓ is connected with the cross ratio $\overline{XY,X'Y'}$ by

$$\cos^2 \frac{2\delta}{r} = \overline{XY,X'Y'}, \tag{4.14}$$

the distance δ between two points X and Y in H^n or H^n_ℓ is connected with the cross ratio $\overline{XY,X'Y'}$ by

$$\cosh^2 \frac{2\delta}{q} = \overline{XY,X'Y'}, \tag{4.15}$$

where the points X' and Y' are obtained from X and Y by inversion with respect to corresponding hypersphere.

The groups of motions in S^n, H^n, S^n_ℓ, and H^n_ℓ are isomorphic to the subgroups of the groups of conformal transformations in C^n or C^n_ℓ preserving these hyperspheres.

PROOF. In **3.8.2** we have represented the space C^n by the hypersphere in R^{n+1} connected with the space R^n via (3.69). If we replace in this formula the coordinates x^i of a point X by the coordinates $'x^i = -x^i/\sum_i (x^i)^2$ of the point X' inverse to X with respect to the imaginary hypersphere $\sum_i (x^i)^2 = -1$, we obtain a point in R^{n+1} with coordinates $-X^i$, $-X^{n+1}$, that is, an antipodal point of the hypersphere in R^{n+1}. Formula (4.14) can be proved by a calculation similar to the calculation in the proof of (4.10).

Since motions in S^n transform points inverse with respect to the imaginary hypersphere into inverse points, they preserve this hypersphere.

The assertions of the theorem concerning the spaces H^n, S^n_ℓ, and H^n_ℓ can be proved analogously. ■

4.4.2. Interpretations of Quadratic and Hermitian Hyperbolic Spaces in Euclidean Spaces. If we regard the oval absolute hyperquadric of the projective interpretation of H^n or the space C^{n-1} of the conformal interpretation of this space as a hypersphere in R^n, we obtain *projective and conformal interpretations of H^n inside a hypersphere in R^n.*

These interpretations were proposed for $n = 2$ and 3 by Eugenio Beltrami [Bel] and Henri Poincaré [Poi1], respectively. Therefore these interpretations are called the *interpretations of Beltrami and Poincaré.*

Theorem 4.16. *The Beltrami interpretation of H^n can be obtained by projecting the hypersphere of radius qi in R^{n+1}_1 from its center onto a tangent hyperplane to this hypersphere (Fig. 4.1a).*

The Poincaré interpretation of H^n can be obtained by stereographic projection of the same hypersphere from one of its poles onto the tangent hyperplane to this hypersphere at the opposite pole (Fig. 4.1b).

PROOF. Since the tangent hyperplane at a point of a hypersphere is perpendicular to the position vector of the point of tangency, tangent hyperplanes to

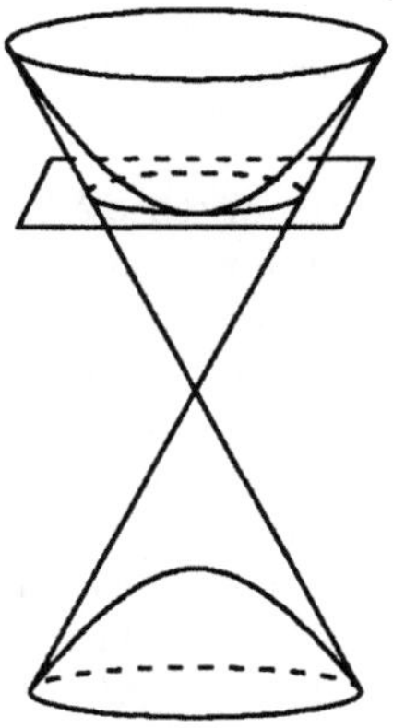

FIG. 4.1A

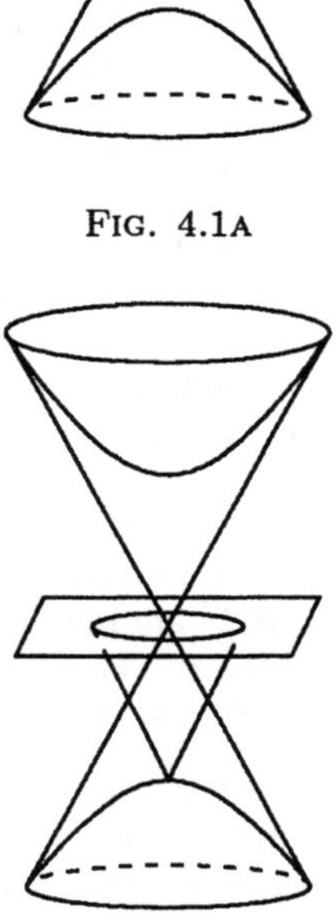

FIG. 4.1B

hyperspheres of imaginary radius in R_1^{n+1} are Euclidean spaces R^n. The projections of points on this hypersphere from its center fall inside the intersection of the hyperplane of the projection with the asymptotic hypercone of this hypersphere, that is, the isotropic hypercone in R_1^{n+1}; this intersection is a hypersphere in R^n.

Under this projection, lines in H^n are projected to planes through the center of the hypersphere. Therefore, the projections of lines in H^n are lines in R^n and the map from H^n to the interior domain of the hypersphere in R^n is projective.

Under stereographic projection of this hypersphere, the projections of points on the hypersphere also fall inside the intersection of the hyperplane of the projection with the isotropic hypercone in R_1^{n+1} with apex at the pole; this intersection is also a hypersphere in R^n. Lines in H^n, that is, great circles on the hypersphere, are projected to circles that are preserved under reflection in the center of the hypersphere. Therefore, the projections of lines in H^n are circles in R^n orthogonal to the hypersphere or its diameters. Since stereographic projection in R_1^{n+1}, as in R^{n+1}, preserves angles between curves, this map from H^n to the interior domain of the hypersphere in R^n is conformal. ∎

Note that in formula (4.5) for the distance δ between two points in H^n, in the projective interpretation of this space we can write the cross ratio $\overline{XY,\alpha\beta}$ in the same form as in (4.15), if X' and Y' are the points of intersection of the line XY with the hyperplanes α and β, while the factor 2 in the second formula is explained by the fact that the central angle leaning on an arc of a great circle on a hypersphere in R_1^{n+1}, like in R^{n+1}, is equal to twice the inscribed angle leaning on the same arc.

If a hypersphere in C^n is mapped by a conformal transformation to a hyperplane, we obtain another form of the Poincaré interpretation of H^n in R^n, the *Poincaré interpretation of H^n in a halfspace of R^n*. In this interpretation the lines in H^n are interpreted as halfcircles or halflines in R^n orthogonal to the hyperplane, and the distance δ between two points X and Y of H^n is connected with the cross ratio $\overline{XY,X'Y'}$ by the correlation (4.15), where X' and Y' are obtained from X and Y by reflection in the hyperplane. The group of motions in H^n is interpreted as the subgroup of the group of conformal transformations in C^n preserving this hyperplane.

The spaces $\mathbb{C}\overline{H}^n$ and $\mathbb{H}\overline{H}^n$ also admit interpretations analogous to the Beltrami interpretation of H^n in R^n. If the space $\mathbb{C}\overline{H}^n$ or $\mathbb{H}\overline{H}^n$ is interpreted inside an oval Hermitian hyperquadric in $\mathbb{C}P^n$, respectively $\mathbb{H}P^n$, and we take this Hermitian hyperquadric as Hermitian hypersphere in $\mathbb{C}\overline{R}^n$, respectively $\mathbb{H}\overline{R}^n$, we obtain the interpretation of $\mathbb{C}\overline{H}^n$ or $\mathbb{H}\overline{H}^n$ inside a Hermitian hypersphere in $\mathbb{C}\overline{R}^n$, respectively $\mathbb{H}\overline{R}^n$. Since the spaces $\mathbb{C}\overline{R}^n$ and $\mathbb{H}\overline{R}^n$ are isometric to the spaces R^{2n} and R^{4n}, respectively, these interpretations can also be regarded as interpretations of $\mathbb{C}\overline{H}^n$ and $\mathbb{H}\overline{H}^n$ inside a hypersphere in R^{2n}, respectively R^{4n}.

Theorem 4.17. *If a point X in the space $\mathbb{C}\overline{H}^n$ or $\mathbb{H}\overline{H}^n$ is interpreted as the center of a hypersphere in R^{2n} or R^{4n} and a point Y is interpreted as an arbitrary point in this hypersphere, the complex or quaternionic line XY can be interpreted as a Poincaré interpretation of the plane H^2, respectively the space H^4, in a sphere in R^2, respectively in a hypersphere in R^4.*

Under the same conditions a normal real n-chain in $\mathbb{C}\overline{H}^n$ or $\mathbb{H}\overline{H}^n$, that is, a set of points in these spaces with real coordinates and all manifolds obtained from it by motions in these spaces, which are isometric to the space H^n of the same curvature and pass through X and Y, can be interpreted as a Beltrami interpretation of H^n in a hypersphere in R^n.

PROOF. The first assertion of this theorem follows from Theorem 4.8, and the center of the hypersphere of radius qi can be taken as the pole of stereographic projection in the hypersphere of radius $\frac{iq}{2}$. The second assertion follows from the fact that a normal real n-chain in $\mathbb{C}\overline{H}^n$ or $\mathbb{H}\overline{H}^n$ of curvature $\frac{1}{-q^2}$ is isometric to the space H^n of the same curvature. ∎

This theorem for $\mathbb{C}\overline{H}^n$ was proved by W.M. Goldman [Gol, pp. 61–63].

Fig. 4.2a,b reproduce two figures from [Gol]. These figures give the interpretations of a line and a normal 2-chain in $\mathbb{C}\overline{H}^2$ in the above mentioned interpretation.

In both figures a normal 1-chain, that is, a line in H^2, is represented by the horizontal diameter of the circle, as well as by the Fermi coordinate lines in H^2, that is, by the perpendiculars to this line and the conics equidistant from it. The perpendiculars are represented in Fig. 4.2a by arcs of circles orthogonal to the circle and in Fig. 4.2b by lines orthogonal to the horizontal diameter. The equidistant conics are represented in Fig. 4.2a by arcs of circles through the extremities of the horizontal diameter, and in Fig. 4.2b by ellipses tangent to the circle at the extremities of the horizontal diameter.

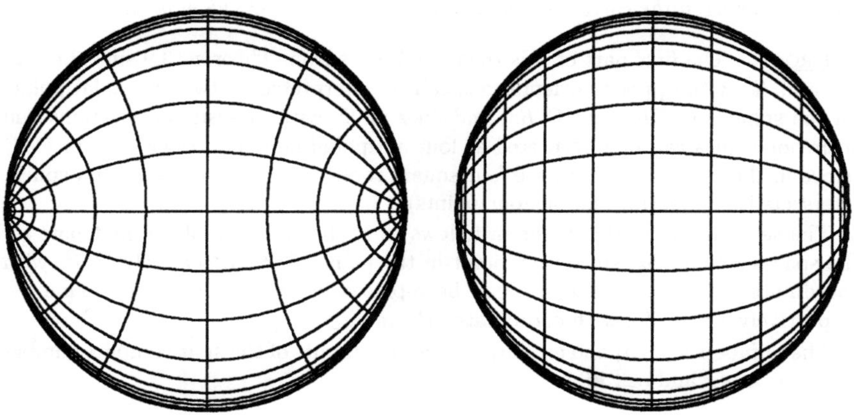

a) b)

FIG. 4.2

Theorem of W. M. Goldman [Gol, p. 77]. *If two points of absolute Hermitian hyperquadric of* $\mathbb{C}\overline{H}^n$ *are interpreted as points* X_1 *and* X_2 *with position vectors* x_1 *and* x_2 *on Hermitian hypersphere* $\overline{x}x = 1$ *in* $\mathbb{C}\overline{R}^n$, *then the geodesic* X_1X_2 *is interpreted in* $\mathbb{C}\overline{R}^n$ *as*

$$x(t) = \frac{(1 - iw)x_1 + e^t x_2}{1 + e^t + iw},$$

where

$$iw = \frac{\overline{x}_1 x_2 - \overline{x}_2 x_1}{2 - \overline{x}_1 x_2 - \overline{x}_2 x_1}.$$

4.4.3. The Interpretations of Kotelnikov-Study-Fubini. We consider some interpretations of quadratic 3-spaces.

Theorem 4.18 (KOTELNIKOV-STUDY-FUBINI [Kot1-2, Stu1, Fu1]). *The planes* $\mathbb{C}'S^2$ *admits interpretation as manifold of pairs of mutually polar lines in the space* S^3.

The plane $\mathbb{C}S^2$ *admits interpretations as manifold of lines in the space* H^3.

The plane $\mathbb{C}^0 S^2$ admits interpretation as manifold of lines in the space R^3.

If the stationary distances between two lines in S^3 are ω_1 and ω_2, the stationary distances between two lines in H^3 are ω_1 and $i\omega_2$, and the shortest distance and the angle between two lines in R^3 are δ and φ, the distances between the corresponding points in $\mathbb{C}' S^2, \mathbb{C}S^2$, and $\mathbb{C}^0 S^2$ are $\omega_1 + e\omega_2$, $\omega_1 + i\omega_2$, and $\varphi + \epsilon\delta$.

The manifold of pairs of mutually polar lines in the space S^3 or S^3_2 admits an interpretation as two planes S^2, respectively H^2.

If the stationary distance between two lines in S^3 or S^3_2 are ω_1 and ω_2, the distances between the corresponding points in S^2 and H^2 are $\omega_1 + \omega_2$ and $|\omega_1 - \omega_2|$.

PROOF. The first part of this theorem follows from Theorem 3.9 since lines in H^3 and R^3 with opposite orientations can be interpreted as two antipodal points on the spheres in $\mathbb{C}R^3$ and $\mathbb{C}^0 R^3$, and lines in S^3 with opposite orientations and their polar lines can be interpreted as four antipodal points on the sphere in $\mathbb{C}' R^3$ (since in the algebra $\mathbb{C}'$ there are four square roots of 1, ± 1 and $\pm e$, such a sphere intersects its diametral lines at four points).

The second part of this theorem follows from the assertion of the first part for the spaces S^3 and S^3_2, since each point in the plane $\mathbb{C}' S^2$, respectively $\mathbb{C}' S^2$, with coordinates $x^i = x^i_+ e_+ + x^i_- e_-$, can be replaced by two points in the plane S^2, respectively S^2_1 or H^2, with coordinates x^i_+ and x^i_-.

The numbers $\omega_1 + \omega_2$ and $\omega_1 - \omega_2$ are the coefficients of the split complex number $\omega_1 + e\omega_2$ at e_+ and e_-. ■

This theorem realizes geometric interpretations of spaces with fundamental groups D_2 established in Theorem 4.10.

4.4.4. Interpretations of Manifolds of Lines and 3-Planes. In Theorem 4.7 we have seen that the space $(\mathbb{C} \otimes \mathbb{C})\, \overset{\simeq}{S}{}^n$ admits a simple interpretation in the space $\mathbb{C}\overline{S}^n$. Let us find similar simple interpretations for the spaces $(\mathbb{C} \otimes \mathbb{H})\overset{\simeq}{S}{}^n$ and $(\mathbb{H} \otimes \mathbb{H})\, \overset{\simeq}{S}{}^n$.

Theorem 4.19. *The spaces $(\mathbb{C} \otimes \mathbb{H})\, \overset{\simeq}{S}{}^n$ and $(\mathbb{H} \otimes \mathbb{H})\, \overset{\simeq}{S}{}^n$ admit interpretations in the manifolds of lines in $\mathbb{C}\overline{S}^{2n+1}$ and of 3-planes in S^{4n+3}, respectively.*

If the distance δ between two points in $(\mathbb{C}\otimes\mathbb{H})\, \overset{\simeq}{S}{}^n$ or $(\mathbb{H}\otimes\mathbb{H})\, \overset{\simeq}{S}{}^n$ is a split complex number $\delta_+ e_+ + \delta_- e_-$, respectively 4-multiple number $\delta_1 e_1 + \delta_2 e_2 + \delta_3 e_3 + \delta_4 e_4$, then the stationary distances of the corresponding lines in $\mathbb{C}\overline{S}^{2n+1}$ are δ_+ and δ_- and the stationary distances of the corresponding 3-planes in S^{4n+3} are $\delta_1, \delta_2, \delta_3$, and δ_4.

PROOF. Since the algebras $\mathbb{C} \otimes \mathbb{H}$ and $\mathbb{H} \otimes \mathbb{H}$ are isomorphic to the matrix algebras $\mathbb{C}_2$ and $\mathbb{R}_4$, respectively, Theorem 2.19 and its complex analog imply that the spaces $(\mathbb{C} \otimes \mathbb{H})P^n$ and $(\mathbb{H} \otimes \mathbb{H})P^n$ admit interpretations in the manifolds of lines in $\mathbb{C}P^{2n+1}$ and of 3-planes in P^{4n+3}, respectively. It is easy to verify that if we introduce in $(\mathbb{C} \otimes \mathbb{H})P^n$ and $(\mathbb{H} \otimes \mathbb{H})P^n$ the metrics of the spaces $(\mathbb{C} \otimes \mathbb{H})\, \overset{\simeq}{S}{}^n$ and $(\mathbb{H} \otimes \mathbb{H})\, \overset{\simeq}{S}{}^n$, then in $\mathbb{C}P^{2n+1}$ and P^{4n+3} the metrics of $\mathbb{C}\overline{S}^{2n+1}$ and S^{4n+3} are introduced, and distances between points in $(\mathbb{C}\otimes\mathbb{H})\,\overset{\approx}{S}{}^n$ and $(\mathbb{H}\otimes\mathbb{H})\,\overset{\approx}{S}{}^n$ are related with stationary distances of lines and 3-planes as mentioned above. ∎

Theorem 4.20 (R. P. VYPLAVINA [RoVZ]). *Denote the division algebras $\mathbb{R},\mathbb{C}$, and $\mathbb{H}$ by $\mathbb{F}_1,\mathbb{F}_2$ and $\mathbb{F}_4$, respectively. The Hermitian elliptic line $(\mathbb{F}_p\otimes\mathbb{F}_q)\,\overset{\approx}{S}{}^1$ admits an interpretation in the manifold of $(p-1)$-planes and their polar $(q-1)$-planes in the real elliptic space S^{p+q-1}.*

PROOF. Note that the line $(\mathbb{F}_p\otimes\mathbb{F}_1)\,\overset{\approx}{S}{}^1$ coincides with $\mathbb{F}_p\overline{S}^1$, and for $p=1$ with S^1. Therefore, the line $(\mathbb{F}_1\otimes\mathbb{F}_1)\,\overset{\approx}{S}{}^1$ is S^1, that is, the manifold of points and their polar points in S^1.

By Theorem 4.8, the lines $\mathbb{C}\overline{S}^1$ and $\mathbb{H}\overline{S}^1$ are isometric to the sphere Σ^2, respectively to the hypersphere Σ^4, therefore, the lines $\mathbb{C}\overline{S}^1$ and $\mathbb{H}\overline{S}^1$ are interpreted as the manifold of points and polar lines of these points in S^2 by the manifold of points and polar hyperplanes of these points in S^4, respectively.

By Theorem 4.7, the line $(\mathbb{C}\otimes\mathbb{C})\,\overset{\approx}{S}{}^1$ is interpreted by pairs of points in two lines $\mathbb{C}\overline{S}^1$. Each of them, as we have seen, can be interpreted as the manifold of points and polar lines of these points in S^2, but by the theorem of Fubini-Study the manifold of lines in S^3 can be interpreted as a pair of planes S^2.

By Theorem 4.15, the lines $(\mathbb{C}\otimes\mathbb{H})\,\overset{\approx}{S}{}^1$ and $(\mathbb{H}\otimes\mathbb{H})\,\overset{\approx}{S}{}^1$ can be interpreted as manifolds of lines in S^5 and of 3-planes of S^7 and, therefore, by manifolds of polar 3-planes of these lines and 3-planes. ∎

4.4.5. Real and Quaternionic Projective Geometries as Hermitian Elliptic Geometries. Since the groups of collineations in the real and quaternionic projective spaces P^n and $\mathbb{H}P^{(n-1)/2}$ are locally isomorphic to the groups SL_{n+1} and $\mathbb{H}SL_{(n+1/2)}$, which are simple Lie groups of the class A_n, it is natural that these groups can be represented in a form similar to the groups of motions in the spaces $\mathbb{C}\overline{S}^n$ or $(\mathbb{C}\otimes\mathbb{H})\,\overset{\approx}{S}{}^{(n-1)/2}$.

Theorem 4.21 [Ro1,5]. *The manifold of 0-pairs in P^n, respectively in $\mathbb{H}P^n$, if to define the distance ω between two 0-pairs by means of their cross ratio w of the form (2.59) by the relation $\cos^2\omega = w$, is isometric to the space $\mathbb{C}'\overline{S}^n$, respectively to the space $(\mathbb{C}'\otimes\mathbb{H})\,\overset{\approx}{S}{}^n$, both of curvature 1.*

The group of projective transformations in P^n, respectively in $\mathbb{H}P^n$ is isomorphic to the group of motions in $\mathbb{C}'\overline{S}^n$, respectively in $(\mathbb{C}\otimes\mathbb{H})\,\overset{\approx}{S}{}^n$.

PROOF. Let a 0-pair in P^n, consisting of a point X with projective coordinates X^i and a hyperplane α with tangential coordinates U_i, correspond to point x with split complex coordinates $x^i = X^i e_+ + U_i e_-$. If X^i and U_i are multiplied by real numbers K and L, respectively, the coordinates x^i are multiplied by the split complex number $k = Ke_+ + Le_-$. The cross ratio (2.59) of the two 0-pairs $\{X,\alpha\}$ and $\{Y,\beta\}$ is equal to

$$w = \frac{U_iY^i \cdot V_iX^i}{U_iX^i \cdot V_iY^i} = \frac{\sum\limits_i \bar{x}^i y^i \cdot \sum\limits_i \bar{y}^i x^i}{\sum\limits_i \bar{x}^i x^i \cdot \sum\limits_i \bar{y}^i y^i}, \tag{4.16}$$

since

$$\sum_i \bar{x}^i y^i = \sum_i (U_i e_+ + X^i e_-) \cdot (Y^i e_+ + V_i e_+) = U_iY^i e_+ + V_iX^i e_-,$$
$$\sum_i \bar{y}^i x^i = \sum_i (V_i e_+ + Y^i e_-) \cdot (X^i e_+ + U_i e_+) = V_iX^i e_+ + U_iY^i e_-,$$
$$\sum_i \bar{x}^i y^i \cdot \sum_i \bar{y}^i x^i = U_iY^i \cdot V_iX^i (e_+ + e_-) = U_iY^i \cdot V_iX^i,$$
$$\sum_i \bar{x}^i x^i = \sum_i (U_i e_+ + X^i e_-) \cdot (X^i e_+ + U_i e_+) = U_iX^i (e_+ + e_-) = U_iX^i,$$
$$\sum_i \bar{y}^i y^i = \sum_i (V_i e_+ + Y^i e_-) \cdot (Y^i e_+ + V_i e_+) = V_iY^i (e_+ + e_-) = V_iY^i.$$

Since under a collineation in P^n of the form $X' = \mathcal{U}X$ covectors representing hyperplanes in P^n are transformed by the law $'U = U\mathcal{U}^{-1}$, the collineations in P^n are represented in $\mathbb{C}'\overline{S}^n$ by the transformations $x' = \mathcal{A}x$, where $\mathcal{A} = \mathcal{U}e_+ + (\mathcal{U}^{-1})^\top e_-$.

The split complex matrices $\mathcal{A}$ are unimodular, since $\det \mathcal{A} = \det \mathcal{U}e_+ + \det (\mathcal{U}^{-1})^\top e_- = e_+ + e_- = 1$, and unitary, since $\overline{\mathcal{A}}^\top \mathcal{A} = (\mathcal{U}^{-1}e_+ + \mathcal{U}^\top e_-)(\mathcal{U}e_+ + (\mathcal{U}^{-1})^\top e_-) = I(e_+ + e_-) = I$. Therefore, the group of collineations in P^n is isomorphic to the group of motions in $\mathbb{C}'\overline{S}^n$.

The second assertion of the theorem can be proved analogously. ■

4.4.6. Interpretation of Pseudoelliptic, Hyperbolic, and Pseudohyperbolic Spaces by Means of Hyperspheres. In **3.8.2** and **3.8.5** we have considered interpretations of the conformal space C^n and the pseudoconformal space C^n_ℓ in the projective space P^{n+1}. Since the right hand sides of formulas (3.75) and (3.81) can be regarded as formulas for determining distances between points in S^{n+1}_1 and $S^{n+1}_{\ell+1}$, respectively, we obtain

Theorem 4.22. *The spaces S^n_1 and S^n_ℓ admit interpretations as manifolds of hyperspheres in the conformal space C^{n-1} and in the pseudoconformal space $C^{n-1}_{\ell-1}$, where the distance ω between two points is equal to the angle between the corresponding hyperspheres.*

The groups of motions in S^n_1 and S^n_ℓ are isomorphic to the groups of conformal transformations in C^{n-1} and $C^{n-1}_{\ell-1}$.

Analogously, we can prove

Theorem 4.23. *The spaces H^n and H^n_ℓ admit interpretations as manifolds of hyperspheres of imaginary radius in the conformal space C^{n-1} and in the pseudoconformal space C^{n-1}_ℓ, where the distance ω between two points is equal to the angle between the corresponding hyperspheres.*

The groups of motions in H^n and H^n_ℓ are isomorphic to the groups of conformal transformations in C^{n-1} and C^{n-1}_ℓ.

4.4.7. The Interpretations of Hesse and Plücker. In Theorem 4.10 we have mentioned an interpretation of the line $\mathbb{C}'\overline{S}^1$ in the plane H^2. Therefore, from

Theorem 4.21 we obtain an interpretation of the line P^1 in the plane S_1^2, or vice versa. It is the classical *Hesse interpretation* [He2], which can be formulated as follows:

Theorem 4.24. *The plane H^2, respectively S_1^2, admits an interpretation as the manifold of conjugate imaginary, respectively real, pairs of points in the line P^1.*

The distance between two points in H^2 or S_1^2 is related with the cross ratio w of the corresponding four points in P^1 by (4.15), respectively (4.14).

The group of motions in H^2 or S_1^2 is isomorphic to the group of collineations in P^1.

PROOF. Since the projective line P^1 can also be regarded as the conformal line C^1, this theorem is a particular case of Theorems 3.28 and 4.15. The map of the absolute conic of H^2 and S_1^2 to P^1 can be realized by stereographic projection. ■

Note that the interpretations of the planes H^2 and S_1^2 on P^2 considered in **4.1.1** and on P^1 considered in this theorem are particular cases of the interpretation of H^2 and S_1^2 on a *norm curve* $x^0 = 1$, $x^1 = t$, $x^2 = t^2, \ldots, x^n = t^n$ in P^n. In this interpretation lines which intersect a norm curve in pairs of real points ("real bisecants") interprete points in S_1^2, lines intersecting a norm curve in pairs of conjugate imaginary points ("imaginary bisecants") interprete points in H^2, the distance between two points in H^2 or S_1^2 is related with the cross ratio $w = \overline{t_1t_2, t_3t_4}$ of the corresponding four points of the norm curve by (4.15) and (4.14), and the group of motions in H^2 or S_1^2 is isomorphic to the subgroup of the group of collineations in P^n preserving the norm curve.

In Theorem 4.10 we have mentioned an interpretation of the space $\mathbb{C}'\overline{S}^3$ in the space S_3^5. Therefore, from Theorem 4.21 we obtain an interpretation of the space P^3 in the space S_3^5, or vice versa. It is the classical *Plücker interpretation* [Pl], which can be formulated as follows:

Theorem 4.25. *The absolute hyperquadric of the space S_3^5 admits an interpretation as the manifold of lines in P^3.*

The group of motions in S_3^5 is isomorphic to the group of projective transformations in P^3.

This theorem can also be proved by means of the equation (2.28) (see **2.2.6**), which connects the Plücker coordinates p^{ij} of lines in P^3, since this equation can be regarded as the equation of the absolute hyperquadric of S_3^5.

The interpretations of Hesse and Plücker are the geometric interpretations of the spaces with fundamental groups A_1 and B_1 and with fundamental groups A_3 and D_3 which were established in Theorem 4.10.

4.4.8. Interpretations of Hermitian 3-Spaces in Real 5-Spaces. Complexification of the spaces mentioned in Theorem 4.25 implies

Theorem 4.26. *The absolute hyperquadric of the space $\mathbb{C}S^5$ admits an interpretation in the manifold of lines in $\mathbb{C}P^3$.*

The group of motions in $\mathbb{C}S^5$ is isomorphic to the group of projective transformations in $\mathbb{C}P^3$.

This theorem implies theorems on two other real spaces with fundamental groups A_3 and D_3:

Theorem 4.27 (Petsko [Pet]). *The manifold of oriented lines in the space S^5, respectively S^5_2, admits an interpretation as the manifold of lines in $\mathbb{C}\overline{S}^3$, respectively $\mathbb{C}\overline{S}^3_2$.*

The group of motions in S^5, respectively S^5_2 is isomorphic to the group of motions in $\mathbb{C}\overline{S}^3$, respectively $\mathbb{C}\overline{S}^3_2$.

Proof. The absolute hyperquadric of S^5 is imaginary and coincides with the absolute hyperquadric of $\mathbb{C}S^5$. A point on the absolute hyperquadric of S^5 together with its conjugate imaginary point determines a real line in S^5 and any real line in S^5 joins two conjugate imaginary points of its absolute hyperquadric. Therefore, any real line in S^5 corresponds to two lines in $\mathbb{C}P^3$ and $\mathbb{C}\overline{S}^3$, and any line in $\mathbb{C}\overline{S}^3$ corresponds to an oriented line in S^5. The isomorphism of the groups of motions in S^5 and $\mathbb{C}\overline{S}^3$ follows from the fact that these groups are maximal compact subgroups of the groups of motions in $\mathbb{C}S^5$ and of the group of projective transformations in $\mathbb{C}P^3$.

The groups of motions in S^5_2 and $\mathbb{C}\overline{S}^3_2$ are obtained from the isomorphic compact groups of motions in S^5 and $\mathbb{C}\overline{S}^3$ by the *Cartan algorithm* (see **1.4.12**) determined by the involutive automorphisms in these groups generated by symmetries in the lines in these spaces, since in the mutual interpretation of these spaces lines in one space correspond to lines in the other space. Therefore, the groups of motions in S^5_2 and $\mathbb{C}\overline{S}^3_2$ are isomorphic. In the spaces S^5_2 and $\mathbb{C}\overline{S}^3_2$ there are elliptic lines which meet their absolute hyperquadrics at imaginary points and hyperbolic lines which meet their absolute hyperquadrics at real points. The stabilizers of elliptic (hyperbolic) lines in these spaces are isomorphic compact (noncompact) subgroups of the groups of motions in these spaces. Therefore, a line in $\mathbb{C}\overline{S}^3_2$ is interpreted as an oriented line in S^5_2. ∎

4.4.9. Connections between Isomorphisms of Simple Lie Groups and Spinor Representations. In **1.6.3** we have defined *spinor representations* of the groups SO_n and $SO_{n,\ell}$. Let us show that the isomorphisms between simple Lie groups and the corresponding geometric interpretations are determined by the spinor representations of the corresponding orthogonal and pseudoorthogonal groups.

Theorem 4.28. *The isomorphisms $A_1 = B_1 = C_1$ are determined by the spinor representation of the group B_1.*

The isomorphisms $D_2 = B_1 \otimes B_1$ and $D_2 = \mathbb{C}B_1$ are determined by the spinor representation of the group D_2.

The isomorphism $B_2 = C_2$ is determined by the spinor representation of the group B_2.

The isomorphism $A_3 = D_3$ is determined by the spinor representation of the group D_3.

Proof. The spinor representations of the groups SO_3 and $SO_{3,1}$ of the class B_1, representing motions in S^2 and H^2, are realized in the algebras $\mathbb{A}_3 = \mathbb{H}$ and $\mathbb{A}_{3,1} = \mathbb{H}' = \mathbb{R}_2$, respectively. The spinor group $\overline{SO}_3$ is isomorphic to the quaternionic group $|\alpha| = 1$ and to the group $\mathbb{C}SU_2$. The first of these groups is the compact group C_1, the second group is the compact group A_1 and represents motions in $\mathbb{C}\overline{S}^1$. The spinor group $\overline{SO}_{3,1}$ is isomorphic to the split quaternionic group $|\alpha| = 1$ and to the groups $SL_2, \mathbb{C}SU_{2,1}$, and $\mathbb{C}'SU_2$. The first of these groups is the noncompact group C_1, the second and third groups are the noncompact group A_1 and represent collineations in P^1 and motions in $\mathbb{C}\overline{S}^1_1$ and $\mathbb{C}'\overline{S}^1$, respectively.

The spinor representations of the groups $SO_4, SO_{4,1}$, and $SO_{4,2}$ of the class D_2, representing motions in S^3, H^3, and S^3_2, are realized in the algebras $\mathbb{A}_4 = \mathbb{H} \oplus \mathbb{H}, \mathbb{A}_{4,1} = \mathbb{C}_2$, and $\mathbb{A}_{4,2} = \mathbb{H}' \oplus \mathbb{H}' = \mathbb{R}_2 \oplus \mathbb{R}_2$, respectively. The spinor group $\overline{SO}_4$ is isomorphic to the group $SO_3 \otimes SO_3$, the spinor group $\overline{SO}_{4,1}$ is isomorphic to the group $\mathbb{C}SL_2$, and the spinor group $\overline{SO}_{4,2}$ is isomorphic to the group $\mathbb{C}'SL_2 = SL_2 \otimes SL_2$. These groups represent motions in $\mathbb{C}'S^2, \mathbb{C}S^2$, and $\mathbb{C}'H^2$, respectively.

The spinor representations of the groups $SO_5, SO_{5,1}$, and $SO_{5,2}$ of the class B_2, representing motions in S^4, H^4, and S^4_2, are realized in the algebras $\mathbb{A}_5 = \mathbb{A}_{5,1} = \mathbb{H}_2$, $\mathbb{A}_{5,2} = \mathbb{H}'_2 = \mathbb{R}_4$, respectively. The spinor groups $\overline{SO}_5$ and $\overline{SO}_{5,1}$ are isomorphic to the groups $\mathbb{H}U_2$ and $\mathbb{H}U_{2,1}$. The spinor group $\overline{SO}_{5,2}$ is isomorphic to the group $\mathbb{H}'U_2 = Sp_4$. These groups represent motions in $\mathbb{H}\overline{S}^1, \mathbb{H}\overline{S}^1_1$, and $\mathbb{H}'S^1$, respectively.

The spinor representations of the groups $SO_6, SO_{6,1}, SO_{6,2}, SO_{6,3}$ of the class D_3, representing motions in S^5, H^5, S^5_2, and S^5_3, are realized in the algebras $\mathbb{A}_6 = \mathbb{C}_4$, $\mathbb{A}_{6,1} = \mathbb{H}_2 \oplus \mathbb{H}_2$, $\mathbb{A}_{6,2} = \mathbb{C}_4$, $\mathbb{A}_{6,3} = \mathbb{R}_4 \oplus \mathbb{R}_4$, respectively. The spinor groups $\overline{SO}_6, \overline{SO}_{6,1}, \overline{SO}_{6,2}$, and $\overline{SO}_{6,3}$ are isomorphic to the groups $\mathbb{C}SU_4, \mathbb{H}SL_2, \mathbb{C}SU_{4,2}$, and $\mathbb{C}'SU_4 = SL_4$, respectively. These groups represent motions in $\mathbb{H}\overline{S}^1$, $(\mathbb{C}' \otimes \mathbb{H})\ \overline{\widetilde{S}}^1$, $\mathbb{H}\overline{S}^1_1$, and $\mathbb{C}'\overline{S}^3$, respectively; the second and fourth of these groups also represent collineations in $\mathbb{H}P^1$ and P^3, respectively.

§4.5. Trigonometry

4.5.1 Trigonometry in the Spaces $\mathbf{S}^n$ and $\mathbf{H}^n$. Since the spaces S^n and H^n are hyperspheres in the spaces R^{n+1} and R^{n+1}_1 with identified antipodal points, the formulas of trigonometry in S^n coincide with the formulas of spherical trigonometry on hyperspheres in R^{n+1}, that is, with the cosine law (3.39), the sine law (3.40), and the dual cosine law (3.41), and the formulas of trigonometry in H^n coincide with the formulas of spherical trigonometry on hyperspheres of imaginary radius in R^{n+1}_1, that is, with the cosine law (3.45), the sine law (3.46), and the dual cosine law (3.47) (see **3.6.3**).

4.5.2. Holomorphy Angles and Holomorphic and Antiholomorphic Real 2-Directions in the Spaces $\mathbb{C}\overline{S}^n$, $\mathbb{C}\overline{H}^n$, $\mathbb{H}\overline{S}^n$ and $\mathbb{H}\overline{H}^n$. In **3.3.6** we have defined *holomorphy angles* and *holomorphic and antiholomorphic 2-directions* in the spaces $\mathbb{C}\overline{R}^n$ and $\mathbb{H}\overline{R}^n$. Since the tangent spaces of $\mathbb{C}\overline{S}^n$ and $\mathbb{C}\overline{H}^n$ coincide with $\mathbb{C}\overline{R}^n$ and the tangent spaces of $\mathbb{H}\overline{S}^n$ and $\mathbb{H}\overline{H}^n$ coincide with $\mathbb{H}\overline{R}^n$, holomorphy angles and holomorphic and antiholomorphic 2-directions in the spaces $\mathbb{C}\overline{S}^n, \mathbb{C}\overline{H}^n, \mathbb{H}\overline{S}^n$, and $\mathbb{H}\overline{H}^n$ can be defined as holomorphy angles and holomorphic and antiholomorphic 2-directions in their tangent spaces $\mathbb{C}\overline{R}^n$ and $\mathbb{H}\overline{R}^n$.

In **3.3.6** we also have defined *holomorphy vectors* in $\mathbb{H}\overline{R}^n$. Since the tangent spaces of $\mathbb{H}\overline{S}^n$ and $\mathbb{H}\overline{H}^n$ coincide with $\mathbb{H}\overline{R}^n$, holomorphy vectors in the spaces $\mathbb{H}\overline{S}^n$ and $\mathbb{H}\overline{H}^n$ can be defined as holomorphy vectors in their tangent spaces $\mathbb{H}\overline{R}^n$.

4.5.3. Trigonometry in the Spaces $\mathbb{C}\overline{S}^n$ and $\mathbb{H}\overline{S}^n$. Consider a geodesic triangle ABC in $\mathbb{C}\overline{S}^n$ or $\mathbb{H}\overline{S}^n$. Denote the lengths of the sides BC, CA, AB of this triangle by a, b, c, respectively, the angles between the lines AB and AC, BC and BA, CA and CB by A, B, C, respectively, the angles between the geodesic segments AB and AC, BC and BA, CA and CB by A_0, B_0, C_0, respectively, and the holomorphy angles in the 2-directions tangent to this triangle at its vertices A, B, C, by α, β, γ, respectively. Formula (3.26) in Theorem 3.13, concerning the geometry in $\mathbb{C}\overline{R}^n$ and $\mathbb{H}\overline{R}^n$, implies

$$\sin A = \sin A_0 \ \sin\alpha, \ \sin B = \sin B_0 \ \sin\beta, \ \sin C = \sin C_0 \ \sin\gamma. \tag{4.17}$$

Theorem 4.29 (P. Shirokov [ShP2], B. Rosenfeld [Ro8], Hsiang Wu-Yi [Hsi]). . *The angles A_0, B_0, C_0, the holomorphy angles α, β, γ at the vertices A, B, C of a geodesic triangle ABC in the spaces $\mathbb{C}\overline{S}^n$ and $\mathbb{H}\overline{S}^n$ of curvature $\frac{1}{r^2}$, and its sides a, b, c are connected by the following laws of trigonometry in these spaces:*

the first cosine law

$$\cos^2\frac{a}{r} = \left(\cos\frac{b}{r}\cos\frac{c}{r} + \sin\frac{b}{r}\sin\frac{c}{r}\cos A_0\right)^2 + \sin^2\frac{b}{r}\sin^2\frac{c}{r}\sin^2 A_0\cos^2\alpha, \tag{4.18}$$

and the second cosine law

$$\cos\frac{2a}{r} = \cos\frac{2b}{r}\cos\frac{2c}{r} + \sin\frac{2b}{r}\sin\frac{2c}{r}\cos A_0 - 2\sin\frac{2b}{r}\sin^2\frac{2c}{r}\sin^2 A_0\sin^2\alpha, \tag{4.19}$$

the first sine law

$$\frac{\sin A_0\sin\alpha}{\sin\frac{a}{r}} = \frac{\sin B_0\sin\beta}{\sin\frac{b}{r}} = \frac{\sin C_0\sin\gamma}{\sin\frac{c}{r}}, \tag{4.20}$$

and the second sine law

$$\frac{\sin A_0\cos\alpha}{\sin\frac{2a}{r}} = \frac{\sin B_0\cos\beta}{\sin\frac{2b}{r}} = \frac{\sin C_0\cos\gamma}{\sin\frac{2c}{r}}. \tag{4.21}$$

Proof. Let us represent the points in $\mathbb{C}\overline{S}^n$ and $\mathbb{H}\overline{S}^n$ by the lines of a bundle in $\mathbb{C}\overline{R}^{n+1}$ and $\mathbb{H}\overline{R}^{n+1}$, respectively, and by the direction vectors x of these lines satisfying the condition $\overline{x}x = 1$. Let the vertex A be represented by the basis vector e_0, and let the vertex B be represented by the vector $e_0\cos\frac{c}{r} + e_1\sin\frac{c}{r}$ $(\overline{e}_0e_1 = 0)$. Then the unit tangent vector to the side AB at A is e_1. Let the unit tangent vector to the side AC at A be f $(\overline{e}_0 f = 0)$. Since the angle between e_1 and f at A is A_0 and the holomorphy angle in the tangent 2-direction at A is α, the Hermitian inner product $\overline{e}_1 f$ is equal to $\cos A_0 + u\sin A_0\cos\alpha$, where $u = i$ for $\mathbb{C}\overline{S}^n$ and u is the holomorphy vector in the tangent 2-direction at A for $\mathbb{H}\overline{S}^n$. The vertex C

is represented by the vector $e_0 \cos\frac{b}{r} + f \sin\frac{b}{r}$ and the distance a between B and C can be determined by

$$\cos\frac{a}{r} = \left|\left(\bar{e}_0\cos\frac{c}{r} + \bar{e}_1\sin\frac{c}{r}\right)\left(e_0\cos\frac{b}{r} + f\sin\frac{b}{r}\right)\right| = \left|\cos\frac{b}{r}\cos\frac{c}{r} + \sin\frac{b}{r}\sin\frac{c}{r}\cos A_0 + u\sin\frac{b}{r}\sin\frac{c}{r}A_0\cos\alpha\right|, \tag{4.22}$$

hence we obtain (4.18). Formulas (4.19), (4.20), and (4.21) can be obtained from (4.18) by analytical transformations. ■

The formulas (4.18) and (4.21) for $\mathbb{C}\overline{S}^n$ were proved by P. A. Shirokov [ShP2], all four formulas (4.18), (4.19), (4.20), and (4.21) for both spaces $\mathbb{C}\overline{S}^n$ and $\mathbb{H}\overline{S}^n$ were proved by the author [Ro8] and independently in other notations by Hsiang Wu-Yi [Hsi].

Note that by (4.17), the formulas (4.20) can be written as (3.38) where A, B, C are the angles between the complex or quaternionic lines containing the sides of the triangle ABC. This form of the sine law for $\mathbb{C}\overline{S}^n$ was proved by J. L. Coolidge [Coo2].

For $\alpha = \beta = \gamma = \frac{\pi}{2}$ the triangle ABC is located in a real normal 2-chain isometric to the space S^n of curvature $\frac{1}{r^2}$, and formulas (4.18) and (4.20) coincide with the formulas (3.39) and (3.40) for S^n of the same curvature. For $\alpha = \beta = \gamma = 0$ the triangle ABC is located in a complex or quaternionic line isometric to a sphere or hypersphere of radius $\frac{r}{2}$, and formulas (4.19) and (4.21) coincide with formulas (3.37) and (3.38) for a sphere of this radius.

Formula (4.18) and the comparison of (4.20) with (4.21) show that the holomorphy angles β and γ can be expressed in terms of a, b, c (or A_0, B_0, C_0), and α. The fourth independent invariant which together with the sides a, b, c and angles A_0, B_0, C_0 determines all elements of a geodesic triangle in $\mathbb{C}\overline{S}^n$ and $\mathbb{H}\overline{S}^n$ can also be written in symmetric form. Such invariants were found by W. Blaschke and H. Terheggen [BlT] and U. Brehm [Bre]. If we denote the vectors representing the vertices A, B, C by x, y, z, respectively, the invariant ω of Blaschke and Terheggen can be expressed by

$$(\overline{x}y)(\overline{y}z)(\overline{z}x) = \cos\frac{a}{r}\cos\frac{b}{r}\cos\frac{c}{r}(\cos\omega + u\sin\omega). \tag{4.23}$$

If we substitute in (4.23) $x = e_0$, $y = e_0\cos\frac{c}{r} + e_1\sin\frac{c}{r}$, $z = e_0\cos\frac{b}{r} + f\sin\frac{b}{r}$ and use (4.22), we obtain

$$\begin{aligned}\cos\omega &= \frac{\cos\frac{b}{r}\cos\frac{c}{r} + \sin\frac{b}{r}\sin\frac{c}{r}\cos A_0}{\cos\frac{a}{r}} \\ &= \frac{\cos\frac{c}{r}\cos\frac{a}{r} + \sin\frac{c}{r}\sin\frac{a}{r}\cos B_0}{\cos\frac{b}{r}} \\ &= \frac{\cos\frac{a}{r}\cos\frac{b}{r} + \sin\frac{b}{r}\cos C_0}{\cos\frac{c}{r}}.\end{aligned} \tag{4.24}$$

If we denote the common value of the quotients (4.21) by S, then

$$\sin\omega = 2\sin\frac{a}{r}\sin\frac{b}{r}\sin\frac{c}{r}\cdot S. \tag{4.25}$$

Formulas (4.23), (4.24), and (4.25) show that $\omega = 0$ for $\alpha = \beta = \gamma = \frac{\pi}{2}$.

In $\mathbb{C}\overline{S}^n$ and $\mathbb{H}\overline{S}^n$ there is also a *dual cosine law*, for $\mathbb{C}\overline{S}^n$ it was found by Blaschke and Terheggen [BlT], for $\mathbb{H}\overline{S}^n$ it has the same form.

Since for $\mathbb{H}\overline{S}^n$ the quaternion (4.23) has the form $a+bu$ where u is the holomorphy vector at any vertex of a geodesic triangle, we obtain

Theorem 4.30 (BREHM [Bre]). *The holomorphy vectors at all vertices of a geodesic triangle in $\mathbb{H}\overline{S}^n$ are equal.*

This theorem can also be formulated as follows: geodesic triangles in $\mathbb{H}\overline{S}^n$ are located in normal complex 2-chains of this space (these 2-chains are isometric to $\mathbb{C}\overline{S}^n$).

4.5.4. Trigonometry in the Spaces $\mathbb{C}\overline{H}^n$ and $\mathbb{H}\overline{H}^n$. Consider a geodesic triangle ABC in $\mathbb{C}\overline{H}^n$ and $\mathbb{H}\overline{H}^n$.

Theorem 4.31 (HSIANG WU-YI [Hsi]). *The angles A_0, B_0, C_0, the holomorphy angles α, β, γ at the vertices A, B, C of a geodesic triangle ABC in the spaces $\mathbb{C}\overline{H}^n$ or $\mathbb{H}\overline{H}^n$ of curvature $-\frac{1}{q^2}$, and its sides a, b, c are related by following laws of trigonometry of these spaces:*

the first cosine law

$$\begin{aligned}\cosh^2\frac{a}{q} = \left(\cosh\frac{b}{q}\cosh\frac{c}{q} - \sinh\frac{b}{q}\sin\frac{c}{q}\cos A_0\right)^2 + \\ + \sinh^2\frac{b}{q}\sinh^2\frac{c}{q}\sin^2 A_0\cos^2\alpha,\end{aligned} \tag{4.26}$$

the second cosine law

$$\begin{aligned}\cosh\frac{2a}{q} = \cosh\frac{2b}{qr}\cosh\frac{2c}{q} - \sinh\frac{2b}{q}\sinh\frac{2c}{q}\cos A_0 - \\ - 2\sinh^2\frac{b}{q}\sinh^2\frac{c}{q}\sin^2 A_0\sin^2\alpha,\end{aligned} \tag{4.27}$$

the first sine law

$$\frac{\sin A_0\sin\alpha}{\sinh\frac{a}{q}} = \frac{\sin B_0\sin\beta}{\sinh\frac{b}{q}} = \frac{\sin C_0\sin\gamma}{\sinh\frac{c}{q}}, \tag{4.28}$$

and the second sine law

$$\frac{\sin A_0\cos\alpha}{\sin\frac{2a}{q}} = \frac{\sin B_0\cos\beta}{\sinh\frac{2b}{q}} = \frac{\sin C_0\cos\gamma}{\sinh\frac{2c}{q}}. \tag{4.29}$$

The holomorphy vectors at all vertices of a geodesic triangle in $\mathbb{H}\overline{H}^n$ are equal.

The proof is analogous to the proofs of Theorems 4.29 and 4.30.

4.5.5. Trigonometry in the Grassmann Manifolds. The trigonometric formulas in the spaces $S^n, H^n, \mathbb{C}\overline{H}^n, \mathbb{H}\overline{S}^n$, and $\mathbb{H}\overline{H}^n$ are particular cases of trigonometric formulas in the Grassmann manifolds of these spaces. Let us formulate the

Theorem of Th. Hangan [Han]. *Let A, B, C be three m-planes in the space S^N $(N \geq 3m+2)$ of curvature 1, let $a_0, a_1, \ldots, a_m$ be the stationary distances between the m-planes B and C, let $b_0, b_1, \ldots, b_m$ be the stationary distances between the m-planes C and A, let $c_0, c_1, \ldots, c_m$ be the stationary distances between the m-planes A and B, let $A_0, A_1, \ldots, A_m$ be stationary distances between the $(2m+1)$-planes which are the sums of the m-planes A, B and A, C, let $B_0, B_1, \ldots, B_m$ be stationary distances between the $(2m+1)$-planes which are the sums of the m-planes B, C and B, A, and let $C_0, C_1, \ldots, C_m$ be stationary distances between the $(2m+1)$-planes which are the sums of the m-planes C, A and C, B. Then these angles are related by the sine law*

$$\frac{\sin A_0 \sin A_1 \ldots \sin A_m}{\sin a_0 \sin a_1 \ldots \sin a_m} = \frac{\sin B_0 \sin B_1 \ldots \sin B_m}{\sin b_0 \sin b_1 \ldots \sin b_m} = \frac{\sin C_0 \sin C_1 \ldots \sin C_m}{\sin c_0 \sin c_1 \ldots \sin c_m}. \tag{4.30}$$

Similar theorems are valid in the spaces $H^n, \mathbb{C}\overline{S}^n, \mathbb{C}\overline{H}^n, \mathbb{H}\overline{S}^n$, and $\mathbb{H}\overline{H}^n$ (see [Ro11]).

Trigonometry in general symmetric Riemannian spaces was investigated by E. Leuzinger [Leu].

§4.6. Sectional Curvature of Hermitian Spaces

4.6.1. Sectional Curvature of Symmetric Spaces. Since the spaces $\mathbb{C}\overline{S}^n$, $\mathbb{C}\overline{H}^n$ and $\mathbb{H}\overline{S}^n$, $\mathbb{H}\overline{H}^n$ are isometric to the Riemannian manifolds V^{2n} and V^{4n}, and the spaces $\mathbb{C}\overline{S}^n_\ell, \mathbb{C}'\overline{S}^n, \mathbb{H}\overline{S}^n_\ell$, and $\mathbb{H}'\overline{S}^n$ are isometric to the pseudo-Riemannian manifolds $V^{2n}_{2\ell}, V^{2n}_n$, $V^{4n}_{4\ell}$, and V^{4n}_{2n}, we can determine the sectional curvature in any real 2-direction at each point in these spaces. Since all these manifolds are symmetric spaces, these manifolds are isometric to totally geodesic submanifolds in the fundamental groups of these spaces, which are simple Lie groups with their invariant metric or pseudometric (see **0.8.5**). Therefore, the Riemann tensors in all these manifolds can be determined by formula (0.70) for the Riemann tensor in semisimple Lie groups. Each symmetric space is determined by an involutive automorphism J in its fundamental group G and by a corresponding involutive automorphism (1.33) in the tangent Lie algebra $\mathbb{G}$ of this group in a basis e_a, e_u whose vectors e_a are invariant under this automorphism while the vectors e_u are multiplied by -1 under this automorphism. The vectors e_a form a basis of the Lie algebra $\mathbb{K}$ of the stabilizer of the symmetric space and the vectors e_u form a basis of a subspace $\mathbb{E}$ in $\mathbb{G}$ which can be regarded as the tangent space to the symmetric space.

Theorem 4.32. *Choose an orthonormal frame in a symmetric Riemannian or pseudo-Riemannian space G/K. Identify its vectors with the vectors e_u in a subspace $\mathbb{E}$ in the Lie algebra $\mathbb{G} = \mathbb{K} \oplus \mathbb{E}$. Then the Riemann tensor can be written as*

$$R^z_{uv,w}e_z = \rho[[e_u e_v]e_w] \tag{4.31}$$

with respect to this frame.

PROOF. In the basis $\{e_a, e_u\}$ the structure formulas (0.65) in $\mathbb{G}$ have the form (1.125). Therefore, formula (0.70) leads to the following Riemann tensor in this symmetric space

$$R^z_{uv,w} = \rho C^a_{uw} C^z_{va}, \tag{4.32}$$

which is equivalent to formula (4.31) for $\rho = \frac{1}{4}$. This coefficient in formula (0.70) corresponds to the Killing-Cartan metric or pseudometric in semisimple Lie groups. Since symmetric Riemannian and pseudo-Riemannian spaces can have various scales, for these spaces the coefficient ρ can have various values. ■

4.6.2. Sectional Curvature in Complex and Split Complex Spaces. The tangent Lie algebra $\mathbb{G}$ of the group G of motions in $\mathbb{C}\overline{S}^n$ consists of the complex skew symmetric $((n+1) \times (n+1))$-matrices $A = (A_{ij})$ with zero trace ($A_{ij} = -A_{ji}$, $\sum_i A_{ii} = 0$, $i, j = 0, 1, \ldots, n$), which can be written in the form $A = \sum_i \sum_j A_{ij} E_{ij}$, where the matrices E_{ij} are the same as in the proof of Theorem 1.29 (see **1.4.3**). The matrices A in the subspace $\mathbb{E}$ of this Lie algebra have the form

$$A = E_{0i} a^i - E_{i0} \bar{a}^i. \tag{4.33}$$

Theorem 4.33 (P. SHIROKOV [ShP2]). *Denote a basis of the algebra $\mathbb{C}$ by $\{i_\alpha\}$, that is, $i_0 = 1$, $i_1 = i$. Then we can write a complex number x as $x^\alpha i_\alpha$. Denote double indices $i\alpha$ by the letters $I, J, \ldots$. Choose in the Riemannian manifold V^{2n} which is isometric to the space $\mathbb{C}\overline{S}^n$ of curvature $\frac{1}{r^2}$ the frame of vectors $e_{i\alpha} = e_i i_\alpha$. Then the Riemann tensor has, in this frame, the form*

$$R^L_{IJ,K} = \frac{1}{r^2}\left(\delta_{IK}\delta^L_J - \delta_{JK}\delta^L_I + \epsilon_{IK}\epsilon^L_J - \epsilon_{JK}\epsilon^L_I + 2\epsilon_{IJ}\epsilon^L_K\right), \tag{4.34}$$

where the matrices (δ_{IJ}), (δ^I_J), (ϵ_{IJ}), and (ϵ^I_J), are $(2n \times 2n)$-matrices with diagonal (2×2)-submatrices $(\delta_{\alpha\beta})$, (δ^β_α), $(\epsilon_{\alpha\beta})$, and (ϵ^β_α), two first of which are identity (2×2)-matrices and two last are skew symmetric (2×2)-matrices with entries $\epsilon_{01} = \epsilon^1_0 = -1$. All entries of $(2n \times 2n)$-matrices outside diagonal $(2 \times 2$-submatrices are equal to zero.

The covariant Riemann tensor $R_{IJ,KL} = R^H_{IJ,K} e_{HL}$ in the same manifold V^{2n}, in the same frame, has the form

$$R_{IJ,KL} = \frac{1}{r^2}\left(\delta_{IK}\delta_{JL} - \delta_{IL}\delta_{JK} + \epsilon_{IK}\epsilon_{JL} - \epsilon_{IL}\epsilon_{JK} + 2\epsilon_{IJ}\epsilon_{KL}\right). \tag{4.35}$$

The sectional curvature K in the manifold V^{2n} in the 2-direction determined by two orthogonal unit vectors a and b is equal to

$$K = \frac{1 + 3\cos\alpha}{r^2}, \tag{4.36}$$

where α is the holomorphy angle in this 2-direction and r is the radius of curvature of $\mathbb{C}\overline{S}^n$.

PROOF. Since arbitrary vectors in the tangent space R^{2n} to the manifold V^{2n} are represented by the matrices (4.33), the basis vectors $e_{i\alpha} = e_i i_\alpha$ are represented by the matrices

$$A_{i\alpha} = E_{0i} i_\alpha = E_{i0} \bar{i}_\alpha. \tag{4.37}$$

Therefore, the double commutator $[[A_{i\alpha} A_{j\beta}] A_{k\gamma}]$ has the form

$$\begin{aligned}[][[A_{i\alpha} A_{j\beta}] A_{k\gamma}] = &- E_{0i}\delta_{jk} i_\gamma (\bar{i}_\beta i_\alpha) + E_{0j}\delta_{ik} i_\gamma (\bar{i}_\alpha i_\beta) + \\ &+ E_{0k}\delta_{ij}(i_\beta \bar{i}_\alpha - i_\alpha \bar{i}_\beta) i_\gamma + E_{0i}\delta_{jk}(\bar{i}_\alpha i_\beta)\bar{i}_\gamma - \\ &- E_{j0}\delta_{ik}(\bar{i}_\beta i_\alpha)\bar{i}_\gamma + E_{k0}\delta_{ij}\bar{i}_\gamma (i_\beta \bar{i}_\alpha - i_\alpha \bar{i}_\beta).\end{aligned} \tag{4.38}$$

Hence, by (4.31),

$$R^{\ell\delta}_{i\alpha,j\beta;k\gamma} e_\ell i_\delta = \rho(-e_i\delta_{jk} i_\gamma (\bar{i}_\beta i_\alpha) + e_j \delta_{ik} i_\gamma (\bar{i}_\alpha i_\beta) + e_k \delta_{ij}(i_\beta \bar{i}_\alpha - i_\alpha \bar{i}_\beta)) i_\gamma. \tag{4.39}$$

Since the field $\mathbb{C}$ is associative, $i_\gamma(\bar{i}_\beta i_\alpha) = (i_\gamma \bar{i}_\beta) i_\alpha$ and $i_\gamma(\bar{i}_\alpha i_\beta) = (i_\gamma \bar{i}_\alpha) i_\beta$. Since $e_i = \delta^\ell_i e_\ell$, at the left hand side of (4.39) we can delete e_ℓ and at the right hand side we can replace the vectors e_i, e_j, and e_k by the Kronecker symbols $\delta^\ell_i, \delta^\ell_j$, and δ^ℓ_k, respectively.

To determine the coefficient ρ we consider the sectional curvature in an antiholomorphic 2-direction in a normal 2-chain $x^i = \bar{x}^i \quad (i = 0, 1, 2)$, $x^j = 0 \quad (j > 2)$. In (4.39) the basis elements i_α can only be equal to $i_0 = 1$, that is, this formula (we omit the indices $\alpha = \beta = \gamma = \delta = 1$) has the form $R^\ell_{ij,k} e_\ell = \rho(-e_i\delta_{jk} + e_j\delta_{ik})$ and, therefore, $R_{ij,k\ell} = R^h_{ij,k} e_{h\ell} = R^h_{ij,k} e_h e_\ell = \rho(\delta_{ik}\delta_{j\ell} - \delta_{i\ell}\delta_{jk})$ (we omit the indices $\alpha = \beta = \gamma = \delta = 1$). Since the basis $\{i_i\}$ is orthonormal, the metric tensor e_{ij} has the form δ_{ij}. By (0.45) the sectional curvature K in this 2-direction is equal to ρ. Therefore, since the sectional curvature of normal n-chains in $\mathbb{C}\overline{S}^n$ of curvature $\frac{1}{r^2}$ is equal to $\frac{1}{r^2}$, we obtain $\rho = \frac{1}{r^2}$.

For the algebra $\mathbb{C}$ formula (0.14) can be written as $i_\alpha i_\beta = c^\gamma_{\alpha\beta} i_\gamma$, where the structure constants $c^\gamma_{\alpha\beta}$ are determined by

$$c^\gamma_{0\beta} = \delta^\gamma_\beta, \quad c^\gamma_{1\beta} = \epsilon^\gamma_\beta. \tag{4.40}$$

Therefore,

$$\bar{i}_\alpha, i_\beta = \delta_{\alpha\beta} i_0 - \epsilon_{\alpha\beta} i_1, \quad i_\alpha \bar{i}_\beta = \delta_{\alpha\beta} i_0 + \epsilon_{\alpha\beta} i_1. \tag{4.41}$$

Substituting (4.41) and ρ in (4.39) gives (4.34); convolution of $R^H_{IJ,K}$ with $e_{HL} = \delta_{HL}$ gives (4.35).

Since $\delta_{IJ} a^I a^J = a^2 = 1$, $\delta_{IJ} b^I b^J = b^2 = 1$, $\delta_{IJ} a^I b^J = ab = 0$, $\epsilon_{IJ} a^I a^J = aJa = 0$, $\epsilon_{IJ} b^I b^J = bJb = 0$, $\epsilon_{IJ} a^I b^J = aJb = \cos\alpha$, substituting (4.35) in (0.45) gives (4.36). ■

Note that for holomorphic 2-directions, for which $\alpha = 0$, the sectional curvature is equal to $\frac{4}{r^2}$, which corresponds to the isometry of lines in $\mathbb{C}\overline{S}^n$ to spheres in R^3 of radius $\frac{r}{2}$.

The sectional curvature in holomorphic 2-directions is called the *holomorphic sectional curvature.* Since the holomorphic curvature in V^{2n}, which is isometric to $\mathbb{C}\overline{S}^n$ of curvature $\frac{1}{r^2}$, is equal to the constant number $\frac{4}{r^2}$, this Riemannian manifold is called a *Riemannian manifold of constant holomorphic sectional curvature* [KoN].

Similarly we can prove

Theorem 4.34 (PRVANOVIĆ [Prv]). *Denote the basis of the algebra $\mathbb{C}'$ by $\{i_\alpha\}$, that is, $i_0 = 1$, $i_1 = e$. Then we can write a split complex number x as $x^\alpha i_\alpha$. Also, denote the double indices $i\alpha$ by letters $I, J, \ldots,$. Take in the pseudo-Riemannian manifold V_n^{2n}, which is isometric to the space $\mathbb{C}'\overline{S}^n$ of curvature $\frac{1}{r^2}$, the frame consisting of the vectors $e_{i\alpha} = e_i i_\alpha$. The the Riemann tensor $R^L_{IJ,K}$ has the form*

$$R^L_{IJ,K} = \frac{1}{r^2}\left(\delta_{IK}\delta^L_J - \delta_{JK}\delta^L_I + \eta_{IK}\eta^L_J - \eta_{JK}\eta^L_I + 2\eta_{IJ}\eta^L_K\right), \tag{4.42}$$

where the matrices (δ_{IJ}), (δ^I_J), (η_{IJ}), and (η^I_J) are $(2n\times 2n)$-matrices with diagonal (2×2)-submatrices $(\delta_{\alpha\beta}), (\delta^\beta_\alpha), (\eta_{\alpha\beta})$, and (η^β_α), two first of which are identity (2×2)-matrices and two last are symmetric (2×2)-matrices with entries $\eta_{00} = \eta^0_0 = \eta_{11} = \eta^1_1 = 0, \eta_{01} = \eta^1_0 = 1$. All entries of $(2n\times 2n)$-matrices outside diagonal $(2\times 2$-subnatrices are equal to zero.

The covariant Riemann tensor $R_{IJ,KL} = R^H_{IJ,K}e_{HL}$ of the same manifold V_n^{2n}, in the same frame, has the form

$$R_{IJ,KL} = \frac{1}{r^2}\left(\delta_{IK}\delta_{JL} - \delta_{IL}\delta_{JK} + \eta_{IK}\eta_{JL} - \eta_{IL}\eta_{JK} + 2\eta_{IJ}\eta_{KL}\right). \tag{4.43}$$

The sectional curvature K of the pseudo-Riemannian manifold V_n^{2n} in the 2-direction determined by two orthogonal vectors a and b with moduli $|a| = 1$, $|b| = i$ is equal to

$$K = \frac{1 + 3\cosh\alpha}{r^2}, \tag{4.44}$$

where α is the modulus of the pure imaginary holomorphy angle in this 2-direction. The sectional curvature K of the same manifold in the 2-direction determined by two unit orthogonal vectors a and b is equal to (4.36). In (4.36) and (4.44) r is the radius of curvature of $\mathbb{C}'\overline{V}^n$.

4.6.3. Sectional Curvature in Quaternionic and Split Quaternionic Spaces. The tangent Lie algebra $\mathbb{G}$ of the group G of motions in $\mathbb{H}\overline{S}^n$ consists of the quaternionic skew Hermitian symmetric $((n+1)\times(n+1))$-matrices $A = (A_{ij})$ $(A_{ij} = -\bar{A}_{ji})$, with zero real part of the trace, $\operatorname{Re}\sum_i A_{ii} = 0$, $(i, j = 0, 1, \ldots, n)$ which can be written as $A = \sum_i\sum_j A_{ij}E_{ij}$, where the matrices E_{ij} are the same as in **4.6.2**; the matrices A in the subspace $\mathbb{E}$ of this Lie algebra have the form (4.33).

Theorem 4.35. *Denote the basis of the algebra* $\mathbb{H}$ *by* $\{i_\alpha\}$, *that is,* $i_0 = 1$, $i_1 = i$, $i_2 = j$, $i_3 = k$. *Then we can write a quaternion* x *as* $x^\alpha i_\alpha$. *Also, denote the double indices* $i\alpha$ *by letters* $I, J, \ldots,$. *Take in the Riemannian manifold* V^{4n}, *which is isometric to* $\mathbb{H}\overline{S}^n$ *of curvature* $\frac{1}{r^2}$, *the frame consisting of vectors* $e_{i\alpha} = e_i i_\alpha$. *Then the Riemann tensor* $R^L_{IJ,K}$ *has the form*

$$R^L_{IJ,K} = \frac{1}{r^2}\left(\delta_{IK}\delta^L_J - \delta_{JK}\delta^L_I + \sum_\lambda \epsilon_{\lambda IK}\epsilon^L_{\lambda J} - \sum_\lambda \epsilon_{\lambda JK}\epsilon^L_{0\lambda I} + 2\sum_\lambda \epsilon_{\lambda IJ}\epsilon^L_{\lambda K}\right), \quad (4.45)$$

where the matrices $(\delta_{IJ}), (\delta^I_J), (\epsilon_{\lambda IJ})$, *and* $(\epsilon^I_{\lambda J})$ *are the* $(4n \times 4n)$*-matrices with diagonal* (4×4)*-submatrices*

$$(\delta_{\alpha\beta}) = (\delta^\alpha_\beta) = \begin{bmatrix} 1 & 0 & 0 & 0 \\ 0 & 1 & 0 & 0 \\ 0 & 0 & 1 & 0 \\ 0 & 0 & 0 & 1 \end{bmatrix}, \quad (\epsilon_{1ij}) = (\epsilon^i_{1j}) = \begin{bmatrix} 0 & -1 & 0 & 0 \\ 1 & 0 & 0 & 0 \\ 0 & 0 & 0 & -1 \\ 0 & 0 & 1 & 0 \end{bmatrix},$$

$$(\epsilon_{2\alpha\beta}) = (\epsilon^\alpha_{2\beta}) = \begin{bmatrix} 0 & 0 & -1 & 0 \\ 0 & 0 & 0 & 1 \\ 1 & 0 & 0 & 0 \\ 0 & -1 & 0 & 0 \end{bmatrix}, \quad (\epsilon_{3\alpha\beta}) = (\epsilon^\alpha_{3\beta}) = \begin{bmatrix} 1 & 0 & 0 & 0 \\ 0 & 1 & 0 & 0 \\ 0 & 0 & 1 & 0 \\ 0 & 0 & 0 & 1 \end{bmatrix}. \quad (4.46)$$

All entries of $(4n \times 4n)$*-matrices outside diagonal* $(4 \times 4$*-subnatrices are equal to zero.*

The covariant Riemann tensor $R_{LJ,KL} = R^H_{LJ,K} e_{HL}$ *in the same* V^{4n}, *and in the same frame, has the form*

$$R_{IJ,KL} = \frac{1}{r^2}\left(\delta_{IK}\delta_{JL} - \delta_{IL}\delta_{JK} + \sum_\lambda \epsilon_{\lambda IK}\epsilon_{\lambda JL} - \sum_\lambda \epsilon_{\lambda IL}\epsilon_{\lambda JK} + 2\sum_\lambda \epsilon_{\lambda IJ}\epsilon_{\lambda KL}\right). \quad (4.47)$$

The sectional curvature K *in the manifold* V^{4n} *in the* 2*-direction determined by two orthogonal unit vectors* a *and* b *is equal to (4.48), where* α *is the holomorphy angle in this* 2*-direction.*

PROOF. Since an arbitrary vector in the tangent space R^{4n} to V^{4n} can be represented by a matrix (4.33), the basis vectors $e_{i\alpha} = e_i i_\alpha$ can be represented by the matrices (4.37). In this case the double commutator $[[A_{i\alpha}A_{j\beta}]A_{k\gamma}]$ has the form (4.38) and, according to (4.31), the tensor $R^{\ell\delta}_{i\alpha,j\beta;k\gamma} e_\ell i_\delta$ has the form (4.39). Since the algebra $\mathbb{H}$ is associative, $i_\gamma(\bar{i}_\beta i_\alpha) = (i_\gamma \bar{i}_\beta) i_\alpha$ and $i_\gamma(\bar{i}_\alpha i_\beta) = (i_\gamma \bar{i}_\alpha) i_\beta$. Since $e_i = \delta^\ell_i e_\ell$, at the left hand side of the equality (4.39), like in the proof of Theorem 4.33, we can replace the vectors e_i, e_j, and e_k by the Kronecker symbols $\delta^\ell_i, \delta^\ell_j$, and δ^ℓ_k, respectively. Like in the proof of Theorem 4.33, we can prove that ρ in (4.39) is equal to $\frac{1}{r^2}$.

The structure constants $c^{\gamma}_{\alpha\beta}$ of the algebra $\mathbb{H}$ are

$$c^{\gamma}_{0\beta} = \delta^{\gamma}_{\beta}, \quad c^{\gamma}_{1\beta} = \epsilon^{\gamma}_{1\beta}, \quad c^{\gamma}_{2\beta} = \epsilon^{\gamma}_{2\beta}, \quad c^{\gamma}_{3\beta} = \epsilon^{\gamma}_{3\beta}. \tag{4.48}$$

Therefore,

$$\bar{i}_{\alpha} i_{\beta} = \delta_{\alpha\beta} i_0 - \sum_{\lambda} \epsilon_{\lambda\alpha\beta} i_1, \; i_{\alpha}\bar{i}_{\beta} = \delta_{\alpha\beta} i_0 + \sum_{\lambda} \epsilon_{\lambda\alpha\beta} i_1. \tag{4.49}$$

Substituting (4.49) and ρ in (4.39) gives formula (4.45). Convolution of $R^{H}_{LJ,K}$ with $e_{HL} = \delta_{HL}$ gives formula (4.47).

Since $\delta_{IJ} a^I a^J = a^2 = 1$, $\delta_{IJ} b^I b^J = b^2 = 1$, $\delta_{IJ} a^I b^J = ab = 0$, $\epsilon_{IJ} a^I a^J = aJa = 0$, $\epsilon_{IJ} b^I b^J = bJb = 0$, $\epsilon_{IJ} a^I b^J = aJb = \cos\alpha$, substituting (4.47) in (0.45), like in Theorem 4.33, gives formula (4.36). ■

Similarly we can prove

Theorem 4.36. *Denote the basis of the algebra $\mathbb{H}'$ by $\{i_{\alpha}\}$, that is, $i_0 = 1$, $i_1 = i$, $i_2 = e$, $i_3 = f$. Then we can write a split quaternion x as $x^{\alpha} i_{\alpha}$. Denote the double indices $i\alpha$ by letters $I, J, \ldots,$. Take in the pseudo-Riemannian manifold V^{4n}_{2n}, which is isometric to $\mathbb{H}'\overline{S}^n$ of curvature $\frac{1}{r^2}$ in the frame consisting of vectors $e_{i\alpha} = e_i i_{\alpha}$. Then the Riemann tensor $R^L_I J, K$ has the form*

$$R^{L}_{IJ,K} = \frac{1}{r^2}\left(\delta_{IK}\delta^{L}_{J} - \delta_{JK}\delta^{L}_{I} + \sum_{\lambda} \eta_{\lambda IK}\eta^{L}_{\lambda J} - \sum_{\lambda} \eta_{\lambda JK}\eta^{L}_{0\lambda I} + 2\sum_{\lambda} \eta_{\lambda IJ}\eta^{L}_{\lambda K}\right), \tag{4.50}$$

where the matrices $(\delta_{\alpha\beta}), (\delta^{\alpha}_{\beta}), (\eta_{\lambda\alpha\beta})$, and $(\eta^{\alpha}_{\lambda\beta})$ are the $(4n \times 4n)$-matrices with diagonal $(4 \times 4$-matrices

$$(\delta_{\alpha\beta}) = (\delta^{\alpha}_{\alpha}) = \begin{bmatrix} 1 & 0 & 0 & 0 \\ 0 & 1 & 0 & 0 \\ 0 & 0 & 1 & 0 \\ 0 & 0 & 0 & 1 \end{bmatrix}, \quad (\eta_{1\alpha\beta}) = (\eta^{\alpha}_{1\beta}) = \begin{bmatrix} 0 & -1 & 0 & 0 \\ 1 & 0 & 0 & 0 \\ 0 & 0 & 0 & -1 \\ 0 & 0 & 1 & 0 \end{bmatrix},$$

$$(\eta_{2\alpha\beta}) = (\eta^{\alpha}_{2\beta}) = \begin{bmatrix} 0 & 0 & 1 & 0 \\ 0 & 0 & 0 & -1 \\ 1 & 0 & 0 & 0 \\ 0 & -1 & 0 & 0 \end{bmatrix}, \quad (\eta_{3\alpha\beta}) = (\eta^{\alpha}_{3\beta}) = \begin{bmatrix} 0 & 0 & 0 & 1 \\ 0 & 0 & 1 & 0 \\ 0 & 1 & 0 & 0 \\ 1 & 0 & 0 & 0 \end{bmatrix}. \tag{4.51}$$

All entries of $(4n \times 4n)$-matrices outside diagonal $(4 \times 4$-submatrices are equal to zero.

The covariant Riemann tensor $R_{IJ,KL} = R^{H}_{IJ,K} e_{HL}$ of the same V^{4n}_{2n}, in the same frame, has the form

$$\begin{aligned} R_{IJ,KL} = \frac{1}{r^2}\Big(&\delta_{IK}\delta_{JL} - \delta_{IL}\delta_{JK} + \sum_{\lambda} \eta_{\lambda IK}\eta_{\lambda JL} \\ &- \sum_{\lambda} \eta_{\lambda IL}\eta_{\lambda JK} + 2\sum_{\lambda} \eta_{\lambda IJ}\eta_{\lambda KL}\Big). \end{aligned} \tag{4.52}$$

The sectional curvature K in the manifold V_{2n}^{4n} in the 2-direction determined by two orthogonal vectors a and b with moduli $|a| = 1$, $|b| = i$ is equal to (4.44), where α is the modulus of the pure imaginary holomorphy angle in this 2-direction. The sectional curvature K of the same manifold in the 2-direction determined by two unit orthogonal vectors a and b is equal to (4.36), where α is the holomorphy angle in this 2-direction.

§4.7. Polyhedra, Hyperquadrics, and Hyperspheres

4.7.1. Polyhedra. Since the quadratic spaces $\mathbb{A}S^n$, $\mathbb{A}H^n$, $\mathbb{A}S^n_\ell$, and $\mathbb{A}H^n_\ell$ can be defined as the projective space $\mathbb{A}P^n$ in which an absolute hyperquadric is given and since the Hermitian spaces $\mathbb{A}\overline{S}^n$, $\mathbb{A}\overline{H}^n$, $\mathbb{A}\overline{S}^n_\ell$, and $\mathbb{A}\overline{H}^n_\ell$ can be defined as the projective space $\mathbb{A}P^n$ in which an absolute Hermitian hyperquadric is given, polyhedra in these spaces are polyhedra in $\mathbb{A}P^n$. In particular, a simplex in $\mathbb{A}P^n$, that is, the convex envelope of $n+1$ points which are not in one hyperplane, is called a *simplex* in these spaces; these $n+1$ points are called the *vertices* of the simplex.

In **3.6.5** we have considered spherical polyhedra on hyperspheres in R^{n+1} and on hyperspheres of pure imaginary radius in R^{n+1}_1, and volumes of spherical simplexes on these hyperspheres. Under the identification of antipodal points of these hyperspheres, these spherical polyhedra become polyhedra in an elliptic space S^n and in a hyperbolic space H^n, respectively.

4.7.2. Centers of Hyperquadrics and Hermitian Hyperquadrics. Each hyperquadric or Hermitian hyperquadric in $\mathbb{A}P^n$ determines *autopolar simplexes*, that is, simplexes whose vertices are the poles of opposite faces. Since the spaces $\mathbb{A}S^n$, $\mathbb{A}H^n$, $\mathbb{A}S^n_\ell$, $\mathbb{A}H^n_\ell$, $\mathbb{A}\overline{S}^n$, $\mathbb{A}\overline{H}^n$, $\mathbb{A}\overline{S}^n_\ell$, or $\mathbb{A}\overline{H}^n_\ell$, are the projective space $\mathbb{A}P^n$, in which an absolute hyperquadric or Hermitian hyperquadric is given, there are the common autopolar simplexes of hyperquadrics and Hermitian hyperquadrics under the consideration in these spaces and of absolute hyperquadrics and Hermitian hyperquadrics of these spaces. The vertices of these simplexes are fixed points of collineations which are products of polar transformations with respect to considered and absolute (Hermitian) hyperquadrics. These vertices are called the *centers* of the hyperquadrics and Hermitian hyperquadrics.

Theorem 4.37. *If the equations of a hyperquadric and of the absolute hyperquadric in $\mathbb{A}S^n, \mathbb{A}H^n, \mathbb{A}S^n_\ell$, or $\mathbb{A}H^n_\ell$ are $xAx = 0$ and $xEx = 0$, respectively, and if the equations of a Hermitian hyperquadric and of the absolute Hermitian hyperquadric in $\mathbb{A}\overline{S}^n$, $\mathbb{A}\overline{H}^n$, $\mathbb{A}\overline{S}^n_\ell$, or $\mathbb{A}\overline{H}^n_\ell$ are $\bar{x}Ax = 0$ and $\bar{x}Ex = 0$, respectively, then the centers of the hyperquadric under consideration are determined by the condition $(A - \lambda E)x = 0$.*

<u>Proof</u>. In quadratic spaces the polar transformations with respect to the hyperquadrics $xAx = 0$ and $xEx = 0$ are $u = Ax$ and $u = Ex$, respectively. In Hermitian spaces the polar transformations with respect to the Hermitian hyperquadrics $\bar{x}Ax$ and $\bar{x}Ex$ are $\bar{u} = Ax$ and $\bar{u} = Ex$, respectively. Therefore, the fixed points of products of these polar transformations are in both cases determined by the condition $Ax = \lambda Ex$, or $(A - \lambda E)x = 0$. ■

For commutative algebras $\mathbb{A}$ this condition can be fulfilled only if det $(A - \lambda I) =$

0, or det $(A - \lambda E) = 0$. If these algebraic equations of degree $(n+1)$ have $n+1$ different roots, the hyperquadric has $n+1$ centers. If these equations have $k+1$ equal roots, the hyperquadric has a k-plane every point of which is a center. For hyperquadrics of index ℓ in the space S^n and for Hermitian hyperquadrics of index ℓ in the space $\mathbb{C}\overline{S}^n$, which divide these spaces into two domains, ℓ centers are located in one of these domains and $n+1-\ell$ centers are located in the other domain.

The case of noncommutative algebras $\mathbb{A}$ is analogous, but instead of the determinants det $(A - \lambda I)$ and det $(A - \lambda E)$ we must consider the semideterminants of the matrices $A - \lambda I$ and $A - \lambda E$.

4.7.3. Hyperspheres and m-Equidistant Hyperquadrics. When the centers of a hyperquadric in a space $\mathbb{A}S^n, \mathbb{A}H^n, \mathbb{A}S^n_\ell$, or $\mathbb{A}H^n_\ell$ or of a Hermitian hyperquadric in space $\mathbb{A}\overline{S}^n, \mathbb{A}\overline{H}^n, \mathbb{A}\overline{S}^n_\ell$, or $\mathbb{A}\overline{H}^n_\ell$ are all points of two mutually polar m-plane and $(n-m-1)$-plane, the (Hermitian) hyperquadric is called an *m-equidistant (Hermitian) hyperquadric.* The m-plane and $(n-m-1)$-plane containing the centers of such a hyperquadric are called its *basis m-plane and $(n-m-1)$-plane.* If the centers of such a hyperquadric are a separate point and all points of its polar hyperplane, the hyperquadric is called a *hypersphere*, respectively a *Hermitian hypersphere*; the separate center of such a hyperquadric is called the *center of the hypersphere.* The names "m-equidistant hyperquadrics" and "hyperspheres" are explained by

Theorem 4.38. *An m-equidistant hyperquadric or Hermitian hyperquadric in a space $\mathbb{A}S^n$, $\mathbb{A}H^n$, $\mathbb{A}S^n_\ell$, $\mathbb{A}H^n_\ell$, $\mathbb{A}\overline{S}^n$, $\mathbb{A}\overline{H}^n$, $\mathbb{A}\overline{S}^n_\ell$, or $\mathbb{A}\overline{H}^n_\ell$ is the set of points in the space that are equidistant from its basis m-plane and $(n-m-1)$-plane.*

A hypersphere or Hermitian hypersphere in such a space is a set of points that are equidistant from the center and from the basis hyperplane.

PROOF. The equation of an m-equidistant hyperquadric in $\mathbb{A}S^n$ or $\mathbb{A}S^n_\ell$ can be reduced to the form $-\lambda_0 \sum_a (x^a)^2 + \lambda_1 \sum_u (x^u)^2 = 0$; the equation of an m-equidistant Hermitian hyperquadric in $\mathbb{A}\overline{S}^n$ or $\mathbb{A}\overline{S}^n_\ell$ can be reduced to the form $-\lambda_0 \sum_a \bar{x}^a x^a + \lambda_1 \sum_u \bar{x}^u x^u = 0$. If the coefficients λ_0 and λ_1 are normalized by the condition $\lambda_0^2 + \lambda_1^2 = 1$, we can put $\lambda_0 = \cos a$ and $\lambda_1 = \sin a$. Therefore, since the equations of the basis m-plane and $(n-m-\ell)$-plane of an equidistant hyperquadric are $x^u = 0$ and $x^a = 0$, respectively, in a space of curvature $+1$ the distances of any point on this hyperquadric to its basis m-plane and $(n-m-1)$-plane are a and $\frac{\pi}{2} - a$, respectively. For $m = 0$ we obtain a (Hermitian) hypersphere. ∎

The distances a and $\frac{\pi}{2} - a$ of points on an m-equidistant hyperquadric from its basis m-plane and $(n-m-\ell)$-plane are called *parameters* of this m-equidistant hyperquadric. A similar theorem is valid also in the spaces $\mathbb{A}H^n, \mathbb{A}H^n_\ell, \mathbb{A}\overline{H}^n$, and $\mathbb{A}\overline{H}^n_\ell$. In H^n the hyperspheres with ideal center are called *equidistant hyperquadrics.*

Theorem 4.39. *A hypersphere of radius R in a space S^n of curvature $\frac{1}{r^2}$ is isometric to the hypersphere of radius $r \sin \frac{R}{r}$ in R^n.*

A hypersphere of radius R in a space H^n of curvature $-\frac{1}{q^2}$ is isometric to the hypersphere of radius $q \sin h \frac{R}{q}$ in R^n.

An equidistant hyperquadric of parameter P in H^n is isometric to the hypersphere of radius $iq \cosh \frac{P}{q}$ in R^n_1.

PROOF. In the first case, let O be the center of the hypersphere of radius r in R^{n+1} whose identification of antipodal points leads to S^n, let C be the center of the $(n-1)$-sphere of spherical radius R on this hypersphere and let X be a point on this $(n-1)$-sphere. Then in the sector OCX the angle $\angle COX$ is equal to $\frac{R}{r}$. The $(n-1)$-sphere is located in the hyperplane perpendicular to the line OC, suppose they meet in C'. Then the Euclidean radius $C'X$ of this $(n-1)$-sphere is equal to $r \sin \frac{R}{r}$.

In the second and third cases, let O be the center of the hypersphere of radius qi in R_1^{n+1} whose identification of antipodal points leads to H^n, let C be the center of the $(n-1)$-sphere of spherical radius R on this hypersphere and let X be a point on this $(n-1)$-sphere. Then in the sector OCX the angle $\angle COX$ is equal to $i\frac{R}{q}$. The $(n-1)$-sphere is located in the Euclidean hyperplane perpendicular to the line OC, suppose they meet in C'. Then the Euclidean radius $C'X$ of this $(n-1)$-sphere is equal to $q \sin h \frac{R}{q}$.

The equidistant $(n-1)$-quadric is located in the pseudo-Euclidean hyperplane parallel to the hyperplane which determines the basis hyperplane of the equidistant $(n-1)$-quadric in H^n. Let the base of the perpendicular which is dropped from O onto the pseudo-Euclidean hyperplane be O' and X be a point on the equidistant hyperquadric. Then in the triangle $OO'X$ the angle $\angle O'OX$ is equal to $i\frac{P}{r}$. The equidistant $(n-1)$-quadric is an $(n-1)$-sphere of imaginary radius located in the pseudo-Euclidean hyperplane perpendicular to the line OO', and the imaginary length of the radius $C'X$ of this $(n-1)$-sphere is equal to $iq \cosh \frac{R}{q}$. ■

A 1-equidistant quadric in S^3 or H^3 is called a *Clifford quadric*, respectively an *equidistant barrel.*

Theorem 4.40. *The intrinsic geometry of a Clifford quadric in S^3 and of an equidistant barrel in H^3 is Euclidean.*

PROOF. Let the basis lines of a Clifford quadric in S^3 of curvature 1 and of an equidistant barrel in H^3 of curvature -1 be the edges E_0E_1 and E_2E_3 of autopolar tetrahedra in these spaces. Then an arbitrary point X on the Clifford quadric can be represented by the vector $x = (e_0 \cos u + e_1 \sin u) \cdot \cos a + (e_2 \cos v + e_3 \sin v) \cdot \sin a$ and the linear element $ds^2 = dx^2$ of this surface is equal to $\cos^2 a\, du^2 + \sin^2 a\, dv^2 = dU^2 + dV^2$.

Analogously, an arbitrary point X on the equidistant barrel can be represented by the vector $x = (e_0 \cosh u + e_1 \sinh u) \cdot \cosh a + (e_2 \cos v + e_3 \sin v) \cdot \sinh a$ and the linear element $ds^2 = dx^2$ of this surface is equal to $\cosh^2 a\, du^2 + \sinh^2 a\, dv^2 = dU^2 + dV^2$. But $dU^2 + dV^2$ is the linear element of R^2. ■

This property of Clifford quadrics was proved by W. K. Clifford [Cli1]; this property of equidistant barrels was proved by A. Volman [Vol]. It is easy to verify that a Clifford quadric of parameter a in S^3 of curvature 1 is isometric to the rhombus with acute angle $2a$ and side π and with identified opposite sides, and that an equidistant barrel of parameter a in H^3 of curvature -1 is isometric to the strip of width π on R^2 with identified boundary lines.

4.7.4. Horohyperspheres. An intermediate position between hyperspheres and equidistant hyperquadrics in H^n is taken up by *horohyperspheres* or *limit hy-*

perspheres: a hypersphere and an equidistant hyperquadric are determined by sections of a hypersphere of radius qi in R_1^{n+1} by a Euclidean or pseudo-Euclidean hyperplane, respectively, whereas a horohypersphere is determined by a section of this hypersphere by an isotropic hyperplane. For $n = 2$ and 3 the role of horohyperspheres is played by ***horocycles*** and ***horospheres***, respectively.

Theorem 4.41. *The intrinsic geometry of a horohypersphere in H^n is the Euclidean geometry of R^{n-1}.*

PROOF. Let the isotropic hyperplane determining a horohypersphere in H^n be $x^0 + x^n = 0$. Then the vectors of this hyperplane are linear combinations of the vectors $e_0 - e_n, e_1, \ldots, e_{n-1}$, and the vector representing a point X on this horohypersphere has the form $x^0 = (e_0 - e_n) + x^i e_i \quad (i = 1, 2, \ldots, n-1)$. Therefore, the linear element $ds^2 = dx^2$ of this hypersurface is equal to $ds^2 = \sum_i (dx^i)^2$. But $\sum_i (dx^i)^2$ is the linear element of R^{n-1}. ■

Horohyperspheres in H^n can be obtained by a limit transition from hyperspheres and equidistant hyperquadrics in this space if the center of the hypersphere tends to a point of the absolute hyperquadric of H^n or if the basic hyperplane of an equidistant hyperquadric tends to a tangent hyperplane to the absolute hyperquadric.

4.7.5 Hermitian Hyperspheres, Equidistant Hyperquadrics, and Horohyperspheres. In **4.7.3** we have defined Hermitian hyperspheres and m-equidistant hyperquadrics in Hermitian spaces. Here we will consider Hermitian hyperspheres in $\mathbb{C}\overline{S}^n$ and Hermitian hyperspheres and equidistant Hermitian hyperquadrics, as well as Hermitian horohyperspheres, in $\mathbb{C}\overline{H}^n$.

Note that a Hermitian horohypersphere in $\mathbb{C}\overline{H}^n$ can be obtained by a limit transition from Hermitian hyperspheres and equidistant hyperquadrics in this space by letting the center of a Hermitian hypersphere tend to a point of the absolute Hermitian hyperquadric of H^n or by letting the basis hyperplane of an equidistant Hermitian hyperquadric tend to a tangent hyperplane to the absolute hyperquadric.

Theorem 4.42. *Hermitian hyperspheres in $\mathbb{C}\overline{S}^n$ and $\mathbb{C}\overline{H}^n$ are isometric to a hypersphere Σ^{2n-1} in R^{2n} and are fibrations of this real hypersphere with great circles as fibers and the space $\mathbb{C}\overline{S}^{n-1}$ as base.*

An equidistant Hermitian hyperquadric in $\mathbb{C}\overline{H}^n$ is isometric to a equidistant hyperquadric in H^{2n} and is a fibration of this real equidistant hyperquadric with geodesic equidistant conics as fibers and the space $\mathbb{C}\overline{H}^{n-1}$ as base.

A horohypersphere in $\mathbb{C}\overline{H}^n$ is isometric to a horohypersphere in H^{2n} and is a fibration of this real horohypersphere with geodesic horocycles as fibers and the space $\mathbb{C}\overline{R}^{n-1}$ as base.

PROOF. Analogously to Theorem 4.39 we can prove that a Hermitian hypersphere in $\mathbb{C}\overline{S}^n$ or $\mathbb{C}\overline{H}^n$ is isometric to a Hermitian hypersphere in $\mathbb{C}\overline{R}^n$ which, as we have seen in **3.6.1**, is isometric to a hypersphere in R^{2n}. In **4.1.2** we have seen that this hypersphere is a fibration with its great circles as fibers and $\mathbb{C}\overline{S}^{n-1}$ as base. These great circles are cut out from a Hermitian hypersphere by lines in $\mathbb{C}\overline{S}^n$ or $\mathbb{C}\overline{H}^n$ through the center of the hypersphere.

The second and third assertions can be proved analogously. The geodesic equidistant conics on the equidistant hyperquadric are cut out from it by lines in $\mathbb{C}\overline{H}^n$ orthogonal to the basic hyperplane of this equidistant hyperquadric, that is, lines passing through the pole of this hyperplane.

The geodesic horocycles on the horohypersphere are cut out from it by lines in $\mathbb{C}\overline{H}^n$ through the point of tangency of the horohypersphere with the absolute hyperquadric of the space. Since a horohypersphere in H^{2n} is isometric to R^{2n-1}, geodesic horocycles on this horohypersphere are interpreted as straight lines in R^{2n-1}. Since these horocycles are cut out by complex lines through one point, the real straight lines in R^{2n-1} have no common points and, are therefore parallel, and the set of these parallel lines is isometric to R^{2n} and $\mathbb{C}\overline{R}^n$.

The complex structure in this R^{2n} follows from the form of the horocyclic turns on the horohypersphere in $\mathbb{C}\overline{H}^n$ which introduce translations on $\mathbb{C}\overline{R}^{n-1}$. A. Korányi [Kor] has proved that the horocyclic turn corresponding to the translation $x' = x+a$ is the motion with $((n+1) \times (n+1))$-matrix

$$A(a,t) = \begin{bmatrix} 1+\{a,t\} & i\bar{a} & \{a,t\} \\ -ia & I & -ia \\ -\{a,t\} & -i\bar{a} & 1-\{a,t\} \end{bmatrix}, \tag{4.53}$$

where $\{a,t\} = \frac{-t+\bar{a}a}{2}$.

Note that the horocyclic turns on a horohypersphere form a noncommutative group. For the horohypersphere tangent to the absolute hyperquadric of $\mathbb{C}H^n$ at the point represented by the vector $e_0 - e_n$, multiplication of the matrices $A(a,t)$ has the form

$$A(a,t)A(b,u) = A(a+b, t+u-2\ \mathrm{Im}\ \bar{a}b). \tag{4.54}$$

Therefore a hyperhorosphere in $\mathbb{C}\overline{H}^n$ is a fibration of a horohypersphere in H^{2n} with horocycles as fibers and $\mathbb{C}\overline{R}^{n-1}$ as base. ∎

4.7.6. Heisenberg Group. The group (4.54) is called the *Heisenberg group*. This group is noncommutative since (4.54) implies that

$$[A(a,t)A(b,u)] = A(a,t)B(b,u) - B(b,u)A(a,t) = -2(\bar{a}b - \bar{b}a). \tag{4.55}$$

The Heisenberg group is solvable since it has a commutative invariant subgroup isomorphic to the group of translations in $\mathbb{C}\overline{R}^{n-1}$. This group is related to the particular cases of solvable groups called *nilpotent groups.*

K. Iwasawa [Iw] has proved that a connected semisimple Lie group G admits a representation as a product $G = KAN$ of the maximal compact subgroup K, a commutative subgroup A isomorphic to the group R_s and a nilpotent group N. This representation is called the *Iwasawa decomposition.* In **1.4.15** we have seen that G is homeomorphic to the topological product of K and a Euclidean space, therefore N is also homeomorphic to a Euclidean space. For the group of motions in $\mathbb{C}\overline{H}^n$ the group K is the group $\mathbb{C}U_{n-1}$, $s = 1$, that is, $R_s = R$ and N is the Heisenberg group of $\mathbb{C}\overline{H}^n$.

This group is called the *Heisenberg group* since the commutation relations in the tangent Lie algebra of this group, which coincide with (4.55), are similar to W. Heisenberg's commutation relations in Quantum Mechanics which are the base for his Uncertainty Principle. In Quantum Mechanics the role of coordinates q_i and momenta p_i of particles or systems of particles is played by Hermitian symmetric linear operators Q_i and P_i acting in the infinite-dimensional complex linear space of wave functions and the commutation relations of these operators have the form

$$[Q_i P_j] = i\frac{h}{2\pi}\delta_{ij} I,$$

where I is the identity operator and h is the Planck constant.

Since the Heisenberg group can be interpreted as a one-transitive group of transformations in a hyperhorosphere in $\mathbb{C}\overline{H}^n$, it can also be interpreted as a one-transitive group of transformations in the absolute hyperquadric of this space without one point.

Analogously to Theorem 4.43 we can prove that a Hermitian hypersphere in $\mathbb{H}\overline{S}^n$ or $\mathbb{H}\overline{H}^n$ is isometric to a hypersphere Σ^{4n-1} in R^{4n} and is a fibration of this real hypersphere with great 3-spheres as fibers and the space $\mathbb{H}\overline{S}^{n-1}$ as base. We can also prove that an equidistant Hermitian hyperquadric in $\mathbb{H}\overline{H}^n$ is isometric to an equidistant hyperquadric in H^{4n} and is a fibration of this real equidistant hyperquadric with totally geodesic equidistant 3-quadrics as fibers and the space $\mathbb{H}\overline{H}^{n-1}$ as base, and that a horohypersphere in $\mathbb{C}\overline{H}^n$ is isometric to a hyperhorosphere in H^{4n} and is a fibration of this real hyperhorosphere with totally geodesic 3-horospheres as fibers and the space $\mathbb{C}\overline{R}^{n-1}$ as base.

In the last case there is also a generalization of the Heisenberg group, represented by the matrices (4.53) in which the pure imaginary number it must be replaced by a quaternion t satisfying the condition $\bar{t} = -t$ and in the rule (4.54) $2\,i$ Im $\bar{a}b$ must be replaced by $\overline{a}b - \overline{b}a$.

If, in (4.53), we put $t = 0$ we obtain a commutative group isomorphic to the group of horocyclic turns on a hyperhorosphere in H^n. Note that for the plane H^2 this group corresponds to the group of complex (2×2)-matrices $\begin{bmatrix} 1 & a \\ 0 & 1 \end{bmatrix}$ in the group $\mathbb{C}SL_2$, which is the spinor group for the group $SO_{3,1}$.

4.7.7. Classification of Conics in the Hyperbolic Plane. The classification of hyperquadrics $xAx = 0$ and Hermitian hyperquadrics $\bar{x}Ax = 0$ in quadratic or Hermitian spaces with absolute (Hermitian) hyperquadrics $xEx = 0$ or $\overline{x}Ex = 0$ can be reduced to the classification of the collineations $x' = AEx$, whose fixed points, as we have seen in Theorem 4.37, are centers of these hyperquadrics. Note that the matrices AE are equivalent to diagonal matrices if these hyperquadrics and absolute hyperquadrics do not touch and can be reduced to Jordan forms with Jordan cells if they touch. A horohypersphere is an example of a hyperquadric of the last type.

For the hyperbolic plane H^2 this method implies

Theorem 4.43 (H. LIEBMANN [Lib, p. 142]). *In the plane H^2 there are the following 12 types of conics:*

1) ellipse, a conic which meets the absolute conic in 4 imaginary points and has with it 4 common imaginary tangents (Fig. 4.3a),

2) convex hyperbola, a conic which meets the absolute conic in 4 real points and has with it 4 common imaginary tangents (Fig. 4.3b),

3) concave hyperbola, a conic which meets the absolute conic in 4 real points and has with it 4 common real tangents (Fig. 4.3c),

4) semihyperbola, a conic which meets the absolute conic in 2 real and 2 imaginary points and has with it 2 common real and 2 common imaginary tangents (Fig. 4.3d),

5) circle, a conic which meets the absolute conic in two couples of coincident imaginary points (Fig. 4.3e),

6) equidistant conic, a conic which meets the absolute conic in two couples of coincident real points (Fig. 4.3f),

7) elliptic parabola, a conic which touches the absolute conic at one point, meets it in 2 imaginary points, and has with it 2 common imaginary tangents (Fig. 4.3g),

8) convex hyperbolic parabola, a conic which touches the absolute conic at one point, meets it in 2 real points, and has with it 2 common imaginary tangents (Fig.4.3h),

9) concave hyperbolic parabola with one branch, a conic which touches the absolute conic at one point, meets it in 2 real points, has with it 2 common real tangents,

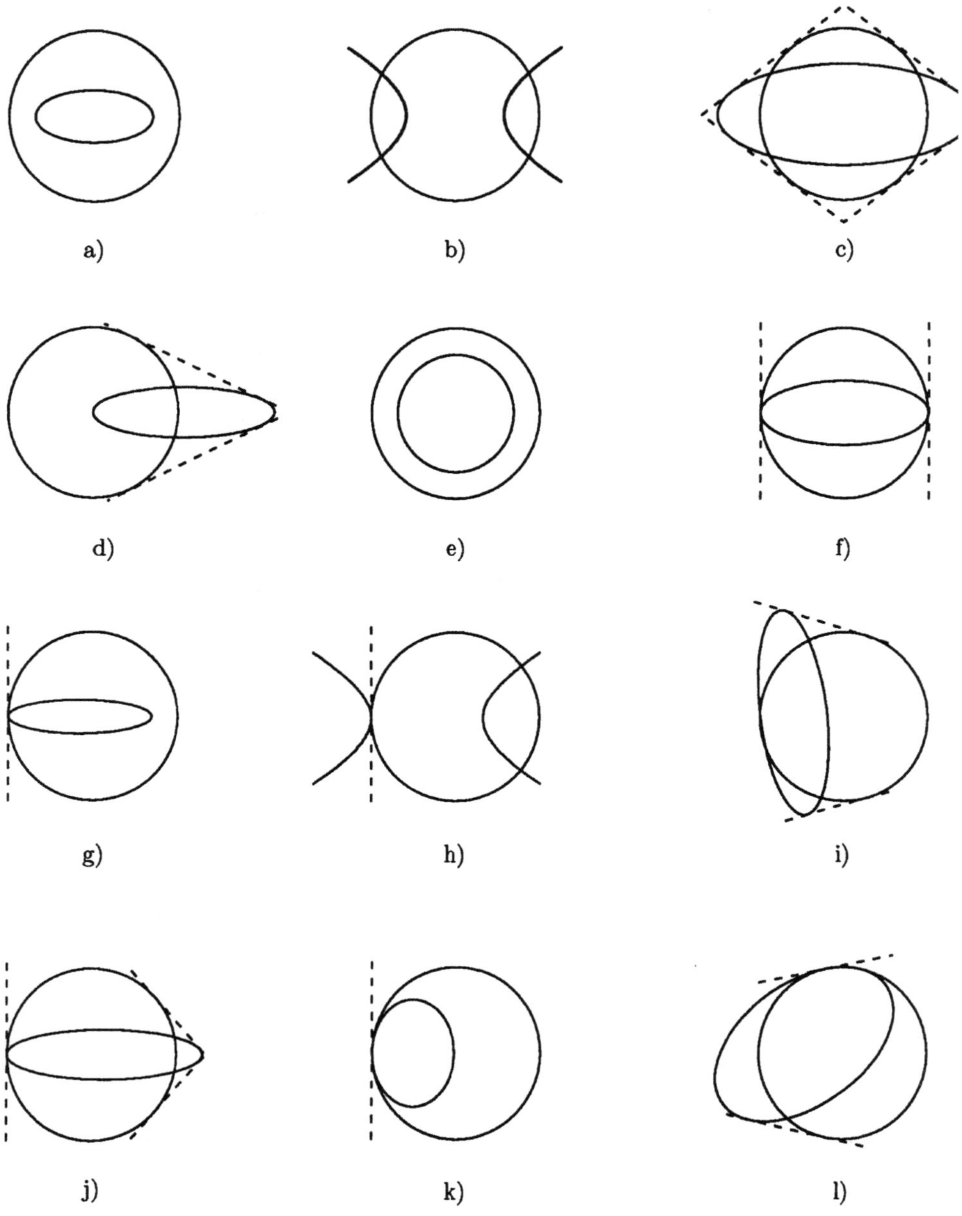

FIG. 4.3

and consists of one branch (Fig. 4.3i),

10) concave hyperbolic parabola with two branches, a conic which touches the absolute conic at one point, meets it in 2 real points, has with it 2 common real tangents, and consists of two branches (Fig. 4.3j),

11) horocycle, a conic which touches the absolute conic at one quadruple point (Fig. 4.3k),

12) osculating parabola, a conic which touches the absolute conic at one triple point, meets it in one point, and has with it one common tangent (Fig. 4.3l).

PROOF. The classification of conics $xAx = 0$ in H^2 with absolute conic $xEx = 0$ is based on the classification of collineations $x' = AEx$, which are products of polar transformations with respect to the conics $xAx = 0$ and $xEx = 0$. ■

The classification of hyperquadrics $xAx = 0$ in an n-space with absolute hyperquadric $xEx = 0$ is also based on the classification of collineations $x' = AEx$, which are products of polar transformations with respect to the hyperquadrics $xAx = 0$ and $xEx = 0$. We can analogously obtain the classification of Hermitian hyperquadrics $\overline{x}Ax = 0$ in an n-space with absolute Hermitian hyperquadric $\overline{x}Ex = 0$.

4.7.8. Topology of Manifolds of Generators of Maximal Dimension of Real Hyperquadrics. Formula (2.90) shows that the dimension of the manifold of $(n-1)$-planar generators of a real hyperquadric in the space P^{2n} or P^{2n+1}, that is, of the manifold of $(n-1)$-planar generators of the absolute hyperquadric in S^{2n}_n or S^{2n+1}_{n+1}, is equal to $\frac{n(n+1)}{2}$, that is, to the dimension of the group of motions in S^n. These coincidences are explained by

Theorem 4.44. *The manifold of $(n-1)$-planar generators of a hyperquadric in P^{2n} or P^{2n+1} is homeomorphic to the group of motions in S^n.*

PROOF. Consider n-equidistants in S^{2n} and S^{2n+1}, which are hyperquadrics with $(n-1)$-planar generators, the bases of the first n-equidistant being an $(n-1)$-plane and an n-plane, both bases of the second equidistant being n-planes. The $(n-1)$-planar generators of these n-equidistants are paratactic to their bases. Let us choose an autopolar $(n-1)$-simplex in a basic $(n-1)$-plane of the first n-equidistant and in one basic n-plane of the second one and drop from the vertices of this $(n-1)$-simplex the perpendiculars on each $(n-1)$-planar generator of the n-equidistant. These perpendiculars are common perpendiculars of the $(n-1)$-plane of the $(n-1)$-simplex and of the $(n-1)$-planar generator and therefore are mutually polar and meet the other basic n-plane of both n-equidistants in the vertices of an autopolar $(n-1)$-simplex. These vertices, together with the pole of its $(n-1)$-plane, are the vertices of an autopolar n-simplex in the basic n-plane. Thus we obtain a correspondence between the manifold of $(n-1)$-planar generators of a hyperquadric in P^{2n} or P^{2n+1} and the manifold of autopolar simplexes in S^n. This correspondence is a continuous bijection between autopolar n-simplexes in S^n with ordered vertices and $(n-1)$-planar generators of n-equidistants. However, the manifold of n-simplexes in S^n with ordered vertices is homeomorphic to the group of motions in S^n. ■

4.7.9. Triality Principle in Real Quadratic 7-Spaces. In **2.4.6** we have defined the *duality principle* in $\mathbb{A}P^n$. In **2.7.1** and **2.7.2** we have considered dual and autodual projective configurations in P^2 and P^3. There is an analogous duality principle in quadratic and Hermitian elliptic, hyperbolic, pseudoelliptic, and pseudohyperbolic spaces over algebras. In **4.5.1** we have considered dual and autodual formulas of trigonometry in S^2 and H^2.

Note that the complete duality principle holds only in the elliptic spaces $\mathbb{A}S^n$ and $\mathbb{A}\overline{S}^n$ and in the pseudoelliptic spaces $\mathbb{A}S_m^{2m-1}$ and $\mathbb{A}\overline{S}_m^{2m-1}$.

É. Cartan [Car11] has generalized the duality principle to homogeneous spaces whose group of transformations is a semisimple Lie group with two-fold symmetry of the fundamental roots, which is clearly seen from the symmetry of the Dynkin and Satake diagrams. For the groups of collineations in P^n and $\mathbb{C}P^n$ this symmetry can be seen in Fig. 1.3, A_n, for the group of collineations of $\mathbb{H}P^n$ it is in Fig. 1.9, AII.

Two-fold symmetry of the Dynkin diagrams can also be observed for groups of the classes D_n and E_6. The first of these symmetries corresponds to the equivalence of the families of imaginary or real $(n-1)$-planar generators of absolute hyperquadrics in the space S^{2n-1} or S_n^{2n-1}, the second of these symmetries corresponds to the projective duality principle in the plane $\mathbb{O}P^2$ which we will consider in **7.2.2**.

In the same paper [Car11], É. Cartan discovered also the ***triality principle*** for the group D_4, whose system of fundamental roots has three-fold symmetry, for the Dynkin diagram of this group (see Fig. 1.5). The fundamental roots $\alpha_1, \alpha_2, \alpha_3$, and α_4 of this group represent points on the absolute hyperquadric Q^6 of S^7 or S_4^7, as well as the rectilinear generators of Q^6 and 3-planar generators of both families of Q^6. The triality principle in S^7 or S_4^7 consists in the equivalence of points on Q^6 in S^7 or S_4^7 and of 3-planar generators of both families of Q^6.

Theorem 4.44 implies that the manifolds of 3-planar generators of Q^6 are homeomorphic to the group of motions in S^3, but in the **1.6.3** we have seen that the spinor group $\overline{SO}_4$ can be interpreted as the intersection of the hypercone (1.164) and a hypersphere (1.165) in R^8, and, therefore, the group of motions in S^3 itself is also homeomorphic to Q^6.

The triality principle in the spaces S^7 and S_4^7 can be clearly illustrated by means of the algebras $\mathbb{O}$ and $\mathbb{O}'$ of octonions and split octonions: if we consider in $\mathbb{O}$ the metric of R^8 and in $\mathbb{O}'$ the pseudometric of R_4^8 introduced in **1.3.1** and **1.3.2**, the spaces S^7 and S_4^7 can be regarded as the hyperspheres $|x| = 1$ in these spaces with identified antipodal points; the points on the imaginary and real absolute hyperquadrics in these spaces are represented by the rectilinear generators of the imaginary hypercone $|x| = 0$ in R^8, that is, by the bioctonions with zero modulus, respectively by the rectilinear generators of the real hypercone $|x| = 0$ in R_4^8, that is, by the split octonions with zero modulus; these bioctonions and split octonions with zero modulus are determined up to a real multiplier.

Theorem 4.45 (E. Weiss [Wes]). *The 3-planar generators of both families of the absolute hyperquadric of S^7 are determined by the equations*

$$ax = 0, \; xb = 0, \tag{4.56}$$

where a and b are bioctonions with zero modulus.

The two 3-planar generators of different families determined by bioctonions a and b with product $ab = c \neq 0$ intersect at the point determined by the bioctonion $x = \bar{c}$; if $ab = 0$, the 3-planar generators intersect at a 2-planar generator.

The two points determined by bioctonions x and y satisfying the condition

$$\overline{x}y + \overline{y}x = 0 \tag{4.57}$$

are on a rectilinear generator. If x and y do not satisfy this condition and are conjugate imaginary, these points are on a real line.

The two 3-planar generators of the same family and determined by bioctonions a and b satisfying the analogous condition

$$\overline{a}b + \overline{b}a = 0 \tag{4.58}$$

intersect in a rectilinear generator, while if a and b do not satisfy this condition and are conjugate imaginary, the conjugate imaginary points of these 3-planar generators are joined by lines of a paratactic congruence of lines.

All assertions of this theorem can be checked by simple calculations.

This theorem shows that the points on the absolute hyperquadric of S^7, as well as the 3-planar generators of it, are determined by bioctonions with zero modulus, and real lines in S^7 correspond in the triality principle to paratactic congruences determined by two conjugate imaginary 3-planar generators of the first or of the second family.

Similarly we can prove

Theorem 4.46. *The 3-planar generators of both families of the absolute hyperquadric of S^7_4 are determined by the equations (4.56), where a and b are split octonions with zero modulus.*

The two 3-planar generators of different families determined by split octonions a and b with product $ab = c \neq 0$ intersect at the point determined by the split octonion $x = \bar{c}$; if $ab = 0$, the 3-planar generators intersect at a 2-planar generator.

The two points determined by split octonions x and y satisfying the condition (4.57) are on a rectilinear generator. If x and y do not satisfy this condition, these points are on a hyperbolic real line.

The two 3-planar generators of the same family determined by split octonions a and b which satisfy the analogous condition (4.58) intersect at a rectilinear generator; if a and b do not satisfy this condition, the points of these 3-planar generators are joined by lines of a hyperbolic paratactic congruence.

4.7.10. Circles, Equidistant Conics, and Horocycles. Consider the most important kinds of conics: circles in S^2 and circles, equidistant conics, and horocycles in H^2.

Theorem 4.47. *Circles in S^2 and circles, equidistant conics, and horocycles in H^2 are trajectories of motions, orthogonal trajectories of bundles of lines, and curves of constant geodesic curvature.*

PROOF. Theorem 4.38 implies that circles in S^2 and H^2 are equidistant from their centers and, therefore, they are preserved under rotations around these points and are orthogonal to lines through these points.

The same theorem also implies that equidistant conics in H^2 are equidistant from their bases and, therefore, they are preserved under shifts along these lines and are orthogonal to perpendiculars to these lines; these lines are lines through an ideal point in H^2.

Since horocycles in H^2 can be obtained by a limiting process from circles and equidistant conics, they have analogous properties, a motion preserving such a curve is called a *horocyclic turn*; they are orthogonal to lines of a bundle of parallel lines; these lines are lines through the point on the absolute conic of H^2 which is the point of tangency of the horocycle and the absolute conic.

The centers of these bundles are fixed points of the motions preserving the circles, equidistant conics, and horocycles. Since these conics are preserved under motions in S^2 and H^2, they are curves of constant geodesic curvature. ■

Analogously, we can prove that hyperspheres in S^n and hyperspheres, equidistant hyperquadrics, and horohyperspheres in H^n are preserved under rotations around centers of hyperspheres and under motions preserving the basis hyperplane of an equidistant hyperquadric and the point of tangency of a horohypersphere with the absolute hyperquadric, and that these hyperquadrics are orthogonal to lines through the center of a hypersphere, through the pole of the basis hyperplane of an equidistant hyperconic, and through the point of tangency of a horohypersphere with the absolute hyperquadric.

Theorem 4.39 implies that the length of the arc of a circle in S^2 or H^2 corresponding to the central angle α is equal to $\alpha r \sin \frac{R}{r}$, respectively $\alpha q \sinh \frac{R}{q}$, and that the length of the arc of an equidistant conic in H^2 of parameter P corresponding to the segment with length a of its basic line is equal to $a \cosh \frac{P}{q}$.

4.7.11. The Quadrangles of Khayyam-Saccheri and Ibn al-Haytham-Lambert. In their attempts to prove the Fifth postulate of Euclid, Ibn al-Haytham in the 10-th century, Omar Khayyam in the 11-th century and G. Saccheri and J. H. Lambert in the 18-th century considered certain quadrangles and three hypotheses about some of their angles.

Khayyam and Saccheri considered quadrangles $ABCD$ with right angles A and B at one side and with equal sides AD and BC, and considered three hypotheses regarding their angles C and D: these angles can be acute, obtuse, or right (Fig. 4.4a). Ibn al-Haytham and Lambert considered quadrangles $ABCD$ with right angles A, B, and C and the same three hypotheses about the angle D (Fig. 4.4b). These three hypotheses are called the "hypothesis of the acute angle", the "hypothesis of the obtuse angle", and the "hypothesis of the right angle", respectively (see [Ro4, pp. 59-70 and 98-101]). In R^2 both types of quadrangles are rectangles, that is, in R^2 the hypothesis of the right angle is valid.

Theorem 4.48. *In the plane S^2, for the quadrangles of Khayyam-Saccheri*

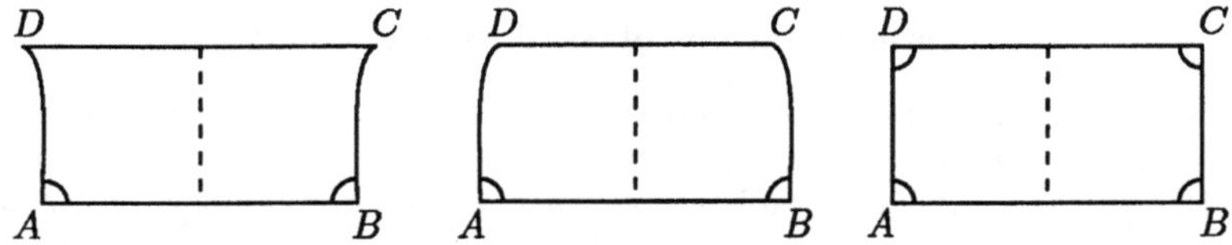

FIG. 4.4A

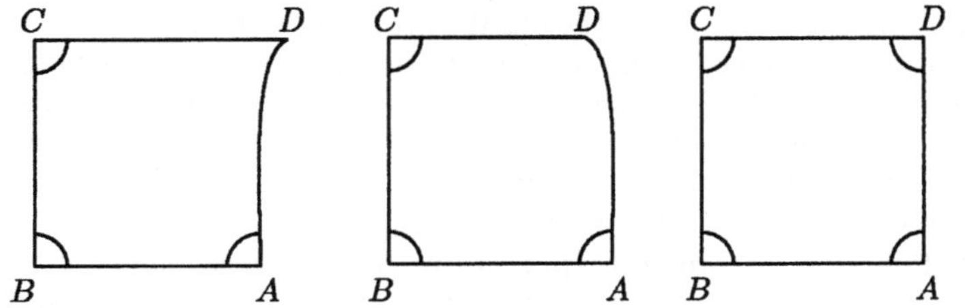

FIG. 4.4B

and Ibn al-Haytham-Lambert the hypothesis of the obtuse angle is valid.

In the plane H^2, the hypothesis of the acute angle is valid for these quadrangles.

In S^2, respectively H^2, the side CD is shorter, respectively longer, than AB.

PROOF. Since the quadrangles of Khayyam-Saccheri are symmetric with respect to the line joining the midpoints of the sides AB and CD, the angles C and D of these quadrangles are equal. From (3.64) and (3.66) it is clear that the sum of the angles of a triangle in S^2 or H^2 is more, respectively less, than π. Therefore, the sum of the angles of a quadrangle in these planes is more, respectively less, than 2π. Therefore, the angles C and D of the quadrangles of Khayyam-Saccheri and the angle D of the quadrangles of Ibn al-Haytham-Lambert are more, respectively less, than $\frac{\pi}{2}$.

If the vertices C and D of a quadrangle $ABCD$ of Khayyam-Saccheri with sides $AB = a$ and $AD = BC = P$ are joined in S^2 by the arc of the circle with center at the pole of AB and in H^2 by the arc of the equidistant conic with basis line AB, the lengths of these arcs are equal to $a \cos \frac{P}{r}$, respectively $a \cosh \frac{P}{r}$. Therefore, in S^2 the side CD is shorter than the arc CD, which is shorter than a, therefore it is shorter than a. In H^2 the side CD is also shorter than the arc CD, but since the arc CD is longer than AB, the side CD is also longer than AB: this can be proved by the construction of the quadrangle $ABC'D'$ symmetric for $ABCD$ with respect to the line AB. ■

This theorem implies that in S^2 two lines converge on both sides from their common perpendicular which is the polar line of the point of their intersection, and in H^2 two lines diverge on both sides from their common perpendicular which is the polar line of the ideal point of their intersection (in H^2 parallel lines meet at a point on the absolute conic).

§4.8. Interpretations of Skopets and Popović

4.8.1. Skopets' First Interpretation. Skopets' first interpretation of the space H^3 is equivalent to the interpretation of the pseudoelliptic space S_1^3 on the plane R^2 by circles, which, as we have seen in **4.4.6**, interprets points in S_1^3 and, therefore, planes in H^3. Z. A. Skopets [Sk1] obtained his interpretation from a horosphere in H^3, which is isometric to a plane R^2 (see Theorem 4.41). Any plane in H^3 is interpreted on this horosphere by the circle consisting of the points such that all lines through them orthogonal to the horosphere are parallel to this plane. Therefore, this circle can be regarded as the projection onto the horosphere of the intersection of this plane with the absolute quadric of H^3 from the point of tangency of this quadric with this horosphere. Since horospheres are isometric to R^2, the circles on a horosphere interpreting planes in H^3 can be regarded as circles in R^2. Skopets proved that the angle between two of these circles in R^2 is equal to the angle between the planes in H^3 which are interprete by these circles.

4.8.2. Skopets' Second Interpretation. An analogous map of planes in H^3 to a single plane Π in this space is proposed in Skopets' second interpretation of H^3 on a plane H^2 [Sk2]. Any plane in H^3 can be interpreted by a circle, an equidistant conic, or a horocycle in the plane Π in this space consisting of the bases of the perpendiculars to the plane Π that are parallel to the mapped plane. Therefore, these circles, equidistant conics, and horocycles can be regarded as projections of the intersection of the mapped plane with the absolute quadric of H^3 from the pole of the plane Π onto this plane. In this interpretation, circles in the plane Π interpret planes which diverge with Π (the centers of the circles are the bases of common perpendiculars of Π and the mapped plane), equidistant conics in the Π interpret planes intersecting Π (the bases of the equidistant conics are the lines of intersections of Π with the mapped plane), horocycles in Π interpret planes parallel to Π (the point of tangency of a horocycle with the absolute conic of H^2 is the common point of the line of intersection of Π and the mapped plane with the absolute quadric of H^3).

Both interpretations of Skopets can be generalized to interpretations of the space H^n in R^{n-1} and H^{n-1}.

4.8.3. Interpretation of Popović. The interpretation of Milanka Popović [Pop] is analogous to the interpretation of hyperspheres in R^n in the space S_1^{n+1} considered in **4.4.6**. By Theorems 4.22 and 4.23 (see **4.4.6**) the metric of the space S_1^{n+1} of curvature $+1$ can be introduced also in the manifold of hyperspheres, hyperplanes, equidistant hyperquadrics, and horohyperspheres in H^n, if for the distance between two such hypersurfaces we take the angle between them. M. Popović proved that points, that is, hyperspheres of radius 0, in H^n are interpreted as points of the absolute hyperquadric of S_1^{n+1}, the absolute hyperquadric of H^n is interpreted as a point P in S_1^{n+1} outside its absolute hyperquadric, hyperplanes in H^n are interpreted as points in a hyperplane in S_1^{n+1} polar to P, horohyperspheres in H^n are interpreted as points on a hypercone in S_1^{n+1} with apex at P and tangent to the absolute hyperquadric of S_1^{n+1}, hyperspheres of real radius and elliptic hyperplanes in H^n are interpreted as points in S_1^{n+1} outside this hypercone, equidistant hyperquadrics in H^n are interpreted as points in S_1^{n+1} inside this hypercone, but outside the absolute hyperquadric, hyperspheres of imaginary radius and hyperbolic hyperplanes in H^n are interpreted as points in S_1^{n+1} inside the absolute hyperquadric, isotropic hyperplanes in H^n (that is, hyperplanes tangent the absolute hyperquadric) are interpreted as points of tangency of this hypercone and the absolute hyperquadric of S_1^{n+1}.

§4.9. Regular Polyhedra and Honeycombs

4.9.1. Regular Polyhedra and Honeycombs in Elliptic Space. In the elliptic space S^n, like in R^n, we can define regular n-polyhedra and n-honeycombs. However, unlike in R^n, regular n-honeycombs in S^n consist of a finite number of cells. These n-honeycombs can be defined by means of regular $(n+1)$-polyhedra in R^{n+1}: any regular $(n+1)$-polyhedron in R^{n+1} can be inscribed in a hypersphere, which can be transformed into the space S^n by identification of its antipodal points. Projection of the faces of a regular $(n+1)$-polyhedron in R^{n+1} from the center of the hypersphere circumscribed around it determines an n-honeycomb in S^b if this projection is preserved under reflection in the center of the hypersphere.

The regular n-honeycombs in S^n are denoted by the same symbols $\{p_1, p_2, \ldots, p_n\}$ as the corresponding regular $(n+1)$-polyhedra in R^{n+1}.

Theorem 4.49 (Coxeter [Cox6]). *In S^2 there are four regular 2-honeycombs, $\{3,4\}, \{4,3\}, \{3,5\}$, and $\{5,3\}$. They consist of 4 autopolar triangles, 3 squares, 10 equilateral triangles and 6 regular pentagons, respectively.*

In S^3 there are five regular 3-honeycombs $\{3,3,4\}$, $\{4,3,3\}$, $\{3,4,3\}$, $\{3,3,5\}$, and $\{5,3,3\}$. They consist of 8 autopolar tetrahedra, 6 cubes, 12 regular octahedra, 300 regular tetrahedra, and 60 regular dodecahedra, respectively.

In S^n for $n > 3$ there are two regular n-honeycombs $\{3,3,\ldots,3,4\}$ and $\{4,3,\ldots,3,3\}$. They consist of the 2^n autopolar n-simplexes and the $3 \cdot 2^{n-1}$ n-cubes, respectively.

Proof. Among the 5 regular polyhedra in R^3 only an octahedron $\{3,4\}$, a cube $\{4,3\}$, an icosahedron $\{3,5\}$, and a dodecahedron $\{5,3\}$ are symmetric with respect to the center of the circumscribed sphere. Therefore, in S^2 there are 4 regular 2-honeycombs with the same notations.

Among the 6 regular 4-polyhedra in R^4, determined by the Theorem of Schläfli (see **3.4.2**), only a 4-cocube $\{3,3,4\}$, a 4-cube $\{4,3,3\}$, a bicube $\{3,4,3\}$, a 600-hedron $\{3,3,5\}$, and a 120-hedron $\{5,3,3\}$ are symmetric with respect to the center of the circumscribed hypersphere. Therefore, in S^3 there are 5 regular 3-honeycombs, with the same notations.

Among the 3 regular n-polyhedra of R^n with $n > 4$, determined by the Theorem of Schläfli (see **3.4.2**), only an n-cocube $\{3,3,\ldots,3,4\}$, and an n-cube $\{4,3,\ldots,3,3\}$, are symmetric with respect to the center of the circumscribed hypersphere. Therefore, in S^n for $n > 3$ there are two regular n-honeycombs, with the same notations. ■

4.9.2. Regular Honeycombs in the Hyperbolic Plane. In H^n, like in R^n and S^n, we can define regular polyhedra and honeycombs.

Note that for $n = 3$ the determinant in (3.27) vanishes if $\cos^2\frac{\pi}{p} + \cos^2\frac{\pi}{q} = 1$, that is, $\frac{\pi}{p} + \frac{\pi}{q} = \frac{\pi}{2}$ or

$$\frac{1}{p} + \frac{1}{q} = \frac{1}{2} \tag{4.59}$$

(in **3.4.3** we have seen that the p and q satisfying this condition are $p = q = 4$, $p = 3$ and $q = 6$, $p = 6$ and $q = 3$). Therefore, for $n = 3$ the inequality (3.27) can be written as $\cos^2 \frac{\pi}{p} + \cos^2 \frac{\pi}{q} < 1$, or

$$\frac{1}{p} + \frac{1}{q} > \frac{1}{2}. \tag{4.60}$$

(in **3.4.2** and **4.9.1** we have seen that the p and q satisfying this condition are $p = q = 3$, $p = 3$ and $q = 4$ or 5, $p = 4$ or 5 and $q = 3$).

Theorem 4.50. *In H^2 there are regular 2-honeycombs $\{p, q\}$. They consist of regular p-gons with q p-gons glued at each vertex for all values of p and q satisfying the inequality*

$$\frac{1}{p} + \frac{1}{q} < \frac{1}{2}. \tag{4.61}$$

PROOF. Join two neighboring vertices A and B of a p-gon of a regular 2-honeycomb in H^2 with the center O of the circle circumscribed around this p-gon. In the triangle ABO the angle at O is equal to $\frac{2\pi}{p}$ and the angles at A and B are equal to $\frac{\pi}{q}$. Therefore, the sum of the angles of this triangle is equal to $\frac{2\pi}{p} + \frac{2\pi}{q}$. Since the sum of the angles of a triangle in H^2 is less than π, we obtain inequality (4.61). This condition is sufficient for the existence of regular 2-honeycombs $\{p, q\}$ since the plane H^2 is infinite and the covering of p-gons (like the covering of the triangle 2-honeycomb $\{3, 3\}$ in S^2) in H^2 is impossible. ■

Note that similar proofs based on the sum of the angles of plane triangles in R^2, spherical triangles in R^3, and triangles in S^2 are valid for the equality (4.59) for regular honeycombs in R^2, and for the inequality (4.60) for regular polyhedra in R^3 and regular honeycombs in S^2.

In H^2 there are also regular 2-honeycombs $\{p, \infty\}$ (for them condition (4.61) is fulfilled for $p \geq 3$). A 2-honeycomb $\{3, \infty\}$ consists of triangles all vertices of which are on the absolute conics, and all angles of these triangles being equal to 0; this honeycomb can be obtained by reflection of such a triangle in its sides and by analogous reflections of the resulting triangles. The 2-honeycombs $\{p, \infty\}$ consist of regular p-gons all vertices of which are on the absolute conic, all angles of these p-gons also being equal to 0; these 2-honeycombs can be obtained by reflection of such a p-gon in its sides and by analogous reflections of the resulting p-gons.

The plane H^2 can be partitioned also by congruent p-gons with unequal angles $\frac{\pi}{n_i}$, where the n_i are arbitrary integers ≥ 2 satisfying the condition $\frac{1}{n_1} + \frac{1}{n_2} + \cdots + \frac{1}{n_p} < p - 2$.

4.9.3. Regular Polyhedra and Honeycombs in Hyperbolic Space. In H^n there are three kinds of regular polyhedra: *elliptic regular polyhedra*, which can be inscribed in hyperspheres, *parabolic regular polyhedra*, which can be inscribed in horohyperspheres, and *hyperbolic regular polyhedra* which can be inscribed in equidistant hyperquadrics. Elliptic regular n-polyhedra in H^n have the same shape forms as regular n-polyhedra in R^n; the vertices of such an n-polyhedron in the Beltrami interpretation inside the hypersphere in R^n, with as center the center of the hypersphere circumscribed around this polyhedron, are interpreted as the vertices of a regular n-polyhedron in R^n inscribed in a hypersphere concentric with the hypersphere which interprets the absolute hyperquadric of H^n. These regular n-polyhedra in H^n and R^n are determined by the same Schläfli symbols $\{p_1, p_2, \ldots, p_{n-1}\}$.

If a regular n-polyhedron in R^n is inscribed in a hypersphere which interprets the absolute hyperquadric of H^n, we obtain an elliptic regular n-polyhedron all vertices of which are at infinity and all angles at the vertices of this polyhedron are equal to 0. In the general case the angles at the vertices of an elliptic regular n-polyhedron in H^n are less than the angles at the vertices of the corresponding regular n-polyhedron in R^n.

Vertices of parabolic regular n-polyhedra in H^n are such points of horohyperspheres in this space which under the isometric map of this horohypersphere to R^{n-1} are interpreted as vertices of cells of a regular $(n-1)$-honeycomb in R^{n-1}.

The vertices of a hyperbolic regular n-polyhedron in H^n are points on two sheets of equidistant hyperquadrics in this space located symmetrically with respect to its basic hyperplane such that under an isometric map of these sheets to H^{n-1} they are interpreted as vertices of a cell of a regular $(n-1)$-honeycomb in H^{n-1}. **3.4.2**, **3.4.3** and Theorem 4.46 imply

Theorem 4.51. *In H^3 there are five types of elliptic regular 3-polyhedra analogous to the regular 3-polyhedra in R^3: a tetrahedron $\{3,3\}$, a cube $\{4,3\}$, an octahedron $\{3,4\}$, a dodecahedron $\{5,3\}$, and an icosahedron $\{3,5\}$, there are three types of parabolic regular 3-polyhedra determined by regular 2-honeycombs in R^2: the regular polyhedra $\{4,4\}, \{6,3\}$, and $\{3,6\}$, and there are many types of hyperbolic regular 3-polyhedra $\{p,q\}$, where p and q satisfy the condition (4.61), determined by regular 2-honeycombs in H^2.*

Theorem of H.S.M.Coxeter [Cox7]. *In H^3 there are 15 types of regular honeycombs: $\{3,5,3\}$, $\{4,3,5\}$, $\{5,3,4\}$, $\{5,3,5\}$, $\{3,3,6\}$, $\{6,3,3\}$, $\{3,4,4\}$, $\{4,4,3\}$, $\{3,6,3\}$, $\{4,3,6\}$, $\{6,3,4\}$, $\{4,4,4\}$, $\{5,3,6\}$, $\{6,3,5\}$, and $\{6,3,6\}$.*

In H^4 there are 7 types of regular honeycombs: $\{3,3,3,5\}$, $\{5,3,3,3\}$, $\{4,3,3,5\}$, $\{5,3,3,4\}$, $\{5,3,3,5\}$, $\{3,4,3,4\}$, and $\{4,3,4,3\}$.

In H^5 there are 5 types of regular honeycombs: $\{3,3,3,4,3\}$, $\{3,4,3,3,3\}$, $\{3,3,4,3,3\}$, $\{3,4,3,3,4\}$, and $\{4,3,3,4,3\}$.

In H^n for $n > 5$ there are no regular honeycombs.

The Schläfli symbols for regular honeycombs in H^3, H^4, and H^5 show that in H^3 there are regular 3-honeycombs consisting of all 5 types of elliptic regular 3-polyhedra and of 2 types of parabolic regular 3-polyhedra, in H^4 there are regular 4-honeycombs consisting of 4 types of elliptic regular 4-polyhedra, 4-simplexes, 4-cubes, bicubes, and 120-hedra, and of one type of parabolic regular 4-polyhedra, and in H^5 there are regular 5-honeycombs consisting of 5-cocubes and of all 3 types of parabolic regular 5-polyhedra.

4.9.4. Fuchsian Groups. Regular 2-honeycombs in H^2 are related with *Fuchsian groups*, that is, discrete subgroups of the group of motions in H^2 preserving the orientation of figures in this plane.

Fuchsian groups in H^2 can be represented by matrices in the group $SO_{3,1}$ and, since the spinor group of this group is the group SL_2, Fuchsian groups in H^2 can also be represented by matrices in SL_2. These matrices can be obtained as follows: in **4.4.2** we have defined a Poincaré interpretation of H^n in the halfspace of R^n. For $n = 2$ this interpretation is an interpretation of H^2 in the halfplane of R^2. If this halfplane is regarded as the upper halfplane of the complex plane bounded by the real axis, then motions in H^2 are represented by fractional linear transformations of the complex plane with real coefficients; the matrices of these transformations belong to SL_2.

Examples of Fuchsian groups are the *modular group of F. Klein*, that is, the group of fractional linear transformations in the complex plane with integer coefficients and determinant 1, the group of fractional linear transformations in the complex plane with rational coefficients and determinant 1, and, more general, *arithmetic Fuchsian groups*, determined by subalgebras of the algebra $\mathbb{Q}_2$ of (2×2)-matrices with rational entries and of analogous algebras of (2×2)-matrices whose entries are elements of algebraic number fields.

Fuchsian groups were investigated by H. Poincaré [Poi1]. For Fuchsian groups see the book by Svetlana Katok [Kat].

4.9.5. Kleinian Groups. Regular 3-honeycombs in H^3 are related with *Kleinian groups*, that is, discrete subgroups of the group of motions in H^3 preserving the orientation of figures in this space. Therefore, Kleinian groups in H^3 can be represented by matrices of the group $SO_{4,1}$ and, since the spinor group of this group is the group $\mathbb{C}SL_2$, Kleinian groups in H^3 can also be represented by matrices in $\mathbb{C}SL_2$. These matrices can be obtained as follows: in **3.8.2** we have defined an interpretation of the conformal space C^n on an oval hyperquadric in P^{n+1}, which can be regarded as the absolute hyperquadric of H^{n+1}. In this interpretation collineations preserving this hyperquadric, that is, motions in H^{n+1} are represented by conformal transformations in C^n. For $n = 2$ this interpretation is an interpretation of H^3 in the plane C^2 or in the complex plane. In this interpretation motions in H^3 are represented by fractional linear transformations in the complex plane, whose matrices belong to matrices in $\mathbb{C}SL_2$.

Kleinian groups were investigated by H. Poincaré [Poi2], see also [Kl3].

4.9.6. Groups of Motions of Hyperbolic Spaces Generated by Reflections. In H^n there are also groups of motions generated by reflections in hyperplanes. These groups are characterized by the *hyperbolic Coxeter diagrams*, analogous to the Coxeter diagrams in R^n (see **1.4.7**). Fig. 4.5 presents the hyperbolic Coxeter diagrams of the groups of motions in H^n generated by reflections in the faces of simplexes with dihedral angles equal to $\frac{\pi}{n}$; these groups were found by F. Lanner [Lan] and are called *Lanner groups*. Fig. 4.6a-d present the hyperbolic Coxeter diagrams of the groups of motions in H^n generated by reflections in the faces of unbounded simplexes with finite volume all bihedral angles of which, except one, are equal to $\frac{\pi}{n}$ while one pair of faces is parallel. These groups are called *quasi-Lanner groups*.

The Lanner and quasi-Lanner groups coincide with the groups of symmetries of the regular n-honeycombs in H^n whose cells or vertex n-polyhedra are parabolic regular n-polyhedra. For instance, the Lanner group generated by reflections in the faces of one of 9 tetrahedra in H^3 found by Lanner, whose bihedral angles at the edges AC, AD, and BC are equal to $\frac{\pi}{2}$, the bihedral angle at the edge BD is equal to $\frac{\pi}{3}$, and the bihedral angles at the edges AB and CD are equal to $\frac{\pi}{5}$, coincides with the group of symmetries of the regular 3-honeycomb $\{5,2,5\}$.

Note that the groups of symmetries of the regular n-honeycombs in H^n correspond to the Coxeter diagrams of the Lanner and quasi-Lanner groups that do not contain bifurcations and cycles. The Lanner groups exist only in H^n for $n \le 4$, and quasi-Lanner groups exist only in H^n for $n \le 9$.

§4.10. Symmetry and Parabolic Figures

4.10.1. Symmetry Figures. Since motions in $\mathbb{A}S^n$, $\mathbb{A}H^n$, $\mathbb{A}S^n_\ell$, $\mathbb{A}H^n_\ell$, $\mathbb{A}\overline{S}^n$, $\mathbb{A}\overline{H}^n$, $\mathbb{A}\overline{S}^n_\ell$, and $\mathbb{A}\overline{H}^n_\ell$ are particular cases of collineations in $\mathbb{A}P^n$, symmetry figures in $\mathbb{A}S^n$, $\mathbb{A}H^n$, $\mathbb{A}S^n_\ell$, $\mathbb{A}H^n_\ell$, $\mathbb{A}\overline{S}^n$, $\mathbb{A}\overline{H}^n$, $\mathbb{A}\overline{S}^n_\ell$, and $\mathbb{A}\overline{H}^n_\ell$ are particular cases of symmetry figures in $\mathbb{A}P^n$.

Since reflections in 0-pairs and m-pairs in $\mathbb{A}P^n$ are motions in $\mathbb{A}S^n$, $\mathbb{A}H^n$, $\mathbb{A}S^n_\ell$, $\mathbb{A}H^n_\ell$, $\mathbb{A}\overline{S}^n$, $\mathbb{A}\overline{H}^n$, $\mathbb{A}\overline{S}^n_\ell$, and $\mathbb{A}\overline{H}^n_\ell$ only if points and hyperplanes of 0-pairs, and m-planes and $(n-m-1)$-planes of m-pairs, are mutually polar with respect to the absolute (Hermitian) hyperquadrics in these spaces, points in these spaces are symmetry figures together with their polar hyperplanes, and m-planes in these spaces are symmetry figures together with their polar $(n-m-1)$-planes.

Symmetries in n-chains in complex and split complex spaces and complex and split complex n-chains in quaternionic and split quaternionic spaces are motions in these spaces only if these n-chains are *normal*, that is, are obtained from n-chains $'x^i = \bar{x}^i$, $'x^i = i^{-1}x^i i$, and $'x^i = ex^i e$, respectively, by motions in these spaces.

Symmetries in linear congruences of lines in S^{2n+1}, $\mathbb{C}S^{2n+1}$, and $\mathbb{C}'S^{2n+1}$, and symmetries in Hermitian linear congruences of lines in $\mathbb{C}\overline{S}^{2n+1}$ and $\mathbb{C}'\overline{S}^{2n+1}$, are motions in these spaces only if these congruences consist of paratactic lines, that is, are *paratactic congruences of lines.*

Theorem 2.61 implies

Theorem 4.52. *The symmetry figures in S^n, $\mathbb{C}S^n$, $\mathbb{C}\overline{S}^n$, and $\mathbb{H}\overline{S}^n$ are points with their polar hyperplanes, lines with their polar $(n-2)$-planes, m-planes with their polar $(n-m-1)$-planes $\left(m = 2, 3, \ldots, \left[\frac{n-1}{2}\right]\right)$; for odd n in S^n, $\mathbb{C}S^n$, and $\mathbb{C}\overline{S}^n$ we have to add paratactic congruences of lines, and in $\mathbb{C}\overline{S}^n$ and $\mathbb{H}\overline{S}^n$, normal n-chains and complex n-chains, respectively.*

A reflection in a point, line, or m-plane in these spaces can be reduced to the form (2.118), a reflection in a paratactic congruence of lines in S^{2n+1} or $\mathbb{C}\overline{S}^{2n+1}$ can be reduced to the form (2.6), respectively (2.13), a reflection in a normal n-chain in $\mathbb{C}\overline{S}^n$ can be reduced to the form (2.119), a reflection in a normal complex n-chain in $\mathbb{H}\overline{S}^n$ can be reduced to the form (2.120).

Analogously we can define symmetry figures in the spaces H^n, S^n_ℓ, H^n_ℓ, $\mathbb{C}'S^n$, $\mathbb{C}H^n$, $\mathbb{C}S^n_\ell$, $\mathbb{C}H^n_\ell$, $\mathbb{C}\overline{H}^n$, $\mathbb{C}\overline{S}^n_\ell$, $\mathbb{C}\overline{H}^n_\ell$, $\mathbb{C}'\overline{S}^n$, $\mathbb{C}^0S^n$, $\mathbb{C}^0H^n$, $\mathbb{C}^0S^n_\ell$, $\mathbb{C}^0H^n_\ell$, $\mathbb{C}^0\overline{S}^n$, $\mathbb{C}^0\overline{H}^n$, $\mathbb{C}^0\overline{S}^n_\ell$, $\mathbb{C}^0\overline{H}^n_\ell$, $\mathbb{H}\overline{H}^n$, $\mathbb{H}\overline{S}^n_\ell$, $\mathbb{H}\overline{H}^n_\ell$, $\mathbb{H}'\overline{S}^n$, $\mathbb{H}^0\overline{S}^n$, $\mathbb{H}^0\overline{H}^n$, $\mathbb{H}^0\overline{S}^n_\ell$, $\mathbb{H}^0\overline{H}^n_\ell$, and $\mathbb{H}'^0\overline{S}^n$.

Since the space $\mathbb{C}'\overline{S}^n$ admits an interpretation in P^n and the fundamental groups of these spaces are isomorphic,the symmetry figures in each of these spaces can

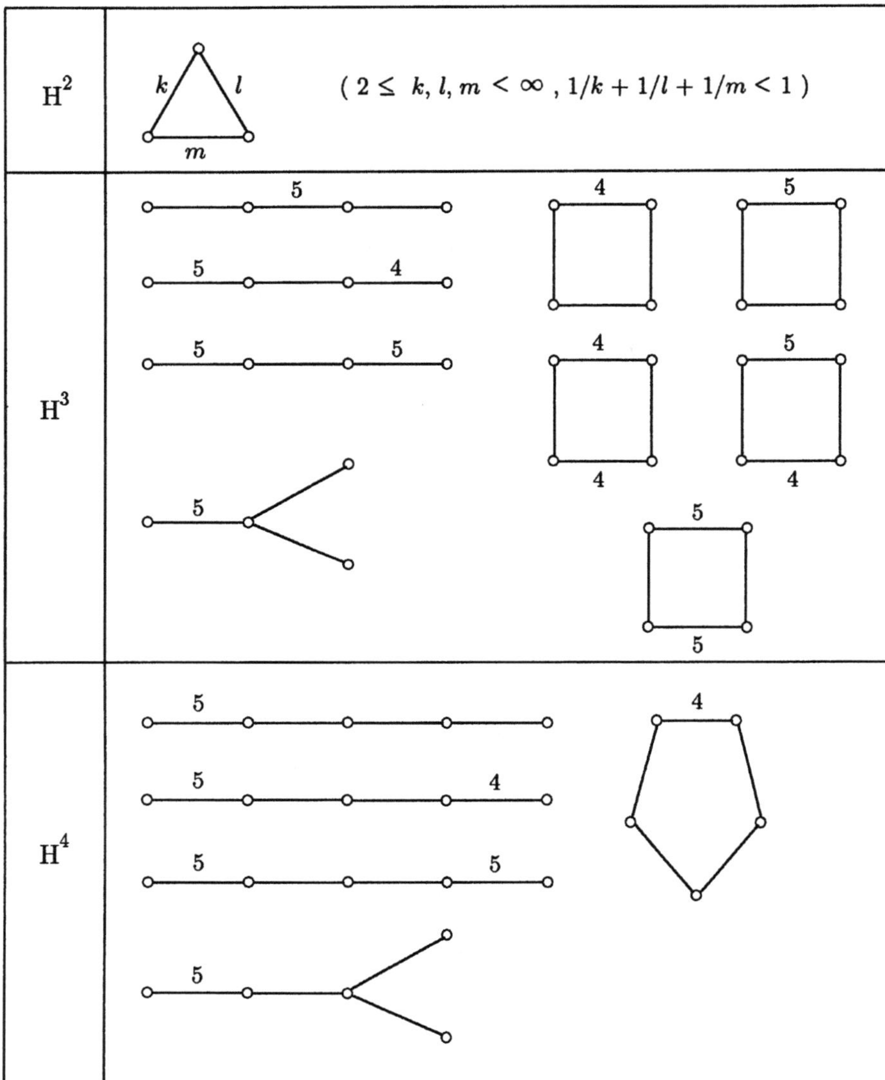

FIG. 4.5

be interpreted as the symmetry figures in the other space. In particular, points, lines, and m-planes in $\mathbb{C}'\overline{S}^n$ are interpreted as 0-pairs, 1-pairs, and m-pairs in P^n, respectively, pairs of conjugate imaginary $\frac{n-1}{2}$-planes on the absolute Hermitian hyperquadric in $\mathbb{C}'\overline{S}^n$ can be interpreted as elliptic linear congruences in P^n,

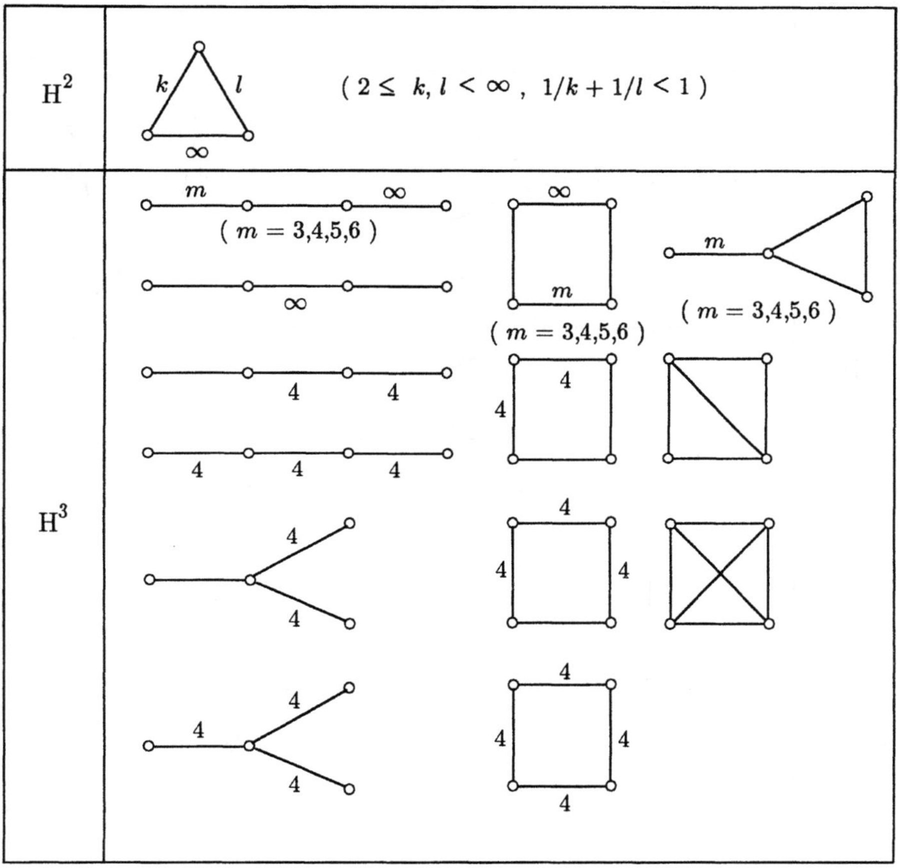

FIG. 4.6A

normal n-chains in $\mathbb{C}'\overline{S}^n$ can be interpreted as imaginary hyperquadrics in P^n, pseudonormal n-chains in $\mathbb{C}'\overline{S}^n$, reflections in which can be reduced to a product of the transformations (2.118) and (2.119), can be interpreted as real hyperquadrics in P^n of various indexes, Hermitian linear congruences of lines in $\mathbb{C}'\overline{S}^n$ can be interpreted as linear complexes of lines in P^n.

4.10.2. Absolutes and Superabsolutes of Symmetric Spaces. In **2.5.10** we have mentioned that Cartan called the Veronesean Υ_n in P^n the *absolute* of the symmetric space represented by a variety in P^n; this absolute can be represented by degenerate hyperquadrics which are double hyperplanes. Therefore we call the variety of all points in P^n representing other degenerate hyperquadrics in P^n the *superabsolute* of this symmetric space.

For Riemannian and pseudo-Riemannian symmetric spaces, represented by m-planes in H^n, S^n_ℓ, and H^n_ℓ and in analogous spaces over algebras, the role of abso-

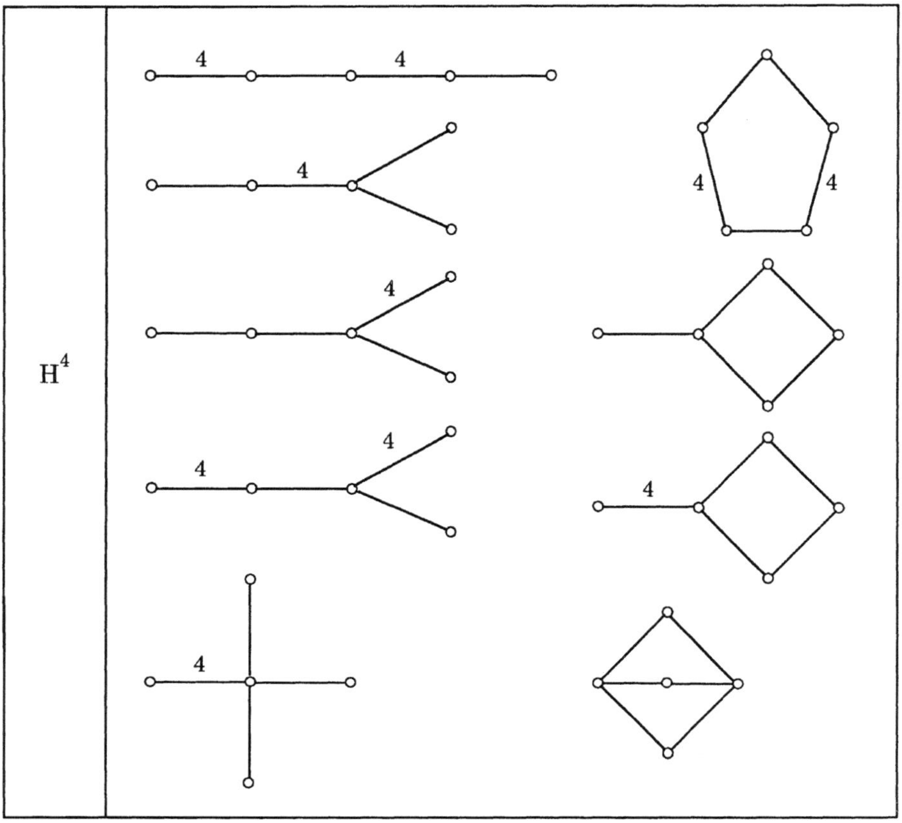

FIG. 4.6B

lutes is played by varieties representing manifolds of m-planar generators of absolute hyperquadrics of these spaces and the role of superabsolutes is played by varieties representing manifolds of m-planes tangent to these absolute hyperquadrics.

In the Riemannian and pseudo-Riemannian symmetric spaces V^n and V^n_ℓ, next to absolutes and superabsolutes there are *local absolutes* and *local superabsolutes*. If these spaces are not of constant curvature, the stabilizers of their points are subgroups of the groups O_n and $O_{n,\ell}$ of rotations in the spaces R^n and R^n_ℓ tangent to these spaces. These subgroups leave invariant cones in R^n or R^n_ℓ. A variety in the hyperplane at infinity of R^n or R^n_ℓ such that the rotations in such a space leaving invariant this variety coincide with the rotations in stabilizers of points in the corresponding symmetric space is called a *local absolute*. Complete varieties in these hyperplanes cut out by invariant cones are called *local superabsolutes* of the symmetric spaces.

Since the tangent spaces of the spaces $\mathbb{C}\overline{S}^n$ and $\mathbb{C}'\overline{S}^n$ are $\mathbb{C}\overline{R}^n$, respectively

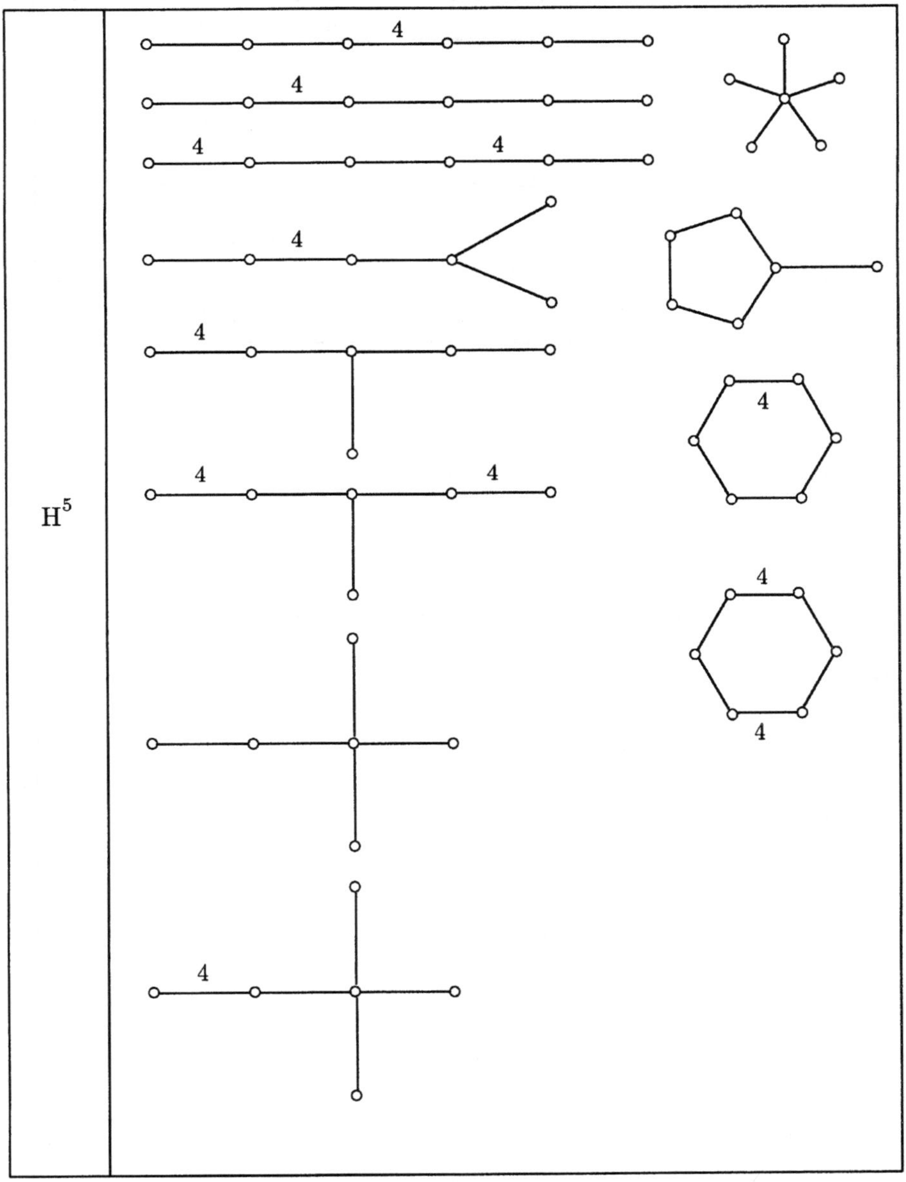

FIG. 4.6C

$\mathbb{C}'\overline{R}^n$, the rectilinear generators of invariant cones in the tangent spaces of the symmetric spaces V^{2n} and V_n^{2n}, which are isometric to $\mathbb{C}\overline{S}^n$ and $\mathbb{C}'\overline{S}^n$, coincide with real lines with singular direction vectors in $\mathbb{C}\overline{R}^n$ and $\mathbb{C}'\overline{R}^n$ and, by Theorem

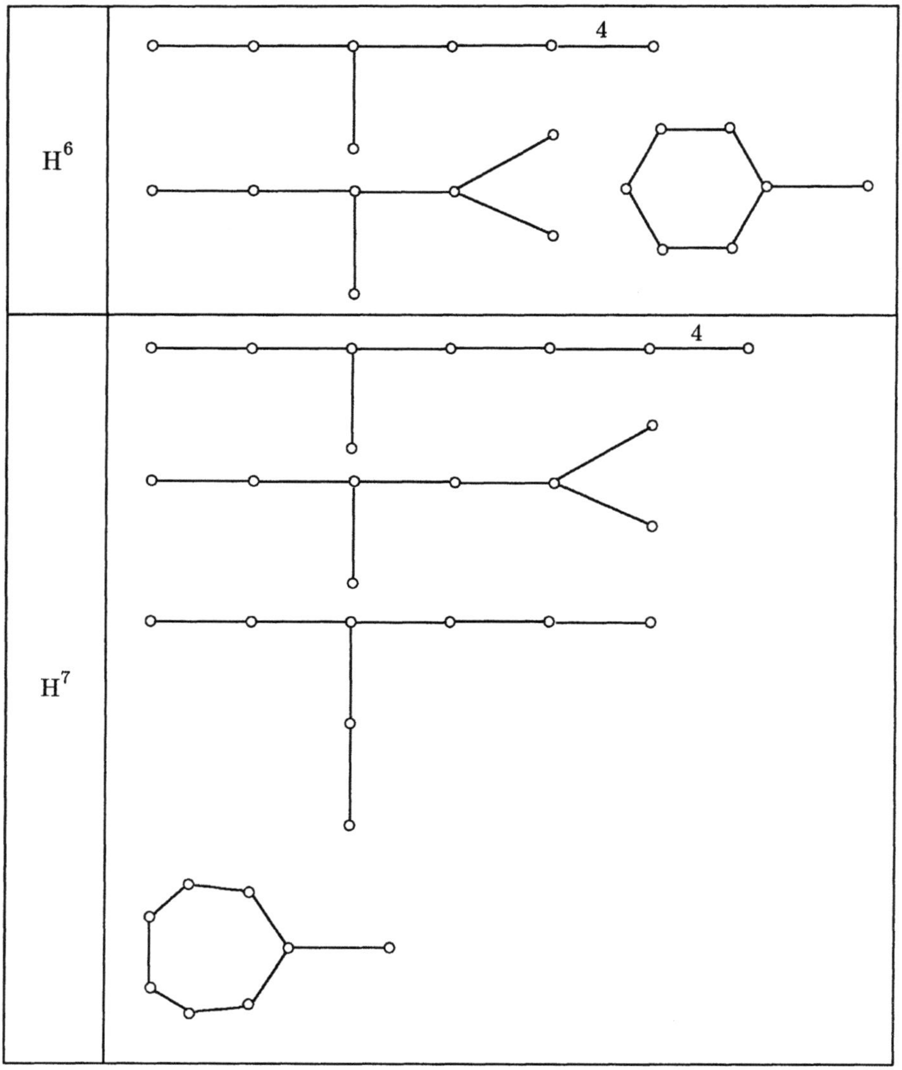

FIG. 4.6D

2.6 (see **2.1.7**), local absolutes in V^{2n} or V^{2n}_n are conjugate imaginary, respectively real, $(n-1)$-planes.

Since the tangent spaces of the spaces $\mathbb{H}\overline{S}^n$ and $\mathbb{H}'\overline{S}^n$ are $\mathbb{H}\overline{R}^n$, respectively $\mathbb{H}'\overline{R}^n$, analogously we find that local absolutes in the symmetric spaces V^{4n} and V^{4n}_{2n}, which are isometric to $\mathbb{H}\overline{S}^n$ and $\mathbb{H}'\overline{S}^n$, by Theorem 2.7 (see **2.1.8**), are imaginary, respectively real, Segreans $\Sigma_{1,2n-1}$.

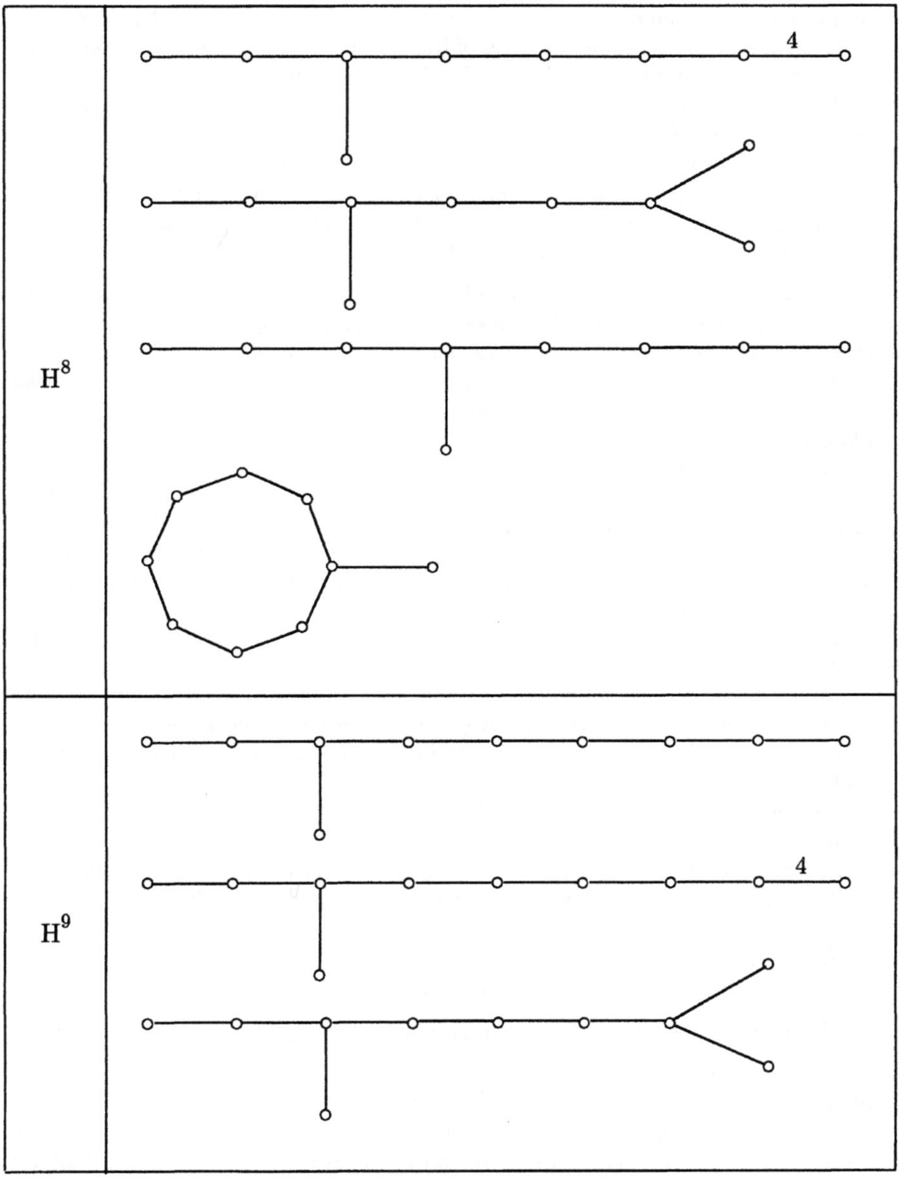

FIG. 4.6E

Since the tangent space of the space $\mathbb{R}_{m+1}\overline{S}^n$ is $\mathbb{R}_{m+1}\overline{R}^n$, we analogously find that local absolutes in the symmetric space $V^{(m+1)(n-m)}$, which is isometric to $\mathbb{R}_{m+1}\overline{S}^n$, by Theorem 2.8 (see **2.1.8**), are Segreans $\Sigma_{m,nm+n-1}$ and local syperabsolutes consist of points in k-planes intersecting this Segrean in $k+1$ points.

Local absolutes and superabsolutes of Riemannian and pseudo-Riemannian symmetric spaces were investigated by A. F. Masagutova [RoMa].

4.10.3. Fundamental and Parabolic Figures. Since the groups of motions in the spaces S^n, H^n, S^n_ℓ, H^n_ℓ, $\mathbb{C}S^n$, $\mathbb{C}\overline{S}^n$, $\mathbb{C}\overline{H}^n$, $\mathbb{C}S^n_\ell$, $\mathbb{C}\overline{H}^n_\ell$, $\mathbb{C}'\overline{S}^n$, $\mathbb{H}\overline{S}^n$, $\mathbb{H}\overline{H}^n$, $\mathbb{H}\overline{S}^n_\ell$, $\mathbb{H}\overline{H}^n_\ell$, and $\mathbb{H}'\overline{S}^n$ are simple and semisimple Lie groups, we can define in these spaces fundamental α_i-figures and parabolic $(\alpha_{i_1}, \alpha_{i_2}, \ldots, \alpha_{i_k})$-figures (see **1.6.4**).

Analogously to the proof of Theorem 2.61, by the rules of **1.6.4** we can prove the following Theorems 4.53, 4.54, and 4.55.

Theorem 4.53. *The fundamental α_i-figures in S^{2n}, H^{2n}, S^{2n}_ℓ, and H^{2n}_ℓ are the $(i-1)$-planar generators of the absolute hyperquadrics of these spaces. For $i=1$ they are points on these hyperquadrics, for $i=2$ their rectilinear generators.*

The parabolic $(\alpha_{i_1}, \alpha_{i_2}, \ldots, \alpha_{i_k})$-figures in these spaces are the flags consisting of incident (i_1-1)-planar, (i_2-1)-planar, ... ,(i_k-1)-planar generators of these absolute hyperquadrics.

The fundamental α_i-figures in S^{2n-1}, H^{2n-1}, S^{2n-1}_ℓ, and H^{2n-1}_ℓ for $i=1,2,\ldots,$ $n-2$ are also the $(i-1)$-planar generators of the absolute hyperquadrics in these spaces. For $i=1$ they are points on these absolute hyperquadrics, for $i=2$ their rectilinear generators; the fundamental α_{n-1}-figures and α_n-figures in these spaces are $(n-1)$-planar generators belonging to two families of $(n-1)$-planar generators of these hyperquadrics.

The parabolic (α_{n-1}, α_n)-figures in these spaces are the $(n-2)$-planar generators of these absolute hyperquadrics (which are intersections of two $(n-1)$-planar generators of two different families), other parabolic $(\alpha_{i_1}, \alpha_{i_2}, \ldots, \alpha_{i_k})$-figures in these spaces are flags consisting of incident generators of these hyperquadrics.

In S^{2n} and S^{2n-1} all fundamental and parabolic figures are imaginary; in S^{2n}_n and S^{2n-1}_n all these figures are real.

The fundamental α_i-figures in S^{2n}_ℓ and S^{2n-1}_ℓ are real for $i=1,2,\ldots,\ell$ and imaginary for $i>\ell$.

In the space S^{2n-1}_{n-2}, α_{n-1}-figures and α_n-figures are imaginary conjugate.

The fundamental α_i-figures and parabolic $(\alpha_{i_1}, \alpha_{i_2}, \ldots, \alpha_{i_k})$-figures in H^n and H^n_ℓ are the same as in S^n_1, respectively $S^n_{\ell+1}$.

The fundamental α_i-figures and parabolic $(\alpha_{i_1}, \alpha_{i_2}, \ldots, \alpha_{i_k})$-figures in $\mathbb{C}S^{2n}$ and $\mathbb{C}S^{2n-1}$ are the same as in S^{2n}_n, respectively S^{2n-1}_n.

The geometric meaning of spinor representations is given by

Theorem of É. Cartan [Car4]. *The spinor representations of the groups of motions in $\mathbb{C}S^{2n-1}$, $\mathbb{C}S^{2n}$, S^{2n-1}_n and S^{2n}_n are linear transformations of the coordinates of the $(n-1)$-planar generators of the absolute hyperquadrics of these spaces. These $(n-1)$- planes are represented by points in Lipschitzians Ω_n that are preserved under transformations in the spinor groups of the groups of motions in $\mathbb{C}S^{2n-1}$, $\mathbb{C}S^{2n}$, S^{2n-1}_n, and S^{2n}_n.*

Theorem 4.54. *The parabolic $\{\alpha_i, \alpha_{n-i}\}$-figures in $\mathbb{C}\overline{S}^n$, $\mathbb{C}\overline{H}^n$, $\mathbb{C}\overline{S}^n_\ell$ and $\mathbb{C}\overline{H}^n_\ell$ are the $(i-1)$-planar generators of the absolute Hermitian hyperquadrics of these spaces. For $i=1$ they are points on these hyperquadrics, for $i=2$ their rectilinear generators.*

The parabolic $(\alpha_{i_1}, \ldots, \alpha_{i_k})$*-figures in these spaces are the flags consisting of incident* $(i_1 - 1)$*-planar,* $(i_2 - 1)$*-planar,...,* $(i_k - 1)$*-planar generators of these Hermitian hyperquadrics.*

All fundamental and parabolic figures in $\mathbb{C}\overline{S}^n$ *are imaginary.*

The fundamental α_i*-figures and* α_{n-i+1}*-figures in* $\mathbb{C}\overline{S}^n_\ell$ *are imaginary conjugate for* $i \leq \ell$ *and imaginary for* $i > \ell$.

The fundamental α_i*-figures and parabolic* $(\alpha_{i_1}, \alpha_{i_2}, \ldots, \alpha_{i_k})$*-figures in* $\mathbb{C}\overline{H}^n$ *and* $\mathbb{C}\overline{H}^n_\ell$ *are the same as in* $\mathbb{C}\overline{S}^n_1$*, respectively* $\mathbb{C}\overline{S}^n_{\ell+1}$.

Note that the (α_1, α_n)-figures in $\mathbb{C}\overline{H}^n$ are points on the absolute oval Hermitian hyperquadric of this space. The stabilizers of these points contain the stabilizers of the Hermitian horohyperspheres in this space tangent to the absolute Hermitian hyperquadric at these points, and therefore these groups are parabolic. The Borel groups of these parabolic groups coincide with the *Heisenberg group* of $\mathbb{C}\overline{H}^n$ (see **4.7.6**).

Theorem 4.55. *The fundamental* α_{2i}*-figures in* $\mathbb{H}\overline{S}^n$*,* $\mathbb{H}\overline{H}^n$*,* $\mathbb{H}\overline{S}^n_\ell$ *and* $\mathbb{H}\overline{H}^n_\ell$ *are the* $(i-1)$*-planar generators of the absolute Hermitian hyperquadrics of these spaces. For* $i = 1$ *they are points on these Hermitian hyperquadrics, for* $i = 2$ *rectilinear generators of these Hermitian hyperquadrics. All fundamental* α_{2i-1}*-figures of these spaces are imaginary* $(i-\frac{1}{2})$*-planar generators of the same Hermitian hyperquadrics, for* $i = 1$ *halfpoints, for* $i = 2$ *halflines.*

The parabolic $(\alpha_{i_1}, \ldots, \alpha_{i_k}, \alpha_{n-i_k} + 1, \ldots, \alpha_{n-i_1} + 1)$*-figures of these spaces are the flags consisting of incident generators of these Hermitian hyperquadrics.*

All fundamental and parabolic figures in $\mathbb{H}\overline{S}^n$ *are imaginary.*

The fundamental α_{2i}*-figures in* $\mathbb{H}\overline{S}^n_\ell$ *are real for* $i < \ell$ *and imaginary for* $i > \ell$.

The fundamental α_i*-figures and parabolic* $(\alpha_{i_1}, \alpha_{i_2}, \ldots, \alpha_{i_k})$*-figures in* $\mathbb{H}\overline{H}^n$ *and* $\mathbb{H}\overline{H}^n_\ell$ *are the same as in* $\mathbb{H}\overline{S}^n_1$*,respectively* $\mathbb{H}\overline{S}^n_{\ell+1}$.

Note that the α_2-figures in $\mathbb{H}\overline{H}^n$ are points on the absolute oval Hermitian hyperquadric of this space. The stabilizers of these points contain the stabilizers of the Hermitian horohyperspheres in this space tangent to the absolute Hermitian hyperquadric at these points, and therefore these groups are parabolic. The Borel groups of these parabolic groups can be regarded as an analog of the Heisenberg group of $\mathbb{C}\overline{H}^n$.

The Satake diagrams of the compact groups of motions in $\mathbb{C}\overline{S}^n$, S^{2n}, $\mathbb{H}\overline{S}^{n-1}$, and S^{2n-1} coincide with the Dynkin diagrams of the simple Lie groups A_n, B_n, C_n, and D_n (Fig. 1.3), respectively, but all dots of these diagrams are black.

The Satake diagrams of the split groups of motions in S^{2n}_n and S^{2n-1}_{n-1} coincide with the Dynkin diagrams of the simple Lie groups B_n and D_n, respectively, but all dots of these diagrams are white.

The Satake diagrams of the groups of motions in $\mathbb{C}\overline{S}^n_\ell$, S^{2n}_ℓ, $\mathbb{H}\overline{S}^{n-1}_\ell$, and S^{2n-1}_ℓ are given in Fig. 1.9 ($AIII, BI, CII$, and DI, respectively).

The Satake diagrams of the groups of motions in $\mathbb{C}\overline{H}^n$, H^{2n}, $\mathbb{H}\overline{H}^{n-1}$ and H^{2n-1} are also given in Fig. 9 (AIV, BII, CII for $\ell = 1$, and DII, respectively).

4.10.4. Local Absolutes and Superabsolutes in Parabolic Spaces. Local absolutes and superabsolutes can be also defined in parabolic spaces. The structure of these figures is determined by the structure $\mathbb{G}_0 \oplus \mathbb{G}_1 \oplus ... \oplus \mathbb{G}_h$ of the tangent Lie algebra of the stabilizer of a point in the parabolic space.

For the manifold of the m-planar generators of the absolute hyperquadric of H^n or S^n_ℓ the role of local absolutes and superabsolutes is played by an m-plane generator and the intersection of its polar $n-m-1$-plane with the absolute hyperquadric, this intersection is a hypercone of the second degree with m-planar apex in P^{n-m-1}, for $m=0$ a hypercone of the second degree with point apex.

For Grassmann manifold $Gr^P_{n,m}$ of m-planes in P^n the role of local absolutes is played by an m-plane and the set of all m-planes intersecting this m-plane, this set of m-planes is interpreted on Grassmannian by the *Segre-cone*, that is, a cone consisting of rectilinear generators of Grassmannian, which projects a Segrean $\Sigma_{m,n-m-1}$.

Local absolutes C^n, C^n_ℓ, and $Gr^P_{n,m}$ were considered by M. A. Akivis and V. V. Goldberg [AG].

4.10.5. Representations of a Simple Lie Group by Fractional Linear Transformations in a Jordan Algebra. In Theorem 1.38 (see **1.6.4**) we have seen that the subspace $\mathbb{G}_{-1}$ in the decomposition (1.170) of the tangent Lie algebra of a simple Lie group G for $h=1$ can be regarded as a Jordan algebra and the simple Lie group G with tangent Lie algebra $\mathbb{G} = \mathbb{G}_{-1} \oplus \mathbb{G}_0 \oplus \mathbb{G} - 1$ can be represented as the group of fractional linear transformations (1.140) in this Jordan algebra.

For hyperbolic or pseudoelliptic space H^n, respectively, S^n_ℓ this Jordan algebra is $\mathbb{J}(\mathbb{A}_{n-1})$, respectively $\mathbb{J}(\mathbb{A}_{n-1,\ell})$. For $n=1$ and 2 the representations of groups of motions in H^2, H^3, and S^3_2 by fractional linear transformations in $\mathbb{R} = \mathbb{A}_1$, $\mathbb{C} = \mathbb{A}_2$, and $\mathbb{C}' = \mathbb{A}_{2,1}$ follow from Theorems 4.23 and 4.24 (see **4.4.6-7**). For general case this representation was obtained by I. I. Kolokol'tseva [RKMV]; in another form this interpretation was found by M. A. Javadov in his geometric interpretation of spinor representations of groups of motions in H^n and S^n_ℓ [Jav1] (see also [Ro1, pp. 523–532]. Another approach to this interpretation was found by A. Blunck [Blu].

§4.11. Space Forms

4.11.1. Space Forms of Clifford-Klein. In his paper [Cli1], W. K. Clifford noted that there are two forms of Riemannian manifold V^n of constant positive curvature: a hypersphere Σ^n in the Euclidean space R^{n+1} and the elliptic space S^n. The term "space form" was introduced by W. Killing in his book [Ki1], where he considered four "space forms": the "space of Euclid" R^n, the "space of Lobachevsky" H^n, the hypersphere Σ^n called the "space of Riemann" by him, and the space S^n, called the "polar form of the space of Riemann" by him (because this space admits an interpretation as the set of great $(n-1)$-spheres in Σ^n).

Another space form, also connected with Clifford (see [Cli1]) is the Clifford quadric in S^3 (see **4.7.3**). In Theorem 4.40 we have seen that the intrinsic geometry of this surface is Euclidean and that this surface is isometric to a rhombus in R^2 with identified opposite sides and, therefore, its area is finite.

As a result of his investigations of Fuchsian and Kleinian groups in H^2 (see **4.9.4** and **4.9.5**), F. Klein [Kl3] came to the problem of discribing all compact Riemannian manifolds V^n of constant curvature, which as particular case is Clifford quadric. In the book [Ki3] Killing proved that the manifolds described by Klein are quotient spaces V^n/G of Riemannian manifolds V^n of constant curvature by a discrete subgroup of the group of motions in V^n, and called these manifolds "space forms of Clifford-Klein".

Nowadays these manifolds are called "space forms of Clifford" when the discrete group of motions is finite (as for the space S^n, where this group consists of two elements) and "space forms of Klein" when this group is infinite (as for Fuchsian and Kleinian groups).

4.11.2. 2-Dimensional Space Forms. The well known classification of discrete subgroups of the group of motions in R^2 implies

Theorem 4.56. *The 2-dimensional Euclidean space forms are the plane R^2 itself and its quotient spaces by discrete subgroups of the group of motions in this plane:*

1° *cylinder, for subgroups in the group of translations consisting of the translations $x' = x + an$, where a is a vector and n is an integer number (Fig. 4.7a),*

2° *torus, for subgroups in the group of translations consisting of the translations $x' = x + am + bn$, where a and b are two linearly independent vectors and m and n are integers (Fig. 4.7b); for $|a| = |b|$ it is the Clifford quadric,*

3° *the Möbius strip (see* **2.2.2***), for subgroups in the group of translations and translation reflections consisting of the translations $'x^1 = x^1 + an$, $'x^2 = (-1)^n x^2$, where a is a real number and n is an integer, and the reflections in lines invariant under these translations(Fig. 4.7c) (see also Fig. 2.4b),*

4° *the Klein bottle, obtained by glueing two pairs of opposite sides of a rectangle, as shown on Fig. 4.7e, for subgroups in the group of translations and translation reflections consisting of the translations in two orthogonal directions $'x^1 = x^1 + am$, $'x^2 = (-1)^m x^2 + bn$, where a and b are two real numbers and m and n are two integers, and the reflections in lines invariant under one of these translations.*

In **2.2.2** we have seen that the projective plane P^2 can be obtained by glueing the boundary of a Möbius strip with the circumference of a circle. As space forms, the cylinder and Möbius strip are considered without their boundaries. A Möbius strip, like the plane P^2, is a one-sided surface; P^2 is also homeomorphic to a rectangle with glued opposite sides; this glueing is shown in Fig. 4.7d.

A Klein bottle, like a Möbius strip and P^2, is a one-sided surface; this surface, which cannot be depicted in R^3 without self-intersection, can be depicted without self-intersection in R^4. It can be obtained from a tube with one end a little narrower than its other end by bending this tube so that the narrow end of the tube passes through its side to its wide end while the circumferences of both ends of the tube will become concentric; subsequently both ends are glued together (Fig. 4.8).

We can analogously define 2-dimensional spherical and hyperbolic space forms. Note that if R^2 is regarded as the complex plane, the sphere Σ^2 in R^3 is regarded as the Riemann sphere, and the plane H^2 is considered in the Poincaré interpretation as the upper halfplane of the complex plane, all three kinds of two-sided 2-dimensional space forms can be realized as Riemann surfaces determined by various functions of a complex variable. These surfaces are closed surfaces of genus p (homeomorphic to complex plane algebraic curves of genus p). Note that compact Riemann surfaces of genus $p \geq 2$ are 2-dimensional hyperbolic space forms, and

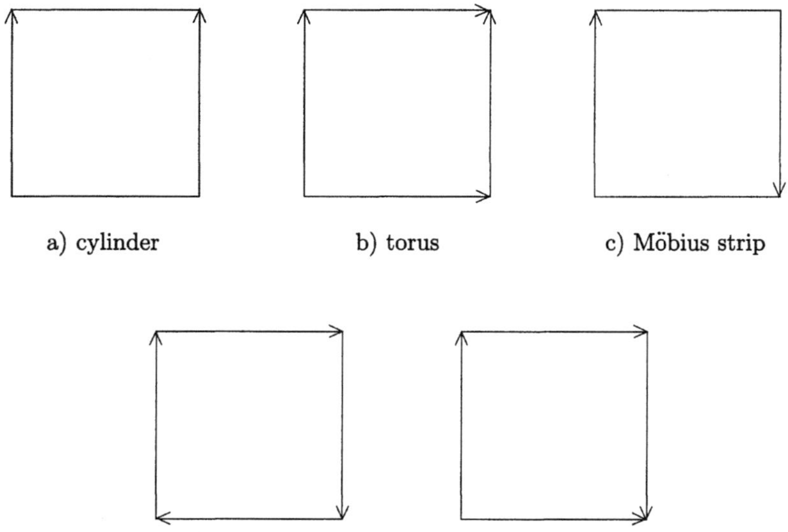

a) cylinder b) torus c) Möbius strip

d) projective plane e) Klein bottle

FIG. 4.7

FIG. 4.8

that two-sided 2-dimensional hyperbolic space forms are determined by Fuchsian groups in H^2.

4.11.3. 3-Dimensional and Multidimensional Space Forms. Dodecahedron Spaces. The 3-dimensional spherical space forms of Klein can be obtained by representing a 3-sphere as the group of quaternions with modulus one ($|x| = 1$) and by considering finite subgroups in this group. When this subgroup consists of the quaternions 1 and -1 we obtain the space S^3. When this subgroup is the complete preimage of the group of symmetries of a dodecahedron we obtain the *dodecahedron space* discovered by Poincaré [Poi4]. This space can be obtained by glueing the opposite faces of the dodeahedron in S^3 which is a cell of the regular 3-honeycomb $\{5, 3, 3\}$ in S^3.

The homology groups of this space are the same as the homology groups of the 3-sphere Σ^3, but this space is not homeomorphic to this 3-sphere.

The analogous dodecahedron space which is a hyperbolic space form of Klein can be obtained by glueing the opposite faces of the dodecahedron in H^3 which is a cell of the regular 3-honeycomb $\{5, 3, 5\}$ in H^3; it was investigated by C. Weber and H. Seifert [WeS].

The classification of space forms in H^3 is connected also with topological problems.

Space forms in multidimensional Riemannian and pseudo-Riemannian manifolds of constant curvature can be defined analogously. These space forms were investigated by J. Wolf in his book [Wol].

§4.12. Sliding Vectors

4.12.1. Sliding Vectors in Elliptic Space. The theory of sliding vectors on the hypersphere in R^{n+1} (see **3.7.5**) entails a theory of sliding vectors in S^n. In particular, each sliding vector $\breve{a}$ in S^n can be represented by a sliding vector a in R^{n+1}, and for each system of sliding vectors $\breve{a}_i$ we can determine a skew symmetric linear operator $M = \sum_i x_i \wedge a_i$, where x_i is the vector representing the initial point of the vector $\breve{a}_i$. M is called the *main moment* of this system.

Theorem 3.27 implies

Theorem 4.57. *Each system of sliding vectors in S^n of curvature $\frac{1}{r^2}$ can be reduced to the canonical form consisting of sliding vectors $\breve{a}_i$ on mutually polar lines that are non-meeting edges of an autopolar simplex in S^n and determined by invariant planes of the main moment M of this system; the modulus $|a_i|$ of the vector a_i representing a sliding vector $\breve{a}_i$ of the canonical system is equal to the modulus of an eigenvalue of the linear operator M divided by r.*

Canonical systems of sliding vectors in S^n are called *screws* in S^n. For $n = 3$ this theory was built by A. P. Kotelnikov [Kot2] and E. Study [Stu1].

Note that the linear operators M can be regarded as elements of the tangent Lie algebra of the group of motions in S^n.

4.12.2. Sliding Vectors in Hyperbolic Space. The theory of sliding vectors on the hypersphere of imaginary radius in R_1^{n+1} (see **3.7.6**) entails a theory of sliding vectors in H^n. Like for S^n, each sliding vector $\breve{a}$ in H^n can be represented by a sliding vector a in R_1^{n+1}, and for each system of sliding vectors $\breve{a}_i$ we can determine a skew symmetric linear operator $M = \sum_i x_i \wedge a_i$, where x_i is the vector representing the initial point of the vector $\breve{a}_i$. M is called the *main moment* of this system. Theorem 3.28 implies

Theorem 4.58. *Each system of sliding vectors in H^n of curvature $-\frac{1}{q^2}$ can be reduced to the canonical form consisting of sliding vectors $\breve{a}_i$ on mutually polar lines that are non-meeting edges of an autopolar simplex in H^n and determined by invariant planes of the linear operator EM that is the product of the operator E*

determining the inner product xEy in R_1^{n+1} by the main moment M of this system; the modulus $|a_i|$ of the vector a_i representing a sliding vector $\breve{a}_i$ of the canonical system is equal to the modulus of an eigenvalue of the linear operator EM divided by q.

Canonical systems of sliding vectors in H^n are called *screws* of H^n. For $n = 3$ this theory was built by A. P. Kotelnikov [Kot2] and E. Study [Stu1].

Note that the linear operators EM can be regarded as elements of the tangent Lie algebra of the group of motions in H^n.

§4.13. Finite Geometries

4.13.1. Finite Quadratic Spaces. Hyperquadrics in the spaces $\mathbb{F}_q P^{2n}$ and elliptic and hyperbolic hyperquadrics in the space $\mathbb{F}_q P^{2n+1}$, whose equations can be reduced to the forms (2.131), (2.133), and (2.132), respectively, can be regarded as absolute hyperquadrics of the *elliptic spaces* $\mathbb{F}_q S^{2n}$ and $\mathbb{F}_q S^{2n+1}$, respectively of the *hyperbolic space* $\mathbb{F}_q H^{2n+1}$, *over the Galois field* $\mathbb{F}_q$.

The collineations in $\mathbb{F}_q P^{2n}$ and $\mathbb{F}_q P^{2n+1}$ that preserve these absolute hyperquadrics are called *motions* in the spaces $\mathbb{F}_q S^{2n}$, $\mathbb{F}_q S^{2n+1}$, and $\mathbb{F}_q H^{2n+1}$. Invariants of pairs of points in these spaces that are preserved under these motions are called *metric invariants* of pairs of points in these spaces. The cardinalities of absolute hyperquadrics of these spaces are expressed by (2.137), (2.139), respectively (2.138).

The motions in $\mathbb{F}_q S^{2n}$ and $\mathbb{F}_q H^{2n-1}$ form finite groups of Lie type $\mathbb{F}_q B^n$, respectively $\mathbb{F}_q D^n$. The motions in $\mathbb{F}_q S^{2n-1}$ form a 2-twisted finite group of Lie type $\mathbb{F}_q D_n^{(2)}$. These groups have invariant subgroups which are simple finite groups $\mathbb{F}_q S\Omega_{2n+1}, \mathbb{F}_q S\Omega_{2n}$, respectively $\mathbb{F}_q S\Omega_{2n}^{(2)}$, which are simple finite groups.

4.13.2. Finite Hermitian Space. Hermitian hyperquadrics in $\mathbb{F}_{q^2} P^n$, whose equation can be reduced to the form (2.143) can be regarded as the absolute hyperquadric of the *Hermitian elliptic space* $\mathbb{F}_{q^2} \overline{S}^n$ *over the Galois field* $\mathbb{F}_q$.

The collineations in $\mathbb{F}_{q^2} P^n$ that preserve this absolute Hermitian hyperquadric are called *motions* in $\mathbb{F}_{q^2} \overline{S}^n$. Invariants of pairs of points in this space that are preserved under these motions are called *metric invariants* of pairs of points in this space. The cardinality of the absolute Hermitian hyperquadric of this space is expressed by (2.145).

The motions in the space $\mathbb{F}_{q^2} \overline{S}^n$ form a 2-twisted finite group $\mathbb{F}_{q^2} SU_{n+1}$ of Lie type $\mathbb{F}_q A_n^{(2)}$ which is a simple finite group.

§4.14. Applications to Physics

4.14.1. Application of Hyperbolic Geometry. Since, as we have seen in **3.10.1**, the group of Lorentz transformations and the group of rotations in the space-time of the Special Relativity of Einstein are isomorphic to the group of rotations in the space R_1^4, these groups are locally isomorphic to the group of motions in the space H^3. In **3.10.2** we have seen that the rule of addition of velocities in Special Relativity is equivalent to the cosine law on one sheet of the hypersphere of imaginary radius in R_1^4, which can be regarded as the space H^3.

In **3.10.3** we have seen that the electromagnetic field in Special Relativity admits an interpretation by means of systems of sliding vectors on the same hypersphere of imaginary radius, that is, it admits an interpretation by means of systems of sliding vectors in H^3.

4.14.2. Twistor Program of R. Penrose. In **3.10.3** we have seen that the electromagnetic field in Special Relativity is invariant under the group of conformal transformations in the pseudoconformal space C_1^4. Since this group is isomorphic to the group of motions in the pseudoelliptic space S_2^5, which is locally isomorphic to the group of motions in the Hermitian pseudoelliptic space $\mathbb{C}\overline{S}_2^3$, the geometry of this space can be applied to the theory of electromagnetic fields in Special Relativity.

The Twistor Program of Roger Penrose [Pen] is based on this isomorphism. *Twistors* are spinors of the group $SO_{6,2}$, whose (6×6)-matrices represent motions in S_2^5. Their coordinates can be regarded as coordinates of points on the absolute Hermitian quadric of $\mathbb{C}\overline{S}_2^3$. Since the groups of motions in S_2^5 and $\mathbb{C}\overline{S}_2^3$ are locally isomorphic, there are correspondences between fundamental figures in these spaces. Since the parabolic α_1-figures and (α_2, α_3)-figures in S_2^5 are points and rectilinear generators of the absolute hyperquadric of this space and the (α_1, α_3)-figures and α_2-figures in $\mathbb{C}\overline{S}_2^3$ are also points and rectilinear generators of the absolute quadric of this space, and since the dimensions of the manifolds of these figures are equal to 4 and 5 for S_2^5 and to 5 and 4 for $\mathbb{C}\overline{S}_2^3$ while their respective stabilizers are isomorphic, points on the absolute hyperquadric of S_2^5 can be interpreted as rectilinear generators of the absolute quadric of $\mathbb{C}\overline{S}_2^3$ and rectilinear generators of the absolute hyperquadric of S_2^5 can be interpreted as points on the absolute Hermitian quadric of $\mathbb{C}\overline{S}_2^3$. Therefore, the points in space-time, that is, points in C_1^4, which can be interpreted as points on the absolute hyperquadric of S_2^5, can also be interpreted as complex rectilinear generators in the absolute Hermitian quadric of $\mathbb{C}\overline{S}_2^3$. For this reason S. G. Gindikin called this interpretation the "Complex Universe of Roger Penrose" [Gin].

Note that the isotropic lines issueing from a point in C_1^4 can be interpreted as rectilinear generators of the absolute hyperquadric of S_2^5. Each of these rectilinear generators corresponds to a point on the absolute Hermitian quadric of $\mathbb{C}\overline{S}_2^3$, and all isotropic lines issueing from a point in C_1^4 can be interpreted as points in a rectilinear generator of the absolute Hermitian quadric. Since the line $\mathbb{C}P^1$ can be regarded as the complex plane, an isotropic hypercone with apex at a point in C_1^4 can be interpreted as the complex plane.

CHAPTER V

QUASIELLIPTIC, QUASIHYPERBOLIC, AND QUASI-EUCLIDEAN GEOMETRIES

§5.1. Quasielliptic, Quasihyperbolic, and Quasi-Euclidean Spaces

5.1.1. Co-Euclidean and Copseudo-Euclidean Spaces. Since the Euclidean space R^n and pseudo-Euclidean space R^n_ℓ can be defined as the projective space P^n with a distinguished hyperplane in which the absolute hyperquadrics of the spaces S^{n-1} or S^{n-1}_ℓ are given, there are spaces corresponding them by the duality principle of P^n. These dual spaces are called *co-Euclidean*, respectively *copseudo-Euclidean, spaces* and are denoted by R^{n*}, respectively R^{n*}_ℓ.

Since the points in P^n correspond by the duality principle to hyperplanes, we define distance between points in R^{n*} and R^{n*}_ℓ as metric invariants of hyperplanes in R^n and R^n_ℓ, respectively: for general hyperplanes we take their angle, for parallel hyperplanes their distance.

Theorem 5.1. *The role of absolute in R^{n*}, respectively, R^{n*}_ℓ, is played by a hypercone of second order with a point apex, imaginary, respectively real.*

If the equations of these hypercones are

$$x^2 = (x^0)^2 + (x^1)^2 + \cdots + (x^{n-1})^2 = 0, \tag{5.1}$$

respectively

$$xEx = -(x^0)^2 - (x^1)^2 - \cdots - (x^{\ell-1})^2 + (x^\ell)^2 \cdots + (x^{n-1})^2 = 0, \tag{5.2}$$

and the projective coordinates x^i of points are normalized by the conditions $x^2 = 1$, respectively $xEx = \pm 1$, then the distance δ between two points x and y in R^{n*}, respectively in R^{n*}_l, is defined by

$$\cos\frac{\delta}{r} = |xy| = |x^0y^0 + x^1y^1 + \cdots + x^{n-1}y^{n-1}|, \tag{5.3}$$

respectively

$$\cos\frac{\delta}{r} = |xEy| = |-x^0y^0 - x^1y^1 - \cdots - x^{\ell-1}y^{\ell-1} + x^\ell y^\ell + \cdots + x^{n-1}y^{n-1}|, \tag{5.4}$$

where r is the radius of curvature of the space. When $\delta = 0$ the points x and y in R^{n*} or R^{n*}_ℓ have a second distance, d, equal to

$$d = |y^n - x^n|. \tag{5.5}$$

PROOF. Since the role of absolute in R^n and R^n_ℓ is played by their hyperplanes at infinity and the absolute hyperquadrics of S^{n-1} or S^{n-1}_ℓ in these hyperplanes, the absolutes of R^{n*} and R^{n*}_ℓ consist of points corresponding to the hyperplanes at infinity and of hyperplanes, imaginary, respectively real, corresponding to points on absolute hyperquadrics of S^{n-1} or S^{n-1}_ℓ on these hyperplanes. Since these hyperquadrics are on hyperplanes at infinity, the hyperplanes corresponding to points on the hyperquadrics pass through the points of the absolutes. Since the projective coordinates of points on the hyperquadrics satisfy equations of the second degree, the tangential coordinates of the imaginary or real hyperplanes corresponding to points on the absolute hyperquadrics of S^{n-1} and S^{n-1}_ℓ also satisfy equations of the second degree. Therefore, these hyperplanes are tangent to hypercones of the second order with a point apex, imaginary, respectively real.

If the normal equation of a hyperplane in R^n is $u_iX^i + u = 0$ $(\sum_i u_i^2 = 1)$, this hyperplane corresponds to a point in R^{n*} with projective coordinates $x^i = u_{i-1}$, $x^n = u$. Therefore, the normalized coordinates x^i coincide with the rectangular coordinates of a unit normal vector of the corresponding hyperplane in R^n and the expression (5.3) for $\cos\frac{\delta}{r}$ coincides with the cosine of the angle between the two hyperplanes corresponding to the two points. The condition $\delta = 0$ coincides with the condition that the two hyperplanes be parallel and the expression (5.5) for d coincides with the distance between two parallel hyperplanes.

The proof for R^{n*}_ℓ is analogous. ■

The requirement that collineations in P^n preserve the distance δ between points in R^{n*} and R^{n*}_ℓ and, when $\delta = 0$, preserve the distance d implies

Theorem 5.2. *The motions in R^{n*} or R^{n*}_ℓ have the form*

$$'x^i = U^i_j x^j,\ 'x^n = a_j x^j + x^n \quad (i, j = 0, 1, \ldots, n-1), \tag{5.6}$$

where (U^i_j) are $(n \times n)$-matrices in the groups O_n, respectively $O_{n,\ell}$, and a_j are arbitrary real numbers.

The groups (5.6) are isomorphic to the groups of motions in R^n, respectively R^n_ℓ.

We can analogously define the quadratic and Hermitian co-Euclidean and co-pseudo-Euclidean spaces $\mathbb{A}R^{n*}$, $\mathbb{A}R^{n*}_\ell$, $\mathbb{A}\overline{R}^{n*}$, and $\mathbb{A}\overline{R}^{n*}_\ell$, corresponding to the spaces $\mathbb{A}R^n$, $\mathbb{A}R^n_\ell$, $\mathbb{A}\overline{R}^n$, and $\mathbb{A}\overline{R}^n_\ell$, respectively, by the duality principle in $\mathbb{A}P^n$.

For $\mathbb{A}\overline{R}^n$ and $\mathbb{A}\overline{R}^n_\ell$, the quadratic and bilinear forms in (5.1), (5.2), (5.3), and (5.4) must be replaced by quadratic, respectively bilinear, Hermitian forms.

5.1.2. Quasielliptic, Quasihyperbolic, Quasipseudoelliptic, and Quasipseudohyperbolic Spaces. The spaces R^n, R^n_ℓ, R^{n*}, and R^{n*}_ℓ are particular cases of the more general *quasielliptic spaces* $S^{m,n}$ and *quasipseudoelliptic spaces* $S^{m,n}_{k\ell}$, that is, the space P^n, in which an imaginary or real hypercone C_0 of the second order with $(n-m-1)$-planar apex A_0, and an imaginary or real nondegenerate $(n-m-2)$-quadric C_1 in A_0 are given. The hypercone C_0, the $(n-m-1)$-plane A_0, and the $(n-m-2)$-quadric C_1 are called the *absolute hypercone*, the *absolute $(n-m-1)$-plane*, and the *absolute $(n-m-2)$ quadric.*

The equations of the absolute hypercone C_0 in $S^{m,n}$ or $S^{m,n}_{k\ell}$ can be reduced to the form

$$x_0^2 = (x^0)^2 + (x^1)^2 + \cdots + (x^m)^2 = 0, \tag{5.7}$$

respectively

$$x_0 E_0 x_0 = -(x^0)^2 - (x^1)^2 - \cdots - (x^{k-1})^2 + (x^k)^2 \cdots + (x^m)^2 = 0. \tag{5.8}$$

The equations of the $(n-m-1)$-planar apex A_0 of this hypercone have the form $x^0 = x^1 = \cdots = x^m = 0$.

The equations of the absolute $(n-m-2)$-quadric C_1 in $S^{m,n}$ or $S^{m,n}_{k\ell}$ can be reduced to the form

$$x_1^2 = (x^{m+1})^2 + (x^{m+2})^2 + \cdots + (x^n)^2 = 0, \tag{5.9}$$

respectively

$$x_1 E_1 x_1 = -(x^{m+1})^2 - (x^{m+2})^2 - \cdots - (x^{m+\ell})^2 + (x^{m+\ell+1})^2 \cdots + (x^n)^2 = 0. \tag{5.10}$$

If the projective coordinates x^i of points are normalized by the condition $x_0^2 = 1$, respectively $x_0 E_0 x_0 = \pm 1$, the distance δ between the points x and y in $S^{m,n}$ or $S^{m,n}_{k\ell}$ is defined by

$$\cos \frac{\delta}{r} = |x_0 y_0| = |x^0 y^0 + x^1 y^1 + \cdots + x^m y^m|, \tag{5.11}$$

respectively,

$$\cos \frac{\delta}{r} = |x_0 E_0 y_0| = |-x^0 y^0 - x^1 y^1 - \cdots - x^{k-1} y^{k-1} + x^k y^k + \cdots + x^m y^m|, \tag{5.12}$$

where r is the radius of curvature of the space.

If $\delta = 0$, the points x and y in $S^{m,n}$ or $S^{m,n}_{k\ell}$ have a second distance, d, defined by

$$d^2 = (y_1 - x_1)^2 = (y^{m+1} - x^{m+1})^2 + (y^{m+2} - x^{m+2})^2 + \cdots + (y^n - x^n)^2, \tag{5.13}$$

respectively

$$\begin{aligned} d^2 = (y_1 - x_1) E_1 (y_1 - x_1) &= -(y^{m+1} - x^{m+1})^2 - \cdots - (y^{m+\ell} - x^{m+\ell})^2 \\ &+ (y^{m+\ell+1} - x^{m+\ell+1})^2 + \cdots + (y^n - x^n)^2. \end{aligned} \tag{5.14}$$

Note that when $\delta = 0$ the points x and y are located in a space R^{n-m}, respectively R^{n-m}_ℓ, and the distance d between these points is equal to their distance in these $(n-m)$-spaces.

The spaces R^n and R^n_ℓ are the spaces $S^{m,n}$ and $S^{m,n}_{k\ell}$ for $m = 0$. Then the hypercone C_0 is the double hyperplane $x^0 = 0$ and the $(n-m-2)$-quadric C_1 is the absolute hyperquadric of S^{n-1} or S^{n-1}_ℓ. The spaces R^{n*} and R^{n*}_k are the spaces $S^{m,n}$ and $S^{m,n}_{k\ell}$ for $m = n-1$. Then the hypercone C_0 is a hypercone with a point apex and the $(n-m-2)$-quadric C_1 is the double point coinciding with this apex.

The requirement that collineations in P^n preserve the distance δ between points in $S^{m,n}$ and $S^{m,n}_{k\ell}$ and, when $\delta = 0$, preserve the distance d between these points, leads to

Theorem 5.3. *The motions in $S^{m,n}$, respectively $S^{m,n}_{k\ell}$, have the form*

$$'x^a = U^a_b x^b,\ 'x^u = T^u_b x^b + V^u_v x^v \quad (a, b = 0, 1, \dots, m,\ u, v = m+1, \dots, n), \tag{5.15}$$

where (U^a_b) are $((m+1) \times (m+1))$-matrices in the groups O_{m+1}, respectively $O_{m+1,k}$, (V^u_v) are $((n-m) \times (n-m))$-matrices in the groups O_{n-m}, respectively $O_{n-m,\ell}$, and (T^u_b) are $((m+1) \times (n-m))$-matrices with arbitrary entries.

The *quasihyperbolic space* $H^{m,n}$ and *quasipseudohyperbolic space* $H^{m,n}_{k\ell}$ have the same absolutes as $S^{m,n}_{10}$, respectively $S^{m,n}_{k+1,\ell}$, but the radius of curvature of these spaces is pure imaginary, $r = iq$, and, therefore, in (5.11) and (5.12) the expression $\cos \frac{\delta}{r}$ must be replaced by $\cosh \frac{\delta}{q}$.

We can analogously define the quadratic and Hermitian quasielliptic, quasihyperbolic, quasipseudoelliptic, and quasipseudohyperbolic spaces $\mathbb{A}S^{m,n}, \mathbb{A}H^{m,n}, \mathbb{A}S^{m,n}_{k\ell}$, $\mathbb{A}H^{m,n}_{k\ell}$, $\mathbb{A}\overline{S}^{m,n}$, $\mathbb{A}\overline{H}^{m,n}$, $\mathbb{A}\overline{S}^{m,n}_{k\ell}$, $\mathbb{A}\overline{H}^{m,n}_{k\ell}$. In $\mathbb{A}\overline{S}^{m,n}$, $\mathbb{A}\overline{H}^{m,n}$, $\mathbb{A}\overline{S}^{m,n}_{k\ell}$, $\mathbb{A}\overline{H}^{m,n}_{k\ell}$ in formulas (5.7), (5.8), (5.9), (5.10), (5.11), (5.12), (5.13), and (5.14) the quadratic and bilinear forms must be replaced by quadratic and bilinear Hermitian forms.

It is well-known that a Euclidean space can be obtained by a limiting process from elliptic or hyperbolic spaces, and that pseudo-Euclidean spaces can be obtained by a limiting process from pseudoelliptic or pseudohyperbolic spaces. These facts are particular cases of

Theorem 5.4. *All quasielliptic, quasihyperbolic, quasipseudoelliptic, and quasipseudohyperbolic spaces can be obtained by a limiting process from elliptic, hyperbolic, pseudoelliptic, and pseudohyperbolic spaces.*

PROOF. Consider the space S^n with the absolute hyperquadric $\sum_i (x^i)^2 = 0$. The tangential coordinates of the imaginary tangential coordinates of this hyperquadric satisfy the equation $\sum_i u_i^2 = 0$, called the tangential equation of this hyperquadric. Replace this hyperquadric by the hyperquadric $\sum_a (x^a)^2 + t\sum_u (x^u)^2 = 0$. For all $t \neq 0$ it is also the absolute hyperquadric of the space S^n. The tangential equation of this hyperquadric is $t\sum_a u_a^2 + \sum_u u_u^2 = 0$. For $t = 0$ the first equation becomes the equation of the absolute hypercone C_0 in the space $S^{m,n}$ and the second equation becomes the tangential equation of the absolute $(n-m-2)$-quadric C_1 in the same space. Thus, the limiting process for $t \to 0$ transforms the absolute hyperquadric of S^n into the absolute of $S^{m,n}$ and, therefore, the space S^n itself into the space $S^{n,m}$.■

We can obtain the same result by the same limiting process from the pseudoelliptic space S^n_m.

We can prove analogous theorems for real quasipseudoelliptic, quasihyperbolic, and quasipseudohyperbolic spaces and for quadratic and Hermitian spaces over algebras.

A space $\mathbb{A}P^n$, in which the only absolute $(n-m-1)$-plane A_0 is given, is called *quasiaffine n-space* and is denoted by $\mathbb{A}P^{m,n}$. The space $\mathbb{A}P^{0,n}$ coincides with the affine space $\mathbb{A}E^n$ supplemented by a hyperplane at infinity, the space $\mathbb{A}P^{n-1,n}$ is called *centroprojective* or *coaffine n-space*, the space $\mathbb{A}P^{n-2,n}$ is called *axial n-space.*

Collineations preserving the absolute $(n-m-1)$-plane A_0 are called *quasiaffine transformations* in $\mathbb{A}P^{m,n}$. These transformations have the form (5.15) where the matrices (U^a_b) and (V^u_v) are arbitrary nonsingular matrices.

5.1.3. Quasi-Euclidean and Quasipseudo-Euclidean Spaces. The affine space E^n in whose hyperplane at infinity the geometry of a space $S^{m,n-1}$ or $S^{m,n-1}_{k\ell}$ is given is called a *quasi-Euclidean space* $R^{m,n}$, respectively a *quasipseudo-Euclidean space* $R^{m,n}_{k\ell}$. The absolutes of $R^{m,n}$ and $R^{m,n}_{k\ell}$ consist of their hyperplanes at infinity and of absolutes of the spaces $S^{m,n-1}$ and $S^{m,n-1}_{k\ell}$ in these hyperplanes.

The distance δ between two points x and y in $R^{m,n}$ or $R^{m,n}_{k\ell}$ is defined by $\delta^2 = (y_0 - x_0)^2$, respectively $\delta^2 = (y_0 - x_0)E_0(y_0 - x_0)$, where the inner products x_0y_0 and $x_0E_0y_0$ have the same form as in (5.11) and (5.12) with one difference: here the only coordinates of x and y whose squares enter in these formulas are $x^1, x^2, \dots, x^m$ and $y^1, y^2, \dots, y^m$ and not $x^0, x^1, \dots, x^m$ and $y^0, y^1 \dots, y^m$. Like in $S^{m,n}$ and $S^{m,n}_{k\ell}$, if $\delta = 0$ the points x and y in $R^{m,n}$ and $R^{m,n}_{k\ell}$ have a second distance d, defined by (5.13), respectively (5.14).

The requirement that affine transformations in E^n preserve the distance δ between these points implies

Theorem 5.5. *The rotations in the spaces $R^{m,n}$ and $R^{m,n}_{k\ell}$ have the form (5.15), where $a, b = 1, 2, \dots, m, \quad u, v = m+1, \dots, n$. The motions in $R^{m,n}$ and $R^{m,n}_{k\ell}$ are products of these rotations and the translations $'x^i = x^i + a^i$.*

The space $R^{1,n}$, respectively $R^{1,n}_{0\ell}$, is the space E^n in whose hyperplane at infinity the geometry of R^{n-1*}, respectively R^{n-1*}_{ℓ}, is given; it is called the *isotropic space* I^n, respectively the *pseudoisotropic space* I^n_ℓ.

The space $R^{n-1,n}$, respectively $R^{n-1,n}_{\ell 0}$, is the space E^n in whose hyperplane at infinity the geometry of the space R^{n-1}, respectively R^{n-1}_{ℓ} is given; it is called the *Galilean space* Γ^n, respectively the *pseudo-Galilean space* Γ^n_ℓ.

The name "isotropic space" is explained by the fact that the geometry of this space exists on isotropic hyperplanes in R^{n+1}_1, the name "Galilean space" is explained by the fact that the formulas of motions in the space Γ^4 coincide with the formulas for transforming the rectangular coordinates x^1, x^2, x^3 and time coordinate t in the space-time of the classical Mechanics of Galilei-Newton (in **5.8.1** we will see that the geometry in this space-time coincides with the geometry in I^4) (see [Ro10]). The plane I^2 coincides with the plane Γ^2 and with the *flag plane* F^2, which is a particular case of a *flag space* F^n which we will consider in **5.2.2.**

We can analogously define the quadratic and Hermitian quasi-Euclidean and quasipseudo-Euclidean spaces $\mathbb{A}R^{m,n}$, $\mathbb{A}R^{m,n}_{k\ell}$, $\mathbb{A}\overline{R}^{m,n}$, $\mathbb{A}\overline{R}^{m,n}_{k\ell}$. For $\mathbb{A}\overline{R}^{m,n}$ and $\mathbb{A}\overline{R}^{m,n}_{k\ell}$ we must replace the quadratic and bilinear forms in (5.11), (5.12), (5.13) and (5.14) by quadratic and bilinear Hermitian forms.

The geometry of the quasielliptic space was founded by W. Blaschke [Bl]. The geometry of the isotropic space and plane was investigated by É. Cartan [Car2, 17] and by K. Strubecker [Str1, 2]. The geometry of the Galilean space was founded by A. P. Kotelnilov [Kot3]. The geometry of quadratic quasielliptic and quasihyperbolic spaces was investigated by T. G. Chakhlenkova [Cha]. For the geometry of quadratic quasielliptic, quasihyperbolic, quasi-Euclidean, quasipseudoelliptic, quasipseudohyperbolic, and quasipseudo-Euclidean spaces see [Ro3]. The geometry of Hermitian quasielliptic and quasihyperbolic spaces was investigated by T. M. Klimanova [Klm] and I. N. Semyonova [Sem]. For the history of the geometry of these spaces see [Ro4, pp. 372–376].

5.1.4. Quasisimple and r-Quasisimple Lie Groups. In **1.1.10** we have defined quasisimple and r-quasisimple associative algebras. We can analogously define a *quasisimple and r-quasisimple Lie algebras.* If in a semisimple Lie algebra $\mathbb{G}$ with basis e_i, there exists an involutive automorphism (1.33), then in this basis the structure equations $[e_i e_j] = C_{ij}^k e_k$ can be written in the form (1.125) (see **1.4.11**). Then the quasisimple Lie algebra $\mathbb{G}^0$ obtained from $\mathbb{G}$ is characterized by the structure equations

$$\begin{gathered}[e_a e_b] = C_{ab}^c e_c, \\ [e_a e_u] = C_{au}^v e_v, \ [e_u e_v] = 0.\end{gathered} \tag{5.16}$$

Since the transition from the Lie algebra $\mathbb{G}$ with structure equations (1.125) to the Lie algebra $\mathbb{G}'$ with structure equations (1.127) is called the Cartan algorithm, the transition from $\mathbb{G}$ to the Lie algebra $\mathbb{G}^0$ with structure equations (5.16) is called the *quasi-Cartan algorithm.*

If the quasi-Cartan algorithm is applied r times to the semisimple Lie algebra, the new Lie algebra obtained is *r-quasisimple,* for $r = 2$ *biquasisimple.*

A *quasisimple,* respectively *r-quasisimple, Lie group* is a Lie group whose tangent Lie algebra is a quasisimple, respectively r-quasisimple, Lie algebra.

Theorem 5.6. *The groups of motions in the quasielliptic, quasihyperbolic, quasi-pseudoelliptic, and quasipseudohyperbolic spaces $S^{m,n}, H^{m,n}, S_{k\ell}^{m,n}$, and $H_{k\ell}^{m,n}$, in particular, in the Euclidean, co-Euclidean, pseudo-Euclidean, and copseudo-Euclidean spaces R^n, R^{n*}, R_ℓ^n, and R_ℓ^{n*}, are quasisimple Lie groups. The groups of motions in the quasi-Euclidean and quasipseudo-Euclidean spaces $R^{m,n}$ and $R_{k\ell}^{m,n}$, in particular, in the isotropic, Galilean, pseudoisotropic, and pseudo-Galilean spaces I^n, Γ^n, I_ℓ^n, and Γ_ℓ^n, are biquasisimple Lie groups.*

<u>Proof</u>. The groups of motions in $S^{m,n}$, $H^{m,n}$, $S_{k\ell}^{m,n}$, and $H_{k\ell}^{m,n}$ which have the form (5.15), are represented by the $((n+1) \times (n+1))$-matrices

$$\mathcal{U} = \begin{bmatrix} U & 0 \\ T & V \end{bmatrix}, \tag{5.17}$$

where U are $((m+1) \times (m+1))$-matrices in the groups O_{m+1} or $O_{m+1,k}$, V are $((n-m) \times (n-m))$-matrices in the groups O_{n-m} or $O_{n-m,\ell}$, and T are $((m+1) \times (n-m))$-matrices with arbitrary entries. These matrices can be obtained from the $((n+1) \times (n+1))$-matrices in the groups O_{n+1} or $O_{n+1,\ell}$ by the quasi-Cartan algorithm since there is an involutive automorphism in O_{n+1} or $O_{n+1,\ell}$, which multiplies all entries U_a^v and U_u^b of an orthogonal or pseudoorthogonal $((n+1) \times (n+1))$-matrix U by -1 and leaves invariant all entries U_a^b and U_u^v of this matrix $(a, b = 0, 1, \dots, m, \ u, v = m+1, \dots, n)$; this involutive automorphism determines an involutive automorphism in the tangent Lie algebra $\mathbb{A}$ of each group. In **1.1.10** we have seen that the quasisimple associative algebra $\mathbb{A}^0$ obtained by the quasi-Cartan algorithm from the semisimple algebra $\mathbb{A} = \mathbb{B} \oplus \mathbb{E}$ with involutive automorphism $\mathbb{A} = \mathbb{B} \oplus \mathbb{E} \to \mathbb{A} = \mathbb{B} \oplus (-\mathbb{E})$ can be represented in the form $\mathbb{A}^0 = \mathbb{B} \oplus \epsilon\mathbb{E}$. This is also true for the quasi-Cartan algorithm applied to semisimple Lie algebras $\mathbb{A}$.

Therefore, if the $((n+1)\times(n+1))$-matrices in the tangent Lie algebra of O_{n+1} or $O_{n+1,\ell}$ can be written in the form

$$\mathcal{A} = \begin{bmatrix} A_0^0 & A_0^1 \\ A_1^0 & A_1^1 \end{bmatrix}, \tag{5.18}$$

where A_0^0 are $((m+1)\times(m+1))$-submatrices, A_0^1 are $((n-m)\times(m+1))$-submatrices, A_1^0 are $((m+1)\times(n-m))$-submatrices, and A_1^1 are $((n-m)\times(n-m))$-submatrices of the matrix (5.18), then the $((n+1)\times(n+1))$-matrices in the Lie algebras $\mathbb{A}^0$ can be obtained from (5.18) by multiplication of the submatrices A_0^1 and A_1^0 by ε. However, this Lie algebra is isomorphic to the Lie algebra of real $((n+1)\times(n+1))$-matrices (5.18) where the submatrices A_0^0 and A_1^1 are $((m+1)\times(m+1))$-matrices and $((n-m)\times(n-m))$-submatrices in the tangent Lie algebras of the groups O_{m+1} or $O_{m+1,k}$, respectively O_{n-m} or $O_{n-m,\ell}$, the submatrix A_0^1 is the zero matrix, and the submatrix A_1^0 consists of arbitrary real numbers. This latter Lie algebra is the tangent Lie algebra of the group (5.17).

The second assertion of this theorem can be similarly proved. ■

We can analogously prove

Theorem 5.7. *The groups of motions in the Hermitian quasielliptic, quasihyperbolic, quasipseudoelliptic, and quasipseudohyperbolic spaces* $\overline{\mathbb{C}S}^{m,n}$, $\overline{\mathbb{C}H}^{m,n}$, $\overline{\mathbb{C}S}^{m,n}_{k\ell}$, $\overline{\mathbb{C}H}^{m,n}_{k\ell}$, $\overline{\mathbb{H}S}^{m,n}$, $\overline{\mathbb{H}H}^{m,n}$, $\overline{\mathbb{H}S}^{m,n}_{k\ell}$, *and* $\overline{\mathbb{H}H}^{m,n}_{k\ell}$, *in particular in the Hermitian Euclidean, co-Euclidean, pseudo-Euclidean, and copseudo-Euclidean spaces* $\overline{\mathbb{C}R}^{n}$, $\overline{\mathbb{C}R}^{n*}$, $\overline{\mathbb{C}R}^{n}_{\ell}$, $\overline{\mathbb{C}R}^{n*}_{\ell}$, $\overline{\mathbb{H}R}^{n}$, $\overline{\mathbb{H}R}^{n*}$, $\overline{\mathbb{H}R}^{n}_{\ell}$, *and* $\overline{\mathbb{H}R}^{n*}_{\ell}$, *are quasisimple Lie groups. The groups of motions in the Hermitian quasi-Euclidean and quasipseudo-Euclidean spaces* $\overline{\mathbb{C}R}^{m,n}$, $\overline{\mathbb{C}R}^{m,n}_{k\ell}$, $\overline{\mathbb{H}R}^{m,n}$, $\overline{\mathbb{H}R}^{m,n}_{k\ell}$, *in particular,in the Hermitian isotropic, Galilean, pseudoisotropic, and pseudo-Galilean spaces* $\overline{\mathbb{C}I}^{n}$, $\overline{\mathbb{C}\Gamma}^{n}$, $\overline{\mathbb{C}I}^{n}_{\ell}$, $\overline{\mathbb{C}\Gamma}^{n}_{\ell}$, $\overline{\mathbb{H}I}^{n}$, $\overline{\mathbb{H}\Gamma}^{n}$, $\overline{\mathbb{H}I}^{n}_{l}$, *and* $\overline{\mathbb{H}\Gamma}^{n}_{l}$, *are biquasisimple Lie groups.*

5.1.5. Quasimetrics of Order One. Spaces in which the points have a distance δ and, when $\delta=0$, have a second distance d are called *quasimetric spaces of the first order* or spaces with a *quasimetric of order one.*

The definitions of the quasielliptic, quasihyperbolic, quasipseudoelliptic, and quasipseudohyperbolic spaces $S^{m,n}$, $H^{m,n}$, $S^{m,n}_{k\ell}$, and $H^{m,n}_{k\ell}$, for $m \neq 0$, in particular, of the co-Euclidean and copseudo-Euclidean spaces R^{n*}, and R^n_ℓ, and the definitions of the quasi-Euclidean and quasipseudo-Euclidean spaces $R^{m,n}$ and $R^{m,n}_{k\ell}$, in particular, of the isotropic, Galilean, pseudoisotropic, and pseudo-Galilean spaces I^n, Γ^n, I^n_ℓ, Γ^n_ℓ, and of their complex and quaternion Hermitian analogues, show that all these spaces are quasimetric spaces of the first order.

In **1.1.3** we mentioned the quasimetric of the flag plane F^2, which coincides with the planes I^2 and Γ^2, in the algebra $\mathbb{C}^0$ of dual numbers. We can define analogous quasimetrics in the algebras $\mathbb{H}^0$ and $\mathbb{H}'^0$ of semiquaternions and split semiquaternions: if we regard the modulus $|y-x|$ of the difference $y-x$ between the elements x and y in one of these algebras as distance δ and, if $\delta=0$, define the second distance d between these elements, which in this case can be written in the form $x_0+\epsilon x_1$, as the modulus $|y_1-x_1|$, we see that the algebras $\mathbb{H}^0$ and $\mathbb{H}'^0$ with these quasimetrics are isometric to the spaces $R^{2,4}$ and $R^{2,4}_{11}$, respectively. These quasimetrics allow us to define in these algebras *metric automorphisms*, analogous to the metric automorphisms in the algebra $\mathbb{C}^0$ (see **1.1.4**).

5.1.6. Invariant Metrics, Pseudometrics, and Quasimetrics in the Groups of Motions of Planes. In **0.8.5** we have defined invariant Riemannian metrics in all compact semisimple Lie groups and invariant pseudo-Riemannian pseudometrics in all noncompact semisimple Lie groups. Since the group SO_3 is isomorphic to the quotient group of the group U of quaternions with unit modulus by its subgroup $\{1,-1\}$, the group U can be regarded as a hypersphere of unit radius in R^4 and the metric of this hypersphere is an invariant metric in U, the group of motions in the plane S^2, which is isomorphic to SO_3 with its invariant metric, is isometric to the elliptic space S^3.

Analogously, since the group $SO_{3,2}$ is isomorphic to the quotient group of the group U' of split quaternions with unit modulus by its subgroup $\{1,-1\}$, the group U' can be regarded as a hypersphere of unit radius in R^4_2 and the pseudometric of this hypersphere is an invariant pseudometric in U'. The connected group of motions in the plane H^2, which is isomorphic to $SO_{3,2}$ with its invariant pseudometric, is isometric to the pseudoelliptic space S^3_2.

Theorem 5.8 (W. Blaschke [Bl]). *The group of motions in the plane R^2 in its invariant quasimetric is isometric to the quasielliptic space $S^{1,3}$ of curvature 1.*

PROOF. Let A and B be two motions in R^2. If the motion BA^{-1} is a rotation over the angle φ, we take this angle to mean the "distance δ" between these motions; if BA^{-1} is a translation $x' = x + a$, we take that $\varphi = 0$ and call the real number $|a|$ the "distance d" between these motions. Since φ and $|a|$ are invariant under the transformations $A \to AC$, $A \to CA$, and $A \to A^{-1}$, this quasimetric is invariant. We assign to each motion $'X^1 = X^1\cos\varphi + X^2\sin\varphi + a^1$, $'X^2 = -X^1\sin\varphi + X^2\cos\varphi + a^2$ in R^2 the point x in $S^{1,3}$ with coordinates $x^0 = \cos\varphi$, $x^1 = \sin\varphi$, $x^2 = a^1$, $x^3 = a^2$. Then the distances δ and d between points x with these coordinates and y with coordinates $y^0 = \cos\psi$, $y^1 = \sin\psi$, $y^2 = b^1$, $y^3 = b^2$ are $\delta = |\psi - \varphi|$, and, when $\varphi = \psi$, $\delta = 0$ and $d = |b - a|$. ■

Note that the term "quasielliptic space" appeared at first for the space $S^{1,3}$ in the paper [Bl], and that the terms "quasihyperbolic space", "quasi-Euclidean space", "quasimetric", "quasisimple groups", etc. were introduced by the analogy with the term of Blaschke.

We can analogously prove

Theorem 5.9. *The group of motions in the plane R^2_1 in its invariant quasimetric is isometric to the quasipseudoelliptic space $S^{1,3}_{11}$ of curvature 1.*

§5.2. r-Quasielliptic, r-Quasihyperbolic, and r-Quasi-Euclidean Spaces.

5.2.1. r-Quasielliptic, r-Quasihyperbolic, r-Quasipseudoelliptic, and r-Quasipseudohyperbolic Spaces. The spaces $S^{m,n}$ and $S^{m,n}_{k\ell}$ are particular cases of the more general *r-quasielliptic spaces*, $S^{m_0,m_1,\dots,m_{r-1},n}$ and *r-quasipseudoelliptic spaces* $S^{m_0,m_1,\dots,m_{r-1},n}_{\ell_0,\ell_1\dots,\ell_r}$, that is, the space P^n in which an imaginary or real hypercone C_0 of the second order with an $(n-m_0-1)$-planar apex A_0, an imaginary or real $(n-m_0-2)$-cone C_1 of the second order with an $(n-m_1-1)$-planar apex

A_1 in A_0, an imaginary or real $(n-m_1-2)$-cone C_2 of the second order with an $(n-m_2-1)$-planar apex A_2 in A_1, etc., up to an imaginary or real nondegenerate $(n-m_{r-1}-2)$-quadric C_r in an $(n-m_{r-1}-1)$-plane A_{r-1}, the apex of the $(n-m_{r-2}-2)$-cone C_{r-1}, are given. These hypercone C_0, $(n-m_{a-1}-2)$-cones C_a, $(n-m_a-1)$-planes A_a, and $(n-m_{r-1}-2)$-quadric C_r are called the *absolute hypercone*, $(n-m_{a-1}-2)$-*cones*, $(n-m_a-1)$-*planes*, and $(n-m_{r-1}-2)$-*quadric*. The equation of the absolute hypercone C_0 of $S^{m_0,m_1,\ldots,m_{r-1},n}$ or $S^{m_0,m_1,\ldots,m_{r-1},n}_{\ell_0,\ell_1\ldots\ell_r}$ can be reduced to the form

$$x_0^2=(x^0)^2+(x^1)^2+\cdots+(x^{m_0})^2=0, \tag{5.19}$$

respectively

$$x_0E_0x_0=-(x^0)^2-(x^1)^2-\cdots-(x^{\ell_0-1})^2+(x^{\ell_0})^2+\cdots+(x^{m_0})^2=0. \tag{5.20}$$

The equations of the $(n-m_0-1)$-planar apex A_0 of this hypercone have the form $x^0=x^1=\cdots=x^{m_0}=0$.

The equation of the absolute $(n-m_{a-1}-2)$-cone C_a of $S^{m_0,m_1,\ldots,m_{r-1},n}$ or $S^{m_0,m_1,\ldots,m_{r-1},n}_{\ell_0,\ell_1\ldots\ell_r}$ in the $(n-m_{a-1}-1)$-plane A_{a-1} can be reduced to the form

$$x_a^2=(x^{m_{a-1}+1})^2+(x^{m_{a-1}+2})^2+\cdots+(x^{m_a})^2=0 \tag{5.21}$$

respectively

$$x_aE_ax_a=-(x^{m_{a-1}+1})^2-\cdots-(x^{m_{a-1}-\ell_a})^2+(x^{m_{a-1}+\ell_a+1})^2\cdots+(x^{m_a})^2=0. \tag{5.22}$$

The equations of the $(n-m_a-1)$-planar apex A_a of C_a have the form $x^0=x^1=\cdots=x^{m_a}=0$.

The equation of the absolute $(n-m_{r-1}-2)$-quadric C_r of this space on the $(n-m_{r-1}-1)$-plane A_{r-1} can be reduced to the form

$$x_r^2=(x^{m_{r-1}+1})^2+(x^{m_{r-1}+2})^2+\cdots+(x^n)^2=0, \tag{5.23}$$

respectively

$$x_rE_rx_r=-(x^{m_{r-1}+1})^2-\cdots-(x^{m_{r-1}+\ell_r})^2+(x^{m_{r-1}+\ell_r+1})^2\cdots+(x^n)^2=0. \tag{5.24}$$

If the projective coordinates x^i of points are normalized by the condition $x_0^2=1$ or $x_0E_0x_0=\pm1$, the distance δ between two points x and y in $S^{m_0,m_1,\ldots,m_{r-1},n}$ or $S^{m_0,m_1,\ldots,m_{r-1},n}_{\ell_0,\ell_1\ldots\ell_r}$ is defined by

$$\cos\frac{\delta}{r}=|x_0y_0|=|x^0y^0+x^1y^1+\cdots+x^{m_0}y^{m_0}|, \tag{5.25}$$

respectively

$$\cos\frac{\delta}{r}=|x_0E_0y_0|=|-x^0y^0-x^1y^1-\cdots-x^{\ell_0-1}y^{\ell_0-1}+x^{\ell_0}y^{\ell_0}+\cdots+x^{m_0}y^{m_0}|, \tag{5.26}$$

where r is the radius of curvature of the space.

If $\delta = 0$, the points x and y in $S^{m_0,m_1,\dots,m_{r-1},n}$ or $S^{m_0,m_1,\dots,m_{r-1},n}_{\ell_0,\ell_1,\dots,\ell_r}$ have a second distance, d_1, defined by

$$d_1^2 = (y_1 - x_1)^2 = (y^{m_0+1} - x^{m_0+1})^2 + (y^{m_0+2} - x^{m_0+2})^2 + \dots + (y^{m_1} - x^{m_1})^2, \quad (5.27)$$

respectively

$$\begin{aligned} d_1^2 &= (y_1 - x_1)E_1(y_1 - x_1) = -(y^{m_0+1} - x^{m_0+1})^2 - \dots \\ &- (y^{m_0+\ell_1} - x^{m_0+\ell_1})^2 + (y^{m_0+\ell_1+1} - x^{m_0+\ell_1+1}) + \dots \\ &+ (y^{m_1} - x^{m_1})^2. \end{aligned} \quad (5.28)$$

If $\delta = d_1 = d_2 = \dots = d_{a-1} = 0$, the points x and y in $S^{m_0,m_1,\dots,m_{r-1},n}$ or $S^{m_0,m_1,\dots,m_{r-1},n}_{\ell_0,\ell_1,\dots,\ell_r}$ have an $(a+1)$-th distance, d_a, defined by

$$\begin{aligned} d_a^2 &= (y_a - x_a)^2 = (y^{m_{a-1}+1} - x^{m_{a-1}+1})^2 \\ &+ (y^{m_{a-1}+2} - x^{m_{a-1}+2})^2 + \dots + (y^{m_a} - x^{m_a})^2, \end{aligned} \quad (5.29)$$

respectively

$$\begin{aligned} d_a^2 &= (y_a - x_a)E_a(y_a - x_a) \\ &= -(y^{m_{a-1}+1} - x^{m_{a-1}+1})^2 - \dots - (y^{m_{a-1}+\ell_a} - x^{m_{a-1}+\ell_a})^2 \\ &+ (y^{m_{a-1}+\ell_a+1} - x^{m_{a-1}+\ell_a+1})^2 + \dots + (y^{m_a} - x^{m_a})^2. \end{aligned} \quad (5.30)$$

If $\delta = d_1 = d_2 = \dots = d_{r-1} = 0$, the points x and y in $S^{m_0,m_1,\dots,m_{r-1},n}$ or $S^{m_0,m_1,\dots,m_{r-1},n}_{\ell_0,\ell_1\dots\ell_r}$ have an $(r+1)$-th distance, d_r, defined by

$$\begin{aligned} d_r^2 &= (y_r - x_r)^2 = (y^{m_{r-1}+1} - x^{m_{r-1}+1})^2 \\ &+ (y^{m_{r-1}+2} - x^{m_{r-1}+2})^2 + \dots + (y^n - x^n)^2, \end{aligned} \quad (5.31)$$

respectively

$$\begin{aligned} d_r^2 &= (y_r - x_r)E_r(y_r - x_r) \\ &= -(y^{m_{r-1}+1} - x^{m_{r-1}+1})^2 - \dots - (y^{m_{r-1}+\ell_r} - x^{m_{r-1}+\ell_r})^2 \\ &+ (y^{m_{r-1}+\ell_r+1} - x^{m_{r-1}+\ell_r+1}) + \dots + (y^n - x^n)^2. \end{aligned} \quad (5.32)$$

Note that if $m_a - m_{a-1} = 1$ the $(n - m_{a-1} - 2)$-cone C_a is the double $(n - m_a - 1)$-plane A_a and if $m_{r-1} = n - 1$, the $(n - m_{r-2} - 2)$-cone C_{r-1} has a point apex and the $(n - m_{r-1} - 2)$-quadric C_r is the double point coinciding with this apex.

The requirement that collineations in P^n preserve the distance δ between points in $S^{m_0,m_1,\dots,m_{r-1},n}$ or $S^{m_0,m_1,\dots,m_{r-1},n}_{\ell_0,\ell_1\dots\ell_r}$ and, when $\delta = d_1 = d_2 = \dots = d_{a-1} = 0$, preserve the distance d_a between these points implies

Theorem 5.10. *The motions in $S^{m_0,m_1,\dots,m_{r-1},n}$ and $S^{m_0,m_1,\dots,m_{r-1},n}_{\ell_0,\ell_1\dots\ell_r}$ have the form*

$$\left.\begin{aligned}
{}'x^{i_0} &= U^{i_0}_{j_0}x^{j_0}\\
{}'x^{i_1} &= T^{i_1}_{j_0}x^{j_0} + U^{i_1}_{j_1}x^{j_1}\\
{}'x^{i_2} &= T^{i_2}_{j_0}x^{j_0} + T^{i_2}_{j_1}x^{j_1} + U^{i_2}_{j_2}x^{j_2}\\
&\dots\dots\dots\dots\dots\dots\dots\dots\dots\dots\dots\dots\\
{}'x^{i_r} &= T^{i_r}_{j_0}x^{j_0} + T^{i_r}_{j_1}x^{j_1} + \dots + T^{i_r}_{j_{r-1}}x^{j_{r-1}} + U^{i_r}_{j_r}x^{j_r}\\
(i_0,j_0 &= 0,1,\dots,m_0,\ i_1,j_1 = m_0+1,\dots,m_1,\\
i_2,j_2 &= m_1+1,\dots,m_2,\dots,\quad i_r,j_r = m_{r-1}+1,\dots,n),
\end{aligned}\right\} \tag{5.33}$$

where $(U^{i_0}_{j_0})$ are $((m_0+1)\times(m_0+1))$-matrices in the groups O_{m_0+1}, respectively O_{m_0+1,ℓ_0}, $(U^{i_a}_{j_a})$ are $((m_a-m_{a-1})\times(m_a-m_{a-1}))$-matrices in the groups $O_{m_a-m_{a-1}}$, respectively $O_{m_a-m_{a-1},\ell_a}$, $(U^{i_r}_{j_r})$ are $((n-m_{r-1})\times(n-m_{r-1}))$-matrices in the groups $O_{n-m_{r-1}}$, respectively O_{n-m_{r-1},ℓ_r} and $(T^{i_a}_{j_b})$ are $((m_a-m_{a-1})\times(m_b-m_{b-1}))$-matrices with arbitrary entries.

An *r-quasihyperbolic space* $H^{m_0,m_1,\dots,m_{r-1},n}$ or *r-quasipseudohyperbolic space* $H^{m_0,m_1,\dots,m_{r-1},n}_{\ell_0,\ell_1\dots\ell_r}$ has the same absolute as the quasipseudoelliptic space $S^{m_0,m_1,\dots,m_{r-1},n}_{10}$, respectively $S^{m_0,m_1,\dots,m_{r-1},n}_{\ell_0+1,\ell_1\dots\ell_r}$, but its radius of curvature is pure imaginary, $r = iq$, and, therefore, in formulas (5.25) and (5.26) the expression $\cos\frac{\delta}{r}$ must be replaced by $\cosh\frac{\delta}{q}$.

We can analogously define the quadratic and Hermitian r-quasielliptic, r-quasihyperbolic, r-quasipseudoelliptic, and r-quasipseudohyperbolic spaces $\mathbb{A}S^{m_0,m_1,\dots,m_{r-1},n}$, $\mathbb{A}H^{m_0,m_1,\dots,m_{r-1},n}$, $\mathbb{A}S^{m_0,m_1,\dots,m_{r-1},n}_{\ell_0,\ell_1,\dots,\ell_r}$, $\mathbb{A}H^{m_0,m_1,\dots,m_{r-1},n}_{\ell_0,\ell_1\dots ell_r}$, $\mathbb{A}\overline{S}^{m_0,m_1,\dots,m_{r-1},n}$, $\mathbb{A}\overline{H}^{m_0,m_1,\dots,m_{r-1},n}$, $\mathbb{A}\overline{S}^{m_0,m_1,\dots,m_{r-1},n}_{\ell_0\ell_1,\dots,\ell_r}$, and $\mathbb{A}\overline{H}^{m_0,m_1,\dots,m_{r-1},n}_{\ell_0,\ell_1,\dots,\ell_r}$.

For $\mathbb{A}\overline{S}^{m_0,m_1,\dots,m_{r-1},n}$, $\mathbb{A}\overline{H}^{m_0,m_1,\dots,m_{r-1},n}$, $\mathbb{A}\overline{S}^{m_0,m_1,\dots,m_{r-1},n}_{\ell_0,\ell_1,\dots,\ell_r}$, and $\mathbb{A}\overline{H}^{m_0,m_1,\dots,m_{r-1},n}_{\ell_0,\ell_1,\dots,\ell_r}$ the quadratic and bilinear forms in formulas (5.19), (5.20), (5.21), (5.22), (5.23), (5.24), (5.25), (5.26), (5.27), (5.28), (5.29), (5.30), (5.31), and (5.32) must be replaced by quadratic and bilinear Hermitian forms.

Theorem 5.11. *All r-quasielliptic, r-quasihyperbolic, r-quasipseudoelliptic, and r-quasipseudohyperbolic spaces can be obtained by the r-fold limiting process from elliptic, hyperbolic, pseudoelliptic, and pseudohyperbolic spaces.*

The proof is analogous to the proof of Theorem 5.4.

Theorem 5.12. *The groups of motions in the r-quasielliptic, r-quasihyperbolic, r-quasipseudoelliptic, and r-quasipseudohyperbolic spaces $S^{m_0,m_1,\dots,m_{r-1},n}$, $H^{m_0,m_1,\dots,m_{r-1},n}$, $S^{m_0,m_1,\dots,m_{r-1},n}_{\ell_0,\ell_1,\dots,\ell_r}$, and $H^{m_0,m_1,\dots,m_{r-1},n}_{\ell_0,\ell_1,\dots,\ell_r}$, and the groups of motions in the Hermitian r-quasielliptic, r-quasihyperbolic, r-quasipseudoelliptic, and r-quasipseudohyperbolic spaces $\mathbb{C}\overline{S}^{m_0,m_1,\dots,m_{r-1},n}$, $\mathbb{C}\overline{H}^{m_0,m_1,\dots,m_{r-1},n}$,*

$\mathbb{C}S^{m_0,m_1,\ldots,m_{r-1},n}_{\ell_0,\ell_1,\ldots,\ell_r}$, $\mathbb{C}H^{m_0,m_1,\ldots,m_{r-1},n}_{\ell_0,\ell_1,\ldots,\ell_r}$, $\mathbb{H}\overline{S}^{m_0,m_1,\ldots,m_{r-1},n}$, $\mathbb{H}\overline{H}^{m_0,m_1,\ldots,m_{r-1},n}_{\ell_0,\ell_1,\ldots,\ell_r}$, $\mathbb{H}\overline{S}^{m_0,m_1,\ldots,m_{r-1},n}_{\ell_0,\ell_1,\ldots,\ell_r}$, *and* $\mathbb{H}\overline{H}^{m_0,m_1,\ldots,m_{r-1},n}_{\ell_0,\ell_1,\ldots,\ell_r}$, *are r-quasisimple Lie groups.*

The proof is analogous to the proofs of Theorems 5.6 and 5.7.

A space $\mathbb{A}P^n$, in which the only absolute $(n-m_a-1)$-planes A_a $(a=0,1,...,r$ are given, is called *r-quasiaffine n-space* and is denoted by $\mathbb{A}P^{m_0,m_1,\ldots,m_r,n}$.

Collineations preserving the absolute $(n-m_a-1)$-planes A_0 in $\mathbb{A}P^{m_0,m_1,\ldots,m_r,n}$ are called *r-quasiaffine transformations* in this space. These transformations have the form (5.33) where the matrices $(U^{i_a}_{j_a})$ consist of arbitrary entries.

5.2.2. r-Quasi-Euclidean and r-Quasipseudo-Euclidean Spaces. The spaces $S^{m_0,m_1,\ldots,m_{r-1},n}$ and $S^{m_0,m_1,\ldots,m_{r-1},n}_{\ell_0,\ell_1,\ldots,\ell_r}$ with $m_0=0$ are called *r-quasi-Euclidean spaces* and *r-quasipseudo-Euclidean spaces*, and are denoted by $R^{m_1,m_2,\ldots,m_{r-1},n}$, respectively $R^{m_1,m_2,\ldots,m_{r-1},n}_{\ell_1,\ell_2,\ldots,\ell_r}$.

These spaces can also be defined as the affine space E^n in whose hyperplane at infinity the geometry of the space $S^{m_1,m_2,\ldots,m_{r-1},n}$ or $S^{m_1,m_2,\ldots,m_{r-1},n}_{\ell_1,\ell_2,\ldots,\ell_r}$ is given.

In these spaces there are only distances $d_1,d_2,\ldots,d_r$.

Analogously, the spaces $\mathbb{A}S^{m_0,m_1,\ldots,m_{r-1},n}$, $\mathbb{A}S^{m_0,m_1,\ldots,m_{r-1},n}_{\ell_0,\ell_1,\ldots,\ell_r}$, $\mathbb{A}\overline{S}^{m_0,m_1,\ldots,m_{r-1},n}$, and $\mathbb{A}\overline{S}^{m_0,m_1,\ldots,m_{r-1},n}_{\ell_0,\ell_1,\ldots,\ell_r}$ with $m_0=0$ are called *quadratic and Hermitian r-quasi-Euclidean spaces* and *r-quasipseudo-Euclidean spaces over the algebra* $\mathbb{A}$ and are denoted by $\mathbb{A}R^{m_1,m_2,\ldots,m_{r-1},n}$, $\mathbb{A}R^{m_1,m_2,\ldots,m_{r-1},n}_{\ell_1,\ell_2,\ldots,\ell_r}$, $\mathbb{A}\overline{R}^{m_1,m_2,\ldots,m_{r-1},n}$, and $\mathbb{A}\overline{R}^{m_1,m_2,\ldots,m_{r-1},n}_{\ell_1,\ell_2,\ldots,\ell_r}$.

The space $R^{12\ldots n}$, that is, n-quasi-Euclidean space whose absolute is a complete *flag* consisting of a hyperplane, an $(n-2)$-plane in this hyperplane, an $(n-3)$-plane in this $(n-2)$-plane, ..., a plane in a 3-plane, a line in this plane, and a point in this line, is called a *flag space* and is denoted by F^n.

The group of motions in F^n is a particular case of the group (5.33) and has the form

$$\left.\begin{aligned}
'x^1 &= \pm x^1 + a^1,\\
'x^2 &= T^2_1 x^1 \pm x^2 + a^2,\\
&\ldots\ldots\ldots\ldots\ldots\ldots\ldots\ldots\ldots\ldots\\
'x^n &= T^n_1 x^1 + T^n_2 x^2 + \cdots + T^n_{n-1} x^{n-1} + \pm x^n + a^n.
\end{aligned}\right\} \tag{5.34}$$

Note that although the absolute of F^n coincides with absolute of n-quasiaffine n-space $P^{0,1,\ldots,n-1,n}$ the group of motions in F^n does not coincide with the group of n-quasiaffine transformations in $P^{0,1,\ldots,n-1,n}$, the last group is represented by triangular $((n+1)\times(n+1))$-matrices, the first group is a subgroup of the last group.

Since r-quasi-Euclidean and r-quasipseudo-Euclidean spaces are particular cases of r-quasielliptic and r-quasipseudoelliptic spaces, Theorem 5.12 implies that the groups of motions in the spaces $R^{m_1,m_2,\ldots,m_{r-1},n}$, $R^{m_1,m_2,\ldots,m_{r-1},n}_{\ell_1,\ell_2,\ldots,\ell,r}$, $\mathbb{C}\overline{R}^{m_1,m_2,\ldots,m_{r-1},n}$, $\mathbb{C}\overline{R}^{m_1,m_2,\ldots,m_{r-1},n}_{\ell_1,\ell_2,\ldots,\ell_r}$, $\mathbb{H}\overline{R}^{m_1,m_2,\ldots,m_{r-1},n}$, and $\mathbb{H}\overline{R}^{m_1,m_2,\ldots,m_{r-1},n}_{\ell_1,\ell_2,\ldots,\ell_r}$ are r-quasisimple Lie groups.

General real r-quasielliptic, r-quasihyperbolic, r-quasipseudoelliptic and r-quasipseudohyperbolic spaces were defined by D. M. Y. Sommerville [Smv] and, independently, by Ch. Müntz [Mü]. These spaces were investigated by I. M. Yaglom and E. U. Yasinskaya [YRY]. For the geometry of these spaces see [Ro3].

For the history of the geometry of quadratic and Hermitian r-quasielliptic, r-quasihyperbolic, r-quasi-Euclidean, r-quasipseudoelliptic, r-quasipseudohyperbolic, and r-quasipseudo-Euclidean spaces see [Ro4, pp. 376–377].

5.2.3. Quasimetrics of higher orders. Spaces in which points have a distance δ and, when $\delta = 0$, have a second distance d_1, and, when $\delta = d_1 = 0$, have a third distance $d_2, \ldots$, and, when $\delta = d_1 = \cdots = d_{r-1} = 0$, have an $(r+1)$-th distance d_r, are called *quasimetric spaces of order* r.

The definitions of the r-quasielliptic, r-quasihyperbolic, r-quasipseudoelliptic, and r-quasipseudohyperbolic spaces $S^{m_0,m_1,\ldots,m_{r-1},n}$, $H^{m_0,m_1,\ldots,m_{r-1},n}$, $S^{m_0,m_1,\ldots,m_{r-1},n}_{\ell_0,\ell_1,\ldots,\ell_r}$, and $H^{m_0,m_1,\ldots,m_{r-1},n}_{l_0,\ell_1,\ldots,\ell_r}$ for $m_0 \neq 0$, and the definitions of the r-quasi-Euclidean and r-quasipseudo-Euclidean spaces $R^{m_1,m_2,\ldots,m_{r-1},n}$ and $R^{m_1,m_2,\ldots,m_{r-1},n}_{\ell_1,\ell_2,\ldots,\ell_r}$ and of their complex and quaternionic Hermitian analogs show that all these spaces are quasimetric spaces of order r.

§5.3. Hyperquadrics, Hyperspheres, and Hypercycles

5.3.1. Centers of Hyperquadrics and Hermitian Hyperquadrics. Like hyperquadrics in S^n and S^n_ℓ and Hermitian hyperquadrics in $\mathbb{C}\overline{S}^n$, $\mathbb{C}\overline{S}^n_\ell$, $\mathbb{H}\overline{S}^n$, and $\mathbb{H}\overline{S}^n_\ell$, hyperquadrics in $S^{m_0,m_1,\ldots,m_{r-1},n}$ and $S^{m_0,m_1,\ldots,m_{r-1},n}_{\ell_0,\ell_1\ldots\ell_r}$ and Hermitian hyperquadrics in $\mathbb{C}\overline{S}^{m_0,m_1,\ldots,m_{r-1},n}$, $\mathbb{C}\overline{S}^{m_0,m_1,\ldots,m_{r-1},n}_{\ell_0\ell_1,\ldots,\ell_r}$, $\mathbb{H}\overline{S}^{m_0,m_1,\ldots,m_{r-1},n}$, and $\mathbb{H}\overline{S}^{m_0,m_1,\ldots,m_{r-1},n}_{\ell_0.\ell_1,\ldots,\ell_r}$ also have *centers*, that is, vertices of a simplex which can also be called *autopolar simplex* of the (Hermitian) hyperquadric.

The centers of (Hermitian) hyperquadrics Q in $S^{m_0,m_1,\ldots,m_{r-1},n}$, $S^{m_0,m_1,\ldots,m_{r-1},n}_{\ell_0,\ell_1,\ldots,\ell_r}$, $\mathbb{C}\overline{S}^{m_0,m_1,\ldots,m_{r-1},n}$, $\mathbb{C}\overline{S}^{m_0,m_1,\ldots,m_{r-1},n}_{\ell_0,\ell_1,\ldots,\ell_r}$, $\mathbb{H}\overline{S}^{m_0,m_1,\ldots,m_{r-1},n}$, $\mathbb{H}\overline{S}^{m_0,m_1,\ldots,m_{r-1},n}_{\ell_0,\ell_1\ldots\ell_r}$, which are not tangent to the absolute $(n-m_a-1)$-planes A_a $(a = 0, 1, \ldots, r-1)$ and to the absolute hypercones C_0, absolute $(n-m_{a-1}-2)$-cones $C_a (a = 1, 2, \ldots, r-1)$ and absolute $(n-m_{r-1}-2)$-quadrics C_r of these spaces can be defined as follows: we denote the polar m_0-plane of A_0 with respect to Q by A'_0, and the polar $(m_a - m_{a-1} - 1)$-plane of A_a in A_{a-1} with respect to the intersection of Q with A_{a-1}, denoted by Q_a, by A'_a $(a = 1, 2, \ldots, r-1)$. We denote the intersection of C_a with A_{a-1} by C'_a. Then the centers of Q are the vertices of the common autopolar simplexes of C'_a and Q_a $(0, 1, \ldots, r-1)$ and the vertices of the common autopolar simplexes of C_r and Q_r.

Note that for $m_0 = 0$ and $m_a - m_{a-1} = 1$, A'_0 and A'_a are points and these points are centers of Q.

Theorem 5.13. *The centers of a hyperquadric $xAx = 0$ or Hermitian hyperquadric $\overline{x}Ax = 0$ in a space $S^{m_0,m_1,\ldots,m_{r-1},n}$, $S^{m_0,m_1,\ldots,m_{r-1},n}_{\ell_0,\ell_1,\ldots,\ell_r}$, $\mathbb{C}\overline{S}^{m_0,m_1,\ldots,m_{r-1},n}$, $\mathbb{C}\overline{S}^{m_0,m_1,\ldots,m_{r-1},n}_{\ell_0,\ell_1,\ldots,\ell_r}$, $\mathbb{H}\overline{S}^{m_0,m_1,\ldots,m_{r-1},n}$, $\mathbb{H}\overline{S}^{m_0,m_1,\ldots,m_{r-1},n}_{\ell_0,\ell_1,\ldots,\ell_r}$, with absolute hypercone $x_0E_0x_0 = 0$ or $\overline{x}_0E_0x_0 = 0$, absolute $(n-m_{a-1}-2)$-cone $x_aE_ax_a = 0$ or $\overline{x}_aE_ax_a = 0$, and absolute $(n-m_{r-1}-2)$-quadric $x_rE_rx_r = 0$ or $\overline{x}_rE_rx_r = 0$, which are not tangent to the absolute $(n-m_a-1)$-planes A_a $(a = 0, 1, \ldots, r-1)$ and to the absolute hypercone C_0, the absolute $(n-m_{a-1}-2)$-cones C_a $(a = 1, 2, \ldots, r-1)$ and the absolute $(n-m_{r-1}-2)$-quadric C_r in the spaces, are determined by the conditions $(A-\lambda\hat{E}_0)x = 0$ and $(\hat{A}_a - \lambda\hat{E}_a)\hat{x}_a = 0$ $(a = 1, 2, \ldots, r)$, where $\hat{E}_0$ is the $((n+1)\times(n+1))$-matrix in whose upper left corner there is the $((m_0+1)\times(m_0+1))$-matrix E_0, $\hat{A}_a$ is the submatrix of the matrix A consisting of the entries of its last*

$n - m_{a-1} + 1$ rows and columns, $\hat{E}_a$ is the $((n-m_{a-1}+1)\times(n-m_{a-1}+1))$-matrix in whose upper left corner there is the $((m_a+1)\times(m_a+1))$-matrix E_a, and $\hat{x}_a$ is the vector in $(n-m_{a-1}+1)$-space whose coordinates are the last $n-m_{a-1}+1$ coordinates of the vector x.

The proof is analogous to the proof of Theorem 4.37.

5.3.2. Hyperspheres. A hypersphere of radius R in the space $S^{m_0,m_1,\ldots,m_{r-1},n}$ with center a can be defined by the condition $(a_0E_0x_0)^2 = a_0E_0a_0 \cdot x_0E_0x_0 \cdot \cos^2 \frac{R}{r}$. Therefore, a hypersphere in $S^{m_0,m_1,\ldots,m_{r-1},n}$ is a particular case of a hyperquadric in this space. Analogously we can prove that a hypersphere in $S^{m_0,m_1,\ldots,m_{r-1},n}_{\ell_0,\ell_1,\ldots,\ell_r}$ is a particular case of a hyperquadric in this space and that a Hermitian hypersphere in $\mathbb{C}\overline{S}^{m_0,m_1,\ldots,m_{r-1},n}$, $\mathbb{C}\overline{S}^{m_0,m_1,\ldots,m_{r-1},n}_{\ell_0,\ell_1,\ldots,\ell_r}$, $\mathbb{H}\overline{S}^{m_0,m_1,\ldots,m_{r-1},n}$, or $\mathbb{H}\overline{S}^{m_0,m_1,\ldots,m_{r-1},n}_{\ell_0,\ell_1,\ldots,\ell_r}$ is a particular case of a Hermitian hyperquadric in this space.

Theorem 5.14. *The spaces $S^{m_0,m_1,\ldots,m_{r-1},n}$ and $S^{m_0,m_1,\ldots,m_{r-1},n}_{\ell_0,\ell_1,\ldots,\ell_r}$ of curvature $\frac{1}{r^2}$ can, for $m_0 \neq 0$, be represented as hyperspheres of radius r in $R^{m_0+1,m_1+1,\ldots,m_{r-1}+1,n+1}$ and $R^{m_0+1,m_1+1,\ldots,m_{r-1}+1,n+1}_{\ell_0,\ell_1,\ldots,\ell_r}$ with identified antipodal points.*

PROOF. The assertion of this theorem follows from the coincidence of distances between points in real r-quasielliptic and r-quasipseudoelliptic spaces with distances between points on hyperspheres in r-quasi-Euclidean and r-quasipseudo-Euclidean spaces with the same coordinates. ■

5.3.3. Hypercycles. In the spaces $S^{m_0,m_1,\ldots,m_{r-1},n}$ and, $S^{m_0,m_1,\ldots,m_{r-1},n}_{\ell_0,\ell_1,\ldots,\ell_r}$, next to hyperspheres there is another important particular case of hyperquadrics. These hyperquadrics can be interpreted on hyperspheres in the spaces $R^{m_0+1,m_1+1,\ldots,m_{r-1}+1,n+1}$ and $R^{m_0+1,m_1+1,\ldots,m_{r-1}+1,n+1}_{\ell_0,\ell_1,\ldots,\ell_r}$ as sections by hyperplanes. Since a section of a hyperquadric in P^{n+1} by a hyperplane is a hyperquadric in P^n, these hypersurfaces are hyperquadrics. We call these hyperquadrics *hypercycles*; for $n = 2$ *cycles*. If a hypersphere in $R^{m_0+1,m_1+1,\ldots,m_{r-1}+1,n+1}$ or $R^{m_0+1,m_1+1,\ldots,m_{r-1}+1,n+1}_{\ell_0,\ell_1,\ldots,\ell_r}$ is defined by the equation $\sum_{i_0} \epsilon_{i_0}(x^{i_0})^2 = r^2$, its section by a hyperplane $p_ix^i = p$ is defined by

$$p^2 \sum_{i_0} \epsilon_{i_0}(x^{i_0})^2 = r^2(p_ix^i)^2. \tag{5.35}$$

We can define hypercycles also in the r-quasi-Euclidean and r-quasipseudo-Euclidean spaces $R^{m_1,m_2,\ldots,m_{r-1},n}$ and $R^{m_1,m_2,\ldots,m_{r-1},n}_{\ell_1,\ell_2,\ldots,\ell_r}$. Let us locate these spaces in $R^{m_0+1,m_1+1,\ldots,m_{r-1}+1,n+1}$ and $R^{m_0+1,m_1+1,\ldots,m_{r-1}+1,n+1}_{\ell_0,\ell_1,\ldots,\ell_r}$ as the hyperplane $x^0 = 0$ and project this hyperplane onto the hypersphere $\sum_{i_0} \epsilon_{i_0}(x^{i_0})^2 = 1$ from its point $(1,0,0,\ldots,0)$. So, a hyperquadric in the r-quasi-Euclidean or r-quasipseudo-Euclidean space whose projection onto a hypersphere corresponds to a hypercycle in the r-quasielliptic or r-quasipseudoelliptic space, is called a hypercycle in the r-quasi-Euclidean or r-quasipseudo-Euclidean space.

Using the formulas of stereographic projection, which maps points $x(x^i)$ in the hyperplane $x^0 = 0$ to points $y(y^0, y^i)$ on the hypersphere

$$y^0 = \frac{1 - \sum\limits_{i_1} \epsilon_{i_1}(x^{i_1})^2}{\sum\limits_{i_1} \epsilon_{i_1}(x^{i_1})^2 + 1}, \qquad y^i = \frac{2x^i}{\sum\limits_{i_1} \epsilon_{i_1}(x^{i_1})^2 + 1}, \tag{5.36}$$

obtained analogously to (3.71), we find that a hypercycle in $R^{m_1,m_2,\ldots,m_{r-1},n}$ and $R^{m_1,m_2,\ldots,m_{r-1},n}_{\ell_1,\ell_2,\ldots,\ell_r}$ is defined by the equation (for details see [Ro3, p. 357])

$$\sum_{i_1} \epsilon_{i_1}(x^{i_1})^2 + 2a_i x^i + a = 0. \tag{5.37}$$

5.3.4. Cycles in the Flag Plane. In particular, in the flag plane F^2 (which coincides with the planes I^2 and Γ^2) formula (5.37) has the form

$$(x^1)^2 + 2a_1x^1 + 2a_2x^2 + a = 0. \tag{5.38}$$

Note that on the complex plane with the metric of R^2 or the split complex plane with the pseudometric of R^n_1 the equation

$$A\overline{x}x + \overline{b}x + \overline{x}b + C = 0 \quad (A = \overline{A}, C = \overline{C}) \tag{5.39}$$

represents circles, and in dual plane with the quasimetric of F^2 this equation represents a pair of isotropic lines. However the equation

$$A\overline{x}x + \overline{b}x - \overline{x}b + C = 0 \quad (A = -\overline{A}, C = -\overline{C}) \tag{5.40}$$

in $\mathbb{C}$ and $\mathbb{C}'$ with the same metric and pseudometric also represents a circle in R^2 and R^2_1, but in $\mathbb{C}^0$ with the same quasimetric (5.40) coincides with (5.38), that is, then it represents a cycle in F^2.

5.3.5. Angles between Hypercycles. Like between two hyperspheres in S^n, S^n_ℓ, R^n or R^n_ℓ, between two hypercycles in $S^{m_0,m_1,\ldots,m_{r-1},n}$, $S^{m_0,m_1,\ldots,m_{r-1},n}_{\ell_0,\ell_1,\ldots,\ell_r}$, $R^{m_1,m_2,\ldots,m_{r-1},n}$ or $R^{m_1,m_2,\ldots,m_{r-1},n}_{\ell_1,\ell_2,\ldots,\ell_r}$ we can define an *angle*.

Analogously to Theorem 4.22 we can prove

Theorem 5.15. *The manifold of hypercycles in* $S^{m_0,m_1,\ldots,m_{r-1},n}$, $S^{m_0,m_1,\ldots,m_{r-1},n}_{\ell_0,\ell_1,\ldots,\ell_r}$, $R^{m_1,m_2,\ldots,m_{r-1},n}$ *or* $R^{m_1,m_2,\ldots,m_{r-1},n}_{\ell_1,\ell_2,\ldots,\ell_r}$, *if we take as distance between two hypercycles the angle between them, is isometric to the space* $S^{n-m_{r-1},\ldots,n-m_0,n+1}_{00,\ldots,01}$, $S^{n-m_{r-1},\ldots,n-m_0,n+1}_{\ell_r,\ell_{r-1},\ldots,\ell_1,\ell_0+1}$, $S^{n-m_{r-1},\ldots,n-m_1,n+1}_{00,\ldots,01}$, *respectively* $S^{n-m_{r-1},\ldots,n-m_1,n+1}_{\ell_r,\ell_{r-1},\ldots,\ell_2,\ell_1+1}$.

For details of the proof see [Ro3, pp. 360-365].

In particular, since the flag plane F^2 is the quasi-Euclidean plane $R^{1,2}$, the manifold of cycles in this plane, if as distance between two cycles we take the angle between them, is isometric to the space $S^{2,3}_{001}$, that is, the copseudo-Euclidean space R^{3*}_1.

5.3.6. Conformal Transformations and Quasiconformal, Quasipseudoconformal, r-Quasiconformal, and r-Quasipseudoconformal Spaces. The transformations in $S^{m_0,m_1,\dots,m_{r-1},n}$, $S^{m_0,m_1,\dots,m_{r-1},n}_{\ell_0,\ell_1,\dots,\ell_r}$, $R^{m_1,m_2,\dots,m_{r-1},n}$ and $R^{m_1,m_2,\dots,m_{r-1},n}_{\ell_1,\ell_2,\dots,\ell_r}$ preserving the hypercycles in these spaces and the angles between them are called *conformal transformations* in these spaces.

The spaces $R^{m_1,m_2,\dots,m_{r-1},n}$, $R^{m_1,m_2,\dots,m_{r-1},n}_{\ell_1,\ell_2,\dots,\ell_r}$, $S^{m_1,m_2,\dots,m_{r-1},n}$, and $S^{m_1,m_2,\dots,m_{r-1},n}_{\ell_1,\ell_2,\dots,\ell_r}$ with as fundamental groups the groups of conformal transformations supplemented by points such that these transformations become bijective, are called the *r-quasiconformal spaces* $C^{m_1,m_2,\dots,m_{r-1},n}$ and *r-quasipseudoconformal spaces* $C^{m_1,m_2,\dots,m_{r-1},n}_{\ell_1,\ell_2,\dots,\ell_r}$, respectively. For $r = 1$ they are the *quasiconformal spaces* $C^{m,n}$ and *quasipseudoconformal spaces* $C^{m,n}_{k\ell}$.

Note that in the interpretation of the manifolds of hypercycles in these spaces as the spaces $S^{n-m_{r-1},\dots,n-m_0,n+1}_{00,\dots,01}$ and $S^{n-m_{r-1},\dots,n-m_1,n+1}_{\ell_r,\ell_{r-1},\dots,\ell_2,\ell_1+1}$, the conformal transformations in $C^{m_1,m_2,\dots,m_{r-1},n}$ and $C^{m_1,m_2,\dots,m_{r-1},n}_{\ell_1,\ell_2,\dots,\ell_r}$ are interpreted as motions in $S^{n-m_{r-1},\dots,n-m_0,n+1}_{00,\dots,01}$ and $S^{n-m_{r-1},\dots,n-m_1,n+1}_{\ell_r,\ell_{r-1},\dots,\ell_2,\ell_1+1}$.

In particular, the flag plane F^2 supplemented by one point at infinity is the *quasiconformal plane* $C^{1,2}$. Like the planes C^2 and C^2_1, which can be interpreted as the extended complex and split complex planes, respectively, the plane $C^{1,2}$ can be interpreted as the extended dual plane, and *circular transformations* in $C^{1,2}$, that is, bijections in this plane preserving its cycles and the angles between them, like circular transformations in C^2 and C^2_1, are represented by the fractional linear transformations (2.46) and (2.47) of a dual variable.

An involutive conformal transformation in $C^{m_1,m_2,\dots,m_{r-1},n}$ or $C^{m_1,m_2,\dots,m_{r-1},n}_{\ell_1,\ell_2,\dots,l_r}$ preserving the points of a hypercycle is called *inversion in the hypercycle.*

For quasiconformal n-space and a hypercycle, which can be regarded as a paraboloid in E^n, inversion in this hypercycle is inversion in paraboloid in E^n (see **2.5.6**). In particular, the inversion in the cycle (5.38) in F^2 is the inversion in a parabola considered by Apollonius (see [Ro4, p. 116]). This inversion maps each point x in this plane to the point x' in this plane such that the line xx' is the isotropic line $x^1 = \text{const}$ and the points x and x' are symmetric with respect to the point of intersection of the line xx' with this cycle.

§5.4. Lines, m-Planes and Symmetry Figures

5.4.1. Elliptic, Hyperbolic, and Parabolic Lines. A line in a space $S^{m_0,m_1,\dots,m_{r-1},n}$ or $S^{m_0,m_1,\dots,m_{r-1},n}_{\ell_0,\ell_1,\dots,\ell_r}$ is called *elliptic* or *hyperbolic of kind zero* if it meets the hypercone C_0 in two conjugate imaginary, respectively real, points. It is called *elliptic* or *hyperbolic of kind a* if it lies in A_{a-1} and meets the $(n-m_{a-1}-2)$-cone C_a in two conjugate imaginary, respectively real, points. It is called *elliptic* or *hyperbolic of kind r* if it lies in A_{r-1} and meets the $(n-m_{r-1}-2)$-quadric C_r in two conjugate imaginary, respectively real, points. It is called *parabolic of kind a* if it meets the $(n-m_{a-1}-1)$-plane A_a and does not meet the $(n-m_a)-1)$-plane A_{a+1}. If the curvature of the space is $\frac{1}{r^2}$, the elliptic and hyperbolic lines in the space are isometric to the lines S^1, respectively S^1_1, of the same curvature, and the parabolic lines are isometric to the lines R^1.

5.4.2. Elliptic, Pseudoelliptic, and Parabolic m-Planes. Analogously, an m-plane in $S^{m_0,m_1,\ldots,m_{r-1},n}$ or $S^{m_0,m_1,\ldots,m_{r-1},n}_{\ell_0,\ell_1,\ldots,\ell_r}$ is called *elliptic* or *pseudoelliptic of kind zero* if its intersection with the hypercone C_0 is an imaginary, respectively real, nondegenerated $(m-1)$-quadric (this case is possible only if $m < m_0$), *elliptic* or *hyperbolic of kind a* if it lies in A_{a-1} and its intersection with the $(n-m_{a-1}-2)$-cone C_a is an imaginary, respectively real, $(m-1)$-quadric, *elliptic* or *hyperbolic of kind r* if it lies in A_{r-1} and its intersection with the $(n-m_{r-1}-2)$-quadric C_r is an imaginary, respectively real, $(m-1)$-quadric, and *parabolic of kind a* if it meets the $(n-m_{a-1}-1)$-plane A_a and does not meet the $(n-m_a-1)$-plane A_{a+1}. If the curvature of the space is $\frac{1}{r^2}$, the elliptic and hyperbolic m-planes in this space are isometric to the spaces S^m and S^m_k of the same curvature, respectively, and the parabolic m-planes are isometric to a quasielliptic, quasipseudoelliptic, s-quasielliptic, or s-quasipseudoelliptic m-spaces.

5.4.3. Matrix Coordinates of m-Planes. Like in S^n and S^n_ℓ (see **4.3.1**), in $S^{m_0,m_1,\ldots,m_{r-1},n}$ and $S^{m_0,m_1,\ldots,m_{r-1},n}_{\ell_0,\ell_1,\ldots,\ell_r}$ we can define the matrix coordinates $A = (a^i_\alpha)$ and $U = (u^\varphi_i)$ $(\alpha = 0,1,\ldots,m,\ \varphi = m+1,\ldots,n)$. We can also define the submatrices $A_a = (a^{i_a}_\alpha)$ and $U_a = (u^\varphi_{i_a})(a = 0,1,\ldots,r)$. If the points X_{i_0} are mutually polar with respect to the hypercone C_0, the matrix coordinate A_0 satisfies the condition $A_0^\top E_0 A_0 = E_0$, if the points X_{i_a} are mutually polar with respect to the $(n-m_{a-1}-2)$-cone C_a, the matrix coordinate A_a satisfies the condition $A_a^\top E_a A_a = E_a$, if the points X_{i_r} are mutually polar with respect to the $(n-m_{r-1}-2)$-quadric C_r, the matrix coordinate A_r satisfies the condition $A_r^\top E_r A_r = E_r$.

5.4.4. Stationary Distances of Two Elliptic or Pseudoelliptic m-Planes. The stationary distance ω_α of two elliptic or pseudoelliptic m-planes in $S^{m_0,m_1,\ldots,m_{r-1},n}$ or $S^{m_0,m_1,\ldots,m_{r-1},n}_{\ell_0,\ell_1,\ldots,\ell_r}$, for $m \neq m_a$ are determined like the stationary distances of two m-planes in S^n and S^n_ℓ (see **4.3.2**): the squares of cosines of the quotients $\frac{\omega_\alpha}{r}$ are the eigenvalues of the $((m+1)\times(m+1))$-matrices (3.13), where A and B are the matrix coordinates of the m-planes. These stationary distances are realized on the $m+1$ common perpendiculars of the m-planes which are determined by the eigenvectors of the $((m+1)\times(m+1))$-matrices (3.13) and are elliptic or hyperbolic lines.

For $m = m_a$ the common perpendiculars and stationary distances ω_α of two elliptic or pseudoelliptic m-planes in $S^{m_0,m_1,\ldots,m_{r-1},n}$ or $S^{m_0,m_1,\ldots,m_{r-1},n}_{\ell_0,\ell_1,\ldots,\ell_r}$ are determined by

Theorem 5.16. *The stationary distances ω_α between two elliptic or pseudoelliptic m_a-planes in $S^{m_0,m_1,\ldots,m_{r-1},n}$ or $S^{m_0,m_1,\ldots,m_{r-1},n}_{\ell_0,\ell_1,\ldots,\ell_r}$ are the distances between the points of intersection of these m_a-planes with the m_a+1 parabolic lines determined by projection of one of these m_a-planes onto the $(n-m_a-1)$-plane A_a in the direction of the second m_a-plane.*

PROOF. If p and q are two m_a-planes, we can determine the projection x' of any point x in p onto $(n-m_a-1)$-plane A_a in the direction of q (see **2.4.3**). If $m_a \leq n-m_a-1$, the projection lines xx' map the m_a-plane p to an m_a-plane p' in A_a. Since the m_a-plane p is the space S^{m_a} or $S^{m_a}_k$, these projection lines map the absolute hyperquadrics of S^{m_a} or $S^{m_a}_k$ to the m_a-plane p'; since the $(n-m_a-1)$-

plane A_a is an $(r-a)$-quasielliptic or $(r-a)$-quasipseudoelliptic $(n-m_a-1)$-space, in general p' is an s-quasielliptic or s-quasipseudoelliptic m_a-space. Therefore, the quadric in p' obtained by the map of the absolute hyperquadric of S^{m_a} or $S_k^{m_a}$ determined by Theorem 5.14 has m_a+1 centers. If we define the angles between the hyperplanes in $S^{m_0,m_1,\ldots,m_{r-1},n}$ or $S^{m_0,m_1,\ldots,m_{r-1},n}_{\ell_0,\ell_1,\ldots,\ell_r}$ as the distance between the points in spaces corresponding to these spaces by the duality principle of P^n and the angles between lines be the particular case of this definition for $n=2$, we see that all projection lines xx' are common perpendiculars of the m_a-planes p and q. It is easy to check that only for m_a+1 of the lines xx' through these centers, like for the common perpendiculars of two m-planes in S^n and S^n_ℓ, the distances ω_α between the points of intersection with the m_a-planes are stationary distances of these m_a-planes. For details see [Ro3, pp. 343–346].

Therefore, we call these m_a+1 common perpendiculars of the two m_a-planes in $S^{m_0,m_1,\ldots,m_{r-1},n}$ or $S^{m_0,m_1,\ldots,m_{r-1},n}_{\ell_0,\ell_1,\ldots,\ell_r}$ the *parabolic common perpendiculars* of the two m_a-planes in these spaces.

For $m_a > n-m_a-1$ we must project the $(n-m_a-1)$-plane in the m_a-plane p in the direction of a m_a-plane q. The determination of parabolic common perpendiculars and stationary distances of the m_a-planes is analogous, but the number of parabolic common perpendiculars and stationary distances is equal to $n-m_a$. ■

5.4.5. Application of Affine Matrix Coordinates for Quasielliptic or Quasipseudoelliptic Space. For m-planes in $S^{m,n}$ or $S^{m,n}_{k\ell}$ it is very convenient to apply the affine matrix coordinates $\overline{A}$ of m-planes (see **2.4.5**). In this case the formulas (5.15) for motions in $S^{m,n}$ and $S^{m,n}_{k\ell}$ imply the formula

$$\overline{A}' = V\overline{A}U^{-1} + \overline{A}_0, \tag{5.41}$$

where $\overline{A}_0 = TU^{-1}$. For $m \le n-m-1$, there are two $(m-1)$-quadrics, in the m-plane its intersection with the $(n-m-2)$-quadric C_1 and the $(m-1)$-quadric obtained by projecting p onto A_0 in the direction of the plane q obtained from the intersection of p with the hypercone C_0. If the first of these $(m-1)$-quadrics is regarded as the absolute $(m-1)$-quadric, the centers of the second $(m-1)$-quadric are fixed points of a product of polar transformations with respect to both these $(m-1)$ quadrics. The calculation of the matrix of the product of two correlations shows that this matrix is

$$M = (\overline{B}^\top - \overline{A}^\top)E_1(\overline{B} - \overline{A}), \tag{5.42}$$

where E_1 is the matrix of the equation $x_1E_1x_1 = 0$ of the $(n-m-2)$-quadric C_1 (for details see [Ro3, pp. 346–348]). Under the transformation (5.41) the matrix M is transformed into the matrix $M' = (U^\top)^{-1}(\overline{B}^\top - \overline{A}^\top)V^\top E_1 V(\overline{B} - \overline{A})U^{-1}$. Since $U^\top E_0 U = E_0$ and $V^\top E_1 V = E_1$, the matrices, M and M' are similar, their eigenvalues coincide and are, therefore, metric invariants of the two m-planes. It is easy to check that these eigenvalues are equal to the squares ω_α^2 of the stationary distances ω_α of these m-planes.

Two m-planes in $S^{m,n}$ or $S^{m,n}_{kl}$, for wich all stationary distances ω_α are equal are called *paratactic m-planes*. Like paratactic m-planes in S^n and S^n_l, paratactic

m-planes in $S^{m,n}$ and $S^{m,n}_{kl}$ have infinity many parabolic common perpendiculars and in this case two $(m-1)$-quadrics in the m-plane p' coincide.

5.4.6. Lines and m-Planes in Hermitian r-Quasielliptic and r-Quasi-pseudoelliptic Spaces. The geometry of lines and m-planes in the Hermitian r-quasielliptic and r-quasipseudoelliptic spaces $\mathbb{C}\overline{S}^{m_0,m_1,\dots,m_{r-1},n}$, $\mathbb{C}\overline{S}^{m_0,m_1,\dots,m_{r-1},n}_{\ell_0,\ell_1,\dots,\ell_r}$, $\mathbb{H}\overline{S}^{m_0,m_1,\dots,m_{r-1},n}$, and $\mathbb{H}\overline{S}^{m_0,m_1,\dots,m_{r-1},n}_{\ell_0,\ell_1,\dots,\ell_r}$ is similar to the geometry of lines and m-planes in the real r-quasielliptic and r-quasipseudoelliptic spaces $S^{m_0,m_1,\dots,m_{r-1},n}$ and $S^{m_0,m_1,\dots,m_{r-1},n}_{\ell_0,\ell_1,\dots,\ell_r}$. Like in real spaces, in complex and quaternionic r-quasielliptic and r-quasipseudoelliptic spaces we can define elliptic, hyperbolic and parabolic lines and elliptic, pseudoelliptic, and parabolic m-planes and determine stationary distances, common perpendiculars, and parabolic common perpendiculars between two m-planes.

Note that in $S^{2m+1,2n+1}$ and $\mathbb{C}\overline{S}^{2m+1,2n+1}$ there are paratactic congruences of lines, and, analogously to Theorem 4.13, we can prove that these paratactic congruences are isometric to $\mathbb{C}\overline{S}^{m,n}$, respectively $\mathbb{H}\overline{S}^{m,n}$; in particular, paratactic congruences of lines in $S^{1,2n+1}$ and $\mathbb{C}\overline{S}^{1,2n+1}$ are isometric to $\mathbb{C}\overline{R}^{n}$, respectively $\mathbb{H}\overline{R}^{n}$.

5.4.7. Interpretations of Manifolds of Lines in 3-Spaces. In **4.4.3** we have defined the interpretations of Kotelnikov-Study and Fubini-Study of the manifold of pairs of mutually polar lines in S^3 as the split complex plane $\mathbb{C}'S^2$ and as two planes S^2 and of the manifolds of lines in H^3 and R^3 as the complex plane $\mathbb{C}S^2$, respectively the dual plane $\mathbb{C}^0S^2$. There are analogous interpretations for manifolds of lines in the quasielliptic space $S^{1,3}$, the quasihyperbolic space $H^{1,3}$, and the Galilean space Γ^3.

Theorem 5.17 (Zhelezina [Zh1-2]). *The planes $\mathbb{C}'R^2$, $\mathbb{C}R^2$, and $\mathbb{C}^0R^2$ admit interpretations as the manifolds of elliptic lines in $S^{1,3}$, of hyperbolic lines in $H^{1,3}$, and of parabolic lines of kind 0 in Γ^3. The manifold of elliptic lines in $S^{1,3}$ also admits an interpretation as two planes R^2.*

Proof. Consider the line $a=xy$ in $S^{1,3}$ through the points $x(1,0,x^2,x^3)$ and $y(0,1,y^2,y^3)$ and the line $b=zw$ through the points $z(1,0,z^2,z^3)$ and $w(0,1,w^2,w^3)$. Map the lines a and b to the points X and Y in the plane $\mathbb{C}'R^2$ with coordinates

$$\begin{aligned} X^1 &= -y^2+ex^3, \quad X^2 = x^2+ey^3, \\ Y^1 &= -w^2+ez^3, \quad Y^2 = z^2+ew^3. \end{aligned} \tag{5.43}$$

Since $x^0=y^1=z^0=w^1=1$, $x^1=y^0=z^1=w^0=0$, the affine matrix coordinates $\overline{A}$ of a and $\overline{B}$ of b are

$$\overline{A}=\begin{bmatrix} x^2 & y^2 \\ x^3 & y^3 \end{bmatrix}, \quad \overline{B}=\begin{bmatrix} z^2 & w^2 \\ z^3 & w^3 \end{bmatrix}. \tag{5.44}$$

The squares ω_0^2 and ω_1^2 of the stationary distances between the lines a and b are equal to the eigenvalues of the matrix (5.42), where $E_1=I$. But the square Ω^2 of the split complex distance Ω between the points X and Y in $\mathbb{C}'R^2$ is equal to

$$\begin{aligned} \Omega^2 &= (Y^1-X^1)^2+(Y^2-X^2)^2 \\ &= (-w^2+y^2+e(z^3-x^3))^2+(z^2-x^2+e(w^3-y^3))^2. \end{aligned} \tag{5.45}$$

Calculation of the eigenvalues of the matrix (5.42), with $\overline{A}$ and $\overline{B}$ of the form (5.44) and $E_1 = I$, and their comparison with (5.45) shows that

$$\Omega = \omega_0 + e\omega_1 \tag{5.46}$$

(for the details see [Ro3, p. 376]). The second and third assertions of this theorem are proved analogously, but in these cases the plane $\mathbb{C}'R^2$ must be replaced by the planes $\mathbb{C}R^2$ and $\mathbb{C}^0R^2$, and in formulas (5.43), (5.45), and (5.46) e must be replaced by i, respectively ϵ.

The last assertion is obtained from the first one if we replace the plane $\mathbb{C}'R^2$ with coordinates of points $X^i = X^i_+ e_+ + X^i_- e_-$ by the planes R^2_+ and R^2_- with coordinates of points X^i_+ and X^i_-, respectively. Formula (5.46) implies that the distances Ω_+ and Ω_- between the points X_+ and Y_+ in R^2_+ and the points X_- and Y_- in R^2_- which correspond to the lines a and b in $S^{1,3}$ are related with the stationary distances ω_0 and ω_1 by

$$\Omega_+ = \omega_0 + \omega_1, \ \Omega_- = \omega_0 - \omega_1. \tag{5.47}$$

The formulas analogous to formulas (5.47) also hold for the interpretation of Fubini for S^3 (see **4.4.3**), these formulas follow from (3.22). ■

The interpretations of Zhelezina as well as the interpretation of Kotelnikov-Study for R^3 can be obtained by limiting transition from the interpretations of Kotelnikov-Study for S^3 and H^3.

The interpretation of Kotelnikov-Study for S^3 is equivalent to the interpretation of the manifold of lines in S^3 as $(\mathbb{C} \otimes \mathbb{C})\, \overset{\approx}{S}{}^1$.

The interpretation of Kotelnikov-Study for R^3 is equivalent to the interpretation of the manifold of lines in R^3 as $(\mathbb{C} \otimes \mathbb{C}^0)\overline{S}^1$.

The interpretation of Zhelezina for Γ^3 is equivalent to the interpretation of the manifold of lines in Γ^3 as $(\mathbb{C}^0 \otimes \mathbb{C}^0)\overline{S}^1$.

The last interpretation was obtained by A. P. Shirokov, who also obtained the interpretation of the manifold of lines in I^3 as $(\mathbb{C}^0 \otimes \mathbb{C}^0)\, \overset{\approx}{S}{}^1$ [ShA1-2]. These interpretations of Shirokov can also be obtaited by limiting transition from the interpretations of Kotelnikov-Study and Zhelezina.

5.4.8. Symmetry Figures in Quasielliptic and Quasipseudoelliptic Spaces. Since motions in $S^{m,n}$ and $S^{m,n}_{kl}$ are particular cases of collineations in P^n, symmetry figures in $S^{m,n}$ and $S^{m,n}_{kl}$ are particular cases of symmetry figures in P^n. Therefore, we obtain

Theorem 5.18. *The symmetry figures in $S^{m,n}$ or $S^{m,n}_{k\ell}$ are p-pairs and elliptic, respectively hyperbolic, congruences of lines in the space. For $p = m$ these p-pairs consist of elliptic, or pseudoelliptic, m-planes and $(n - m - 1)$-planes A_1; for $p < m$ these p-pairs consist of elliptic, or pseudoelliptic, p-planes and quasielliptic, or quasipseudoelliptic, $(n-p-1)$-planes containing A_1; for $p > m$ these p-pairs consist of quasielliptic, or quasipseudoelliptic, p-planes which intersect A_1 and elliptic, or pseudoelliptic, $(n - p - 1)$-planes in A_1 polar to intersections of A_1 with p-planes. There are also p-pairs which consist of mutually polar parabolic p-planes and $(n -$*

p − 1)*-planes. These p-pairs are called polarized parabolic p-planes. In* $S^{2m+1,2n+1}$ *or* $S^{2m+1,2n+1}_{2k+1,2\ell+1}$ *there are also elliptic, respectively hyperbolic, linear congruences of lines in* P^{2n+1} *which are symmetry figures in* $S^{2m+1,2n+1}$*, respectively* $S^{2m+1,2n+1}_{2k+1,2\ell+1}$. *These congruences consist of paratactic lines and are called paratactic congruences of these spaces.*

In **5.4.6** we have mentioned paratactic congruences in $S^{2m+1,2n+1}$, which are isometric to $\mathbb{C}\overline{S}^{m,n}$; we can analogously prove that paratactic congruences in $S^{2m+1,2n+1}_{2k+1,2\ell+1}$ are isometric to $\mathbb{C}\overline{S}^{m,n}_{k\ell}$.

5.4.9. Cosymmetry Figures in Quasielliptic and Quasipseudoelliptic Spaces. In the autodual spaces S^{2m+1} and S^{2m+1}_{m+1} cosymmetry figures do not exist since any involutive correlation in P^{2m+1} which transforms two points X and Y in one of these spaces with distance ω into hyperplanes α and β with angle ω in one of these spaces is the product of an involutive motion by a polar transformation with respect to the absolute hyperquadric of this space.

However, in the autodual spaces $S^{m,2m+1}$ and $S^{m,2m+1}_{kk}$ polar transformations with respect to the absolute hypercones of these spaces are degenerate correlations. Correlations in P^{2m+1} which transforms any pair of points X and Y in $S^{m,2m+1}$ or $S^{m,2m+1}_{kk}$ with distance ω into hyperplanes α and β with angle ω are called *comotions* of these spaces.

Since involutive comotions in $S^{m,2m+1}$ and $S^{m,2m+1}_{kk}$ cannot be reduced to involutive motions in these spaces, they determine *cosymmetry figures* in these autodual spaces. Since involutive correlations in P^{2n+1} are polar transformations with respect to hyperquadrics and null-systems which preserve linear complexes of lines, we obtain

Theorem 5.19. *The cosymmetry figures in* $S^{m,2m+1}$ *and* $S^{m,2m+1}_{kk}$ *are the hyperquadrics and linear complexes of lines, the polar transformations and null-systems with respect to which are comotions of these spaces.*

The hyperquadrics and linear complexes of lines in $S^{m,2m+1}$ and $S^{m,2m+1}_{kk}$ which are cosymmetry figures in these spaces are called *metric hyperquadrics* and *metric linear complexes of lines* in these spaces.

5.4.10. Symmetry and Cosymmetry Figures in Quasielliptic 3-Space in the Interpretation on Two Euclidean Planes. Theorem 5.18 implies that in the interpretation of the manifold of elliptic lines in $S^{1,3}$ as two planes R^2, elliptic lines are interpreted as points in each R^2.

Paratactic congruences of elliptic lines in $S^{1,3}$ are interpreted as a point in one R^2 (the lines of these congruences are interpreted as this fixed point in one R^2 and an arbitrary point in the other R^2).

Polarized parabolic lines in $S^{1,3}$ are interpreted as lines in each R^2 (elliptic lines which meet polarized and polarizing parabolic lines are interpreted as pairs of points in these two lines).

Metric quadrics in $S^{1,3}$ are interpreted as a line in one R^2 and a point on the other R^2 (this quadric is ruled and its rectilinear generators are interpreted as a fixed point in one R^2 and an arbitrary point on the fixed line in the other R^2).

Metric linear complexes of lines in $S^{1,3}$ are interpreted as pairs of points in two copies of R^2, so that a point in one R^2 belongs to a fixed line and a point in the other R^2 is arbitrary.

§5.5 m-Horospheres in Pseudoelliptic and Pseudohyperbolic Spaces

5.5.1. m-Horospheres in Quadratic Spaces. In **4.7.4** we have defined horohyperspheres in the hyperbolic space H^n, which can be obtained by limit transition from equidistant hyperquadrics in this space. In the pseudoelliptic and pseudohyperbolic spaces S^n_{m+1} and H^n_m we can define m-horospheres, which can be obtained from m-equidistant hyperquadrics in these spaces by limit transition which transforms the basis m-plane of this hyperquadric to an m-planar generator of the absolute hyperquadric of the space. A horohypersphere is the particular case of an m-horosphere for $m = 0$.

Theorem of M. I. Gorbunova [Gor]. *An m-horosphere in S^n_{m+1} for $m > 0$ is a fibration with the Euclidean spaces R^m as fibers and the quasielliptic space $S^{m,n-m-1}$ as base.*

The stabilizers of m-horospheres contain the stabilizers of the m-planar generators of the absolute hyperquadrics of these spaces and are therefore parabolic groups.

5.5.2. m-Horospheres in Hermitian Spaces. We can analogously define m-horospheres in $\mathbb{C}\overline{S}^n_{m+1}$, $\mathbb{C}\overline{H}^n_m$, $\mathbb{H}\overline{S}^n_{m+1}$, and $\mathbb{H}\overline{H}^n_m$. The stabilizer of such an m-horosphere contains the stabilizers of the m-planar generators of the absolute hyperquadrics of these spaces and therefore is a parabolic group. The Borel subgroup of such a group can be regarded as a generalization of the *Heisenberg group* of the space $\mathbb{C}\overline{H}^n$.

§5.6. Sliding Vectors

5.6.1. Sliding Vectors in Quasielliptic Spaces. The theory of sliding vectors in the Euclidean space R^n, considered in **3.7.1–3**, is a particular case of the theory of sliding vectors in the quasielliptic spaces $S^{m,n}$. If R^n is regarded together with its hyperplane at infinity, any pair $(a, -a)$ of vectors in R^n must be replaced by the vector in this hyperplane obtained by projection of the vector a of the pair from the initial point of the vector $-a$ of this pair. Since the hyperplane at infinity of R^n is the elliptic space S^{n-1}, the vectors in this hyperplane which replace pairs of vectors in R^n are called elliptic vectors.

In the general spaces $S^{m,n}$ there are also elliptic and Euclidean sliding vectors, whose lines are elliptic and parabolic lines, respectively (Euclidean vectors in $S^{m,n}$ are vectors in the space R^{n-m} whose hyperplane at infinity coincides with the absolute $(n-m-1)$-plane A_0 of $S^{m,n}$).

By Theorem 5.15 the quasielliptic space $S^{m,n}$ can be regarded as a hypersphere in the quasi-Euclidean space $R^{m+1,n+1}$ with identified antipodal points. Therefore, all elliptic vectors $\breve{a}$ in $S^{m,n}$ can be represented by vectors a in $R^{m+1,n+1}$ tangent to this hypersphere, like vectors in S^n can be represented by vectors in R^{n+1}. In the space $R^{m+1,n+1}$, like in R^{n+1}, we can define the exterior product $x \wedge a$ of vectors, the moment of vectors, and the main moment $M = \sum_i x_i \wedge a_i$ of a system of sliding vectors $\breve{a}_i$. Analogously to Theorems 3.23 and 4.50 we can prove

Theorem 5.20. *Any system of sliding vectors in the space $S^{m,n}$ of curvature $\frac{1}{r^2}$ can be reduced to the canonical form consisting of sliding vectors $\breve{a}_i$ on mutually polar lines which are determined by invariant planes of the main moment M of the system. The modulus $|a_i|$ of a Euclidean vector a_i representing a vector $\breve{a}_i$ of the canonical system is equal to the modulus of an eigenvalue of the linear operator M, divided by r for elliptic vectors $\breve{a}_i$.*

Such systems of sliding vectors in $S^{m,n}$ are called *screws* in $S^{m,n}$. In this case the linear operators M can also be regarded as elements in the tangent Lie algebra of the group of motions in $S^{m,n}$.

The screws in $S^{1,3}$ consist of two elliptic vectors, $\breve{a}_1$ on an elliptic line which intersects the imaginary planes C_0 at two conjugate imaginary points and $\breve{a}_2$ on the absolute line A_0, or of two Euclidean vectors on two mutually polar parabolic lines.

5.6.2. Sliding Vectors in Quasihyperbolic, Quasipseudoelliptic, and Quasipseudohyperbolic Spaces. The theory of sliding vectors in the pseudo-Euclidean space R^n_ℓ considered in **3.7.4** is a particular case of the theories of sliding vectors in the quasihyperbolic spaces $H^{m,n}$, quasipseudoelliptic spaces $S^{m,n}_{k\ell}$, and quasipseudohyperbolic spaces $H^{m,n}_{k\ell}$. If R^n_ℓ is considered together with its hyperplane at infinity, any pair $(a, -a)$ of vectors in R^n_ℓ must be replaced by the vector in this hyperplane obtained by projecting the vector a of the pair from the initial point of the vector $-a$ of this pair. Since the hyperplane at infinity of R^n_ℓ is the pseudoelliptic space S^{n-1}_ℓ, the vectors in this hyperplane which replace pairs of vectors in R^n_ℓ are called hyperbolic or elliptic vectors, if their lines meet or do not meet the absolute hyperquadric of S^{n-1}_ℓ.

In the general spaces $H^{m,n}$, $S^{m,n}_{k\ell}$, and $H^{m,n}_{k\ell}$ there are also elliptic, hyperbolic, Euclidean, and pseudo-Euclidean sliding vectors, whose lines are elliptic, hyperbolic, and parabolic lines,

Euclidean vectors in $S^{m,n}$, $S^{m,n}_{k\ell}$, and $H^{m,n}_{k\ell}$ are vectors in the spaces R^{n-m} and R^{n-m}_ℓ whose hyperplanes at infinity coincide with the $(n-m-1)$-planes A_0 of $H^{m,n}, S^{m,n}_{k\ell}$, and $H^{m,n}_{k\ell}$.

By Theorem 5.14, the spaces $H^{m,n}, S^{m,n}_{k\ell}$, and $H^{m,n}_{k\ell}$ can be regarded as hyperspheres in the spaces $R^{m+1,n+1}_1, R^{m+1,n+1}_{k\ell}$, and $R^{m+1,n+1}_{k+1,\ell}$ with identified antipodal points. Therefore, all elliptic and hyperbolic vectors $\breve{a}$ in $H^{m,n}$, $S^{m,n}_{k\ell}$, and $H^{m,n}_{k\ell}$ can be represented by vectors a in $R^{m+1,n+1}_1$, $R^{m+1,n+1}_{k\ell}$, and $R^{m+1,n+1}_{k+1,\ell}$ tangent to these hyperspheres, like vectors in H^n, S^n_ℓ, H^n_ℓ which can be represented by vectors in R^{n+1}_1, R^{n+1}_ℓ, and $R^n_{\ell+1}$. In the spaces $R^{m+1,n+1}_1$, $R^{m+1,n+1}_{k\ell}$, and $R^{m+1,n+1}_{k+1,\ell}$, like in R^{n+1}_1, R^{n+1}_ℓ, and $R^n_{\ell+1}$, we can define the exterior product $x \wedge a$ of vectors, the moment of vectors, and the main moment $M = \sum_i x_i \wedge a_i$ of a system of sliding vectors a_i. Analogously to Theorems 3.23 and 4.51 we can prove

Theorem 5.21. *Any system of sliding vectors in the space $H^{m,n}$ of curvature $-\frac{1}{q^2}$ can be reduced to the canonical form consisting of sliding vectors $\breve{a}_i$ on mutually polar lines which are determined by invariant planes of the linear operator E_0M of the system. The modulus $|a_i|$ of a vector a_i representing a vector $\breve{a}_i$ of the canonical system is equal to the modulus of an eigenvalue of the linear operator E_0M, divided by q for elliptic or hyperbolic vectors $\breve{a}_i$.*

We can prove similar theorems for systems of sliding vectors in $S^{m,n}_{k\ell}$ and $H^{m,n}_{k\ell}$.

Canonical systems of sliding vectors in $H^{m,n}, S^{m,n}_{k\ell}$, and $H^{m,n}_{k\ell}$ are called *screws* of these spaces. In these cases the linear operators E_0M also can be regarded as elements in the tangent Lie algebras of the groups of motions in these spaces.

5.6.3. Sliding Vectors in r-Quasielliptic, r-Quasihyperbolic, r-Quasipseudoelliptic, and r-Quasipseudohyperbolic Spaces. The theories of sliding vectors in the space $S^{m,n}, H^{m,n}, S^{m,n}_{k\ell}$, and $H^{m,n}_{k\ell}$ considered in **5.6.1** and **5.6.2** are particular cases of the theories of sliding vectors in the r-quasielliptic, r-quasihyperbolic, r-quasipseudoelliptic, and r-quasipseudohyperbolic spaces $S^{m_0,m_1,\ldots,m_{r-1},n}$, $S^{m_0,m_1,\ldots,m_{r-1},n}_{\ell_0,\ell_1,\ldots,\ell_r}$, $H^{m_0,m_1,\ldots,m_{r-1},n}$, and $S^{m_0,m_1,\ldots,m_{r-1},n}_{\ell_0,\ell_1,\ldots,\ell_r}$. In these spaces there are elliptic, hyperbolic, Euclidean, and pseudo-Euclidean vectors of various kinds, whose lines are elliptic, hyperbolic, and parabolic lines of these kinds. By Theorem 5.14, these spaces can be regarded as the hyperspheres in the r-quasi-Euclidean and r-quasipseudo-Euclidean $(n+1)$-spaces with identified antipodal points. Therefore, all elliptic and hyperbolic vectors $\breve{a}$ in these spaces can be represented by vectors a in $(n+1)$-spaces tangent to these hyperspheres. In these $(n+1)$-spaces we can define the exterior product $x \wedge a$ of vectors, the moment of vectors, and the main moment $M = \sum_i x_i \wedge a_i$ of a system of sliding vectors a_i. Analogously to Theorems 5.20 and 5.21 we can prove

Theorem 5.22 (T. M. Klimanova and N. D. Petsko [RKP]). *Any system of sliding vectors in a space $S^{m_0,m_1,\ldots,m_{r-1},n}$, $S^{m_0,m_1,\ldots,m_{r-1},n}_{\ell_0,\ell_1,\ldots,\ell_r}$, $H^{m_0,m_1,\ldots,m_{r-1},n}$, or $S^{m_0,m_1,\ldots,m_{r-1},n}_{\ell_0,\ell_1,\ldots,\ell_r}$ of curvature $\frac{1}{r^2}$ or $-\frac{1}{q^2}$ can be reduced to the canonical form consisting of sliding vectors $\breve{a}_i$ on mutually polar lines which are determined by invariant planes of the linear operator E_0M of the system. The modulus $|a_i|$ of the vector a_i representing a vector $\breve{a}_i$ of the system is equal to the modulus of an eigenvalue of E_0M, divided by r or q for of elliptic or hyperbolic vectors.*

Such systems of sliding vectors in $S^{m_0,m_1,\ldots,m_{r-1},n}$, $S^{m_0,m_1,\ldots,m_{r-1},n}_{\ell_0,\ell_1,\ldots,\ell_r}$, $H^{m_0,m_1,\ldots,m_{r-1},n}$, and $S^{m_0,m_1,\ldots,m_{r-1},n}_{\ell_0,\ell_1,\ldots,\ell_r}$ are called *screws* of these spaces. In these cases the linear operators E_0M can also be regarded as elements in the tangent Lie algebras of the groups of motions in these spaces.

§5.7. Quasi-Riemannian, Quasipseudo-Riemannian, r-Quasi-Riemannian, and r-Quasipseudo-Riemannian Manifolds and Symmetric Spaces

5.7.1. Quasi-Riemannian, Quasipseudo-Riemannian, r-Quasi- Riemannian, and r-Quasipseudo-Riemannian Manifolds. The torsion-free manifold A^n with an affine connection such that in each affine tangent space E^n of which the quasimetric of the quasi-Euclidean, quasipseudo-Euclidean, r-quasi-Euclidean, or r-quasipseudo-Euclidean space $R^{m,n}$, $R^{m,n}_{k\ell}$, $R^{m_1,m_2,\ldots,m_{r-1},n}$, or $R^{m_1 m_2 \ldots m_{r-1} n}_{\ell_1,\ell_2,\ldots,\ell_r}$, is invariant under parallel displacement in A^n is called the *quasi-Riemannian, quasipseudo-Riemannian, r-quasi-Riemannian, and r-quasipseudo-Riemannian manifold*, respectively. We denote these manifolds by $V^{m,n}, V^{m,n}_{k\ell}, V^{m_1,m_2,\ldots,m_{r-1},n}$, and $V^{m_1 m_2 \ldots m_{r-1} n}_{\ell_1,\ell_2,\ldots,\ell_r}$, respectively.

The spaces $S^{m,n}$ and $H^{m,n}$ are manifolds $V^{m,n}$ of constant curvature. The spaces $S^{m,n}_{k\ell}$ and $H^{m,n}_{k\ell}$ are manifolds $V^{m,n}_{k\ell}$ of constant curvature. The spaces

$S^{m_1,m_2,\ldots,m_{r-1},n}$, $H^{m_1,m_2,\ldots,m_{r-1},n}$, $S^{m_1,m_2,\ldots,m_{r-1},n}_{\ell_1,\ell_2,\ldots,\ell_r}$, and $H^{m_1,m_2,\ldots,m_{r-1},n}_{\ell_1,\ell_2,\ldots,\ell_r}$ are manifolds $V^{m_1,m_2,\ldots,m_{r-1},n}$ and $V^{m_1,m_2,\ldots,m_{r-1},n}_{\ell_1,\ell_2,\ldots,\ell_r}$ of constant curvature.

Quasi-Riemannian manifolds first were considered by Enea Bortolotti [Bot], Gregore Moisil [Moi], and Alexander Norden [Nor1-2]. Bortolotti called these manifolds "Riemannian manifolds with specialized metric forms". Moisil called them "singular Riemannian spaces", Norden regarded these manifolds as generalizations of the manifold of lines in R^2, that is, of the co-Euclidean plane R^{2*}.

5.7.2. Quasi-Riemannian, Quasipseudo-Riemannian, r-Quasi-Riemannian, and r-Quasipseudo-Riemannian Symmetric Spaces. If the manifolds $V^{m,n}$, $V^{m,n}_{k\ell}$, $V^{m_1,m_2,\ldots,m_{r-1},n}$, and $V^{m_1,m_2,\ldots,m_{r-1},n}_{\ell_1,\ell_2,\ldots,\ell_r}$ are symmetric spaces with an affine connection, they are called *quasi-Riemannian, quasipseudo-Riemannian, r-quasi-Riemannian*, and *r-quasipseudo-Riemannian symmetric spaces*, respectively.

Theorem 5.23 (L. M. KARPOVA [RoK]). *In any quasisimple Lie group we can define an invariant quasimetric, coming from a quasi-Riemannian or quasipseudo-Riemannian symmetric space.*

PROOF. In a quasisimple Lie group G, like in all Lie groups, there is an invariant torsion-free affine connection defined in **0.8.5**. Since quasisimple Lie groups are not semisimple, the quadratic form (0.71) determined by the structure constants C^k_{ij} of the tangent Lie algebra $\mathbb{G}$ of G is degenerate, and in a tangent affine space of this group this quadratic form determines the distance d_1 of a quasi-Euclidean or quasipseudo-Euclidean space. The tangent Lie algebra $\mathbb{G}$ can be represented as the direct sum $\mathbb{K} \oplus \mathbb{E}$ of subspaces $\mathbb{K}$ and $\mathbb{E}$ whose basis vectors e_a and e_u satisfy the conditions (5.16), and the subspace $\mathbb{E}$ is determined by the condition $d_1 = 0$. The subspace $\mathbb{E}$ can be regarded as the tangent space to a symmetric space. Since $C^b_{uv} = 0$, the Riemannian tensor (0.70) in this space is equal to zero and this space is locally isometric to a Euclidean or pseudo-Euclidean space. Since the points of this space represent symmetry figures of the homogeneous space whose group of transformations is G, the metric or pseudometric in this space is defined by metric invariants of the symmetry figures, which are represented by points in this symmetric space.

In particular, in the groups of motions in $S^{m,n}$ and $S^{m,n}_{k\ell}$, if the points in the symmetric space interpret elliptic or pseudoelliptic m-planes, since two elliptic or pseudoelliptic m-planes in $S^{m,n}$ or $S^{m,n}_{k\ell}$ are characterized by affine matrix coordinates $\overline{A}$ and $\overline{B}$, formula (5.41) shows that the trace $\operatorname{Tr} M$ of the matrix (5.42) can be regarded as the square of the distance between two points in $R^{(m+1)(n-m)}$ or $R^{(m+1)(n-m)}_L$, and this distance can be regarded as the distance d_2 in the tangent space to the group. ■

For $m = 0$ these spaces $R^{(m+1)(n-m)}$ and $R^{(m+1)(n-m)}_L$ are the spaces R^n and R^n_ℓ.

For the quasisimple group of motions in $\mathbb{C}^0\overline{S}^n$, the symmetry figures of this space, which are represented as points in a symmetric space of zero curvature, are *normal n-chains* (analogous to the normal n-chains in $\mathbb{C}\overline{S}^n$ defined in **4.10.1**); these normal n-chains are also isometric to S^n.

Note that the group of motions in the line $\mathbb{C}^0\overline{S}^1$ and in the space $C^0\overline{S}^3$ is isomorphic to the group of motions in R^2, respectively in R^5. These interpretations can be obtained by limit transition from the interpretations of $\mathbb{C}\overline{S}^1$ and $\mathbb{C}\overline{S}^3$ in S^2 and S^5 (in Theorems 4.8 and 4.27). In these interpretations points in R^2 or R^5 are represented as normal chains in $\mathbb{C}^0\overline{S}^1$, respectively as normal 3-chains in $\mathbb{C}^0\overline{S}^3$ (this was proved by N. K. Alexeyeva [Alx] and by N. P. Klimentova [RDKM]).

The invariant quasimetrics in $S^{1,3}$ and $S^{1,3}_{11}$ in the groups of motions in R^2 and R^2_1 obtained in Theorems 5.8 and 5.9 are particular cases of the invariant quasimetrics in quasisimple Lie groups obtained in Theorem 5.23.

Analogously, we can define in any r-quasisimple Lie group the invariant quasimetric of an r-quasi-Riemannian or r-quasipseudo-Riemannian symmetric space.

Arbitrary quasi-Riemannian, quasipseudo-Riemannian, r-quasi-Riemannian, or r-quasipseudo-Riemannian symmetric spaces can be represented as totally geodesic submanifolds in a quasisimple or r-quasisimple Lie group endowed with their invariant quasimetrics.

§5.8. Applications to Physics

5.8.1. Space-Time of Classical Mechanics. The transformations (3.88) of spatial orthogonal coordinates x^1, x^2, x^3 and temporal coordinate t under the transition from one inertial system of coordinates x^1, x^2, x^3, t to another inertial system of coordinates $x^{1'}, x^{2'}, x^{3'}, t'$ in the classical Mechanics of Galilei-Newton coincide with the transformations of orthogonal coordinates in the isotropic space I^4, whose absolute consists of the hyperplane A_0 at infinity, of the imaginary hypercone C_1 of the second order, with a point apex A_1 in this hyperplane and of this apex.

The formulas (5.33) for the group of motions in I^4 can be written as

$$'x^i = U^i_j x^j + a^i, \quad 'x^4 = T^4_j x^j \pm x^4 + a^4 \quad (i, j = 1, 2, 3), \tag{5.48}$$

where (U^i_j) is an orthogonal (3×3)-matrix, and T^4_j, a^i, and a^4 are arbitrary real numbers. These motions leave invariant the direction of the temporal axis Ox^4, whose point at infinity is the point A_1 (the formulas (5.33) have the form (3.88) for the group of motions in Γ^4, this is the reason for the name "Galilean space" for Γ^n) (see [Ro10]).

5.8.2. Applications of Pseudo-Riemannian, Quasipseudo-Riemannian and r-Quasipseudo-Riemannian Geometries. In **3.10.1** we have seen that space-time of the Special Relativity of Einstein carries the pseudometric of the pseudo-Euclidean space R^4_1. In Einstein's General Relativity, space-time carries the pseudometric of the pseudo-Riemannian space V^4_1, whose sectional curvature is greater at places with greater density of the matter. Note that the mean sectional curvature of this space is negative, and if we represent the space V^4_1 with homogeneous density of the matter, that is, V^4_1 of constant negative curvature, by a hypersphere of imaginary radius in R^5_2, the hyperplanes t=const cut out from the hypersphere of imaginary radius in R^5_2 models of the universe; for variable t we obtain the model of "extending universe" discovered by A. A. Friedmann, see [Ro4, pp. 311–314].

General Relativity reduces the properties of gravity of matter to the geometric properties of the space V^4_1. After the creation of General Relativity Einstein proposed the problem of reducing all physical properties to geometric properties of the space-time. Since Einstein believed that all Physics can be reduced to gravity and electromagnetism, at first this problem was the problem of finding unified theory of gravitation and the electromagnetic field. This problem led H. Weyl [We4] to a generalization of the space V_1^4, in which the maps of the tangent spaces R_1^4 realized by parallel displacement of vectors are not isometries but similitudes. The term "affine connection" appeared first in [We4]; the Weyl manifold was generalized by É. Cartan and J. A. Schouten, who introduced the general manifolds A^n with an affine connection.

In their attempts to create a unified field theory P. Defrise [Def], C. Jankiewicz [Jan], R. A. Toupin [Tou] and R. I. Pimenov [Pim1–3] replaced the space V_1^4 by quasipseudo-Riemannian and r-quasipseudo-Riemannian 4-manifolds (see also [Ro4, p. 378]).

CHAPTER VI

SYMPLECTIC AND QUASISYMPLECTIC GEOMETRIES

§6.1. Symplectic Spaces

6.1.1. Quadratic Symplectic Spaces. In **4.1.1** we have seen that the absolutes of the spaces S^n, H^n, S_l^n, and H_l^n are imaginary or real hyperquadrics, which, as we have seen in **2.8.3**, are cosymmetry figures in the space P^n. In **2.8.3** we have also seen that, besides hyperquadrics, in P^{2n+1} there are cosymmetry figures of an other kind: linear complexes of lines. The space P^{2n-1} in which a linear complex of lines is given is said to be a *real quadratic symplectic space* and is denoted by Sy^{2n-1}. The linear complex determining this space is called the *absolute linear complex* of Sy^{2n-1}. The collineations in P^{2n-1} preserving the absolute linear complex of Sy^{2n-1} are called *symplectic transformations* in this space. The absolute linear complex of Sy^{2n-1} can be defined by (2.109), where the $(2n \times 2n)$-matrix (a_{ij}) can be reduced to the form (0.62); then (2.109) has the form

$$\sum_{i=0}^{n-1} p^{2i,2i+1} = 0. \tag{6.1}$$

In this case the matrices of symplectic transformations in Sy^{2n-1} are called *symplectic $(2n \times 2n)$-matrices*, and the group of these matrices is called a *real symplectic group* and is denoted by Sp_{2n} (see **1.4.3**). From **1.4.3** it follows that the dimension of this group is equal to $n(2n+1)$.

The group Sp_{2n} was first called the *group of a linear complex* (in the German literature *Komplex-Gruppe*). When H. Weyl moved from Germany to the USA, he first called this group a *complex group*, but, since this term also means any group with complex parameters, he replaced in [Wey2] this term by "symplectic group". The Greek word *symplektikos* is the Greek translation of the Latin word *complexus* (both these words mean "complicate"). Hence the names of symplectic transformations and of symplectic spaces arose.

The null-system (2.111) corresponding to the absolute linear complex (2.109) of the space Sy^{2n-1} is called the *absolute null-system* of Sy^{2n-1}. If the $(2n \times 2n)$-matrix (a_{ij}) is reduced to the form (0.62), the absolute null-system (2.111) of Sy^{2n-1} has the form

$$u_{2i} = x^{2i+1}, \ u_{2i+1} = -x^{2i}. \tag{6.2}$$

The absolute null-system (2.111) can be written in vector form $u = xA, \quad A^\top = -A$.

We denote the bilinear form $xAy = a_{IJ}x^I y^J$ by $[xy]$. If the absolute null-system has the form (6.2), then

$$[xy] = \sum_i (x^{2i}y^{2i+1} - x^{2i+1}y^{2i}). \tag{6.3}$$

Since $A^{\top} = -A$ for every vector x, we obtain $[xx] = 0$. Therefore, each point in Sy^{2n-1} is located in the hyperplane which corresponds to it in the absolute null-system. In general, the m-plane in Sy^{2n-1} does not belong to the $(2n-m-2)$-plane corresponding to it in the absolute null-system. The m-planes in Sy^{2n-1} which do belong to the $(2n-m-2)$-planes corresponding to them in the absolute null-system are called *null m-planes* (for $m = 1$ *null lines*) in Sy^{2n-1}. From the definition of absolute linear complex of lines in Sy^{2n-1} it follows that the null lines in Sy^{2n-1} coincide with the lines of this linear complex.

Theorem 6.1. *There are no symplectic invariants of two points X and Y in Sy^{2n-1}.*

There is one symplectic invariant of two lines XY and ZT in Sy^{2n-1}:

$$w = \frac{[xz][yt] - [xt][yz]}{[xy][zt]}, \tag{6.4}$$

where x, y, z, t are vectors representing the points X,Y,Z,T, respectively.

PROOF. The symplectic invariants of two m-planes in Sy^{2n-1} coincide with the projective invariants of the two m-pairs consisting of these m-planes and of the two $(2n - m - 2)$-planes corresponding to them in the absolute null-system. These invariants are the eigenvalues of the $((m + 1)(m + 1))$-matrix (2.58), where A and B are the projective matrix coordinates of the m-planes and U and V are the tangential matrix coordinates of the $(2n - m - 2)$-planes.

For $m = 0$ the entries in the $(1 \times 2n)$-matrices A and B coincide with the coordinates of the vectors x and y. The entries in the $(2n \times 1)$-matrices U and V coincide with the coordinates of the covectors representing the corresponding hyperplanes. The (1×1)-matrices UA, UB, VB, VA coincide with the real numbers $[xx]$, $[xy]$, $[yy]$, $[yx]$, respectively. Since $[xx] = [yy] = 0$, the real number (2.58) is equal to infinity and there is no symplectic invariant for two points X and Y in Sy^{2n-1}.

For $m = 1$ the entries in the $(2 \times 2n)$-matrices A and B consist of the coordinates of the vectors x, y and z, t, respectively, representing the points X, Y and Z, T. The entries in the $(2n \times 2)$-matrices U and V coincide with the coordinates of the covectors representing the hyperplanes corresponding to the points X, Y, Z, T. The (2×2)-matrices UA, UB, VB, VA are

$$\begin{aligned} UA &= \begin{bmatrix} [xx] & [yx] \\ [xy] & [yy] \end{bmatrix} = \begin{bmatrix} 0 & -[xy] \\ [xy] & 0 \end{bmatrix}, \quad UB = \begin{bmatrix} [zx] & [tx] \\ [zy] & [ty] \end{bmatrix}, \\ VB &= \begin{bmatrix} [zz] & [tz] \\ [zt] & [tt] \end{bmatrix} = \begin{bmatrix} 0 & -[zt] \\ [zt] & 0 \end{bmatrix}, \quad VA = \begin{bmatrix} [xz] & [yz] \\ [xt] & [yt] \end{bmatrix}, \end{aligned} \tag{6.5}$$

and the (2×2)-matrix (2.58) is equal to

$$(UA)^{-1}(UB)(VB)^{-1}(VA) = \frac{[xz][yt] - [xt][yz]}{[xy][zt]} I, \tag{6.6}$$

that is, it is a scalar (2×2)-matrix with single eigenvalue (6.4). ■

Since for two lines in Sy^{2n-1} and the two $(2n-3)$-planes corresponding to them in the absolute null-system of this space, the (2×2)-matrix (2.58) is a scalar matrix with single eigenvalue, the number of transversals of these lines and $(2n-3)$-planes is not two, as for two generic 1-pairs in P^{2n-1}, but it is infinity and the cross ratios of the points in the intersections of all these transversals with the given lines and $(2n-3)$-planes are equal, while the invariant (6.2) is equal to the common value of these cross ratios.

Analogously, we can define the spaces $\mathbb{A}Sy^{2n-1}$ over any commutative algebra, in particular over the field $\mathbb{C}$ and over the algebras $\mathbb{C}'$ and $\mathbb{C}^0$.

6.1.2. Another Quadratic Symplectic Space. Since the spaces $\mathbb{A}Sy^{2n-1}$ are odd-dimensional, we can define an *even-dimensional quadratic symplectic space* $\mathbb{A}Sy^{2n}$ as the space $\mathbb{A}E^{2n}$ in whose hyperplanes at infinity the geometry of $\mathbb{A}Sy^{2n-1}$ is established. The m-planes in $\mathbb{A}Sy^{2n}$ whose $(m-1)$-planes at infinity are null $(m-1)$-planes in $\mathbb{A}Sy^{2n-1}$ are called *isotropic m-planes* in $\mathbb{A}Sy^{2n}$.

The direct sum of the space Sy^{2n} with the field $\mathbb{R}$, in which the product of two elements $\{x, \lambda\}$ and $\{y, \mu\}$ is defined as $\{x+y, \lambda+\mu+2[xy]\}$, is isomorphic to the *Heisenberg group* which is the Borel subgroup of the stabilizer of a point of the absolute hyperquadric of the space $\mathbb{C}\overline{H}^n$ (see **4.7.6** and **4.10.3**, see also [Gol, pp. 41–45]).

The spaces $\mathbb{A}Sy^{2n-1}$ and $\mathbb{A}Sy^{2n}$ are also called "projective symplectic" and "affine symplectic" spaces, respectively. In classical Mechanics, which is closely connected with symplectic geometry (see **6.6.1**), only the spaces $\mathbb{A}Sy^{2n}$ are called "symplectic spaces", and the spaces $\mathbb{A}Sy^{2n-1}$ are called "contact spaces" (see, for instance, [ArG]). Isotropic n-planes in $\mathbb{A}Sy^{2n}$ are called *Lagrangean n-planes* (this name, proposed by V. P. Maslov [Mas], is connected with J. L. Lagrange's works in classical Mechanics).

6.1.3. Hermitian Symplectic Spaces. The spaces $\mathbb{C}P^n, \mathbb{C}'P^n, \mathbb{H}P^n$, and $\mathbb{H}'P^n$ in which a Hermitian null-system (2.112) is given are called *Hermitian symplectic spaces.* These spaces are denoted by $\mathbb{C}\overline{S}y^n, \mathbb{C}'\overline{S}y^n, \mathbb{H}\overline{S}y^n$, respectively $\mathbb{H}'\overline{S}y^n$. If there are various Hermitian symplectic spaces over the same algebra, an index ℓ is added to this notation.

The Hermitian null-system (2.112) is called the *absolute Hermitian null-system* of the space. Unlike quadratic symplectic spaces, in Hermitian symplectic spaces not all points belong to the hyperplanes corresponding to them in the absolute null-system. The points in a Hermitian symplectic space that do belong to the corresponding hyperplane are called *null points.* The set of null points, that is, of the points satisfying the equation $\overline{x}^j a_{ij} x^i = 0$, is called the *absolute linear complex* of this space. The collineations in $\mathbb{C}P^n, \mathbb{C}'P^n, \mathbb{H}P^n$, and $\mathbb{H}'P^n$ preserving the absolute linear complex of the respective Hermitian symplectic space are called *symplectic transformations* in this space.

Theorem 2.58 implies that the complex Hermitian symplectic spaces coincide with the complex Hermitian elliptic and pseudoelliptic spaces and that the absolute linear complexes of these symplectic spaces coincide with the absolute hyperquadrics of the elliptic and pseudoelliptic spaces while the symplectic transformations in these spaces coincide with motions.

Therefore, the distance between two points in a complex Hermitian elliptic or pseudoelliptic space is a symplectic invariant of these two points.

We can analogously prove that a split complex Hermitian symplectic space coincides with the space $\mathbb{C}'\overline{S}^n$.

Denote by $\mathbb{C}\overline{S}y^n$ the complex Hermitian symplectic space which coincides with the space $\mathbb{C}\overline{S}^{2m-1}_m$ or $\mathbb{C}\overline{S}^{2m}_m$, that is, with the Hermitian pseudoelliptic space with m-planar generators of maximal dimension, and by $\mathbb{C}\overline{S}y^n_\ell$ the complex Hermitian symplectic space which coincides with the space $\mathbb{C}\overline{S}^n_{m-\ell}$. Over the algebra $\mathbb{C}'$ there is only the space $\mathbb{C}'\overline{S}y^n$.

Theorem 2.58 implies also that there is only one kind of quaternionic Hermitian symplectic spaces, because all Hermitian null-systems in $\mathbb{H}P^n$ can be reduced to the form (2.113). On the contrary, there are $\left[\frac{n+1}{2}\right]+1$ kinds of split quaternionic Hermitian symplectic spaces, because Hermitian null-systems in $\mathbb{H}'P^n$ can be reduced to the forms (2.113),

$$u_a = \overline{x}^a i, \quad u_u = -\overline{x}^u i, \tag{6.7}$$

and

$$u_i = \overline{x}^i e. \tag{6.8}$$

We denote the quaternionic Hermitian symplectic space by $\mathbb{H}\overline{S}y^n$, and the split quaternionic Hermitian symplectic space whose absolute null-system can be reduced to the form (2.113), (6.7) $(a = 0, 1, \ldots, l-1,\ u = l, \ldots, n)$, or (6.8) by $\mathbb{H}'\overline{S}y^n, \mathbb{H}'\overline{S}y^n_\ell$, and $\mathbb{H}'\tilde{S}y^n$, respectively.

We can analogously define the semiquaternionic Hermitian symplectic space $\mathbb{H}^0\overline{S}y^n$ and other semiquaternionic Hermitian symplectic spaces.

Theorem 6.2. *There are two symplectic invariants of two points in the spaces* $\mathbb{H}\overline{S}y^n, \mathbb{H}'\overline{S}y^n, \mathbb{H}'\overline{S}y^n_l$, *and* $\mathbb{H}'\tilde{S}y^n$.

PROOF. The symplectic invariants of two points X and Y in $\mathbb{H}\overline{S}y^n$, $\mathbb{H}'\overline{S}y^n$, $\mathbb{H}'\overline{S}y^n_l$, or $\mathbb{H}'\tilde{S}y^n$ coincide with the projective invariants of two 0-pairs in $\mathbb{H}P^n$ or $\mathbb{H}'P^n$, which consist of the points X and Y and of the hyperplanes α and β corresponding to them in the absolute null-system of this space. This cross ratio is equal to the cross ratio of the points X and Y and of the points of intersection of the line XY with the hyperplanes α and β. In Theorem 2.33 we have seen that the cross ratio w of four points in $\mathbb{H}P^n$ or $\mathbb{H}'P^n$ are transformed under collineations of these spaces to $'w = a^{-1}wa$, and, therefore, can be reduced to the forms $w_0 + iw_1, w_0 + ew_1$, or $w_0 + (i+e)w_1$, where w_0 and w_1 are real invariants. ∎

Since the left hand side of the equation of the absolute linear complex of $\mathbb{H}\overline{S}y^n$ is a quaternion x satisfying the condition $\overline{x} = -x$, this equation is equivalent to three real condition and, therefore, the dimension of this submanifold is equal to $4n - 3$ (in particular, it is a curve for the line $\mathbb{H}\overline{S}y^1$).

If the absolute null-system of $\mathbb{H}\overline{S}y^n$ is reduced to the form (2.112), the $((n+1) \times (n+1))$-matrices which represent symplectic transformations in this space satisfy the condition

$$\overline{U}^{\top} iU = iI. \tag{6.9}$$

These matrices are called *quaternionic symplectic matrices.* The group of these matrices is denoted by $\mathbb{H}Sq_n$. In Theorem 6.7 we will see that the group $\mathbb{H}Sq_n$ is a noncompact real simple Lie group of the class D_n. Therefore, from **1.4.3** it follows that the real dimension of the group $\mathbb{H}Sq_n$ is equal to $n(2n-1)$.

§6.2. Interpretations of Symplectic Spaces

6.2.1. Interpretations of the Quadratic Line and 3-Space. The simplest interpretation of real quadratic symplectic spaces is given by

Theorem 6.3. *The symplectic line Sy^1 coincides with the projective line P^1.*

PROOF. Since for $n = 1$ the dimension $n(2n+1)$ of the group Sp_{2n} is equal to 3, the group Sp_2 coincides with the group SL_2 and, therefore, the group of symplectic transformations in Sy^1 coincides with the group of projective transformations in P^1 and the line Sy^1 coincides with the line P^1. ■

Note that the absolute null-system of Sy^1 transforms each of its points into itself.

In **4.5.7** we have considered the Plücker interpretation of the manifold of lines in P^3 on the absolute hyperquadric Q^4 of S^5_3. This interpretation implies

Theorem 6.4. *The manifold of lines in Sy^3 with lines corresponding to another in the absolute null-system of Sy^3 identified admits an interpretation as the space S^4_2. The group of symplectic transformations in Sy^3 is isomorphic to the group of motions in S^4_2.*

PROOF. In the Plücker interpretation the linear complex of lines $a_{ij}p^{ij} = 0$ in P^3 is interpreted as the intersection of the hyperquadric Q^4 with a hyperplane in S^5_3. Therefore, the space Sy^3 with this complex as absolute linear complex is interpreted as this hyperplane, which is the space S^4_2. Any point X in this hyperplane is interpreted as two lines in this Sy^3 whose images on Q^4 are the two points of its intersection with the line joining X with the pole of the hyperplane. Since the absolute null-system of Sy^3 is interpreted in S^5_3 as reflection in the hyperplane $a_{ij}p^{ij} = 0$, these two lines correspond to another in the absolute null-system of Sy^3. The group of symplectic transformations in Sy^3 is isomorphic to the subgroup of the group of motions in S^5_3 which preserves the hyperplane, that is, to the group of motions in S^4_2. ■

Note that in **3.8.7** we have seen that the group of collineations in $\mathbb{H}'P^1$, that is, complete group of collineations in P^3, is isomorphic to the group of conformal transformations in C^4_2, and Theorem 4.25 on the Plücker interpretation implies that linear complexes of lines in P^3 are interpreted in C^4_2 as hyperspheres, elliptic and hyperbolic linear congruences of lines in P^3 are interpreted in C^4_2 as oval, respectively ruled, spheres, and families of rectilinear generators of quadrics in P^3 are interpreted in C^4_2 as circles. Therefore, linear complexes of lines, elliptic and hyperbolic linear congruences of lines, and families of rectilinear generators of quadrics in Sy^3 are interpreted in S^4_2 as hyperspheres or hyperplanes, oval and ruled spheres or elliptic and pseudoelliptic planes, and circles or lines, respectively.

6.2.2. Real Symplectic Geometry as Hermitian Elliptic Geometry. Since the group of symplectic transformations in Sy^{2n-1} is locally isomorphic to the group Sp_{2n} which is a simple Lie group of the class C_n, it is natural that this group can be represented analogously to the group of motions in $\mathbb{H}\overline{S}^{n-1}$.

Theorem 6.5 [Ro1,5]. *The manifold of lines in Sy^{2n-1} is isometric to the space $\mathbb{H}'\overline{S}^{n-1}$ if we define the distance ω between two lines by means of their invariant w of the form (6.4), by $\cos^2\omega = w$.*

The group of symplectic transformations in Sy^{2n-1} is isomorphic to the group of motions in $\mathbb{H}'\overline{S}^{n-1}$.

PROOF. In Theorem 2.19 we have seen that the space $\mathbb{R}_{m+1}P^n$ admits an interpretation in the space P^{mn+n+m} such that each point in $\mathbb{R}_{m+1}P^n$ with coordinates $x^i = (x^{\beta i}_\alpha)$ is interpreted as an m-plane through the points x_α with coordinates $x^{\beta i}_\alpha$. Since the algebra $\mathbb{R}_2$ is isomorphic to the algebra $\mathbb{H}'$, this theorem implies that the space $\mathbb{H}'P^n$ admits an interpretation in P^{2n+1} such that each point in $\mathbb{H}'P^n$ with coordinates x^i, represented by (2×2)-matrices $\begin{bmatrix} X^{2i} & Y^{2i} \\ X^{2i+1} & Y^{2i+1}\end{bmatrix}$, is interpreted as a line XY through the points $X(X^{2i}, X^{2i+1})$ and $Y(Y^{2i}, Y^{2i+1})$. If the coordinates x^i are multiplied by a split quaternion, represented by a (2×2)-matrix $\begin{bmatrix} K & M \\ L & N\end{bmatrix}$, then the points X and Y are replaced by the points $X'\,(KX^{2i}+LY^{2i}, KX^{2i+1}+LY^{2i+1})$ and $Y'\,(MX^{2i}+N^{2i}, MX^{2i+1}+N^{2i+1})$ on the same line. The symplectic invariant (6.4) of the lines XY and ZT in Sy^{2n-1} is

$$w = \frac{[xz][yt]-[xt][yz]}{[xy][zt]} = \frac{\sum\limits_i \overline{x}^i y^i \cdot \sum\limits_i \overline{y}^i x^i}{\sum\limits_i \overline{x}^i x^i \cdot \sum\limits_i \overline{y}^i y^i}, \tag{6.10}$$

since

$$\sum_i \overline{x}^i y^i = \sum_i \begin{bmatrix} Y^{2i+1} & -Y^{2i} \\ -X^{2i+1} & X^{2i}\end{bmatrix} \cdot \begin{bmatrix} Z^{2i} & T^{2i} \\ Z^{2i+1} & T^{2i+1}\end{bmatrix} = \begin{bmatrix} [yz] & [yt] \\ [zx] & [tx]\end{bmatrix},$$

$$\sum_i \overline{y}^i x^i = \begin{bmatrix} [tx] & [ty] \\ [xz] & [yz]\end{bmatrix}, \sum_i \overline{x}^i x^i = \begin{bmatrix} [yx] & [yy] \\ [xx] & [yx]\end{bmatrix}, \sum_i \overline{y}^i y^i = \begin{bmatrix} [tz] & [tt] \\ [zz] & [tz]\end{bmatrix}.$$

If we replace each entry in a unitary split quaternionic $(n\times n)$-matrix by the real (2×2)-matrix representing this split quaternion, we obtain a real symplectic $(2n\times 2n)$-matrix. Therefore, the group of symplectic transformations in Sy^{2n-1} is isomorphic to the group of motions in $\mathbb{H}'\overline{S}^{m-1}$. ■

Note that Theorem 6.4 can be obtained from Theorems 4.8 and 6.5: according to the latter theorem the manifold of lines in Sy^3 admits an interpretation on the line $\mathbb{H}'\overline{S}^1$ and according to the first of these theorems this line is isometric to the hypersphere in R^5_2. Since the lines in Sy^3 which correspond to another in the absolute null-system of this space are interpreted as antipodal points of this hypersphere, the manifold of lines in Sy^3 with identified lines is isometric, in the invariant pseudometric of this manifold to the space S^4_2.

For $n = 1$ the space Sy^{2n-1} becomes the line Sy^1, and the split quaternionic group of class C_1 which corresponds to the group of symplectic transformations in Sy^1 is the group of automorphisms of the algebra $\mathbb{H}'$, which is isomorphic to the group of motions in H^2 and to the group of projective transformations in P^1

(the corresponding compact group of class C_1 is the group of automorphisms of $\mathbb{H}$, which is isomorphic to the group of motions in S^2).

6.2.3. Real Interpretation of the Split Quaternionic Hermitian Symplectic Space. The space $\mathbb{H}'\overline{S}y^{n-1}$ also admits a real interpretation.

Theorem 6.6. *The split quaternionic Hermitian symplectic space $\mathbb{H}'\overline{S}y^{n-1}$ admits an interpretation as the manifold of lines in the real elliptic space S^{2n-1}. The symplectic invariants of two points in $\mathbb{H}'\overline{S}y^{n-1}$ are equivalent to the stationary distances of the corresponding lines in S^{2n-1}. The group of symplectic transformations in $\mathbb{H}'\overline{S}y^{n-1}$ is isomorphic to the group of motions in S^{2n-1}.*

PROOF. In the proof of Theorem 6.5 we have seen that the space $\mathbb{H}'P^n$ admits an interpretation in the space P^{2n+1} such that each point of $\mathbb{H}'P^n$ with coordinates x^i, represented by (2×2)-matrices $\begin{bmatrix} X^{2i} & Y^{2i} \\ X^{2i+1} & Y^{2i+1}\end{bmatrix}$, is interpreted by a line XY through the points $X(X^{2i}, X^{2i+1})$ and $Y(Y^{2i}, Y^{2i+1})$.

The metric invariant (3.13) of the lines XY and ZT in S^{2n-1} is

$$\begin{bmatrix} x^2 & xy \\ xy & y^2 \end{bmatrix}^{-1} \begin{bmatrix} xz & xt \\ yz & yt \end{bmatrix} \begin{bmatrix} z^2 & zt \\ zt & t^2 \end{bmatrix}^{-1} \begin{bmatrix} xz & yz \\ xt & yt \end{bmatrix} = \\ = \left(\sum_i \overline{x}^i ix\right)^{-1} \left(\sum_i \overline{x}^i iy^i\right) \left(\sum_i \overline{y}^i iy\right)^{-1} \left(\sum_i \overline{y}^i ix^i\right), \tag{6.11}$$

since

$$\sum_i \overline{x}^i iy^i = \sum_i \begin{bmatrix} Y^{2i+1} & -Y^{2i} \\ -X^{2i+1} & X^{2i} \end{bmatrix} \begin{bmatrix} 0 & 1 \\ -1 & 0 \end{bmatrix} \begin{bmatrix} Z^{2i} & T^{2i} \\ Z^{2i+1} & T^{2i+1} \end{bmatrix} = \begin{bmatrix} yz & yt \\ -xz & -xt \end{bmatrix},$$

$$\sum_i \overline{y}^i ix^i \begin{bmatrix} xt & yt \\ -xz & -yz \end{bmatrix}, \sum_i \overline{x}^i ix^i = \begin{bmatrix} xy & y^2 \\ -x^2 & -xy \end{bmatrix}, \sum_i \overline{y}^i iy^i = \begin{bmatrix} zt & t^2 \\ -z^2 & -zt \end{bmatrix}.$$

If we replace each entry of the symplectic split quaternionic $(n \times n)$-matrix by the real (2×2)-matrix representing this split quaternion, we obtain a real orthogonal $(2n \times 2n)$-matrix. Therefore, the group of symplectic transformations in $\mathbb{H}'\overline{S}y^{n-1}$ is isomorphic to the group of motions in S^{2n-1}. ∎

Theorem 6.7. *The group of symplectic transformations in the space $\mathbb{H}\overline{S}y^{n-1}$ is a noncompact semisimple (for $n \neq 2$, simple) Lie group of class D_n and of type DIII.*

PROOF. The groups of symplectic transformations in the spaces $\mathbb{H}\overline{S}y^{n-1}$ and $\mathbb{H}'\overline{S}y^{n-1}$ are real Lie groups with the same complex form. Therefore, since the group of motions in S^{2n-1} is a compact semisimple (for $n \neq 2$, simple) Lie group of the class D_n, we see that the group of symplectic transformations in the space $\mathbb{H}\overline{S}y^{n-1}$ is a noncompact semisimple (for $n \neq 2$, simple) Lie group of the class D_n. Since the noncompact simple Lie groups of this class of types DI and DII are the groups of motions in S^{2n-1}_ℓ and H^{2n-1}, respectively (see **1.4.12**), the group of symplectic transformations in $\mathbb{H}\overline{S}y^{n-1}$ is a group of type DIII. ∎

Note that the noncompact simple or semisimple Lie groups of the class D_n are determined by the same involutive automorphisms in the compact group of this class corresponding to the following symmetry figures in S^{2n-1}: $(\ell-1)$-planes with their polar $(2n-\ell-1)$-planes, for $\ell = 1$ points with their polar hyperplanes, and paratactic congruences. The compact groups of the class D_n are locally isomorphic to the groups SO_{2n}. The noncompact groups of this class corresponding to $(\ell-1)$-planes and their polar $(2n-\ell-1)$-planes are locally isomorphic to the groups $SO_{2n,\ell}$. The noncompact groups of this class corresponding to paratactic congruences are locally isomorphic to the groups of motions in $\mathbb{H}'\overline{S}^{n-1}$.

6.2.4. Interpretations of the Hermitian Line, Plane, and 3-Space. We consider interpretations of the line $\mathbb{H}\overline{S}y^1$, the plane $\mathbb{H}\overline{S}y^2$, and the space $\mathbb{H}\overline{S}y^3$.

Theorem 6.8 (L. V. RUMYANTSEVA [Rum1]). *The line $\mathbb{H}\overline{S}y^1$ can be interpreted as pairs of points in two planes S^2 and H^2. The group of symplectic transformations in $\mathbb{H}\overline{S}y^1$ is isomorphic to the direct product of the groups of motions in S^2 and H^2.*

PROOF. The equation $\sum_i \overline{x}^i i x^i = 0$ of the absolute linear complex of the line $\mathbb{H}\overline{S}y^1$ can be written in the affine coordinates $x = x^1(x^0)^{-1}$ as $\overline{x}ix + i = 0$. If we write the quaternion x as $x = x_0 + x_1 i + x_2 j + x_3 k$, where x_0, x_1, x_2, x_3 are real numbers, this equation is equivalent to the three equations $x_0^2 + x_1^2 - x_3^2 - x_4^2 = 1$, $x_1x_2 - x_0x_3 = 0$, and $x_1x_3 + x_0x_2 = 0$.

Since the two last equations are equivalent to the systems $x_0 = x_1 = 0$ and $x_2 = x_3 = 0$, the absolute linear complex of $\mathbb{H}\overline{S}y^1$ consists of the real circle $x_0^2 + x_1^2 = 1$ on the plane $x_2 = x_3 = 0$ and the imaginary circle $x_2^2 + x_3^2 = -1$ in the plane $x_0 = x_1 = 0$.

The projective line $\mathbb{H}P^1$ can be interpreted as the conformal space C^4 or as the absolute hyperquadric of the spaces S_1^5 and H^5 and the absolute linear complex of $\mathbb{H}\overline{S}y^1$ can be interpreted as the real and imaginary circles on two totally orthogonal planes in $\mathbb{H}$ with the metric of R^4 or on the two polar planes in H^5 which cut out these two circles from the oval hyperquadric in P^5 representing C^4.

These two polar planes can be regarded as the planes S^2 and H^2, respectively. Each point in $\mathbb{H}\overline{S}y^1$, interpreted as a point on the oval hyperquadric, can be interpreted as two points in these planes, each of which is the projection of this point on the hyperquadric on one plane in the direction of the second plane.

The group of symplectic transformations in $\mathbb{H}\overline{S}y^1$ can be represented by the subgroup of the group of conformal transformations in C^4 or of the group of motions in S_1^5 that is isomorphic to the direct product of the groups of motions in S^2 and H^2. ■

Theorem 6.9 (RUMYANTSEVA [Rum1]). *The plane $\mathbb{H}\overline{S}y^2$ can be interpreted as the manifold of lines in the space $\mathbb{C}\overline{H}^3$. The group of symplectic transformations in $\mathbb{H}\overline{S}y^2$ is isomorphic to the group of motions in $\mathbb{C}\overline{H}^3$.*

PROOF. Lines in the space $\mathbb{C}P^3$, like lines XY in P^3, can be characterized by the Plücker coordinates $p^{ij} = x^i y^j - y^i x^j$, which satisfy the equation (2.28). If we introduce in $\mathbb{C}P^3$ the pseudometric of the space $\mathbb{C}\overline{H}^3$ and the points X and Y are polar conjugate, that is, satisfy the condition

$$-\overline{x}^0 y^0 + \overline{x}^1 y^1 + \overline{x}^2 y^2 + \overline{x}^3 y^3 = 0, \tag{6.12}$$

and the coordinates of these points are normalized by the conditions

$$-\overline{x}^0 x^0 + \overline{x}^1 x^1 + \overline{x}^2 x^2 + \overline{x}^3 x^3 = -1, \quad -\overline{y}^0 y^0 + \overline{y}^1 y^1 + \overline{y}^2 y^2 + \overline{y}^3 y^3 = 1, \tag{6.13}$$

then we assign to the line XY in $\mathbb{C}\overline{H}^3$ the point in the plane $\mathbb{H}\overline{S}y^2$ with coordinates

$$X^0 = p^{01} + j\overline{p}^{23}, \quad X^1 = p^{02} + j\overline{p}^{31}, \quad X^2 = p^{03} + j\overline{p}^{12}. \tag{6.14}$$

The conditions (6.12) and (6.13) imply the condition

$$-\overline{p}^{01}p^{01} - \overline{p}^{02}p^{02} - \overline{p}^{03}p^{03} + \overline{p}^{23}p^{23} + \overline{p}^{31}p^{31} + \overline{p}^{12}p^{12} = 0. \tag{6.15}$$

The conditions (2.28) and (6.15) imply the condition

$$\overline{X}^0 iX^0 + \overline{X}^1 iX^1 + \overline{X}^2 iX^2 = i. \tag{6.16}$$

Therefore, formulas (6.14) map the manifold of lines in $\mathbb{C}\overline{H}^3$ on the plane $\mathbb{H}\overline{S}y^2$. If the points X, Y, Z, T in $\mathbb{C}\overline{H}^3$ are the points of intersection of the lines XY and ZT with their common perpendiculars, we choose as basis points in this space the points X, Y and points $\widetilde{X}$ and $\widetilde{Y}$ on the common perpendiculars that are polar conjugate with X and Y. Then the coordinates of the points X, Y, Z, T are $1, 0, 0, 0;\ \ 0, 1, 0, 0;\ \ \alpha, 0, \beta, 0;\ \ 0, \gamma, 0, \delta$, respectively, where $|\alpha| = \cosh\omega_1$, $|\beta| = \sinh\omega_1$, $|\gamma| = \cos\omega_0$, $|\delta| = \sin\omega_0$ (ω_1 and $i\omega_0$ are the stationary distances of the lines XY and ZT). Therefore, according to (6.14), $X^0 = 1$, $X^1 = X^2 = 0$, $Y^0 = \alpha\gamma + j\overline{\beta}\overline{\delta}$, $Y^1 = 0$, $Y^2 = \alpha\overline{\delta} - j\beta\overline{\gamma}$, and the symplectic invariants of these two points in $\mathbb{H}\overline{S}y^2$ are determined by the quaternion

$$\begin{aligned} w &= \left(\sum_i \overline{Y}^i iY^i\right)^{-1} \left(\sum_i \overline{Y}^i iX^i\right) \left(\sum_i \overline{X}^i iX^i\right)^{-1} \left(\sum_i \overline{X}^i iY^i\right) = \\ &= |\alpha|^2|\gamma|^2 - |\beta|^2|\delta|^2 + 2j(\alpha\gamma\overline{\beta}\,\overline{\delta}). \end{aligned} \tag{6.17}$$

The real symplectic invariants of these points are equal to

$$\begin{aligned} \frac{w + \overline{w}}{2} &= \cosh^2\omega_1 \cos^2\omega_0 - \sinh^2\omega_1 \sin^2\omega_0, \\ \left|\frac{w - \overline{w}}{2}\right| &= 2\cosh\omega_1 \cos\omega_0 \sinh\omega_1 \sin\omega_0. \end{aligned} \tag{6.18}$$

The invariants (6.18) are the real and imaginary parts of the complex invariant $\cosh^2(\omega_0 + i\omega_1)$ of the lines XY and ZT in $\mathbb{C}\overline{H}^3$. Therefore, the group of symplectic transformations in $\mathbb{H}\overline{S}y^2$ is isomorphic to the group of motions in $\mathbb{C}\overline{H}^3$. ∎

Theorem 6.10 (L. V. Shestyreva and L. S. Nikitina [ShN]). *The space $\mathbb{H}\overline{S}y^3$ can be interpreted as the manifold of lines in the space S^7_2. The group of symplectic transformations in $\mathbb{H}\overline{S}y^3$ is isomorphic to the group of motions in S^7_2.*

PROOF. The Cartan algorithm, which transforms the compact group of motions in S^7 to the noncompact group of motions in S^7_2 and to the noncompact group of symplectic transformations in $\mathbb{H}\overline{S}y^3$, corresponds to involutive automorphisms in the group of motions in S^7 determined by reflection in a line and by symmetry in a paratactic congruence of lines, respectively. According to the triality principle in S^7, considered in **4.7.9**, these symmetry figures in S^7 correspond one to another. Therefore, these noncompact groups are isomorphic. This isomorphism determines the interpretation of the space $\mathbb{H}\overline{S}y^3$ in S^7_2. Since the stabilizer of a point in $\mathbb{H}\overline{S}y^3$ is isomorphic to the stabilizer of a plane in $\mathbb{H}\overline{S}y^3$, which, by Theorem 6.9, is interpreted as the space $\mathbb{C}\overline{H}^3$, the stabilizer of a point in $\mathbb{H}\overline{S}y^3$ is isomorphic to the stabilizer of an elliptic paratactic congruence in S^7_2. Therefore, in this interpretation, points in $\mathbb{H}\overline{S}y^3$ are interpreted as elliptic paratactic congruences in S^7_2. Both groups are isomorphic to the direct product of the group of motions in $\mathbb{C}\overline{H}^3$ by the group of motions in the line S^1. The real dimension of the manifold of points in $\mathbb{H}\overline{S}y^3$ and of the manifold of lines in S^7_2 is 12. ■

6.2.5. Connections between the Interpretations of Symplectic Spaces and Spinor Representations. In Theorem 4.28 we have seen that all interpretations of elliptic and pseudoelliptic spaces based on isomorphisms of simple Lie groups are connected with spinor representations of groups of motions in real spaces. Some interpretations of symplectic spaces are also connected with spinor representations of groups of pseudoorthogonal matrices.

Theorem 6.11. *The matrices of the symplectic transformations in Sy^3 are the matrices of the spinor representations of motions in S^4_2.*

The matrices of the symplectic transformations in $\mathbb{H}\overline{S}y^3$ are the matrices of the spinor representations of motions in S^7_2.

PROOF. The spinor representation of the group $SO_{5,2}$ of the class B_2, representing motions in S^4_2, is realized in the algebra $\mathbb{A}_{5,2} = \mathbb{H}'_2$ isomorphic to the algebra $\mathbb{R}_4$. The spinor group of $SO_{5,2}$ is isomorphic to the group $\mathbb{H}'U_2$, which is isomorphic to the group Sp_4, representing the group of symplectic transformations in Sy^3.

The spinor representation of the group $SO_{8,2}$ of the class D_4, representing motions in S^2_7, is realized in the algebra $\mathbb{A}_{8,2} = \mathbb{H}_4 \oplus \mathbb{H}_4$. The spinor group of $SO_{8,2}$ is the direct product of two groups $\mathbb{H}Sq_4$, representing the group of symplectic transformations in $\mathbb{H}\overline{S}y^3$. ■

The isomorphism of the groups $SL_2, SO_{3,1}$, and Sp_2 of classes A_1, B_1, and C_1, representing the groups of projective transformations in P^1, the group of motions in H^2, and the group of symplectic transformations in Sy^1, the isomorphism of the groups $\mathbb{H}Sq_2$ and $SO_3 \otimes SO_{3,1}$ of class $D_2 = B_1 \otimes B_1$, representing the group of symplectic transformations in $\mathbb{H}\overline{S}y^1$ and the direct product of the groups of motions in S^2 and S^2_1, and the isomorphism of the groups $\mathbb{H}Sq_3$ and $\mathbb{C}SU_{4,1}$ of class $D_3 = A_3$, representing the group of symplectic transformations in $\mathbb{H}\overline{S}y^2$ and the group of motions in $\mathbb{C}\overline{H}^3$, are not connected with spinor representations.

§6.3 Quasisymplectic and r-Quasisymplectic Spaces

6.3.1. Quasisymplectic Spaces. Similarly to the quasielliptic and quasihyperbolic spaces $S^{m,n}$ and $H^{m,n}$, we can define *quasisymplectic spaces* $Sy^{2m-1,2n-1}$. In **6.1.1** and **6.1.2** we have considered only nondegenerate null-systems $u_i = a_{ij}x^j$ and $u_i = \overline{x}^j a_{ij}$, that is, we supposed that the matrices $A = (a_{ij})$ have inverses A^{-1}. For degenerate null-systems in P^{2n-1}, when the matrix A of the null-system has rank $2m$, there is an $(2n-2m-1)$-plane A_0 in P^{2n-1} such that the hyperplane corresponding to a point X in this null-system passes through X and through A_0. We define the *quasisymplectic space* $Sy^{2m-1,2n-1}$ as the space P^{2n-1} in which a degenerate null-system, with matrix (a_{ij}) of rank $2m$, and a nondegenerate null-system in the $(2n-2m-1)$-plane A_0 are given. These two null-systems are called the *absolute null-systems* of $Sy^{2m-1,2n-1}$. We denote the bilinear forms determined by these absolute null-systems by $[xy]_0$ and $[xy]_1$, respectively.

Collineations in $Sy^{2m-1,2n-1}$ that preserve its $(2n-2m-1)$-plane A_0 and commute with its absolute null-systems are called *quasisymplectic transformations*.

Theorem 6.12. *There is no quasisymplectic invariant for two points X and Y in $Sy^{2m-1,2n-1}$.*

There is one quasisymplectic invariant for two lines XY and ZT in $Sy^{2n-1,2n-1}$:

$$w = \frac{[xz]_0[yt]_0 - [xt]_0[yz]_0}{[xy]_0[zt]_0}, \tag{6.19}$$

where x, y, z, t are vectors representing the points X, Y, Z, T. If $w = 1$, these lines have another quasisymplectic invariant:

$$w_1 = [x-z, y-t]_1 = [xy]_1 - [xt]_1 - [zy]_1 + [zt]_1. \tag{6.20}$$

The proof is analogous to the proof of Theorem 6.6.

We can analogously define the quaternionic Hermitian quasisymplectic spaces $\mathbb{H}\overline{S}y^{m,n}$, split quaternionic Hermitian quasisymplectic spaces $\mathbb{H}'\overline{S}y^{m,n}$, and other split quaternionic Hermitian quasisymplectic spaces.

6.3.2. Interpretations of Quasisymplectic Spaces. Similarly to Theorems 6.5 and 6.6 we can prove

Theorem 6.13. *The manifold of lines in $Sy^{2m-1,2n-1}$ is isometric to the space $\mathbb{H}'\overline{S}^{m-1,n-1}$, if we define the distances ω and ω_1 between two lines by their invariants w and w_1 of the forms (6.19) and (6.20), as $\cos^2\omega = w$ and $\omega_1^2 = w_1$. The group of quasisymplectic transformations in $Sy^{2m-1,2n-1}$ is isomorphic to the group of motions in $\mathbb{H}'\overline{S}^{m-1,n-1}$.*

Theorem 6.14. *The split quaternionic Hermitian quasisymplectic space $\mathbb{H}'\overline{S}y^{m-1,n-1}$ admits an interpretation as the manifold of lines in the real quasielliptic space $S^{2m-1,2n-1}$. The quasisymplectic invariants of two points in $\mathbb{H}'\overline{S}y^{m-1,n-1}$ are equivalent to the stationary distances of the corresponding lines in $S^{2m-1,2n-1}$.*

The group of quasisymplectic transformations in $\mathbb{H}'\overline{S}^{m-1,n-1}$ *is isomorphic to the group of motions in* $S^{2m-1,2n-1}$.

Since for $n = 2$, $m = 1$ the space $\mathbb{H}'\overline{S}^{m-1,n-1}$ coincides with the Hermitian Euclidean line $\mathbb{H}'\overline{R}^1$ which according to Theorem 3.1 admits an interpretation as the real space R^4_2, Theorem 6.14 implies

Theorem 6.15. *The manifold of lines in* $Sy^{1,3}$ *is isometric to* R^4_2, *if we define the distance* ω *between two lines by their invariant* w_1 *of the form (6.20) as* $\omega^2 = w_1$. *The group of quasisymplectic transformations in* $Sy^{1,3}$ *is isomorphic to the group of motions in* R^4_2.

This Theorem can also be obtained from Theorem 6.4 by limit transition. Note that the $(2n-2m-1)$-plane A_0 in the space $Sy^{1,3}$ is a line, the absolute null-systems of this space transform any point in the space which is not on A_0 to the plane through this point and A_0 and any point on A_0 into itself. The complete group of collineations in P^3 which preserve the line A_0 is isomorphic to the group of similitudes in R^4_2, the dimension of this group is 11. The complete group of collineations in P^3, as we have seen in **6.2.1**, is isomorphic to the group of conformal transformations in C^4_2, we have seen that linear complexes of lines in P^3 are interpreted in C^4_2 as hyperspheres, elliptic and hyperbolic linear congruences of lines in P^3 are interpreted in C^4_2 as oval and ruled spheres, respectively, and families of rectilinear generators of quadrics in P^3 are interpreted in C^4_2 as circles. Therefore, linear complexes of lines, elliptic and hyperbolic linear congruences of lines, and families of rectilinear generators of quadrics in $Sy^{1,3}$ are interpreted in R^4_2 as hyperspheres or hyperplanes, oval and ruled spheres or Euclidean and pseudo-Euclidean planes, and circles or lines, respectively.

Note that the space $Sy^{1,3}$ has the same absolute (consisting of single line), as axial space $P^{1,3}$ (see **5.1.2**) but the group of quasiaffine transformations in $P^{1,3}$ is isomorphic to the group of similitudes in R^4_1 and the group of quasisymplectic transformations in $Sy^{1,3}$ is a subgroup of this group (the dimensions of these groups are, respectively, 11 and 10).

6.3.3. Semiquaternionic Hermitian Symplectic Spaces. Similarly to the spaces $\mathbb{H}\overline{S}y^n$ and $\mathbb{H}'\overline{S}y^n$ we can define the semiquaternionic Hermitian symplectic spaces $\mathbb{H}^0\overline{S}y^n$. Similarly to Theorems 6.8, 6.9, and 6.10 we can prove

Theorem 6.16. *The line* $\mathbb{H}^0\overline{S}^1$ *can be interpreted as the set of pairs of points on two planes* S^2 *and* R^2. *The group of symplectic transformations in* $\mathbb{H}^0\overline{S}y^1$ *is isomorphic to the direct product of the groups of motions in* S^2 *and* R^2.

Theorem 6.17. *The plane* $\mathbb{H}^0\overline{S}y^2$ *can be interpreted as the manifold of lines in* $\mathbb{C}\overline{R}^3$. *The group of symplectic transformations in* $\mathbb{H}\overline{S}y^2$ *is isomorphic to the group of motions in* $\mathbb{C}\overline{R}^3$.

Theorem 6.18. *The space* $\mathbb{H}^0\overline{S}y^3$ *can be interpreted as the manifold of lines in the space* $S^{1,7}$. *The group of symplectic transformations in* $\mathbb{H}^0\overline{S}y^3$ *is isomorphic to the group of motions in* $S^{1,7}$.

6.3.4. r-Quasisymplectic Spaces. Analogously to the r-quasielliptic and r-quasihyperbolic spaces $S^{m_0m_1\dots m_{r-1}n}$ and $H^{m_0m_1\dots m_{r-1}n}$ we can define the r-*quasisymplectic spaces* $Sy^{2m_0-1,\dots,2m_{r-1}-1,2n-1}$ as the space P^{2n-1} in which are given: a degenerate null-system, determining a $(2n - 2m_0 - 1)$-plane A_0, a second degenerate null-system in A_0, determining a $(2n - 2m_1 - 1)$-plane A_1, a third degenerate null-system in A_1, determining a $(2n - 2m_2 - 1)$-plane $A_2, \ldots$, and a nondegenerate null-system in the $(2n - 2m_{r-1} - 1)$-plane A_{r-1}. These $r+1$ null-systems are called the ***absolute null-systems*** of $Sy^{2m_0-1,\ldots,2m_{r-1}-1,2n-1}$. We denote the bilinear forms determined by the $(a+1)$-th absolute null-system of $Sy^{2m_0-1,\ldots,2m_{r-1}-1,2n-1}$ by $[xy]_a$.

Collineations in $Sy^{2m_0-1,\ldots,2m_{r-1}-1,2n-1}$ preserving its $(2n - 2m_a - 1)$-planes A_{a-1} and commuting with all absolute null-systems of this space are called *r-quasisymplectic transformations.*

Theorem 6.19. *There is no r-quasisymplectic invariant for two points X and Y in* $Sy^{2m_0-1,\ldots,2m_{r-1}-1,2n-1}$.

There is one r-quasisymplectic invariant, (6.19), for two lines XY and ZT in $Sy^{2m_0-1,\ldots,2m_{r-1}-1,2n-1}$. *Here x, y, z, t are vectors representing the points X, Y, Z, T. If $w = 1$ and the r-quasisymplectic invariants*

$$w_a = [x - z, y - t]_a = [xy]_a - [xt]_a - [zy]_a + [zt]_a \tag{6.21}$$

for $a = 1, 2, \ldots, b-1$ are equal to 0, these lines have another r-quasisymplectic invariant, (6.21), for $a = b$.

The proof is analogous to the proof of Theorems 6.6 and 6.11.

We can analogously define the quaternionic Hermitian r-quasisymplectic spaces $\mathbb{H}\overline{S}y^{m_0m_1\ldots m_{r-1}n}$, the split quaternionic Hermitian r-quasisymplectic space $H'\overline{S}y^{m_0m_1\ldots m_{r-1}n}$, and other split quaternionic Hermitian r-quasisymplectic spaces.

Analogously to Theorems 6.12 and 6.13 we can prove theorems on the interpretations of the spaces $Sy^{2m_0-1,\ldots,2m_{r-1}-1,2n-1}$ and $\mathbb{H}'\overline{S}y^{m_0m_1\ldots m_{r-1}n}$ in the r-quasielliptic spaces $\mathbb{H}'\overline{S}^{m_0m_1\ldots m_{r-1}n}$ and $S^{2m_0-1,\ldots,2m_{r-1}-1,2n-1}$

Quasisymplectic spaces and symplectic spaces with quasisimple fundamental groups were investigated by L. P. Andreyeva and L. V. Shestyreva [ASh].

§6.4. Symmetry and Parabolic Figures

6.4.1. Symmetry Figures. Since symplectic transformations in the spaces $\mathbb{A}Sy^n$ and $\mathbb{A}\overline{S}^n$ are particular cases of collineations in $\mathbb{A}P^n$, symmetry figures in $\mathbb{A}Sy^n$ and $\mathbb{A}\overline{S}^n$ are particular cases of symmetry figures in $\mathbb{A}P^n$.

A reflection in an m-pair in $\mathbb{A}P^n$ is a symplectic transformation in $\mathbb{A}Sy^n$ or $\mathbb{A}\overline{S}^n$ only if the m-plane and $(n-m-1)$-plane of this m-pair correspond one to another in the absolute null-system of this space, or, for $m = n - m - 1$, that is, $m = \frac{n-1}{2}$, if both m-planes of this m-pair are null m-planes. In the space $\mathbb{H}'\overline{S}y^{n-1}$, next to reflections in m-planes, which in the interpretation of this space in S^{2n-1} are interpreted as reflections in $(2m+1)$-planes in S^{2n-1}, there exist reflections in $(m-\frac{1}{2})$-planes, interpreted as reflections in $2m$-planes in S^{2n-1}. If the symmetry in a linear congruence of lines in Sy^{2n-1} is a symplectic transformation in this space, this congruence is called a *symplectic congruence* in Sy^{2n-1}. A symmetry in a complex $(n-1)$-chain in $\mathbb{H}\overline{S}y^{n-1}$ and $\mathbb{H}'\overline{S}y^{n-1}$ is a symplectic transformation only if this $(n-1)$-chain is *normal $(n-1)$-chain,* that is, is obtained by a symplectic transformation from a set of points with coordinates of the form $x_0^i + ix_1^i$, a *normal*

$(n-1)$-*pseudochain*, that is, is obtained by a symplectic transformation from a set of points with coordinates of the form $x_0^a + ix_1^a$ and $jx_0^u + kx_1^u$, or an *antinormal* $(n-1)$-*chain*, that is, is obtained by a symplectic transformation from a set of points with coordinates of the form $x_0 + jx_1$. We can analogously, define normal and antinormal complex and split complex $(n-1)$-chains in $\mathbb{H}'\overline{S}y^{n-1}$. Therefore, Theorem 2.61 implies

Theorem 6.20. *The symmetry figures in Sy^{2n-1} are the $(2m-1)$-planes together with the $(2n-2m-1)$-planes corresponding them in the absolute null-system, and elliptic and hyperbolic symplectic congruences of lines which are determined by pairs of conjugate imaginary or real null $(n-1)$-planes.*

The symmetry figures in $\mathbb{H}\overline{S}y^{n-1}$ are the m-planes together with the $(n-m-1)$-planes corresponding them in the absolute null-system, normal complex $(n-1)$-chains and $(n-1)$-pseudochains, and the antinormal complex $(n-1)$-chains.

The symmetry figures in $\mathbb{H}'\overline{S}y^{n-1}$ are the m-planes and $(m-\frac{1}{2})$-planes together with the $(n-m-2)$-planes or $(n-m-\frac{3}{2})$-planes corresponding them in the absolute null-system, and the normal complex $(n-1)$-chains and antinormal split complex $(n-1)$-chains.

Since the line Sy^1 coincides with the line P^1 and admits an interpretation on the plane the H^2, the symmetry figures in Sy^1, that is, conjugate imaginary and real pairs of points, are interpreted as conjugate imaginary and real pairs of points on P^1 and as points and hyperbolic lines in H^2, respectively.

Since the space Sy^3 admits an interpretation in the space S_2^4, the symmetry figures in Sy^3, that is, pairs of lines corresponding in the absolute null-system and elliptic and hyperbolic symplectic congruences of lines, are interpreted as points in S_2^4 together with their polar hyperplanes and by elliptic and hyperbolic lines together with their polar planes, respectively.

Since the space $\mathbb{H}'\overline{S}y^{n-1}$ admits an interpretation in the space S^{2n-1}, the symmetry figures in $\mathbb{H}'\overline{S}y^{n-1}$, that is, m-planes and $(m-\frac{1}{2})$-planes, normal complex $(n-1)$-chains, and antinormal split complex $(n-1)$-chains, are interpreted as $(2m+1)$-planes and $2m$-planes in S^{2n-1} and as paratactic congruences of lines and $(n-1)$-planes, respectively.

6.4.2. Fundamental and Parabolic Figures. Since the groups of symplectic transformations in Sy^{2n-1} and $\mathbb{H}\overline{S}y^{n-1}$ are simple Lie groups, in these spaces we can define fundamental α_i-figures and parabolic $(\alpha_{i_1}, \alpha_{i_2}, \ldots, \alpha_{i_k})$-figures (see **1.6.4**).

Theorem 6.21. *The fundamental α_i-figures in the space Sy^{2n-1} are the null $(i-1)$-planes in this space, for $i=1$ points, for $i=2$ null lines. The parabolic $(\alpha_{i_1}, \alpha_{i_2}, \ldots, \alpha_{i_k})$-figures in this space are the flags consisting of an incident null (i_1-1)-plane, (i_2-1)-plane, ..., (i_k-1)-plane.*

The fundamental α_{2i}-figures in the space $\mathbb{H}\overline{S}y^{n-1}$ are the real null $(i-1)$-planes in this space, for $i=1$ null points, for $i=2$ null lines. The fundamental α_{2i-1}-figures in $\mathbb{H}\overline{S}y^{n-1}$ are the imaginary $(i-\frac{1}{2})$-planes, for $i=1$ halfpoints, for $i=2$ halflines, for $i=3$ halfplanes. The parabolic $(\alpha_{i_1}\alpha_{i_2}, \ldots, \alpha_{i_3})$-figures in this space are the flags consisting of an incident null (i_1-1)-plane, (i_2-1)-plane, ..., (i_k-1)-plane, where $i_1, i_2, \ldots, i_k$ are integers or half-integers.

The proof is analogous to the proof of Theorem 2.63.

Note that the Satake diagrams of the group of symplectic transformations in the space Sy^{2n-1} and of the group Sp_{2n} locally isomorphic to this group coincide with the Dynkin diagram of simple Lie group C_n (Fig. 1.3, C_n).

The Satake diagram of the group of symplectic transformations in the space $\mathbb{H}\overline{S}y^{n-1}$ and of the group $\mathbb{H}Sq_n$ locally isomorphic to this group is given in Fig. 1.9, $DIII$.

Consider the manifolds of null $(n-1)$-planes in $Sy^{2n-1}, \mathbb{C}\overline{S}y^{2n-1}$, and $\mathbb{H}\overline{S}y^{2n-1}$. The affine matrix coordinates of these $(n-1)$-planes are $(n \times n)$-matrices in the algebras $\mathbb{R}_n, \mathbb{C}_n$, and $\mathbb{H}_n$, respectively.

Theorem 6.22 (C. L. SIEGEL [Sie]). *The real $(n \times n)$-matrices representing null $(n-1)$-planes in Sy^{2n-1} are symmetric. The symplectic transformations in this space are represented by the fractional linear transformations (1.140) in the Jordan algebra $\mathbb{J}s(\mathbb{R}_n)$ of these matrices.*

PROOF. In Theorem 2.44 (see **2.4.5**) we have seen that the collineations in P^n have the form (2.63) in affine matrix coordinates, that is, (1.140). We write the $(2n \times 2n)$-matrices U of the Lie group Sp_{2n} as $U = \begin{bmatrix} A & B \\ C & D \end{bmatrix}$ and the $(2n \times 2n)$-matrices $\mathcal{A}$ of the tangent Lie algebra $\mathbb{S}p_{2n}$ of this group in the same form. In **0.8.3** and **0.8.4** we have seen that $(2n \times 2n)$-matrices U of the group Sp_{2n} satisfy the condition (0.61) and that the $(2n \times 2n)$-matrices $\mathcal{A}$ of the Lie algebra $\mathbb{S}p_{2n}$ satisfy the condition (0.68). For the $(2n \times 2n)$-matrices $\mathcal{A} = \begin{bmatrix} A & B \\ C & D \end{bmatrix}$ the last condition can be written as

$$\begin{bmatrix} A^\top & C^\top \\ B^\top & D^\top \end{bmatrix} \begin{bmatrix} 0 & I \\ -I & 0 \end{bmatrix} \begin{bmatrix} A & B \\ C & D \end{bmatrix} = \begin{bmatrix} 0 & I \\ -I & 0 \end{bmatrix}, \qquad (6.22)$$

that is,

$$A^\top = -D,\ B^\top = B,\ C^\top = C. \qquad (6.23)$$

The Lie algebra $\mathbb{S}p_{2n}$ of $(2n \times 2n)$-matrices $\mathcal{A}$ can be written as a direct sum (1.170), where $h = 1$ and the subspaces $\mathbb{G}_{-1}, \mathbb{G}_0$, and $\mathbb{G}_1$ consist of the matrices $\mathcal{A}$ with $A = B = D = 0$, $B = C = 0$, and $A = C = D = 0$, respectively. Therefore, the subspace $\mathbb{G}_{-1}$ can be regarded as the manifold of symmetric $(n \times n)$-matrices B. This subspace can also be regarded as the tangent space to the corresponding homogeneous space, that is, the manifold of null $(n-1)$-planes in Sy^{2n-1}. But since the space of $(n \times n)$-matrices representing this manifold is itself a linear space, it coincides with the tangent space of this manifold, that is, it coincides with the Jordan algebra $\mathbb{J}s(\mathbb{R}_n)$ of real symmetric $(n \times n)$-matrices. ■

Analogously we can prove

Theorem 6.23. *The complex and quaternionic $(n \times n)$-matrices representing null $(n-1)$-planes in $\mathbb{C}\overline{S}y^{2n-1}$ and $\mathbb{H}\overline{S}y^{2n-1}$ are Hermitian symmetric. Transformations in these spaces are represented by the fractional linear transformations (1.140) in the Jordan algebras $\mathbb{J}\overline{s}(\mathbb{C}_n)$, respectively $\mathbb{J}\overline{s}(\mathbb{H}_n)$, of these matrices.*

§6.5. Symplectic and Quasisymplectic Connections

6.5.1. Symplectic Connections. A manifold A^{2n} with a torsion-free affine connection such that in each affine tangent space E^{2n} the geometry of a symplectic space Sy^{2n} (defined in **6.1.2**) which is invariant under parallel displacement of A^{2n} is given, is called a *manifold with a symplectic connection* and we denote this manifold by Vy^{2n}. Since the geometry of Sy^{2n} is determined by a skew symmetric tensor $a_{ij} = -a_{ij}$ by means of which the absolute null-system in the space Sy^{2n-1} in the hyperplane at infinity of Sy^{2n} can be written as $u_i = a_{ij}x^j$, the covariant derivative $\nabla_k a_{ij}$ corresponding to the affine connection in A^{2n} is equal to zero in the manifold Vy^{2n}.

If the exterior differential form $\omega = a_{ij}dx^i \wedge dy^j$ defined by the tensor a_{ij} at each point in Vy^{2n} is closed, that is, the exterior differential $d\omega$ of this form is equal to zero (see **0.7.9**), the manifold Sy^{2n} is called a *symplectic manifold.*

6.5.2. Hermitian Riemannian and Kählerian Manifolds. The manifolds Vy^{2n} are closely connected with complex Hermitian analogs of the Riemannian manifolds V^n and pseudo-Riemannian manifolds V^n_ℓ. These manifolds are complex differentiable (at all points) manifolds $\mathbb{C}X^n$ in which a complex Hermitian symmetric tensor $e_{ij} = \overline{e}_{ji}$ is given. If the differential form $e_{ij}dx^i d\overline{x}^j$ is positive definite, we call this manifold a *Hermitian Riemannian manifold* and denote it by $\mathbb{C}\overline{V}^n$. If this differential form is indefinite of index ℓ, we call this manifold a *Hermitian pseudo-Riemannian manifold of index* ℓ and denote it by $\mathbb{C}\overline{V}^n_\ell$. For $\mathbb{C}\overline{V}^n$ with coordinates $x^i = X^i + iX^{n+i}$ this form can be regarded as the form ds^2 of the real Riemannian manifold V^{2n} with coordinates X^i and X^{n+i}, for $\mathbb{C}\overline{V}^n_\ell$ this form can be regarded as the form ds^2 of the real pseudo-Riemannian manifold $V^{2n}_{2\ell}$. The imaginary part of the bilinear form $e_{ij}dx^i d\overline{y}^j$ is an exterior differential form ω which is invariant under parallel displacement of the manifold V^{2n} or $V^{2n}_{2\ell}$. Therefore, each Hermitian Riemannian and pseudo-Riemannian manifold is also a manifold Vy^{2n}.

If the manifold Vy^{2n} determined by $\mathbb{C}\overline{V}^n$ or $\mathbb{C}\overline{V}^n_\ell$ is a symplectic manifold, that is, if $d\omega = 0$, the Hermitian Riemannian or pseudo-Riemannian manifold $\mathbb{C}\overline{V}^n$ or $\mathbb{C}\overline{V}^n_\ell$ is called a *Kählerian*, respectively *pseudo-Kählerian manifold*, and is denoted by $\mathbb{C}K^n$,respectively $\mathbb{C}K^n_\ell$. Note that the spaces $\mathbb{C}\overline{R}^n$, $\mathbb{C}\overline{S}^n$, and $\mathbb{C}\overline{H}^n$ are particular cases of Kählerian manifold $\mathbb{C}K^n$, while the spaces $\mathbb{C}\overline{R}^n_\ell, \mathbb{C}\overline{S}^n_\ell$, and $\mathbb{C}\overline{H}^n_\ell$ are particular cases of the pseudo-Kählerian manifold $\mathbb{C}K^n_\ell$.

6.5.3. Quasisymplectic Connections. A manifold A^{2n} with a torsion-free affine connection such that in each affine tangent space E^{2n} the geometry of the quasisymplectic space $Sy^{2m,2n}$ determined by the geometry of $Sy^{2m-1,2n-1}$ in the hyperplane at infinity of the tangent space P^{2n-1} of E^{2n} is given and is invariant under parallel displacement of A^{2n} is called a *manifold with a quasisymplectic connection*, and we denote it by $Vy^{2m,2n}$.

We can analogously define *manifolds with an r-quasisymplectic connection* $Vy^{2m_0,2m_1,\dots,2m_{r-1},2n}$.

§6.6. Finite Geometry

6.6.1. Symplectic Space over a Galois Field. A null-system in the finite projective space $\mathbb{F}_q P^{2n+1}$ can be regarded as the absolute null-system of the *symplectic space* $\mathbb{F}_q Sy^{2n+1}$ *over the Galois field* $\mathbb{F}_q$. The collineations in $\mathbb{F}_q P^{2n+1}$

commuting with this null-system are called *symplectic transformations* in $\mathbb{F}_q Sy^{2n+1}$. In this space, like in the real space Sy^{2n+1}, there is no symplectic invariant for two points, and there is one symplectic invariant, (6.4), for two lines, which is an element in $\mathbb{F}_q$. In $\mathbb{F}_q Sy^{2n+1}$, like in Sy^{2n+1}, there are null lines, which form a linear complex of lines, and null m-planes. Theorem 2 of Wan Zhe-xian (see **2.9.5**) implies that the cardinality of a linear complex of null lines in $\mathbb{F}_q Sy^{2n+1}$ is equal to (2.148) and the cardinality of the set of all null m-planes in this space is equal to (2.147).

The "symplectic geometry" in the manifold of null $(n-1)$-planes in $\mathbb{F}_q Sy^{2n-1}$ by means of the Jordan algebra $\mathbb{J}s((\mathbb{F}_q)_n)$ of symmetric $(n \times n)$-matrices whose entries are elements in $\mathbb{F}_q$ was considered by Hua Loo-Keng [Hua].

§6.7. Applications to Physics

6.7.1. Hamilton Equations. If a dynamical system is characterized by *generalized coordinates* q_i and *generalized momenta* p_i, these magnitudes are connected by the *Hamilton canonical equations*, introduced by W. R. Hamilton,

$$\frac{dq_i}{dt} = \frac{\partial \mathcal{H}}{\partial p_i}, \quad \frac{dp_i}{dt} = -\frac{\partial \mathcal{H}}{\partial q_i}, \tag{6.24}$$

where $\mathcal{H}$ is the Hamilton function equal to the sum $T + U$ of the kinetic energy T and the potential energy U of the system. If the coordinates q_i and p_i are regarded as coordinates of vectors q and p in the space R^n, the Hamilton equations (6.24) can be written in vector form

$$\frac{dq}{dt} = \frac{d\mathcal{H}}{dp}, \quad \frac{dp}{dt} = -\frac{d\mathcal{H}}{dq}. \tag{6.25}$$

Theorem 6.24 (C. L. Siegel [SiM], [ArG, p.8]). *If the coordinates q_i and p_i are regarded as coordinates x^i and x^{n+1} of a vector x in the space R^{2n}, the Hamilton equations (6.24) can be written in vector form:*

$$\frac{dx}{dt} = Hx, \tag{6.26}$$

where H is a $(2n \times 2n)$-matrix in the tangent Lie algebra $\mathbb{S}p_{2n}$ of the group Sp_{2n} representing the group of symplectic transformation in the space Sy^{2n-1}.

<u>Proof</u>. In **0.8.3** and **0.8.4** we have seen that the $(2n \times 2n)$-matrices U of the Lie group Sp_{2n} satisfy the condition (0.61) and that the $(2n \times 2n)$-matrices A of the Lie algebra $\mathbb{S}p_{2n}$ satisfy the condition (0.68). The last condition for the $(2n \times 2n)$-matrices $H = \begin{bmatrix} A & B \\ C & D \end{bmatrix}$ can be written as (6.22), that is, the submatrices A, B, C, D of these $(2n \times 2n)$-matrices satisfy the conditions (6.23). The equation (6.26) is equivalent to the system

$$\frac{dq}{dt} = Aq + Bp, \quad ; \frac{dp}{dt} = Cq + Dp. \tag{6.27}$$

Therefore, if the Hamilton function $\mathcal{H}$ is written as

$$\mathcal{H} = \frac{1}{2} xJHx, \tag{6.28}$$

we obtain

$$\mathcal{H} = \frac{1}{2}(qCq - pBp - 2pAq), \tag{6.29}$$

that is, the function $\mathcal{H}$ satisfies the equations (6.25), which are equivalent to the equations (6.24). ■

Therefore, for general dynamical systems the coordinates q_i and p_i can be regarded as coordinates of points in a manifold Vy^{2n} with a symplectic connection.

6.7.2. Duality and Stability. A very important particular case of Hamiltonian dynamical systems is when the $(n \times n)$-matrices A and D are the zero matrix and equations (6.23) have the form

$$\frac{dq}{dt} = -Bp, \quad \frac{dp}{dt} = -Cq. \tag{6.30}$$

In this case the kinetic energy T of the system is

$$T = -\frac{1}{2} pBp = -\frac{1}{2} \sum_i \sum_j b_{ij} p_i p_j \tag{6.31}$$

and the potential energy U of the system is

$$U = -\frac{1}{2} qCq = -\frac{1}{2} \sum_i \sum_j c_{ij} q_i q_j \tag{6.32}$$

and both quadratic forms (6.31) and (6.32) are positive definite. For a general Hamiltonian dynamical system the kinetic energy T also has the form (6.31), but the $(n \times n)$-matrix $B = (b_{ij})$ is not a submatrix of the $(2n \times 2n)$-matrix H having the form $\begin{bmatrix} A & B \\ C & D \end{bmatrix}$ and the entries b_{ij} in this $(2n \times 2n)$-matrix are certain functions of the generalized coordinates q_i. In this general case the potential energy U of the system is a function of general kind of the coordinates q_i. If T and U have the form (6.31) and (6.32) and all coefficients b_{ij} and c_{ij} are constant, the system (6.30) is equivalent to the equation

$$\frac{d^2 q}{dt^2} - W^2 q = 0, \tag{6.33}$$

where the $(n \times n)$-matrix W^2 is equal to the product $-BC$. The equation (6.33) is integrable and the solution of this equation is

$$q = (\cos Wt) q_0 + (W^{-1} \sin Wt) \left(\frac{dq}{dt}\right)_0, \tag{6.34}$$

where q_0 and $\left(\frac{dq}{dt}\right)_0$ are the values of the vectors q and $\frac{dq}{dt}$ at $t = t_0$ and cos Wt and sin Wt are the functions of a matrix argument defined by the series

$$\cos Wt = \frac{(Wt)^2}{2} - \frac{(Wt)^4}{4!} + - \dots,$$
$$\sin Wt = Wt - \frac{(Wt)^3}{3!} + - \dots$$

This dynamical system is the ***system of n mechanical harmonic oscillators***. The coefficients b_{ij} and c_{ij} characterize the *inertia* and *elasticity* of this system, respectively. For $n = 1$ this system represents a single mechanical harmonic oscillator, and its motion is periodic. When $n > 1$, the motion of the system is periodic only if the eigenvalues of the matrix W are commensurable, and it is quasiperiodic in general.

So, complete duality between the kinetic and potential energies T and U of a dynamical system, and between its generalized coordinates and momenta q_i and p_i imply a stable, periodic or quasiperiodic, motion of this system.

Equations of the types (6.30) and (6.33) also occur in other domains of physics: in electrodynamics the equations (6.30), where the matrices $B = (b_{ij})$ and $C = (c_{ij})$ characterize inductivity and electric capacity of the system, determine a *system of n electrodynamical oscillators*.

The analogous equations

$$\frac{\partial \psi_1}{\partial t} = E\psi_2, \quad \frac{\partial \psi_2}{\partial t} = -E\psi_1 \tag{6.35}$$

are equivalent to the *Schrödinger equation*

$$i\,\frac{\partial \psi}{\partial t} = E\psi \tag{6.36}$$

for the atom of Hydrogen, where $\psi = \psi_1 + i\psi_2$ is the complex *wave function* describing the inner motion in this atom.

Physically, processes described by equations of these types consist in the accumulation of potential energy in a mechanical string or in an electric condenser (this role in the atom of Hydrogen is played by a "positronium", the pair electron+positron), in liberation of this energy and its transference to the kinetic energy of inner motions in the system, and in the return of this energy to the place of its accumulation. This return is realized by an inertial part of the system, a heavy weight in a mechanical system, a self-induction coil in an electromagnetic system, a heavy neutron connected with the positron of the "positronium" in an atom (without neutron the "positronium" is unstable). An analogous interaction of inertial parts and parts which are able to accumulate energy can be found in all stable real world phenomena, including in living and social organisms.

6.7.3. Dualities in Philosophy and Mathematics. Already in Antiquity the stability of phenomena was connected with symmetry and duality. Various kinds of duality lie at the bases of many religions and philosophical systems. In certain systems symmetry between dual objects was complete, in other systems this symmetry was not complete. For instance, soul and body in many religions, the ideas and "space" of Plato, the active form and passive matter of Aristotle, whose junction creates living beings and other stable phenomena, are not symmetric, whereas good and evil in the soul are symmetric. Pythagoras considered ten "opposites": finite – infinite, odd – even, one – many, right – left, male – female, resting – moving, straight – curved, light – darkness, good – bad, square – rectangular; five of these "opposites" are symmetric and five asymmetric. The stable phenomena considered in **6.7.2** also contain both symmetric and asymmetric dualities: the atom of Hydrogen is an asymmetric junction of a "positronium" and a neutron, but "positronium" itself is a symmetric junction of positron and electron.

In his book "Symmetry" [We3] H. Weyl has discussed various kinds of symmetry in the history of human culture and has drawn attention to the extremely high symmetry in the images of gods, saints, and divine animals in Babylonian, Greek, Roman, and medieval monuments, statues, and drawings. Since the idea of divinity has always been closely connected with the idea of eternity and stability, in their works of art the ancient and medieval sculptors and artists symbolized stability by symmetry.

In **2.4.6** we have defined the symmetric duality principle in projective geometry, in **2.7.1** and **2.7.2** we have considered dual and autodual projective configurations, in **3.6.3** we have considered symmetric duality in the spherical trigonometry, in **4.7.9** we have mentioned generalizations of this principles, proposed by É. Cartan who also established the triality principle; in **0.2.1**, **0.3.2**, and **0.6.1** we have seen analogous symmetries in topological spaces, linear spaces, and lattices.

In mathematics there are also asymmetric dualities: next to the asymmetric "opposites" of Pythagoras we can mention the duality between ellipse and hyperbola, between elliptic and hyperbolic spaces, between pseudoelliptic and pseudohyperbolic spaces, between quasielliptic and quasihyperbolic spaces, etc. and between compact and noncompact simple Lie groups with common complex form, "dual in the sense of Cartan". There is also a triality: between the dual objects ellipse and hyperbola, elliptic and hyperbolic spaces, etc. there are the parabola, Euclidean, pseudo-Euclidean, quasi-Euclidean spaces, and quasisimple Lie groups, "trial in the sense of Cartan". Since mathematical laws reflect the laws in the real world, mathematical dualities also are reflections of real world dualities.

CHAPTER VII

GEOMETRIES OF EXCEPTIONAL LIE GROUPS. METASYMPLECTIC GEOMETRIES

§7.1. Geometry of the Groups G_2

7.1.1. The Simple Lie Groups G_2. There are two locally non-isomorphic simple real Lie groups G_2: a compact group with character $\delta = -14$ (the Dynkin diagram is given in Fig. 1.3) and a split group with character $\delta = 2$ (the Satake diagram is given in Fig. 1.9).

In Theorem 1.32 (see **1.4.11**) we have seen that the compact simple Lie group G_2 is the group of automorphisms in the alternative division algebra $\mathbb{O}$ of octonions. We can analogously prove that the split simple Lie group G_2 is the group of automorphisms in the alternative algebra $\mathbb{O}'$ of split octonions.

In Theorem 1.27 we have seen that the groups of automorphisms in $\mathbb{O}$ and $\mathbb{O}'$ are subgroups in the groups SO_7 and $SO_{7,3}$, respectively. This was proved using the metric of the space R^8 and the pseudometric of the space R^8_4 introduced in the algebras $\mathbb{O}$ and $\mathbb{O}'$ in **1.3.1** and **1.3.2**. The groups SO_7 and $SO_{7,3}$ are subgroups in the groups of SO_8 and $SO_{8,4}$ of rotations in these 8-spaces.

Note that the Dynkin diagrams of the groups SO_8 and $SO_{8,4}$, that is, of the compact and split groups D_4, of the groups SO_7 and $SO_{7,3}$, that is, the compact and split groups B_3, and of the compact and split groups G_2 are given in Fig. 1.5, 1.3 B_3, and 1.3 G_2. The second of these diagrams can be obtained from the first by joining the points α_3 and α_4, and the third of these diagrams can be obtained from the second by joining the points α_2 and α_3.

In **4.7.9** we have considered the triality principles in the spaces S^7 and S^7_4, whose groups of motions are the compact and split groups D_4. In Theorems 4.45 and 4.46 we have seen that these principles can be proved by means of the algebras $\mathbb{O}$ and $\mathbb{O}'$, respectively. The three-fold symmetry of the Dynkin diagram of the compact and split groups D_4 determines an 3-involutive automorphism σ in these groups (σ^3 is the identity). The compact and split groups G_2 can be defined as the groups of elements in the compact and split groups D_4, respectively, which are invariant under these automorphisms.

7.1.2. G-Elliptic, G-Pseudoelliptic, and G-Pseudohyperbolic 6-Spaces. Since the compact and split groups G_2 are subgroups of the groups of rotations in R^7 and R^7_3, respectively, these groups preserve hyperspheres in these spaces with centers at the elements 0 in the algebras $\mathbb{O}$ and $\mathbb{O}'$. These groups are transitive on these hyperspheres and can be regarded as groups of transformations in homogeneous spaces. The 6-sphere in R^7 with identified antipodal points whose group of transformations is the compact group G_2 is called the *G-elliptic space* Sg^6 and the 6-sphere of real, or pure imaginary, radius in R^7_3 with identified antipodal points whose group of transformations is the split group G_2 is called the *G-pseudoelliptic space* Sg_3^6, respectively the *G-pseudohyperbolic space* Hg_2^6. The spaces Sg^6, Sg_3^6, and Hg_2^6 were investigated by N. N. Adamushko [Ad] and R. G. Tlupova [Tl]. The absolute hyperquadrics of Sg_3^6 and Hg_2^6 were investigated by G. J. Schellekens [Sck]. The 6-sphere whose identification of antipodal points leads to Sg^6 was investigated by A. Fröhlicher [Frö].

Theorem 7.1 (TLUPOVA [Tl]). *The tangent 6-spaces of the spaces* Sg^6, Sg_3^6, *and* Hg_2^6 *at a point are the spaces* $\mathbb{C}\overline{R}^3, \mathbb{C}\overline{R}_1^3$, *and* $\mathbb{C}'\overline{R}^3$, *respectively.*

PROOF. The tangent 6-space to the space Sg^6 at the point represented by the element i in $\mathbb{O}$ can be regarded as the hyperplane in R^7, with the metric of R^6, parallel to the hyperplane through the points represented by the octonions $0, j, k, l, p, q$, and r. In this R^6, for each point represented by an octonion x there is a point represented by the product ix. If we denote the vectors j, k, l, p, r, and q by $e_1, e_2, \ldots, e_6$, the transformation $x \to ix$ can be written as $'x^{2i-1} = x^{2i}$, $'x^{2i} = -x^{2i-1}$, that is, as multiplication of the complex coordinates $X^i = x^{2i-1} + ix^{2i}$ by $-i$. Therefore, this space R^6 can be considered as the complex space $\mathbb{C}\overline{R}^3$, which is isometric to R^6.

The assertions of this theorem concerning tangent 6-spaces to Sg_3^6 and Hg_2^6 can be proved analogously. For Sg_3^6 we must consider the tangent 6-space at the point represented by the element i in $\mathbb{O}'$ with the pseudometric of R_2^6 and for Hg_2^6 the tangent 6-space at the point represented by the element e in $\mathbb{O}'$ with the pseudometric of R_3^6. ∎

Note that in the hyperplanes at infinity of R^6, R_2^6, and R_3^6, which are the spaces S^5, S_2^5, and S_3^5, respectively, the transformations $x \to ix$ in R^6 and R_2^6 and $x \to ex$ in R_3^6 induce a symmetry in a paratactic congruence of lines in S^5, in an elliptic paratactic congruence of lines in S_2^5, and in a hyperbolic paratactic congruence of lines in S_3^5, respectively. These congruences are isometric to the planes $\mathbb{C}\overline{S}^2, \mathbb{C}\overline{S}_1^2$, and $\mathbb{C}'\overline{S}^2$, respectively. These planes are the planes at infinity of $\mathbb{C}\overline{R}^3, \mathbb{C}\overline{R}_1^3$, and $\mathbb{C}'\overline{R}^3$, respectively.

Note also that the complex structures in the tangent space at a point in Sg^6 or Sg_3^6 and the split complex structure in the tangent space at a point in Hg_2^6 do not determine complex and split complex structures in the spaces Sg^6, Sg_3^6, and Hg_2^6, that is, in these spaces it is impossible to introduce complex or split complex coordinates. Therefore, we can only assert that these spaces have an *almost complex* or *almost split complex structure.* On the 6-sphere in $\mathbb{O}$ this almost complex structure was first discovered by Fröhlicher [Frö].

7.1.3. The Quasisimple Lie Groups G_2. G-quasielliptic, G-quasipseudoelliptic and G-quasipseudohyperbolic 6-Spaces. Since the split group G_2 can be obtained from the compact group G_2 by the Cartan algorithm, we can apply also the quasi-Cartan algorithm to these groups. It leads to the two quasisimple groups of this class. Analogously to Theorem 1.32 we can prove that these two quasisimple groups are isomorphic to the groups of *metric automorphisms*, defined just like metric automorphisms in $\mathbb{C}^0$ (see **1.1.4**), in the quasisimple algebras $\mathbb{O}^0$ of semioctonions and $\mathbb{O}'^0$ of split semioctonions (see **1.3.7**). The algebras $\mathbb{O}^0$ and

$\mathbb{O}'^0$ are endowed with the quasimetrics of $R^{4,8}$ and $R^{4,8}_{22}$, in which the distance d between the elements x and y is equal to $|y-x|$, while if $d=0$ and $x=\varepsilon x'$, $y=\varepsilon y'$, where x' and y' are quaternions or split quaternions, these elements have a second distance $d'=|y'-x'|$. The metric automorphisms in $\mathbb{O}^0$ and $\mathbb{O}'^0$, like the metric automorphisms in $\mathbb{H}^0$ and $\mathbb{H}'^0$, are rotations in the spaces $R^{4,8}$ and $R^{4,8}_{22}$ preserving the real axes in $\mathbb{O}^0$ and $\mathbb{O}'^0$ as well as the hyperplanes in these spaces orthogonal to these axes. These hyperplanes are spaces $R^{3,7}$ and $R^{3,7}_{22}$, and the metric automorphisms in $\mathbb{O}^0$ and $\mathbb{O}'^0$ preserve the hyperspheres in these spaces and are transitive on them. Therefore, the groups of metric automorphisms in $\mathbb{O}$ and $\mathbb{O}'^0$ can be regarded as the groups of transformations in the spaces $S^{2,6}, S^{2,6}_{12}$, and $H^{2,6}_{02}$ obtained from these hyperspheres by identification of their antipodal points. The spaces $S^{2,6}, S^{2,6}_{12}$, and $H^{2,6}_{02}$ with these groups of transformations are called *G-quasielliptic, G-quasipseudoelliptic, and G-quasipseudohyperbolic 6-spaces*, and we denote them by $Sg^{2,6}, Sg^{2,6}_{12}$, and $Hg^{2,6}_{02}$, respectively. These spaces were investigated by N. N. Adamushko [Ad].

§7.2. Geometry of the Groups $\mathbf{F}_4$ and $\mathbf{E}_6$

7.2.1. The Simple Lie Groups $\mathbf{F}_4$ and $\mathbf{E}_6$. There are three locally non-isomorphic simple real Lie groups F_4: a compact group with character $\delta=-52$ (the Dynkin diagram is given in Fig. 1.3), and two noncompact groups FI and FII with characters $\delta=4$ and -20, respectively (the Satake diagrams are given in Fig. 1.9).

There are five locally non-isomorphic simple real Lie groups E_6: a compact group with character $\delta=-78$ (the Dynkin diagram is given in Fig. 1.3), and four noncompact groups EI, EII, $EIII$, and EIV, with characters $\delta=6,2,-14$, and -26, respectively (the Satake diagrams are given in Fig. 1.9).

Since each real noncompact simple Lie group can be obtained from the compact real simple group with the same complex form by Cartan algorithm determined by an involutive automorphism in the compact group (see **1.4.12**), these noncompact groups are determined by the subgroups K of the compact group that are invariant under its involutive automorphisms.

Theorem of Cartan on the groups $\mathbf{F}_4$ and $\mathbf{E}_6$. [Car12, pp. 657–658] *The subgroups of the compact real simple Lie group F_4 that determine the noncompact real groups FI and FII with the same complex form are the groups $C_3\otimes A_1$ and B_4, respectively.*

The subgroups of the compact real simple Lie group E_6 that determine the noncompact real groups EI, EII, EIII, and EIV with the same complex form are the groups $C_4, A_5\otimes A_1, D_5\otimes D_1$, and F_4, respectively.

7.2.2. Octonionic and Split Octonionic Projective Planes. In **2.2.1** we have seen that in the projective spaces $P^n, \mathbb{C}P^n, \mathbb{C}'P^n, \mathbb{H}P^n$, and $\mathbb{H}'P^n$ over an associative algebra the points of these spaces can be characterized by coordinates x^i determined up to a multiplication $x^i\to x^i\lambda$ where λ is a nonzero element in the algebras $\mathbb{R}, \mathbb{C}, \mathbb{C}', \mathbb{H}$ and $\mathbb{H}'$, while in $\mathbb{C}'P^n$ and $\mathbb{H}'P^n$ the multiplier λ can not be a zero divisor. This definition is not suitable for the planes $\mathbb{O}P^2$ and $\mathbb{O}'P^2$ over the alternative algebras $\mathbb{O}$ and $\mathbb{O}'$ since here $(x^i\lambda)\mu\neq x^i(\lambda\mu)$. Therefore, the plane $\mathbb{O}P^2$ was defined by H. Freudenthal [Fr1] by means of the *Jordan algebra* $\mathbb{J}\overline{\mathbb{s}}(\mathbb{O}_3)$ of Hermitian symmetric octonionic (3×3)-matrices (x^{ij}) for which $x^{ij} = \overline{x}^{ji}$. Freudenthal characterized the points in $\mathbb{O}P^2$ by such (3×3)-matrices satisfying the condition

$$x^{ij}x^{jk} = x^{jj}x^{ik}, \tag{7.1}$$

and these (3×3)-matrices are determined up to a real multiplier.

Theorem 7.2. *The points in the projective planes $\mathbb{O}P^2$ and $\mathbb{O}'P^2$ can be characterized by the octonions or split octonions x^0, x^1, x^2 in an associative subalgebra of $\mathbb{O}$ or $\mathbb{O}'$ and are determined up to a multiplication $x^i \to x^i\lambda$ with λ a nonzero element in the same subalgebra in $\mathbb{O}$ or $\mathbb{O}'$; for $\mathbb{O}'$, λ cannot be a zero divisor.*

PROOF. Let the points in $\mathbb{O}P^2$ be characterized by the (3×3)-matrices in the Jordan algebra $\mathbb{J}\overline{\mathbb{s}}(\mathbb{O}_3)$ satisfying the condition (7.1). This condition implies that the diagonal entries $x^{ii} = \overline{x}^{ii}$ in these (3×3)-matrices are real numbers while the entries x^{01}, x^{02}, and x^{12} are related by $x^{01}x^{12} = x^{11}x^{02}$. Therefore, since any two elements in an alternative algebra generate an associative subalgebra, all entries in a (3×3)-matrix in $\mathbb{J}\overline{\mathbb{s}}(\mathbb{O}_3)$ satisfying the condition (7.1) belong to an associative division algebra. In this algebra we always can choose elements x^0, x^1, x^2 such that

$$x^{ij} = x^i\overline{x}^j. \tag{7.2}$$

If the points in $\mathbb{O}'P^2$ are characterized by the (3×3)-matrices in the Jordan algebra $\mathbb{J}\overline{\mathbb{s}}(\mathbb{O}'_3)$ satisfying the same condition (7.1), we can analogously choose in $\mathbb{O}'$ elements x^0, x^1, x^2 related with the entries x^{ij} in the (3×3)-matrix by (7.2). The elements x^0, x^1, x^2 are determined up to a multiplication $x^i \to x^i\lambda$ where λ is an element in the same associative subalgebra. ■

The projective plane $\mathbb{O}P^2$ was investigated by various methods by R. Moufang [Mou], G. Hirsch [Hir], and A. Borel [Bor]. Since the algebra $\mathbb{O}$ is not associative, in $\mathbb{O}P^2$ the general Desargues theorem (see **2.2.4**) is not fulfilled, while in this plane the "small Desargues theorem", that is, the Desargues theorem for the case when the point of intersection of the lines AA', BB', and CC' is on the line of points of intersection of the corresponding lines AB and $A'B', BC$ and $B'C'$, and CA and $C'A'$, holds. This theorem was proved by Moufang [Mou]; she called this theorem a "theorem on the quadrilateral".

Since the Desargues theorem in $\mathbb{A}P^n$, where $\mathbb{A}$ is a division algebra and $n > 2$, follows from the incidence axioms (0.17) in $\mathbb{A}P^n$, the spaces $\mathbb{O}P^n$ are possible only for $n = 1$ and 2.

Theorem 7.3. [Ro9] *The collineations in the planes $\mathbb{O}P^2$ and $\mathbb{O}'P^2$ have the form*

$$'x^i = \bar{a}^i_j f(x^j), \tag{7.3}$$

where x^j belong to an associative subalgebra of $\mathbb{O}$ or $\mathbb{O}'$, the transformation $x \to f(x)$ is an automorphism in $\mathbb{O}$ or $\mathbb{O}'$, a^i_j are the entries in a (3×3)-matrix over

$\mathbb{O}$ or $\mathbb{O}'$, and the transformation $x \to \bar{\bar{x}}$ is the "projection" onto the associative subalgebra of $\mathbb{O}$ or $\mathbb{O}'$ containing the elements $f(x^0), f(x^1)$, and $f(x^2)$.

PROOF. The proof is analogous to the proof of Theorem 2.27.

In **1.3.7** we have defined the involutive automorphisms $x \to \hat{x}$ and $x \to \check{x}$ in $\mathbb{O}$ and $\mathbb{O}'$. These automorphisms preserve the associative subalgebras $\mathbb{H}$ in $\mathbb{O}$ and $\mathbb{H}$ and $\mathbb{H}'$ in $\mathbb{O}'$. Therefore, if we denote the involutive automorphism in $\mathbb{O}$ or $\mathbb{O}'$ preserving an associative subalgebra in this algebra by $x \to \hat{x}$, we can define the "projection" onto this subalgebra as

$$x \to \bar{\bar{x}} = \frac{1}{2}(x + \hat{x}). \tag{7.4}$$

■

Theorem 7.4 (H. FREUDENTHAL [Frl]). *The group of collineations in the plane $\mathbb{O}P^2$ is a noncompact real simple Lie group E_6.*

PROOF. The tangent Lie algebra of the group of collineations in $\mathbb{O}P^2$ is the direct sum of the algebra of octonionic (3×3)-matrices with zero traces and the tangent Lie algebra for the group of automorphisms in $\mathbb{O}$. Commutators of elements in this Lie algebra are the same as in the Theorem of Vinberg (see **1.4.11**). Therefore, the dimension of this group, which is equal to the dimension of the tangent Lie algebra, is equal to $8 \cdot 8 + 14 = 78$, that is, to the dimension of the group E_6.

A Cartan subalgebra of the tangent Lie algebra for the group of collineations in $\mathbb{O}P^2$ consists of diagonal octonionic (3×3)-matrices with entries belonging to a commutative subalgebra of $\mathbb{O}$ and with zero trace, and of a Cartan subalgebra of the tangent Lie algebra of the compact group G_2, that is, of the group of automorphisms in $\mathbb{O}$. A commutative subalgebra in $\mathbb{O}$ is $\mathbb{C}$, the independent diagonal entries of complex (3×3)-matrices are two complex numbers, therefore the dimension of the linear space of (3×3)-matrices, which are in a Cartan subalgebra, is equal to 4. The dimension of a Cartan subalgebra in the tangent Lie algebra of the group G_2 is equal to the rank of G_2, that is, 2, so the dimension of whole Cartan subalgebra in the group of collineations in $\mathbb{O}P^2$ is equal to $4 + 2 = 6$. By means of this Cartan subalgebra we can calculate the root system of this group, as in Theorem 1.29. ■

We can analogously prove that the group of collineations in $\mathbb{O}'P^2$ is also a noncompact real simple Lie group E_6. In Theorem 7.14 we will see that this group is a split group, that is, character of this group is equal to 6, and that character of the group of collineations in $\mathbb{O}P^2$ is equal to -26.

Analogously to Theorem 2.34 we can prove

Theorem 7.5. *The projective lines $\mathbb{O}P^1$ and $\mathbb{O}'P^1$ admit real interpretations as the conformal and pseudoconformal spaces C^8 and C^8_4, respectively.*

The collineations and correlations in these lines can be represented by conformal transformations in these 8-spaces.

Since the spaces C^8 and C^8_4 are homeomorphic to hyperspheres in R^9 and R^9_4, respectively, the lines $\mathbb{O}P^1$ and $\mathbb{O}'P^1$ are also homeomorphic to these hyperspheres.

Since the groups of conformal transformations in C^8 and C^8_4 are isomorphic to the groups of motions in the spaces H^9 and H^9_3, respectively, the groups of projective transformations in $\mathbb{O}P^1$ and $\mathbb{O}'P^1$ are also isomorphic to these groups of motions.

7.2.3. Octonionic and Split Octonionic Hermitian Planes. In the plane $\mathbb{O}P^2$, like in the plane $\mathbb{H}P^2$, there are imaginary and real Hermitian conics, whose equations can be written as

$$\bar{x}x = \bar{x}^0x^0 + \bar{x}^1x^1 + \bar{x}^2x^2 = 0, \tag{7.5}$$

respectively

$$\bar{x}Ex = -\bar{x}^0x^0 + \bar{x}^1x^1 + \bar{x}^2x^2 = 0. \tag{7.6}$$

In the plane $\mathbb{O}'P^2$, like in the plane $\mathbb{H}'P^2$, there are Hermitian conics, whose equations can be written as (7.5).

The plane $\mathbb{O}P^2$ in which an imaginary or real Hermitian conic with equation reducible to (7.5) and (7.6) is given is called the *octonionic Hermitian elliptic plane* $\mathbb{O}\overline{S}^2$ and *octonionic Hermitian hyperbolic plane* $\mathbb{O}\overline{H}^2$, respectively. The plane $\mathbb{O}'P^2$ in which a Hermitian conic with equation reducible to (7.5) is given is called the *split octonionic Hermitian elliptic plane* $\mathbb{O}'\overline{S}^2$. These Hermitian conics are called the *absolute Hermitian conics* of these planes.

We determine the distance δ between two points x and y in $\mathbb{O}\overline{S}^2$ or $\mathbb{O}'\overline{S}^2$ by (4.7) and the distance δ between two points x and y in $\mathbb{O}\overline{H}^2$ by (4.8). The *radii of curvature* of these planes are the numbers r in (4.7) and iq in (4.8). The number $\frac{1}{r^2}$ or $-\frac{1}{q^2}$ is called the *curvature* of these planes.

Motions in the planes $\mathbb{O}\overline{S}^2, \mathbb{O}\overline{H}^2$, and $\mathbb{O}'\overline{S}^2$ are the collineations (7.3) in these planes preserving their absolute Hermitian conics (7.5) and (7.6), and, therefore, preserving the distances between points in these planes.

Theorem 7.6 (H. FREUDENTHAL [Fr1]). *The group of motions in the octonionic Hermitian elliptic plane $\mathbb{O}\overline{S}^2$ is the compact real simple Lie group F_4.*

PROOF. Since the motions in $\mathbb{O}\overline{S}^2$ are the collineations (7.3) preserving the imaginary Hermitian conic (7.5), the group of these motions is locally homeomorphic to the topological product of the group $\mathbb{O}SU_3$ of octonionic unimodular unitary (3×3)-matrices (a^i_j), that is, (3×3)-matrices satisfying the condition (1.115), while the condition of unimodularity is equivalent to vanishing of the traces of the matrices in the tangent Lie algebra of this group, and of the group of automorphisms in $\mathbb{O}$. This group is compact, since we can prove the compactness of the group of octonionic unitary (3×3)-matrices (a^i_j) analogously to the proof of the compactness of the group $\mathbb{H}U_n$ in Theorem 1.31 (see **1.4.11**) and since the group of automorphisms in $\mathbb{O}$ is the compact group G_2.

The tangent Lie algebra of this group is the direct sum of the algebra of octonionic skew Hermitian (3×3)-matrices $(a_{ij} = -\bar{a}_{ji})$ with zero trace and of the tangent Lie algebra of the group of automorphisms in $\mathbb{O}$. Commutators of elements of this Lie algebra are the same as in the Theorem of Vinberg (see **1.4.11**). Therefore, the dimension of this group, which is equal to the dimension of its tangent Lie algebra, is equal to $8+8+8+7+7+14 = 52$, that is, to the dimension of the group F_4.

A Cartan subalgebra of the tangent Lie algebra of the group of motions in $\mathbb{O}\overline{S}^2$ consists of the diagonal skew Hermitian octonionic (3×3)-matrices with entries belonging to a commutative subalgebra of $\mathbb{O}$ and with zero trace and of a Cartan subalgebra in the tangent Lie algebra of the compact group G_2, that is, the group of automorphisms in $\mathbb{O}$. A commutative subalgebra in $\mathbb{O}$ is $\mathbb{C}$, the diagonal entries of skew Hermitian octonionic (3×3)-matrices have the form $a = -\bar{a}$, therefore these entries are pure imaginary numbers. The independent diagonal entries of skew Hermitian complex (3×3)-matrices are two pure imaginary numbers, therefore the dimension of the linear space of (3×3)-matrices, which are in a Cartan subalgebra, is equal to 2. The dimension of a Cartan subalgebra of the tangent Lie algebra of the group G_2 also is equal to 2, so the dimension of the whole Cartan subalgebra of the group of motions in $\mathbb{O}\overline{S}^2$ is $2 + 2 = 4$. By means of this Cartan subalgebra we can calculate the root system of this group, like in Theorem 1.29. ■

This theorem follows also from the Theorem of Vinberg.

We can analogously prove the Theorem of J. Tits [Ti1-2] stating that the group of motions in $\mathbb{O}\overline{H}^2$ with real absolute Hermitian conic (7.6) is a noncompact real simple Lie group F_4, and the theorem that the group of motions in $\mathbb{O}'\overline{S}^2$ with real absolute Hermitian conic (7.5) is another noncompact real simple Lie group F_4.

Analogously to Theorem 4.8 we can prove

Theorem 7.7. *The Hermitian elliptic line $\mathbb{O}\overline{S}^1$ of curvature $\frac{1}{r^2}$ is isometric to a hypersphere of radius $\frac{r}{2}$ in R^9.*

The Hermitian hyperbolic line $\mathbb{O}\overline{H}^1$ of curvature $-\frac{1}{q^2}$ is isometric to a hypersphere of radius $\frac{qi}{2}$ in R^9_1.

The Hermitian elliptic line $\mathbb{O}'\overline{S}^1$ of curvature $\frac{1}{r^2}$ is isometric to a hypersphere of radius $\frac{r}{2}$ in R^9_4.

Theorem 7.8. *The group of motions in $\mathbb{O}\overline{H}^2$ is the noncompact real simple Lie group F_4 with character $\delta = -20$. The group of motions in the plane $\mathbb{O}'\overline{S}^2$ is the split real simple Lie group F_4 with character $\delta = 4$.*

PROOF. The group of motions in $\mathbb{O}\overline{H}^2$ is obtained from the compact group of motions in $\mathbb{O}\overline{S}^2$ by the Cartan algorithm corresponding to the involutive automorphism generated by the reflection

$$'x^0 = -x^0,\ 'x^1 = x^1,\ 'x^2 = x^2, \tag{7.7}$$

that is, by reflection in a point and the polar line of this point. Therefore, in this case, it follows from Theorem 7.7 that the subgroup K is locally isomorphic to the group of rotations of a hypersphere in R^9, or to the group of motions in S^8. This subgroup is the group B_4. Therefore, the first assertion of this theorem follows from the Theorem of Cartan on the groups F_4 and E_6.

The group of motions in $\mathbb{O}'\overline{S}^2$ is obtained from the compact group of motions in $\mathbb{O}\overline{S}^2$ by the Cartan algorithm corresponding to the involutive automorphism generated by the involutive automorphism generated by the involutive motion

$$'x^i = \hat{x}^i, \tag{7.8}$$

where $x \to \hat{x}$ is an involutive automorphism (1.94) in $\mathbb{O}$. Since the transformation (7.8) preserves the points in $\mathbb{O}\overline{S}^2$ with quaternionic coordinates, the manifold consisting of such points and all submanifolds obtained from it by motions in $\mathbb{O}\overline{S}^2$

are called *normal quaternionic* 2-*chains* in this plane. Therefore, the motion (7.8) is called a *symmetry in a normal quaternionic* 2-*chain.* Since normal quaternionic 2-chains in $\mathbb{O}\overline{S}^2$ are isometric to $\mathbb{H}\overline{S}^2$, the stabilizers of these 2-chains are isometric to the direct product of the group $\mathbb{H}U_3$ by the group of quaternions with unit modulus, that is, to the group $C_3 \otimes A_1$. Therefore, the second assertion of theorem follows from the Theorem of Cartan on the groups F_4 and E_6. ■

Theorem 7.9. *The plane $\mathbb{O}\overline{S}^2$ is the compact symmetric Riemannian space V^{16}.*

PROOF. Since the group of motions in $\mathbb{O}\overline{S}^2$ is the compact simple Lie group F_4 and the stabilizers of the points in this plane are the compact simple Lie groups B_4, $\mathbb{O}\overline{S}^2$ is the homogeneous space F_4/B_4. Cartan has proved in [Car12] that this space is a irreducible compact symmetric Riemannian space of the rank 1 and has denoted this space by FII. ■

Note that the extended plane $\mathbb{O}\overline{H}^2$ is divided by the absolute Hermitian conic into two domains, one of which is the noncompact Riemannian symmetric space V^{16} and the other is the pseudo-Riemannian symmetric space V_1^{16}. The plane $\mathbb{O}'\overline{S}^2$ is also divided by the absolute Hermitian conic into two domains, which are both pseudo-Riemannian symmetric spaces V_8^{16}. All these symmetric spaces are of rank 1. The stabilizer of a point in these spaces is locally isomorphic to the group of motions in the line $\mathbb{O}\overline{S}^1$ or $\mathbb{O}'\overline{S}^1$, that is, to the group of rotations in R^9 or R_4^9, or to the groups of motions in the spaces S^8 or S_4^8.

The plane $\mathbb{O}\overline{H}^2$ was defined by J. Tits [Til-2]. For this plane see also [Mos]. The Riemannian part of this plane admits an interpretation inside a hypersphere in R^8. For this plane a theorem analogous to Theorem 4.17 is valid.

Analogously to Theorem 4.42 we can prove that Hermitian circles in $\mathbb{O}\overline{S}^2$ and $\mathbb{O}\overline{H}^2$ are isometric to a hypersphere Σ^{15} in R^{16} and are fibrations of this real hypersphere with great 7-spheres as fibers and a line $\mathbb{O}\overline{S}^1$ as base, that an equidistant Hermitian conic in $\mathbb{O}\overline{H}^2$ is isometric to an equidistant hyperquadric in H^{16} and is a fibration of this real equidistant hyperquadric with totally geodesic equidistant 7-quadrics as fibers and a line $\mathbb{O}\overline{H}^1$ as base, and that a Hermitian horocycle in $\mathbb{O}\overline{H}^2$ is isometric to a hyperhorosphere in H^{16} and is a fibration of this real hyperhorosphere with totally geodesic 7-horospheres as fibers and the line $\mathbb{O}\overline{R}^1$, that is R^8, as base.

In the last case there is also a generalization of the Heisenberg group, represented by the matrices (4.53) in which the pure imaginary number it must be replaced by an octonion t satisfying the condition $\bar{t} = -t$ and in the rule (4.54), $2i \operatorname{Im}\bar{a}b$ must be replaced by $\bar{a}b - \bar{b}a$.

7.2.4. The Quasisimple Lie Groups $\mathbf{F_4}$. Since the groups of motions in the planes $\mathbb{O}\overline{H}^2$ and $\mathbb{O}'\overline{H}^2$ can be obtained from the compact group of motions in the plane $\mathbb{O}\overline{S}^2$ by the Cartan algorithm, to these groups we can also apply the quasi-Cartan algorithm, which leads to the quasisimple groups of the class F_4. The quasisimple groups of this class can also be obtained by the quasi-Cartan algorithm from the noncompact groups of this class. The involutive automorphisms in the group of motions in $\mathbb{O}\overline{S}^2$ have the form $x \to \sigma x \sigma$, where σ is an involutive motion in this group. There are two kinds of involutive motions in this group: the motions that can be reduced to the form (7.7) and those that can be reduced to (7.8). If we apply to the group of motions in $\mathbb{O}\overline{S}^2$ the quasi-Cartan algorithm corresponding to the involutive automorphism in this group generated by a reflection (7.7), we obtain the group of motions in the *octonionic Hermitian Euclidean plane* $\mathbb{O}\overline{R}^2$. If we apply to the group of motions in $\mathbb{O}\overline{S}^2$ the quasi-Cartan algorithm corresponding to the involutive automorphism of this group generated by a symmetry (7.8), we obtain the semioctonionic Hermitian elliptic plane $\mathbb{O}^0\overline{S}^2$. If we apply the quasi-Cartan algorithm to the groups of motions in $\mathbb{O}\overline{H}^2$ and $\mathbb{O}'\overline{S}^2$, we obtain the groups of motions in $\mathbb{O}\overline{R}^2_1, \mathbb{O}'\overline{R}^2, \mathbb{O}^0\overline{S}^2_1$, and $\mathbb{O}'^0\overline{S}^2$.

The plane $\mathbb{O}\overline{R}^2$ can be defined as the affine plane $\mathbb{O}E^2$ with Hermitian inner product of vectors $\bar{x}y = \bar{x}^1x^1 + \bar{x}^2x^2$. This plane is isometric to the space R^{16}, but the group of motions has the same dimension as the group of motions in $\mathbb{O}\overline{S}^2$, that is, 52. Therefore, since the dimension of the group of motions in R^{16} is 136, the differential forms of the plane $\mathbb{O}\overline{R}^2$ can be obtained from the differential forms ω^i_j of the space R^{18} by 84 linear correlations (see [Bur1]). The plane $\mathbb{O}\overline{R}^2$ can be regarded as the tangent space of the plane $\mathbb{O}\overline{S}^2$.

7.2.5. Holomorphy Angles and Holomorphic and Antiholomorphic Real 2-Directions in the Planes $\mathbb{O}\overline{\mathbf{S}}^2, \mathbb{O}\overline{\mathbf{H}}^2$, and $\mathbb{O}'\overline{\mathbf{S}}^2$, and Trigonometry in These Planes. In the octonionic and split octonionic Hermitian planes, like in complex, split complex, quaternionic and split quaternionic Hermitian spaces, we can define *holomorphy angles* of real 2-directions, and *holomorphic and antiholomorphic real 2-directions.* Like in quaternionic and split quaternionic Hermitian spaces, we can also define *holomorphy vectors* of 2-directions (see **4.5.2**). These vectors are the elements in $\mathbb{O}$ and $\mathbb{O}'$ satisfying the condition $\bar{x} = -x$.

Analogously to Theorems 4.29, 4.30, and 4.31 we can prove

Theorem 7.10 [Ro8], [Hsi], [Bre]. *The angles A_0, B_0, C_0, the holomorphy angles α, β, γ at the vertices A, B, C of a geodesic triangle ABC in the plane $\mathbb{O}\overline{S}^2$ or $\mathbb{O}\overline{H}^2$ of curvature $\frac{1}{r^2}$, respectively $-\frac{1}{q^2}$, and their sides a, b, c are related by the laws of trigonometry (4.18–21), respectively (4.26–29).*

The holomorphy vectors at all vertices of a geodesic triangle in each of these planes are equal.

7.2.6. Sectional Curvature of Octonionic and Split Octonionic Hermitian Planes. Analogously to Theorems 4.33–4.35 we can prove

Theorem 7.11 (Burtseva [RBDMY]). *If we denote the basis elements in the algebras $\mathbb{O}$ and $\mathbb{O}'$ by i_α ($\alpha = 0, 1, \ldots, 7$), the elements in these algebras by $x = x^\alpha e_\alpha$, and the double indices $i\alpha$ by letters $I, J, \ldots$, then the Riemann tensors $R^L_{IJ,K}$ and $R_{IJ,KL} = R^H_{IJ,K}e_{HL}$ in the Riemannian manifold V^{16} isometric to $\mathbb{O}\overline{S}^2$ and in the pseudo-Riemannian manifold V^{16}_8 isometric to $\mathbb{O}'\overline{S}^2$ of curvature $\frac{1}{r^2}$ have, in the frame consisting of vectors $e_{i\alpha} = e_i i_\alpha$, the forms (4.45) and (4.47) for $\mathbb{O}\overline{S}^2$, where the matrices $(\delta_{IJ}), (\delta^l_J), (\varepsilon_{\lambda IJ})$, and $(\varepsilon^l_{\lambda J})$ are (8×8)-matrices analogous to the (4×4)-matrices (4.46), and the forms (4.50) and (4.52) for $\mathbb{O}'\overline{S}^2$, where the matrices $(\delta_{IJ}), (\delta^l_J), (\eta_{\lambda IJ})$, and $(\varepsilon^l_{\lambda J})$ are (8×8)-matrices analogous to the (4×4)-matrices (4.51).*

Like for $\mathbb{C}\overline{S}^n$, $\mathbb{H}\overline{S}^n$, $\mathbb{C}'\overline{S}n$, and $\mathbb{H}'\overline{S}^n$ the sectional curvature K of V^{16} or V^{16}_8, which are isometric to $\mathbb{O}\overline{S}^2$ or $\mathbb{O}'\overline{S}^2$ of curvature $\frac{1}{r^2}$, in a real 2-direction with angle of holomorphy α or $i\alpha$ is equal to (4.36) and (4.44), respectively.

§7.3. Geometry of the Groups $\mathbf{E_6, E_7}$, and $\mathbf{E_8}$

7.3.1. The Simple Lie Groups $\mathbf{E_6, E_7}$, and $\mathbf{E_8}$. There are four locally non-isomorphic simple real Lie groups E_7: a compact group with character $\delta = -133$ (the Dynkin diagram is given in Fig. 1.3), and three noncompact groups EV, EVI, and $EVII$ with characters $\delta = 7, -5$, and -25, respectively (the Satake diagrams are given in Fig. 1.9).

There are three locally non-isomorphic simple real Lie groups E_8: a compact group with character $\delta = -248$ (the Dynkin diagram is given in Fig. 1.3), and two noncompact groups $EVIII$ and EIX with characters $\delta = 8$ and -24, respectively (the Satake diagrams are given in Fig. 1.9).

Theorem of Cartan on the groups $\mathbf{E_7}$ and $\mathbf{E_8}$ [Car12, pp. 656-657]. *The subgroups in the compact real simple Lie group E_7 that determine the noncompact real groups EV, EVI, and $EVII$ with the same complex form are the groups $A_7, D_6 \otimes A_1$, and $E_6 \otimes D_1$, respectively.*

The subgroups in the compact real simple Lie group E_8 that determine the noncompact real groups $EVIII$ and EIX with the same complex form are the groups D_8 and $E_7 \otimes A_1$, respectively.

7.3.2. Geometry of all Simple Groups $\mathbf{E_6}$. In **7.2.2** we have considered the geometry of the two simple groups E_6, the geometry in $\mathbb{O}P^2$ and in $\mathbb{O}'P^2$. In **4.4.5** we have seen that the manifolds of 0-pairs, that is, pairs point + hyperplane, in P^n and $\mathbb{H}P^n$ admit interpretations in the spaces $\mathbb{C}'\overline{S}^n$ and $(\mathbb{C}'\otimes\mathbb{H})\overline{\overline{S}}^n$, respectively. By analogy we obtain new interpretations for the groups E_6, considered in **7.2.2**, and the planes having the same complex forms as these spaces give us interpretations of all groups of class E_6.

Theorem 7.12 [Ro6]. *The group of motions in the bioctonionic Hermitian elliptic plane $(\mathbb{C} \otimes \mathbb{O})\overline{\overline{S}}^2$ is the compact real simple Lie group E_6.*

<u>PROOF</u>. Analogously to Theorem 7.3 we can prove that the collineations in the bioctonionic projective plane $(\mathbb{C} \otimes \mathbb{O})P^2$ have the form (7.3). Since the motions in $(\mathbb{C} \otimes \mathbb{O})\overline{\overline{S}}^2$ are the collineations in $(\mathbb{C} \otimes \mathbb{O})P^2$ preserving the imaginary absolute Hermitian conic

$$\tilde{\bar{x}}\, x = \tilde{\bar{x}}^0 x^0 + \tilde{\bar{x}}^1\, x^1 + \tilde{\bar{x}}^1 x^2 = 0, \tag{7.9}$$

the group of these motions is locally homeomorphic to the topological product of the group $(\mathbb{C} \otimes \mathbb{O})SW_3$ of bioctonionic unimodular unitary (3×3)-matrices (a^i_j) determined by the condition

$$\tilde{\overline{U}}^{\top} U = I, \tag{7.10}$$

analogous to the condition (1.115), and the condition of unimodularity, which is equivalent to vanishing of the traces of the matrices of the tangent Lie algebra of this group, and of the group of continuous automorphisms in the algebra $\mathbb{C} \otimes \mathbb{O}$, which is isomorphic to the group of automorphisms in $\mathbb{O}$. This group is compact, since compactness of the group $(\mathbb{C}\otimes\mathbb{O})SW_3$ can be proved analogously to the proof of compactness of the group $\mathbb{O}SU_3$ in Theorem 7.6 (see **7.2.3**), and since the group of automorphisms in $\mathbb{O}$ is the compact group G_2.

The tangent Lie algebra of this group is the direct sum of the algebra of bioctonionic skew Hermitian (3×3)-matrices ($a_{ij} = \tilde{\bar{a}}_{ji}$) with zero trace and the tangent Lie algebra of the group of automorphisms in $\mathbb{O}$; commutators of elements in this Lie algebra are the same as in the Theorem of Vinberg (see **1.4.11**). Therefore, the dimension of this group, which is equal to the dimension of the tangent Lie algebra, is equal to $16+16+16+8+8+14 = 78$, that is, to the dimension of the group E_6.

The Cartan subalgebra of the tangent Lie algebra of the group of motions in the plane $(\mathbb{C} \otimes \mathbb{O})\tilde{\bar{S}}^2$ consists of the diagonal skew Hermitian bioctonionic (3×3)-matrices with entries from a commutative subalgebra of $\mathbb{C} \otimes \mathbb{O}$ and with zero trace and of a Cartan subalgebra of the tangent Lie algebra of the compact group G_2. A commutative subalgebra of $\mathbb{C} \otimes \mathbb{O}$ is $\mathbb{C} \otimes \mathbb{C}$, the diagonal entries of skew Hermitian bioctonionic (3×3)-matrix have the form $a_i = -\tilde{\bar{a}}_i$, therefore, these entries are bicomplex numbers of the form $\alpha i + \beta I$. The independent diagonal entries of bioctonionic (3×3)-matrices are two bicomplex numbers of this form, Therefore the dimension of the linear space of (3×3)-matrices, whose entries are in a Cartan subalgebra, is equal to 4. The dimension of a Cartan subalgebra of the tangent Lie algebra of the group G_2 is 2, so the dimension of whole Cartan subalgebra of the group of motions in $(\mathbb{C} \otimes \mathbb{O})\tilde{\bar{S}}^2$ is equal to $4 + 2 = 6$. By means of this Cartan subalgebra we can calculate the root system of this group, like in Theorem 1.29. ■

This theorem follows also from the Theorem of Vinberg.

We can analogously prove that the group of motions in the bioctonionic Hermitian hyperbolic plane $(\mathbb{C} \otimes \mathbb{O})\tilde{\bar{H}}^2$, that is, the plane $(\mathbb{C} \otimes \mathbb{O})P^2$ in which the real absolute Hermitian conic

$$\tilde{\bar{x}}Ex = -\tilde{\bar{x}}^0 x^0 + \tilde{\bar{x}}^1 x^1 + \tilde{\bar{x}}^2 x^2 = 0 \tag{7.11}$$

is given, and the groups of motions in the Hermitian elliptic planes $(\mathbb{C}' \otimes \mathbb{O})\tilde{\bar{S}}^2$, $(\mathbb{C} \otimes \mathbb{O}')\tilde{\bar{S}}^2$, $(\mathbb{C}' \otimes \mathbb{O}')\tilde{\bar{S}}^2$, that is, the projective planes $(\mathbb{C}' \otimes \mathbb{O})P^2$, $(\mathbb{C} \otimes \mathbb{O}')P^2$, and $(\mathbb{C}' \otimes \mathbb{O}')P^2$ in which the real absolute Hermitian conic (7.9) is given, are noncompact real simple Lie groups E_6.

Theorem 7.13. *The Hermitian elliptic plane* $(\mathbb{C} \otimes \mathbb{O})\tilde{\bar{S}}^2$ *admits an interpretation as the compact symmetric Riemannian space* V^{32}.

<u>PROOF</u>. The group of motions in $(\mathbb{C} \otimes \mathbb{O})\tilde{\bar{S}}^2$ is the compact simple Lie group E_6 and the stabilizer of a point in this plane is the compact semisimple Lie groups $D_5 \otimes A_1$. Therefore, $(\mathbb{C} \otimes \mathbb{O})\tilde{\bar{S}}^2$ is the homogeneous space $E_6/(D_5 \otimes A_1)$. Cartan [Car13] proved, that this space can be interpreted as the irreducible compact symmetric Riemannian space of rank 2. Since this space is determined by the same involutive automorphism in the group of motions as the noncompact group $EIII$, this space also is denoted by $EIII$. ■

We can analogously prove that the planes $(\mathbb{C}\otimes\mathbb{O})\widetilde{\overline{H}}^2, (\mathbb{C}'\otimes\mathbb{O})\widetilde{\overline{S}}^2, (\mathbb{C}\otimes\mathbb{O}')\widetilde{\overline{S}}^2$, and $(\mathbb{C}'\otimes\mathbb{O}')\widetilde{\overline{S}}^2$ can be interpreted as the noncompact Riemannian and pseudo-Riemannian symmetric 32-spaces of rank 2, and that the stabilizer of a point in one of these spaces is the group $D_5\otimes A_1$.

Theorem 7.14. *The Hermitian elliptic line* $(\mathbb{C}\otimes\mathbb{O})\widetilde{\overline{S}}^1$ *admits an interpretation as the manifold of lines in the space* S^9.

PROOF. Theorem 7.13 implies that lines in the plane $(\mathbb{C}\otimes\mathbb{O})\widetilde{\overline{S}}^2$ can be interpreted as totally geodesic submanifolds in the symmetric Riemannian space V^{32}. Since the dimension of these lines is equal to the dimension of the algebra $\mathbb{C}\otimes\mathbb{O}$, these submanifolds are the symmetric Riemannian spaces V^{16}. Since these spaces V^{16} have rank 2 and their groups of motions are isomorphic to the group D_5, these symmetric spaces are spaces BI, which admit an interpretation as the Grassmann manifold $Gr^S_{9,1}$ of lines in S^9. ■

This theorem implies that the distance ω between two points X and Y in $(\mathbb{C}\otimes\mathbb{O})\widetilde{\overline{S}}^2$ can be interpreted as follows: these points, regarded as two points in the line XY, are interpreted as two lines in S^9 and if the stationary distances of these lines are ω_0 and ω_1, we have $\omega^2=\omega_0^2+\omega_1^2$.

We can analogously prove that the lines in the planes $(\mathbb{C}\otimes\mathbb{O})\widetilde{\overline{H}}^2, (\mathbb{C}'\otimes\mathbb{O})\widetilde{\overline{S}}^2, (\mathbb{C}\otimes\mathbb{O}')\widetilde{\overline{S}}^2$, and $(\mathbb{C}'\otimes\mathbb{O}')\widetilde{\overline{S}}^2$ admit interpretations as manifolds of lines in real hyperbolic and pseudoelliptic 9-spaces.

Theorem 7.15. *The groups of motions in the planes* $(\mathbb{C}\otimes\mathbb{O})\widetilde{\overline{H}}^2, (\mathbb{C}'\otimes\mathbb{O})\widetilde{\overline{S}}^2$, *and* $(\mathbb{C}\otimes\mathbb{O}')\widetilde{\overline{S}}^2$ *are the noncompact real simple Lie groups* E_6 *with characters* $\delta=-14,-26,2$, *respectively.*

The group of motions in the plane $(\mathbb{C}'\otimes\mathbb{O}')\widetilde{\overline{S}}^2$ *is the split real simple Lie group* E_6 *with character* $\delta=6$.

PROOF. The group of motions in $(\mathbb{C}\otimes\mathbb{O})\widetilde{\overline{H}}^2$ is obtained from the compact group of motions in $(\mathbb{C}\otimes\mathbb{O})\widetilde{\overline{S}}^2$ by the Cartan algorithm corresponding to the involutive automorphism generated by the reflection (7.7), that is, by reflection in a point and the polar line of this point. Therefore, Theorem 7.14 implies that in this case the subgroup K is locally isomorphic to the group of motions in S^9, that is, this subgroup is the group D_5. Therefore, the Theorem of Cartan on the groups F_4 and E_6 implies that the group of motions in $(\mathbb{C}\otimes\mathbb{O})\widetilde{\overline{H}}^2$ has character $\delta=-14$.

The group of motions in $(\mathbb{C}'\otimes\mathbb{O})\widetilde{\overline{S}}^2$ is obtained from the compact group of motions in $(\mathbb{C}\otimes\mathbb{O})\widetilde{\overline{S}}^2$ by the Cartan algorithm corresponding to the involutive automorphism generated by the involutive motion

$$'x^i=\tilde{x}^i, \tag{7.12}$$

where $x\to\tilde{x}$ is conjugation in the field $\mathbb{C}$. The transformation (7.12) preserves the points in $(\mathbb{C}\otimes\mathbb{O})\widetilde{\overline{S}}^2$ with octonionic coordinates. Therefore, the manifold consisting of such points and all submanifolds obtained from it by motions in $(\mathbb{C}\otimes\mathbb{O})\widetilde{\overline{S}}^2$ are called *normal octonionic 2-chains* in this plane. The motion (7.12) is called a *symmetry in the normal octonionic 2-chain.* Since normal octonionic 2-chains in $(\mathbb{C}\otimes\mathbb{O})\widetilde{\overline{S}}^2$ are isometric to $\mathbb{O}\overline{S}^2$, the stabilizers of these 2-chains are isometric to the group of motions in $\mathbb{O}\overline{S}^2$, that is, this group is F_4. Therefore, the Theorem of Cartan on the groups F_4 and E_6 implies that the group of motions in $(\mathbb{C}'\otimes\mathbb{O})\widetilde{\overline{S}}^2$ has character $\delta = -26$.

The group of motions in $(\mathbb{C}\otimes\mathbb{O}')\widetilde{\overline{S}}^2$ is obtained from the compact group of motions in $(\mathbb{C}\otimes\mathbb{O})\widetilde{\overline{S}}^2$ by the Cartan algorithm corresponding to the involutive automorphism generated by the involutive motion (7.8). The transformation (7.8) preserves the points in $(\mathbb{C}\otimes\mathbb{O})\widetilde{\overline{S}}^2$ with biquaternionic coordinates. Therefore, the manifold consisting of such points and all submanifolds obtained from it by motions in $(\mathbb{C}\otimes\mathbb{O})\widetilde{\overline{S}}^2$ are called *normal biquaternionic 2-chains* in this plane. The motion (7.8) in $(\mathbb{C}\otimes\mathbb{O})\widetilde{\overline{S}}^2$ is called a *symmetry in the normal biquaternionic 2-chain.* Since normal biquaternionic 2-chains in $(\mathbb{C}\otimes\mathbb{O})\widetilde{\overline{S}}^2$ are isometric to $(\mathbb{C}\otimes\mathbb{H})\widetilde{\overline{S}}^2$, which, by Theorem 4.19, admits an interpretation as the manifold of lines in $\mathbb{C}\overline{S}^5$, the stabilizers of these 2-chains are isometric to the direct product of the group $\mathbb{C}SU_6$ by the group of quaternions with unit modulus, that is, this group is $A_5\otimes A_1$. Therefore, the Theorem of Cartan on the groups F_4 and E_6 implies that the group of motions in $(\mathbb{C}\otimes\mathbb{O}')\widetilde{\overline{S}}^2$ has character $\delta = 2$.

The group of motions in the plane $(\mathbb{C}'\otimes\mathbb{O}')\widetilde{\overline{S}}^2$ is obtained from the compact group of motions in the plane $(\mathbb{C}\otimes\mathbb{O})\widetilde{\overline{S}}^2$ by the Cartan algorithm corresponding to the involutive automorphism generated by the involutive motion

$$'x^i = \hat{\tilde{x}}, \tag{7.13}$$

where $x \to \tilde{x}$ and $x \to \hat{x}$ are involutive automorphisms in the division algebras $\mathbb{C}$ and $\mathbb{O}$, respectively. Therefore, the manifold consisting of the points with coordinates $x^i = \hat{\tilde{x}}$ and all submanifolds obtained from it by motions in $(\mathbb{C}\otimes\mathbb{O})\widetilde{\overline{S}}^2$ are called *normal 2-bichains* in this plane. The motion (7.13) is called a *symmetry in the normal 2-bichain.* Normal 2-bichains in $(\mathbb{C}\otimes\mathbb{O})\widetilde{\overline{S}}^2$ are isometric to the *split octonionic bi-Hermitian plane* $\mathbb{O}'\overline{\breve{S}}^2$, defined by N. M. Zablotskikh [Zab] as the projective plane $\mathbb{O}'P^2$ in which the absolute *bi-Hermitian conic*

$$\overline{\breve{x}}x = \overline{\breve{x}}^0 x^0 + \overline{\breve{x}}^1 x^1 + \overline{\breve{x}}^2 x^2 = 0, \tag{7.14}$$

where $x \to \breve{x}$ is an involutive automorphism in the algebra $\mathbb{O}'$, is given. The motions in this plane are the collineations in $\mathbb{O}'P^2$ preserving the bi-Hermitian conic (7.14). Zablotskikh [Zab] proved that the plane $\mathbb{O}'\overline{\breve{S}}^2$ admits an interpretation as the manifold of lines in the space $\mathbb{H}\overline{S}^3$: in this interpretation a point $X(X^0, X^1, X^2, X^3)$ on the absolute Hermitian quadric of $\mathbb{H}\overline{S}^3$ is interpreted as the point x on the absolute bi-Hermitian conic in $\mathbb{O}'\overline{\breve{S}}^2$ with coordinates

$$x^i = \frac{1}{\sqrt{3}}X^0 + X^{i+1}e \quad (i = 1, 2, 3) \tag{7.15}$$

and a point x in $\mathbb{O}'\tilde{\overline{S}}^2$ interprets the line in $\mathbb{H}\overline{S}^3$ corresponding to the two points on the absolute bi-Hermitian conic of $\mathbb{O}'\tilde{\overline{S}}^2$ cut out from it by the line polar to the point x. Note that the dimension of $\mathbb{O}'\tilde{\overline{S}}^2$ and of the manifold of lines in $\mathbb{H}\overline{S}^3$ is equal to 12, and the dimension of the absolute bi-Hermitian conic of $\mathbb{O}'\tilde{\overline{S}}^2$ and of the Hermitian quadric of $\mathbb{H}\overline{S}^3$ is equal to 11. Therefore, the stabilizers of these 2-bichains are isomorphic to the group of motions in $\mathbb{H}\overline{S}^3$, that is, this group is the group C_4, and the Theorem of Cartan on the groups F_4 and E_6 implies that the group of motions in $(\mathbb{C}' \otimes \mathbb{O}')\overset{\approx}{\overline{S}}^2$ has character $\delta = 6$ and that it is the split group E_6. ■

7.3.3. Geometry of the Simple Groups $\mathbf{E_7}$ and $\mathbf{E_8}$. The geometries of the simple groups E_7 and E_8 differ from the geometry of the groups E_6 by replacing the algebra $\mathbb{C} \otimes \mathbb{O}$ of bioctonions and its analogs by the algebras $\mathbb{H} \otimes \mathbb{O}$ and $\mathbb{O} \otimes \mathbb{O}$ of quateroctonions and octooctonions and their analogs.

Theorem 7.16 [Ro7]. *The groups of motions in the quateroctonionic Hermitian elliptic plane $(\mathbb{H}\otimes\mathbb{O})\overset{\approx}{\overline{S}}^2$ and in the octooctonionic Hermitian elliptic plane $(\mathbb{O}\otimes\mathbb{O})\overset{\approx}{\overline{S}}^2$ are the compact real simple Lie groups E_7 and E_8, respectively.*

<u>PROOF</u>. The planes $(\mathbb{H} \otimes \mathbb{O})\overset{\approx}{\overline{S}}^2$ and $(\mathbb{O} \otimes \mathbb{O})\overset{\approx}{\overline{S}}^2$ are defined analogously to the plane $(\mathbb{C} \otimes \mathbb{O})\overset{\approx}{\overline{S}}^2$ with an imaginary absolute Hermitian conic (7.9); the motions in these planes have the same form (7.3).

The group of motions in $(\mathbb{H} \otimes \mathbb{O})\overset{\approx}{\overline{S}}^2$ is locally homeomorphic to the topological product of the group $(\mathbb{H} \otimes \mathbb{O})SW_3$ of quateroctonionic unimodular unitary (3×3)-matrices (a^i_j), determined by the condition (7.10) and by the condition of unimodularity, which is equivalent to the vanishing of the traces of the matrices in the tangent Lie algebra of this group, by the group of automorphisms in the algebra $\mathbb{H}\otimes\mathbb{O}$ which is isomorphic to the direct product of the groups of automorphisms in $\mathbb{O}$ and $\mathbb{H}$. The group of motions in $(\mathbb{H}\otimes\mathbb{O})\overset{\approx}{\overline{S}}^2$ is compact, since compactness of the group $(\mathbb{H} \otimes \mathbb{O})SW_3$ can be proved analogously to the proof of the compactness of the group $\mathbb{O}SU_3$ in Theorem 7.6 (see **7.2.3**), and since the groups of automorphisms in $\mathbb{O}$ and $\mathbb{H}$ are the compact groups G_2 and A_1 and the group of automorphisms in $\mathbb{H} \otimes \mathbb{O}$ is the compact group $A_1 \otimes G_2$.

The group of motions in $(\mathbb{O} \otimes \mathbb{O})\overset{\approx}{\overline{S}}^2$ is locally homeomorphic to the topological product of the group $(\mathbb{O} \otimes \mathbb{O})SW_3$ of octooctonionic unimodular unitary (3×3)-matrices (a^i_j), determined by the condition (7.10) and by the condition of unimodularity, which is equivalent to the vanishing of the traces of the matrices in the tangent Lie algebra of this group, by the group of automorphisms in the algebra $(\mathbb{O}\otimes\mathbb{O})$, which is isomorphic to the direct product of two groups of automorphisms in $\mathbb{O}$. The group of motions in $(\mathbb{O} \otimes \mathbb{O})\overset{\approx}{\overline{S}}^2$ is compact, since compactness of the group $(\mathbb{O} \otimes \mathbb{O})SW_3$ can be proved analogously to the proof of the compactness of the group $\mathbb{O}SU_3$ in Theorem 7.6, and since the group of automorphisms in $\mathbb{O} \otimes \mathbb{O}$ is the compact group $G_2 \otimes G_2$.

The tangent Lie algebras of these groups of motions are the direct sums of the algebras of quateroctonionic and octooctonionic skew Hermitian (3×3)-matrices $(a_{ij} = -\bar{\bar{a}}_{ji})$ with zero trace and the tangent Lie algebra of the group of automorphisms in $\mathbb{H} \otimes \mathbb{O}$ or $\mathbb{O} \otimes \mathbb{O}$, respectively. Commutators of elements in these Lie algebras are the same as in the Theorem of Vinberg (see **1.4.11**). Therefore, the dimensions of these groups, which are equal to the dimensions of their tangent Lie algebras, are equal to $32 + 32 + 32 + 10 + 10 + 14 + 3 = 133$ and $64 + 64 + 64 + 14 + 14 + 14 + 14 = 248$, respectively, that is, to the dimensions of the groups E_7 and E_8.

A Cartan subalgebra in the tangent Lie algebra of the group of motions in $(\mathbb{H}\otimes\mathbb{O})\overline{\overline{S}}^2$ or $(\mathbb{O}\otimes\mathbb{O})\overline{\overline{S}}^2$ consists of the diagonal skew Hermitian quateroctonionic or octooctonionic (3×3)-matrices with entries belonging to a commutative subalgebra in $(\mathbb{H}\otimes\mathbb{O})$ or $\mathbb{O}\otimes\mathbb{O}$ with zero trace and of a Cartan subalgebras of the tangent Lie algebra of the compact groups $A_1 \otimes G_2$ or $G_2 \otimes G_2$, respectively. The commutative subalgebras in $\mathbb{H} \otimes \mathbb{O}$ and $\mathbb{O} \otimes \mathbb{O}$ are isomorphic to $\mathbb{C} \otimes \mathbb{C}$, and the diagonal entries of a skew Hermitian quateroctonionic or octooctonionic (3×3)-matrix have the form $a_i = \tilde{\bar{a}}_i$. The independent diagonal entries of bioctonionic (3×3)-matrices are two bicomplex numbers of this form. Therefore the dimension of the linear space of (3×3)-matrices, which are in a Cartan subalgebra, is equal to 4. The dimension of a Cartan subalgebra of the tangent Lie algebra of the group $A_1 \otimes G_2$, respectively $G_2 \otimes G_2$, that is, the rank of this group, is $2 + 1 = 3$, respectively $2 + 2 = 4$, so the dimension of whole Cartan subalgebra of the group of motions in $(\mathbb{C} \otimes \mathbb{O})\overline{\overline{S}}^2$, respectively $(\mathbb{C} \otimes \mathbb{O})\overline{\overline{S}}^2$, is equal to $4 + 3 = 7$, respectively $4 + 4 = 8$. By means of this Cartan subalgebra we can calculate the root systems of these groups, like in Theorem 1.29. ■

This theorem follows also from the Theorem of Vinberg.

We can analogously prove that the group of motions in the quateroctonionic Hermitian hyperbolic plane $(\mathbb{H}\otimes\mathbb{O})\overline{\overline{H}}^2$ with a real absolute Hermitian conic (7.11) and the groups of motions in the Hermitian elliptic planes $(\mathbb{H}' \otimes \mathbb{O})\overline{\overline{S}}^2$, $(\mathbb{H} \otimes \mathbb{O}')\overline{\overline{S}}^2$, and $(\mathbb{H}' \otimes \mathbb{O}')\overline{\overline{S}}^2$ with real absolute Hermitian conics (7.9) are the noncompact real simple Lie groups E_7.

We can analogously also prove that the group of motions in the octooctonionic Hermitian hyperbolic plane $(\mathbb{O}\otimes\mathbb{O})\overline{\overline{H}}^2$ with real absolute Hermitian conic (7.11) and the groups of motions in the Hermitian elliptic planes $(\mathbb{O}' \otimes \mathbb{O})\overline{\overline{S}}^2$ and $(\mathbb{O}' \otimes \mathbb{O}')\overline{\overline{S}}^2$ with real absolute Hermitian conics (7.9) are the noncompact real simple Lie groups E_8.

Theorem 7.17. *The Hermitian elliptic planes $(\mathbb{H} \otimes \mathbb{O})\overline{\overline{S}}^2$ and $(\mathbb{O} \otimes \mathbb{O})\overline{\overline{S}}^2$ admit interpretations as the compact symmetric Riemannian spaces V^{64} and V^{128}, respectively.*

PROOF. Since the groups of motions in $(\mathbb{H}\otimes\mathbb{O})\overline{\overline{S}}^2$ and $(\mathbb{O}\otimes\mathbb{O})\overline{\overline{S}}^2$ are the compact simple Lie group E_7 and E_8 and the stabilizer of a point in these planes is the compact semisimple Lie group $D_6 \otimes A_1$, respectively the compact simple Lie group D_8, the planes $(\mathbb{H} \otimes \mathbb{O})\overline{\overline{S}}^2$ and $(\mathbb{O} \otimes \mathbb{O})\overline{\overline{S}}^2$ are the homogeneous spaces $E_7/D_6 \otimes A_1$ and E_8/D_8. Cartan in [Car13] proved that these spaces can be interpreted as the irreducible compact symmetric Riemannian spaces of ranks 4 and 8, respectively. Since these spaces are determined by the same involutive automorphisms in their groups of motions, as the noncompact groups EVI and $EVIII$, these spaces are also denoted EVI and $EVIII$, respectively. ■

Note that the planes $(\mathbb{H}\otimes\mathbb{O})\tilde{\tilde{H}}^2, (\mathbb{H}'\otimes\mathbb{O})\tilde{\tilde{S}}^2, (\mathbb{H}\otimes\mathbb{O}')\tilde{\tilde{S}}^2$, and $(\mathbb{H}'\otimes\mathbb{O}')\tilde{\tilde{S}}^2$ can be interpreted as the noncompact Riemannian and pseudo-Riemannian symmetric 64-spaces of rank 4; the stabilizer of a point in one of these spaces is also the group $D_6 \otimes A_1$.

Note also that the planes $(\mathbb{O}\otimes\mathbb{O})\tilde{\tilde{H}}^2, (\mathbb{O}'\otimes\mathbb{O})\tilde{\tilde{S}}^2$; $(\mathbb{O}'\otimes\mathbb{O}')\tilde{\tilde{S}}^2$ can be interpreted as the noncompact Riemannian and pseudo-Riemannian symmetric 128-spaces of rank 8; the stabilizer of a point in one of these spaces is the group D_8.

Theorem 7.18. *The Hermitian elliptic lines* $(\mathbb{H}\otimes\mathbb{O})\tilde{\tilde{S}}^1$ *and* $(\mathbb{O}\otimes\mathbb{O})\tilde{\tilde{S}}^1$ *admit interpretations as the manifold of 3-planes in the space* S^{11}*, respectively as the manifold of 7-planes in the space* S^{15}*.*

PROOF. Theorem 7.13 implies that lines in the planes $(\mathbb{H}\otimes\mathbb{O})\tilde{\tilde{S}}^2$ and $(\mathbb{O}\otimes\mathbb{O})\tilde{\tilde{S}}^2$ can be interpreted as totally geodesic submanifolds in the symmetric Riemannian spaces V^{64} and V^{128}. Since the dimensions of these lines are equal to the dimensions of the algebras $\mathbb{H}\otimes\mathbb{O}$ and $\mathbb{O}\otimes\mathbb{O}$, these submanifolds can be interpreted as the symmetric Riemannian spaces V^{32} and V^{64}, respectively. Since the symmetric spaces V^{32} and V^{64} have ranks 4 and 8, respectively, and their groups of motions are isomorphic to the groups D_6 and D_8, respectively, these symmetric spaces are spaces DI, which admit interpretations as the Grassmann manifolds $Gr^S_{11,3}$ of 3-planes in S^{11} and $Gr^S_{15,7}$ of 7-planes in S^{15}, respectively. ■

Note that Theorems 4.20, 7.6, 7.14, and 7.18 can be united in a single theorem: if we denote the division algebras $\mathbb{R}, \mathbb{C}, \mathbb{H}$, and $\mathbb{O}$ by $\mathbb{F}_1, \mathbb{F}_2, \mathbb{F}_4$, and $\mathbb{F}_8$, respectively, the Hermitian elliptic line $(\mathbb{F}_p\otimes\mathbb{F}_q)\tilde{\tilde{S}}^1$ admits an interpretation as the manifold of $(p-1)$-planes and their polar $(q-1)$-planes in the elliptic space S^{p+q-1}.

Theorem 7.18 implies that the distance ω between two points X and Y in $(\mathbb{H}\otimes\mathbb{O})\tilde{\tilde{S}}^2$ and $(\mathbb{O}\otimes\mathbb{O})\tilde{\tilde{S}}^2$ can be interpreted as follows: these points, regarded as two points in the line XY, are interpreted as two 3-planes in S^{11} or two 7-planes in S^{15}, and if the stationary distances of these 3-planes or 7-planes are ω_i ($i = 0,1,2,3$ or $0,1,\dots,7$), we have $\omega^2 = \sum_i \omega_i^2$.

We can analogously prove that a line in one of the planes $(\mathbb{H}\otimes\mathbb{O})\tilde{\tilde{H}}^2, (\mathbb{H}'\otimes\mathbb{O})\tilde{\tilde{S}}^2, (\mathbb{H}\otimes\mathbb{O}')\tilde{\tilde{S}}^2$, and $(\mathbb{H}'\otimes\mathbb{O}')\tilde{\tilde{S}}^2$ admits an interpretation as the manifold of 3-planes in a real hyperbolic or pseudoelliptic 11-space, and that a line in one of the plane $(\mathbb{O}\otimes\mathbb{O})\tilde{\tilde{H}}^2, (\mathbb{O}'\otimes\mathbb{O})\tilde{\tilde{S}}^2$, or $(\mathbb{O}'\otimes\mathbb{O}')\tilde{\tilde{S}}^2$ admits an interpretation as the manifold of 3-planes in a real hyperbolic or pseudoelliptic 15-space.

Theorem 7.19. *The groups of motions in the planes* $(\mathbb{H}\otimes\mathbb{O})\tilde{\tilde{H}}^2, (\mathbb{H}'\otimes\mathbb{O})\tilde{\tilde{S}}^2$*, and* $(\mathbb{H}\otimes\mathbb{O}')\tilde{\tilde{S}}^2$ *are the noncompact real simple Lie groups* E_7 *with characters* $\delta = -5, -25, -5$*, respectively. The first and third of these groups are isomorphic.*

The group of motions in the plane $(\mathbb{H}'\otimes\mathbb{O}')\tilde{\tilde{S}}^2$ *is the split real simple Lie group* E_7 *with character* $\delta = 7$*.*

PROOF. The group of motions in $(\mathbb{H}\otimes\mathbb{O})\widetilde{\overline{H}}^2$ is obtained from the compact group of motions in $(\mathbb{H}\otimes\mathbb{O})\widetilde{\overline{S}}^2$ by the Cartan algorithm corresponding to the involutive automorphism in this group generated by the reflection (7.7), that is, by reflection in a point and the polar line of this point. Therefore, Theorem 7.18 implies that in this case the subgroup K is locally isomorphic to the direct product of the group of motions in S^{11} by the group of quaternions with unit modulus; this subgroup is the group $D_6 \otimes A_1$. Therefore, the Theorem of Cartan on the groups E_7 and E_8 implies that the group of motions in $(\mathbb{H}\otimes\mathbb{O})\widetilde{\overline{H}}^2$ has character $\delta = -5$.

The group of motions in the plane $(\mathbb{H}'\otimes\mathbb{O})\widetilde{\overline{S}}^2$ is obtained from the compact group of motions in $(\mathbb{H}\otimes\mathbb{O})\widetilde{\overline{S}}^2$ by the Cartan algorithm corresponding to the involutive automorphism in this group generated by the involutive motion

$$ {}'x^i = ix^ix^{-i}, \tag{7.16}$$

where $x \to ixi^{-1}$ is an involutive automorphism in $\mathbb{H}$. The transformation (7.16) preserves the points in $(\mathbb{H}\otimes\mathbb{O})\widetilde{\overline{S}}^2$ with bioctonionic coordinates. The manifold of such points and all submanifolds obtained from it by motions in $(\mathbb{H}\otimes\mathbb{O})\widetilde{\overline{S}}^2$ are called *normal bioctonionic 2-chains* in this plane. The motion (7.16) is called a *symmetry in the normal bioctonionic 2-chain.* Since normal bioctonionic 2-chains in $(\mathbb{C}\otimes\mathbb{O})\widetilde{\overline{S}}^2$ are isometric to $(\mathbb{C}\otimes\mathbb{O})\widetilde{\overline{S}}^2$, the stabilizer of such a 2-chain is isomorphic to the direct product of the group of motions in $(\mathbb{C}\otimes\mathbb{O})\widetilde{\overline{S}}^2$ by the group of complex numbers with unit modulus, that is, to the group $E_6 \otimes D_1$. Therefore, Theorem of Cartan on the groups E_7 and E_8 implies that the group of motions in $(\mathbb{H}'\otimes\mathbb{O})\widetilde{\overline{S}}^2$ has character $\delta = -25$.

The group of motions in $(\mathbb{H}\otimes\mathbb{O}')\widetilde{\overline{S}}^2$ is obtained from the compact group of motions in $(\mathbb{H}\otimes\mathbb{O})\widetilde{\overline{S}}^2$ by the Cartan algorithm corresponding to the involutive automorphism in this group generated by the involutive motion (7.8). The transformation (7.8) preserves the points in $(\mathbb{H}\otimes\mathbb{O})\widetilde{\overline{S}}^2$ with quaterquaternionic coordinates. The manifold of such points and all submanifolds obtained from it by motions in $(\mathbb{H}\otimes\mathbb{O})\widetilde{\overline{S}}^2$ are called *normal quaterquaternionic 2-chains* in this plane. The motion (7.8) in $(\mathbb{H}\otimes\mathbb{O})\widetilde{\overline{S}}^2$ is called a *symmetry in the normal quaterquaternionic 2-chain.* Since normal quaterquaternionic 2-chains in $(\mathbb{H}\otimes\mathbb{O})\widetilde{\overline{S}}^2$ are isometric to $(\mathbb{H}\otimes\mathbb{H})\widetilde{\overline{S}}^2$, which by Theorem 4.19 admits an interpretation as the manifold of 3-planes in S^{11}, the stabilizer of such a 2-chain ia isomorphic to the direct product of the group SO_{12} by the group of quaternions with unit modulus, that is, to the group $D_6 \otimes A_1$. Therefore, the group of motions in $(\mathbb{H}\otimes\mathbb{O}')\widetilde{\overline{S}}^2$ is isomorphic to the group of motions in $(\mathbb{H}\otimes\mathbb{O})\widetilde{\overline{H}}^2$ and has character $\delta = -5$.

The group of motions in $(\mathbb{H}'\otimes\mathbb{O}')\widetilde{\overline{S}}^2$ is obtained from the compact group of motions in $(\mathbb{H}\otimes\mathbb{O})\widetilde{\overline{S}}^2$ by the Cartan algorithm corresponding to the involutive automorphism in this group generated by the involutive motion

$$ {}'x^i = i\hat{x}^i i^{-1}, \tag{7.17}$$

where $x \to ixi^{-1}$ and $x \to \hat{x}$ are involutive automorphisms in $\mathbb{H}$ and $\mathbb{O}$, respectively. Therefore, the manifold of points with coordinates $x^i = i\hat{x}i^{-1}$ and all submanifolds obtained from it by motions in $(\mathbb{H}\otimes\mathbb{O})\widetilde{\overline{S}}^2$ are called *normal 2-bichains* in this plane. The motion (7.17) is called *a symmetry in the normal 2-bichain.* Normal 2-bichains in $(\mathbb{H}\otimes\mathbb{O})\widetilde{\overline{S}}^2$ are isometric to the *bi-Hermitian plane* which is the analog of $\mathbb{O}'\hat{\overline{S}}^2$ over the 16-algebra with basis $1, i, j, k, I\ell, Ip, Iq, Ir, J\ell, Jp, Jq, Jr, K\ell, Kp, Kq, Kr$, where $1, i, j, \dots, r$ are the basis elements in $\mathbb{O}$ and $1, I, J, K$ are the basis elements in $\mathbb{H}$, while the elements in these two bases commute. Analogously to the proof of Zablotskikh [Zab], we can prove that this bi-Hermitian plane admits an interpretation as the manifold of lines in $(\mathbb{C}\otimes\mathbb{H})\widetilde{\overline{S}}^3$, with group of motions isomorphic to the group of motions in $\mathbb{C}\overline{S}^7$. Therefore, the stabilizer of such a 2-bichain is isomorphic to the group of motions in $\mathbb{C}\overline{S}^7$, that is, to the group A_7. Therefore, the Theorem of Cartan on the groups F_4 and E_6 implies that the group of motions in $(\mathbb{H}'\otimes\mathbb{O}')\widetilde{\overline{S}}^2$ has character $\delta = 7$ and is the split group E_7. ■

Theorem 7.20. *The groups of motions in the planes* $(\mathbb{O}\otimes\mathbb{O})\widetilde{\overline{H}}^2, (\mathbb{O}'\otimes\mathbb{O})\widetilde{\overline{S}}^2$, *and* $(\mathbb{O}'\otimes\mathbb{O}')\widetilde{\overline{S}}^2$ *are the noncompact real simple Lie groups* E_8 *with characters* $\delta = 8, -24, 8$, *respectively. The first and third groups are isomorphic; these two groups are the split groups* E_8.

PROOF. The group of motions in $(\mathbb{O}\otimes\mathbb{O})\widetilde{\overline{H}}^2$ is obtained from the compact group of motions in $(\mathbb{O}\otimes\mathbb{O})\widetilde{\overline{S}}^2$ by the Cartan algorithm corresponding to the involutive automorphism in this group generated by the reflection (7.7), that is, by reflection in a point and the polar line of this point. Theorem 7.18 implies that in this case the subgroup K is locally isomorphic to the group of motions in S^{15}, that is, to the group D_6. Therefore, the Theorem of Cartan on the groups E_7 and E_8 implies that the group of motions in $(\mathbb{O}\otimes\mathbb{O})\widetilde{\overline{H}}^2$ has character $\delta = 8$ and is the split group E_8.

The group of motions in $(\mathbb{O}'\otimes\mathbb{O})\widetilde{\overline{S}}^2$ is obtained from the compact group of motions in $(\mathbb{O}\otimes\mathbb{O})\widetilde{\overline{S}}^2$ by the Cartan algorithm corresponding to the involutive automorphism in this group generated by the involutive motion (7.8). This transformation preserves the points in $(\mathbb{H}\otimes\mathbb{O})\widetilde{\overline{S}}^2$ with quateroctonionic coordinates. The manifold of such points and all submanifolds obtained from it by motions in $(\mathbb{O}\otimes\mathbb{O})\widetilde{\overline{S}}^2$ are called *normal quateroctonionic 2-chains* in this plane. The motion (7.8) in $(\mathbb{O}\otimes\mathbb{O})\widetilde{\overline{S}}^2$ is called a *symmetry in the normal quateroctonionic 2-chain.* Since normal quateroctonionic 2-chains in $(\mathbb{O}\otimes\mathbb{O})\widetilde{\overline{S}}^2$ are isometric to $(\mathbb{H}\otimes\mathbb{O})\widetilde{\overline{S}}^2$, the stabilizer of such a 2-chain is isomorphic to the direct product of the group of motions in $(\mathbb{H}\otimes\mathbb{O})\widetilde{\overline{S}}^2$ by the group of quaternions with unit modulus, that is, to the group $E_7 \otimes A_1$. Therefore, the Theorem of Cartan on the groups E_7 and E_8 implies that the group of motions in $(\mathbb{C}'\otimes\mathbb{O})\widetilde{\overline{S}}^2$ has character $\delta = -24$.

The group of motions in $(\mathbb{O}'\otimes\mathbb{O}')\widetilde{\overline{S}}^2$ is obtained from the compact group of motions in $(\mathbb{O}\otimes\mathbb{O})\widetilde{\overline{S}}^2$ by the Cartan algorithm corresponding to the involutive automorphism in this group generated by the involutive motion

$$'x^i = \hat{\hat{x}}^i, \tag{7.18}$$

where $x \to \hat{\hat{x}}$ is an involutive automorphism in $\mathbb{O}\otimes\mathbb{O}$, which, in turn, is determined by involutive automorphisms $x \to \hat{x}$ in both tensor factors of this algebra. Therefore, the manifold of points with coordinates $x^i = \hat{\hat{x}}^i$ and all submanifolds obtained from it by motions in $(\mathbb{O}\otimes\mathbb{O})\widetilde{\overline{S}}^2$ are called *normal 2-bichains* in this plane. The motion (7.18) is called a *symmetry in the normal 2-bichain.* Normal 2-bichains in $(\mathbb{H}\otimes \mathbb{O})\widetilde{\overline{S}}^2$ are isometric to the *bi-Hermitian plane*, which is the analog of $\mathbb{O}'\widetilde{\overline{S}}^2$ over the 32-algebra with basis $1, i, j, k, I\ell, Ip, Iq, Ir, J\ell, Jp, Jq, Jr, K\ell, Kp, Kq, Kr, L\ell, Lp, Lq, Lr, P\ell, Pp, Pq, Pr, Q\ell, Qp, Qq, Qr, R\ell, Rp, Rq, Rr$, where $1, i, j, \dots, r$ and $1, I, J, \dots, R$ are bases of two algebras $\mathbb{O}$, while the elements in these two bases commute. Analogously to the proof of Zablotskikh [Zab], we can prove that this bi-Hermitian plane admits an interpretation as the manifold of lines in $(\mathbb{H}\otimes\mathbb{H})\widetilde{\overline{S}}^3$, with group of motions isomorphic to the group of motions in S^{15}. Therefore, the Theorem of Cartan on the groups E_7 and E_8 implies that the group of motions in $(\mathbb{O}'\otimes\mathbb{O}')\widetilde{\overline{S}}^2$ is isomorphic to the group of motions in $(\mathbb{O}\otimes\mathbb{O})\widetilde{\overline{H}}^2$. Therefore, this group has character $\delta = 8$ and is the split group E_8. ■

7.3.4. The Quasisimple Lie Groups $\mathbf{E_6}, \mathbf{E_7}$, and $\mathbf{E_8}$. Since the groups of motions in the planes $(\mathbb{C}\otimes\mathbb{O})\widetilde{\overline{H}}^2, (\mathbb{C}'\otimes\mathbb{O})\widetilde{\overline{S}}^2, (\mathbb{C}\otimes\mathbb{O}')\widetilde{\overline{S}}^2, (\mathbb{C}'\otimes\mathbb{O}')\widetilde{\overline{S}}^2, (\mathbb{H}\otimes\mathbb{O})\widetilde{\overline{H}}^2, (\mathbb{H}'\otimes\mathbb{O})\widetilde{\overline{S}}^2, (\mathbb{H}\otimes\mathbb{O}')\widetilde{\overline{S}}^2, (\mathbb{H}'\otimes\mathbb{O}')\widetilde{\overline{S}}^2, (\mathbb{O}\otimes\mathbb{O})\widetilde{\overline{H}}^2, (\mathbb{O}'\otimes\mathbb{O})\widetilde{\overline{S}}^2$, and $(\mathbb{O}'\otimes\mathbb{O}')\widetilde{\overline{S}}^2$ can be obtained from the compact groups of motions in the planes $(\mathbb{C}\otimes\mathbb{O})\widetilde{\overline{S}}^2, (\mathbb{H}\otimes\mathbb{O})\widetilde{\overline{S}}^2$, and $(\mathbb{O}\otimes\mathbb{O})\widetilde{\overline{S}}^2$ by Cartan algorithms, to these groups we can apply also quasi-Cartan algorithms, which define quasisimple groups of the classes E_6, E_7, and E_8.

The quasisimple groups of these classes can also be obtained by the quasi-Cartan algorithm from noncompact groups of these classes.

Like for the groups F_4, the involutive automorphisms in the compact groups E_6, E_7, and E_8 generated by the reflections (7.7) in the points of elliptic planes lead to the groups of motions in the *bioctonionic, quateroctonionic, and octooctonionic Hermitian Euclidean planes* $(\mathbb{C}\otimes\mathbb{O})\widetilde{\overline{R}}^2, (\mathbb{H}\otimes\mathbb{O})\widetilde{\overline{R}}^2$, and $(\mathbb{O}\otimes\mathbb{O})\widetilde{\overline{R}}^2$.

If we apply to the compact groups the quasi-Cartan algorithms corresponding to the involutive automorphisms in these groups generated by the symmetries in normal 2-chains or 2-bichains, we obtain the Hermitian elliptic planes $(\mathbb{C}^0\otimes\mathbb{O})\widetilde{\overline{S}}^2, (\mathbb{C}\otimes\mathbb{O}^0)\widetilde{\overline{S}}^2, (\mathbb{C}^0\otimes\mathbb{O}^0)\widetilde{\overline{S}}^2, (\mathbb{H}^0\otimes\mathbb{O})\widetilde{\overline{S}}^2, (\mathbb{H}\otimes\mathbb{O}^0)\widetilde{\overline{S}}^2, (\mathbb{H}^0\otimes\mathbb{O}^0)\widetilde{\overline{S}}^2, (\mathbb{O}^0\otimes\mathbb{O})\widetilde{\overline{S}}^2$, and $(\mathbb{O}^0\otimes\mathbb{O}^0)\widetilde{\overline{S}}^2$.

If we apply quasi-Cartan algorithms to the noncompact groups E_6, E_7, and E_8 we obtain the groups of motions in the *bioctonionic, quateroctonionic, and octooctonionic Hermitian pseudo-Euclidean planes* $(\mathbb{C}\otimes\mathbb{O})\widetilde{\overline{R}}^2_1, (\mathbb{H}\otimes\mathbb{O})\widetilde{\overline{R}}^2_1$ and $(\mathbb{O}\otimes\mathbb{O})\widetilde{\overline{R}}^2_1$, and in the planes over tensor products containing the algebras $\mathbb{C}^0, \mathbb{C}'^0, \mathbb{H}^0, \mathbb{H}'^0, \mathbb{O}^0$, and $\mathbb{O}'^0$.

The planes $(\mathbb{C}\otimes\mathbb{O})\widetilde{\overline{R}}^2, (\mathbb{H}\otimes\mathbb{O})\widetilde{\overline{R}}^2$, and $(\mathbb{O}\otimes\mathbb{O})\widetilde{\overline{R}}^2$ can be defined as the affine planes over the same algebras with Hermitian inner product of vectors $\bar{\bar{x}}y = \bar{\bar{x}}^1 y^1 +$

$\bar{\bar{x}}^2y^2$. These planes admit interpretations as the Euclidean spaces R^{32}, R^{64}, and R^{128}, respectively, but their groups of motions have the same dimension as the respective groups of motions in $(\mathbb{C} \otimes \mathbb{O})\bar{\bar{S}}^2, (\mathbb{H} \otimes \mathbb{O})\bar{\bar{S}}^2$, and $(\mathbb{O} \otimes \mathbb{O})\bar{\bar{S}}^2$, that is, 78, 133, and 248. Therefore, the differential forms of these planes can be obtained from the differential forms ω^i and ω^i_j of the spaces R^{32}, R^{64}, and R^{128} by linear correlations, whose numbers are equal to the differences of the dimensions of the groups of motions in R^{32}, R^{64}, and R^{128}, and in these planes. The planes $(\mathbb{C} \otimes \mathbb{O})\bar{\bar{R}}^2, (\mathbb{H} \otimes \mathbb{O})\bar{\bar{R}}^2$, and $(\mathbb{O} \otimes \mathbb{O})\bar{\bar{R}}^2$ can be regarded as the tangent spaces of the planes $(\mathbb{C} \otimes \mathbb{O})\bar{\bar{S}}^2, (\mathbb{H} \otimes \mathbb{O})\bar{\bar{S}}^2$, and $(\mathbb{O} \otimes \mathbb{O})\bar{\bar{S}}^2$.

§7.4. Symplectic and Metasymplectic Geometries

7.4.1. Symplectic 5-Spaces. In Theorem 6.5 (see **6.2.2**) we have seen that the group of symplectic transformations in the space Sy^{2n-1} is isomorphic to the group of motions in the space $\mathbb{H}'\bar{S}^{n-1}$. In Theorem 4.19 (see **4.4.4**) we have seen that the group of motions in the space $\mathbb{C}\bar{S}^{2n-1}_n$, which can be regarded as the group of symplectic transformations in $\mathbb{C}\bar{S}y^{2n-1}$, is isomorphic to the group of motions in the space $(\mathbb{C} \otimes \mathbb{H})\bar{\bar{S}}^{n-1}$, which can be regarded as the space $(\mathbb{C} \otimes \mathbb{H}')\bar{\bar{S}}^{n-1}$. In the same theorem we have seen that the group of motions in the space S^{4n-1}, which can be regarded as the group of symplectic transformations in $\mathbb{H}'\bar{S}y^{2n-1}$, is isomorphic to the group of motions in $(\mathbb{H} \otimes \mathbb{H})\bar{\bar{S}}^{n-1}$. We can analogously prove that the group of symplectic transformations in $\mathbb{H}\bar{S}y^{2n-1}$ is isomorphic to the group of motions in $(\mathbb{H} \otimes \mathbb{H}')\bar{\bar{S}}^{n-1}$.

Therefore, the plane $(\mathbb{H}' \otimes \mathbb{O})\bar{\bar{S}}^2$ can also be interpreted as the *octonionic symplectic space* $\mathbb{O}\bar{S}y^5$. Since the projective spaces $\mathbb{O}P^n$ are impossible for $n > 2$, it is also impossible to define the space $\mathbb{O}\bar{S}y^5$ as the space $\mathbb{O}P^5$ with a restricted group of collineations, but it is possible to define the space $\mathbb{O}\bar{S}y^5$ as a manifold of null planes.

In Theorems 6.22 and 6.23 (see **6.4.2**) we have seen that the analogous manifolds of null $(n-1)$-planes in $Sy^{2n-1}, \mathbb{C}\bar{S}y^{2n-1}$, and $\mathbb{H}\bar{S}y^{2n-1}$ can be represented by the $(n \times n)$-matrices in the Jordan algebras $\mathbb{J}s(\mathbb{R}_n)$, $\mathbb{J}\bar{s}(\mathbb{C}_n)$, and $\mathbb{J}\bar{s}(\mathbb{H}_n)$, respectively. The null planes in $\mathbb{O}\bar{S}y^5$ can, analogously, be represented by (3×3)-matrices in the Jordan algebra $\mathbb{J}\bar{s}(\mathbb{O}_3)$, as was proved by H. Freudenthal [Fr2]. He has also proved that the group of symplectic transformations in $\mathbb{O}\bar{S}y^5$ can be represented by fractional linear transformations in $\mathbb{J}\bar{s}(\mathbb{O}_3)$, and that this group is the noncompact group E_7 with character $\delta = -25$. We can analogously define the *split octonionic symplectic space* $\mathbb{O}'\bar{S}y^5$ whose null planes are represented by the (3×3)-matrices in $\mathbb{J}\bar{s}(\mathbb{O}'_3)$. We can also prove that the group of symplectic transformations in this space can be represented by fractional linear transformations in $\mathbb{J}\bar{s}(\mathbb{O}'_3)$, and that this group is the split group E_7 with character $\delta = 7$.

7.4.2. Metasymplectic Geometries and the Freudenthal Magic Square. In [Fr2], Freudenthal has defined also a new class of geometries, whose fundamental groups are noncompact groups of all four classes F_4, E_5, E_7 and E_8. Freudenthal found that these geometries can be combined in a "magic square", which now is called the *Freudenthal Magic Square.* This square consists of 16 Lie groups

$$\begin{array}{cccc} B_1 & A_2 & C_3 & F_4 \\ A_2 & A_2 \otimes A_2 & A_5 & E_6 \\ C_3 & A_5 & D_6 & E_7 \\ F_4 & E_6 & E_7 & E_8. \end{array} \tag{7.19}$$

The groups of the first line of this square are the groups of motions in the elliptic planes

$$S^2 \quad \mathbb{C}\overline{S}^2 \quad \mathbb{H}\overline{S}^2 \quad \mathbb{O}\overline{S}^2. \tag{7.20}$$

The groups of the second line are the groups of collineations in the projective planes

$$P^2 \quad \mathbb{C}P^2 \quad \mathbb{H}P^2 \quad \mathbb{O}P^2. \tag{7.21}$$

The groups of the third line are the groups of symplectic transformations in the symplectic 5-spaces

$$Sy^5 \quad \mathbb{C}\overline{S}y^5 \quad \mathbb{H}\overline{S}y^5 \quad \mathbb{O}\overline{S}y^5. \tag{7.22}$$

Therefore, Freudenthal called the geometries of the fourth line *metasymplectic geometries.* We denote these geometries by

$$Ms \quad \mathbb{C}\overline{M}s \quad \mathbb{H}\overline{M}s \quad \mathbb{O}\overline{M}s, \tag{7.23}$$

respectively. Freudenthal regarded these geometries as the geometries of manifolds of *symplecta,* which are the manifolds of null planes in the spaces of the third line. These null planes are the planes of the second line and contain the projective lines and points. Freudenthal found that characters of the fundamental groups of the metasymplectic geometries (7.23) are equal to $4, 2, -5$, and -24, respectively. Therefore, if we use the interpretation of projective planes and symplectic 5-spaces by elliptic planes, we obtain

Theorem 7.21. *The groups in the Freudenthal magic square are the groups of motions in the following elliptic planes:*

$$\begin{array}{cccc} S^2 & \mathbb{C}\overline{S}^2 & \mathbb{H}\overline{S}^2 & \mathbb{O}\overline{S}^2 \\ \mathbb{C}'\overline{S}^2 & (\mathbb{C}'\otimes\mathbb{C})\widetilde{\overline{S}}^2 & (\mathbb{C}'\otimes\mathbb{H})\widetilde{\overline{S}}^2 & (\mathbb{C}'\otimes\mathbb{O})\widetilde{\overline{S}}^2 \\ \mathbb{H}'\overline{S}^2 & (\mathbb{H}'\otimes\mathbb{C})\widetilde{\overline{S}}^2 & (\mathbb{H}'\otimes\mathbb{H})\widetilde{\overline{S}}^2 & (\mathbb{H}'\otimes\mathbb{O})\widetilde{\overline{S}}^2 \\ \mathbb{O}'\overline{S}^2 & (\mathbb{O}'\otimes\mathbb{C})\widetilde{\overline{S}}^2 & (\mathbb{O}'\otimes\mathbb{H})\widetilde{\overline{S}}^2 & (\mathbb{O}'\otimes\mathbb{O})\widetilde{\overline{S}}^2. \end{array} \tag{7.24}$$

If we replace in the square (7.24) the algebras $\mathbb{C}'$, $\mathbb{H}'$, and $\mathbb{O}'$ by the division algebras $\mathbb{C}$, $\mathbb{H}$, and $\mathbb{O}$, we obtain

Theorem 7.22. *The compact groups in the Freudenthal magic square are the groups of motions in the following elliptic planes:*

$$\begin{array}{cccc} S^2 & \mathbb{C}\overline{S}^2 & \mathbb{H}\overline{S}^2 & \mathbb{O}\overline{S}^2 \\ \mathbb{C}\overline{S}^2 & (\mathbb{C}\otimes\mathbb{C})\widetilde{\overline{S}}^2 & (\mathbb{C}\otimes\mathbb{H})\widetilde{\overline{S}}^2 & (\mathbb{C}\otimes\mathbb{O})\widetilde{\overline{S}}^2 \\ \mathbb{H}\overline{S}^2 & (\mathbb{H}\otimes\mathbb{C})\widetilde{\overline{S}}^2 & (\mathbb{H}\otimes\mathbb{H})\widetilde{\overline{S}}^2 & (\mathbb{H}\otimes\mathbb{O})\widetilde{\overline{S}}^2 \\ \mathbb{O}\overline{S}^2 & (\mathbb{O}\otimes\mathbb{C})\widetilde{\overline{S}}^2 & (\mathbb{O}\otimes\mathbb{H})\widetilde{\overline{S}}^2 & (\mathbb{O}\otimes\mathbb{O})\widetilde{\overline{S}}^2. \end{array} \tag{7.25}$$

If we replace in the square (7.24) the algebras $\mathbb{C}, \mathbb{H}$, and $\mathbb{O}$ by the algebras $\mathbb{C}', \mathbb{H}'$, and $\mathbb{O}'$, we obtain

Theorem 7.23. *The split groups in the Freudenthal magic square are the groups of motions in the following elliptic planes:*

$$\begin{array}{llll} S^2 & \mathbb{C}'\overline{S}^2 & \mathbb{H}'\overline{S}^2 & \mathbb{O}'\overline{S}^2 \\ \mathbb{C}'\overline{S}^2 & (\mathbb{C}'\otimes\mathbb{C}')\tilde{\overline{S}}^2 & (\mathbb{C}'\otimes\mathbb{H}')\tilde{\overline{S}}^2 & (\mathbb{C}'\otimes\mathbb{O}')\tilde{\overline{S}}^2 \\ \mathbb{H}'\overline{S}^2 & (\mathbb{H}'\otimes\mathbb{C}')\tilde{\overline{S}}^2 & (\mathbb{H}'\otimes\mathbb{H}')\tilde{\overline{S}}^2 & (\mathbb{H}'\otimes\mathbb{O}')\tilde{\overline{S}}^2 \\ \mathbb{O}'\overline{S}^2 & (\mathbb{O}'\otimes\mathbb{C}')\tilde{\overline{S}}^2 & (\mathbb{O}'\otimes\mathbb{H}')\tilde{\overline{S}}^2 & (\mathbb{O}'\otimes\mathbb{O}')\tilde{\overline{S}}^2. \end{array} \tag{7.26}$$

If we replace the elliptic planes in the square (7.25) by lines in these planes and use the interpretation of these lines in real elliptic spaces (Theorems 4.20, 7.6, 7.14, and 7.18), we obtain

Theorem 7.24. *The groups of motions in lines in the planes whose groups of motions are the compact groups in the Freudenthal magic square are locally isomorphic to the groups of motions in the following real elliptic spaces:*

$$\begin{array}{llll} S^1 & S^2 & S^4 & S^8 \\ S^2 & S^3 & S^5 & S^9 \\ S^4 & S^5 & S^7 & S^{11} \\ S^8 & S^9 & S^{11} & S^{15}. \end{array} \tag{7.27}$$

In [Fr2], Freudenthal found the dimensions of the manifolds of points, lines, planes, and symplecta in the metasymplectic geometries $Ms, \overline{\mathbb{C}M}s, \overline{\mathbb{H}M}s$, and $\overline{\mathbb{O}M}s$.

Theorem I of Freudenthal [Fr2]. *The dimensions of the manifolds of points in the geometries $Ms, \overline{\mathbb{C}M}s, \overline{\mathbb{H}M}s$, and $\overline{\mathbb{O}M}s$ are equal to* 15, 24, 42, *and* 78, *respectively.*

The dimensions of the manifolds of lines in these geometries are equal to 20,31,53, *and* 97, *respectively.*

The dimensions of the manifolds of planes in these geometries are equal to 20, 29, 47, *and* 83, *respectively.*

The dimensions of the manifolds of symplecta in these geometries are equal to 15, 21, 33, *and* 57, *respectively.*

§7.5. Symmetry Figures and Symmetric Spaces

7.5.1. Symmetry Figures. Since in the compact group G_2 of automorphisms in $\mathbb{O}$ there is only one kind of involutive automorphism, which can be reduced to the form (1.94), in the space Sg^6 there is only one kind of symmetry figures.

Theorem 7.25. *The symmetry figures in Sg^6 are the special planes obtained from the intersections of the 6-sphere, which is the intersection of the hyperplane $x = -\bar{x}$ with the hypersphere $|x| = 1$, with associative subalgebras in $\mathbb{O}$ isomorphic to the skew field $\mathbb{H}$.*

The stabilizers of these symmetry figures are locally isomorphic to the group of motions in S^3.

The theorem follows from the fact that the involutive automorphism (1.94) can be regarded as a reflection in the subalgebra $\mathbb{H}$ in $\mathbb{O}$.

The manifold of these planes forms an interpretation of the symmetric Riemannian space V^8, which is the space G in the classification of Cartan [Car12].

Since in the split group G_2 of automorphisms in $\mathbb{O}'$ there are two kinds of involutive automorphisms, which can be reduced to the forms (1.94) and (1.95), in Sg_3^6 there are two kinds of symmetry figures, which are the special planes determined by the associative subalgebras $\mathbb{H}$ and $\mathbb{H}'$ in $\mathbb{O}$.

Since in the compact group F_4 of motions in the plane $\mathbb{O}\overline{S}^2$ there are two kinds of involutive motions, which can be reduced to the forms (7.7) and (7.8), we obtain

Theorem 7.26. *The symmetry figures in $\mathbb{O}\overline{S}^2$ are points together with the polar lines of these points, and normal quaternionic 2-chains.*

The stabilizers of these symmetry figures are locally isomorphic to the group of motions in S^8, respectively to the direct product of the group of motions in $\mathbb{H}\overline{S}^2$ by the group of quaternions with unit modulus.

The manifolds of these figures form interpretations of the symmetric Riemannian spaces V^{16} and V^{28}, which are the spaces FII and FI in the classification of Cartan [Car12].

Since in the compact group E_6 of motions in the plane $(\mathbb{C}\otimes\mathbb{O})\widetilde{\overline{\overline{S}}}^2$ there are four kinds of involutive motions, which can be reduced to the forms (7.7), (7.8), (7.12), and (7.13), we obtain

Theorem 7.27. *The symmetry figures in the plane $(\mathbb{C}\otimes\mathbb{O})\widetilde{\overline{\overline{S}}}^2$ are points together with the polar lines of these points, normal octonionic 2-chains, normal biquaternionic 2-chains, and normal 2-bichains.*

The stabilizers of these symmetry figures are locally isomorphic to the direct product of the group of motions in S^9 by the group of complex numbers with unit modulus, to the group of motions in $\mathbb{O}\overline{S}^2$, to the direct product of the group of motions in $\mathbb{C}S^5$ by the group of quaternions with unit modulus, and to the group of motions in $\mathbb{H}\overline{S}^3$.

The manifolds of these figures form interpretations of the symmetric Riemannian spaces V^{32}, V^{26}, V^{49}, and V^{42}, which are the spaces $EIII, EIV, EII$ and EI in the classification of Cartan [Car12].

Since in the compact group E_7 of motions in the plane $(\mathbb{H}\otimes\mathbb{O})\widetilde{\overline{\overline{S}}}^2$ there are four kinds of involutive motions, which can be reduced to the forms (7.7), (7.8), (7.16), and (7.17), we obtain

Theorem 7.28. *The symmetry figures in $(\mathbb{H}\otimes\mathbb{O})\widetilde{\overline{\overline{S}}}^2$ are points together with the polar lines of these points, normal bioctonionic 2-chains, normal quaterquaternionic 2-bichains, and normal 2-bichains.*

The stabilizers of these symmetry figures are locally isomorphic to the direct product of the group of motions in S^{11} by the group of quaternions with unit modulus, to the direct product of the group of motions in $(\mathbb{C}\otimes\mathbb{O})\widetilde{\overline{\overline{S}}}^2$ by the group of complex numbers with unit modulus, to the direct product of the group of motions in $(\mathbb{H}\otimes\mathbb{H})\widetilde{\overline{\overline{S}}}^2$, which itself is isomorphic to the group of motions in S^{11}, by the group of quaternions with unit modulus, and to the group of motions in $(\mathbb{C}\otimes\mathbb{O})\widetilde{\overline{\overline{S}}}^3$, which itself is isomorphic to the group of motions in $\mathbb{H}\overline{S}^7$.

The manifolds of these figures form interpretations of the symmetric Riemannian spaces V^{64}, V^{54}, V^{64}, and V^{70}, which are the spaces $EVI, EVII, EVI$, and EV in the classification of Cartan [Car12].

Since in the compact group E_8 of motions in the plane $(\mathbb{O}\otimes\mathbb{O})\widetilde{\overline{S}}^2$ there are three kinds of involutive motions, which can be reduced to the forms (7.7), (7.8), and (7.18), we obtain

Theorem 7.29. *The symmetry figures in $(\mathbb{O}\otimes\mathbb{O})\widetilde{\overline{S}}^2$ are points together with the polar lines of these points, normal quateroctonionic 2-chains, and normal 2-bichains.*

The stabilizers of these symmetry figures are locally isomorphic to the group of motions in S^{15}, to the direct product of the group of motions in $(\mathbb{H}\otimes\mathbb{O})\widetilde{\overline{S}}^2$ by the group of quaternions with unit modulus, and to the group of motions in $(\mathbb{H}\otimes\mathbb{H})\ \overset{\approx}{S}{}^3$, which itself is isomorphic to the group of motions in S^{15}.

The manifolds of these figures form interpretations of the symmetric Riemannian spaces V^{128}, V^{112}, and V^{128}, which are the spaces $EVIII, EIX$, and $EVIII$ in the classification of Cartan [Car12].

We can analogously define the symmetry figures in geometries with noncompact fundamental groups. Let us mention the symmetry figures in the plane $\mathbb{O}P^2$. There are the *0-pairs*, i.e. pairs point+line, which form an interpretation of the symmetric pseudo-Riemannian 32-space. The cosymmetry figures in this plane are the *Hermitian conics*, which form an interpretation of the symmetric pseudo-Riemannian 26-space.

7.5.2. Local Absolutes in Symmetric Spaces. In **4.10.2** we have defined local absolutes and superabsolutes in Riemannian symmetric spaces.

Theorem 7.30. *The local absolute in the symmetric space V^8 with fundamental compact group G_2 consists of a line with fixed points in S^7 and of the quadric with 2-planar generators in the polar 5-plane of this line.*

PROOF. Since the symmetric space V^8 with fundamental compact group G_2 can be interpreted as the manifold of special 2-planes in Sg^6 and since through each line in this space only one special 2-plane of this space passes, the space V^8 can be interpreted as the quotient space $Gr^S_{6,1}/Gr^S_{2,1} = V^{10}/S^2$. Therefore, the local absolute in this space V^8 can be obtained from the Segrean $\Sigma_{1,4}$ by corresponding factorization and consists of a line with fixed points in S^7 and of the quadric with 2-planar generators, i.e. the Grassmannian $\Gamma\rho_{3,1}$, in the polar 5-plane of this line. ■

Theorem 7.31 (T. A. BURTSEVA [RoB1]). *The local absolutes in the symmetric spaces V^{16} and V^{16}_8, isometric to $\mathbb{O}\overline{S}^2$ and $\mathbb{O}'\overline{S}^2$, are the imaginary and real Lipschitzians Ω_5, respectively.*

PROOF. In **1.6.3** we have defined the Lipschitzian Ω_n, as the algebraic $\frac{n(n-1)}{2}$-surface with equations (1.162) in the space $P^{2^{n-1}-1}$. In particular, the real Lipschitzian Ω_5 is a 10-surface in S^{15} or S^{15}_8, which are hyperplanes at infinity of the spaces R^{16} and R^{16}_8 tangent to V^{16} and V^{16}_8. Since the stabilizers of points in V^{16}

and V_8^{16} are isomorphic to the groups of motions in $\mathbb{O}\overline{S}^1$ and $\mathbb{O}'\overline{S}^1$, they are locally isomorphic to the groups SO_9 and $SO_{9,4}$ of rotations of hyperspheres in R^9 and R_4^9, respectively. These 10-surfaces are the focal surfaces of the paratactic congruences of 7-planes in S^{15} and S_8^{15} which are cut out from the hyperplanes at infinity of the tangent spaces R^{16} and R_8^{16} to the symmetric spaces V^{16} and V_8^{16} by the tangent 8-planes to the 8-spheres in the spaces V^{16} and V_8^{16}, which are isometric to straight lines in $\mathbb{O}\overline{S}^2$ and $\mathbb{O}'\overline{S}^2$. The groups of rotations in R^{16} and R_8^{16} which preserve these paratactic congruences are the ***spinor groups*** $\overline{SO}_9$ and $\overline{SO}_{9,4}$ of the groups SO_9 and $SO_{9,4}$: in **1.6.3** we have seen that the groups $\overline{SO}_9$ and $\overline{SO}_{9,4}$ are represented by alternions and pseudoalternions in the algebras $\mathbb{A}_9$ and $\mathbb{A}_{9,4}$ which, by Theorems 1.23 and 1.24 (see **1.2.8** and **1.2.9**), are isomorphic to subalgebras of the algebra $\mathbb{R}_{16}$ of real (16×16)-matrices.

In the Theorem of Cartan (see **4.10.3**) we have seen that in the linear spaces of the spinor representations of the groups $SO_{2n+1,n}$ and $SO_{2n,n}$ the Lipschitzian Ω_n is preserved by transformations of the spinor groups, and points in Ω_n represent $(n-1)$-planar generators of the absolute hyperquadrics of S_n^{2n} and S_{n-1}^{2n-1}, whose motions are represented by matrices in $SO_{2n+1,n}$ and $SO_{2n,n}$. We have also seen that points in Ω_{n+1} represent these $(n-1)$-planar generators. The analogous theorem is valid also for the groups SO_{2n+1} and SO_{2n} and for the spaces S^{2n} and S^{2n-1} with imaginary absolute hyperquadrics; here the variety Ω_n is imaginary. In our case there are real and imaginary varieties Ω_5 preserved by transformations in the spinor groups $\overline{SO}_9$ and $\overline{SO}_{9,4}$, and the points in these varieties represent the imaginary and real 3-planar generators of the absolute hyperquadrics of S^8 and S_4^8, respectively. In Theorem 4.44 (see **4.7.8**) we have seen that the manifold of 3-planar generators of a hyperquadric in P^8 is homeomorphic to the group of motions in S^4, that is, to the group SO_5. Since the Lipschitzian Ω_5 can be obtained from the intersection of the hypersphere in R^{16} with the hypercone (1.162) by identification of the antipodal points on this hypersphere, and since the group SO_5 is isomorphic to the quotient group of its spinor group by a subgroup consisting of two elements, the local absolute in the symmetric pseudo-Riemannian space V_8^{16} is the Lipschitzian Ω_5 and the local absolute in the symmetric Riemannian space V^{16} is the imaginary Lipschitzian Ω_5. ■

Analogously to Theorem 7.28 we can prove

Theorem 7.32 [RoB2, Ro9]. *The intersections of the local absolutes in the symmetric spaces V^{32}, V^{64}, and V^{126}, which interpret the planes $(\mathbb{C}\otimes\mathbb{O})\widetilde{\overline{S}}^2$, $(\mathbb{H}\otimes\mathbb{O})\widetilde{\overline{S}}^2$, and $(\mathbb{O}\otimes\mathbb{O})\widetilde{\overline{S}}^2$, respectively, with the imaginary absolutes of the elliptic spaces S^{31}, S^{63}, and S^{127}, which are the hyperplanes at infinity of the spaces R^{32}, R^{64}, and R^{128} tangent to V^{32}, V^{64}, and V^{128}, are the imaginary Segrean $\Sigma_{1,15}$ in whose 15-planar generators the imaginary Lipschitzians Ω_5 are given, the imaginary Segrean $\Sigma_{1,31}$ in whose 31-planar generators the imaginary Lipschitzians Ω_6 are given, and the imaginary Lipschitzian Ω_8, respectively.*

Theorem 7.33. *The local superabsolutes in the symmetric spaces V^{32}, V^{64} and V^{128}, which represent the planes $(\mathbb{C}\otimes\mathbb{O})\widetilde{\overline{S}}^2$, $(\mathbb{H}\otimes\mathbb{O})\widetilde{\overline{S}}^2$, and $(\mathbb{O}\otimes\mathbb{O})\widetilde{\overline{S}}^2$, respectively, are the focal varieties of the congruences of 15-planes in S^{31}, of 31-planes in S^{63}, and of 63-planes in S^{127}, respectively, which are cut out from the hyperplanes at* *infinity of the spaces R^{32}, R^{64}, and R^{128} tangent to V^{32}, V^{64}, and V^{128} by the tangent 16-planes, 32-planes, and 64-planes to the submanifolds in V^{32}, V^{64}, and V^{128} determined by lines in the corresponding Hermitian planes.*

The intersections of the local superabsolutes in V^{32}, V^{64}, and V^{128} with the 15-planes, 31-planes, and 63-planes of the above mentioned congruences are isometric to the Segreans $\Sigma_{1,7}, \Sigma_{3,7}$, and $\Sigma_{7,7}$, which are local absolutes in the symmetric spaces modeled by the Grassmann manifolds $Gr^S_{9,1}, Gr^S_{11,3}$, and $Gr^S_{15,7}$, respectively. These local superabsolutes are real.

PROOF. Cartan [Car12] has proved that the stabilizer of a point in one of these symmetric spaces can be represented by the direct product of the spinor representation of the group D_5 and SO_2, by the direct product of the spinor representation of the group D_6 and $\mathbb{C}SU_2$, and by the spinor representation of the group D_8, respectively. The Lipschitzians mentioned in Theorem 7.32 are homeomorphic to the families of generators of maximal dimension on the absolute hyperquadrics of S^9, S^{11} and S^{15}, respectively, whose groups of motions are isomorphic to the groups of motions in lines in the corresponding Hermitian planes.

Since the families of 4-planar generators on the absolute hyperquadrics of S^9, of 5-planar generators on the absolute hyperquadrics of S^{11}, and of 7-planar generators on the absolute hyperquadrics of S^{15} are homeomorphic to the Lipschitzians Ω_5, Ω_6 and Ω_8, respectively, the stabilizers of points in V^{32}, V^{64}, and V^{128} represent transformations in these families of generators under spinor representations of the groups SO_{10}, SO_{12}, and SO_{16}, respectively. ■

§7.6. Parabolic Figures and Fundamental Representations

7.6.1. Fundamental Figures. In **1.6.4** we have defined fundamental figures in a homogeneous space with a simple fundamental Lie group as parabolic figures determined by one simple root α_i of this group. We have called these figures α_i-figures. Note that the stabilizers of these figures are non-semisimple groups. They are maximal non-semisimple subgroups of the fundamental groups of these spaces, but the semisimple parts of the stabilizers of these figures are isomorphic to the direct product of simple groups whose Dynkin diagrams are obtained by removing the root α_i from the Dynkin diagram of the fundamental group.

7.6.2. Parabolic Figures of the Groups $\mathbf{G_2}$. Since the rank of the simple group G_2 is 2, this group has two simple roots, α_1 and α_2. The expressions of these roots as linear combination of the basis vectors E_0, E_1, E_2 are shown by formula (1.109, G_2). All positive roots of this group are linear combinations of these roots, which can be written as

$$0 \lessdot\!\equiv 1 \quad 1 \lessdot\!\equiv 0 \quad 1 \lessdot\!\equiv 1 \quad 2 \lessdot\!\equiv 1 \quad 3 \lessdot\!\equiv 1 \quad 3 \lessdot\!\equiv 2 \tag{7.28}$$

[FdV, p. 531], where instead of the root in the Dynkin diagram of this group the corresponding coefficient of the linear combination is written. The linear combinations written in the form (7.28) admit us to write down the decompositions (1.170) for both simple roots α_1 and α_2. If we remove the roots α_1 and α_2 from the Dynkin diagram of this group, we obtain in both cases the simple group A_1, that is, the group of collineations in the line P^1. From (7.28) we find that the dimensions

$\dim[\alpha_1]$ and $\dim[\alpha_2]$ of the manifolds of α_1-figures and α_2-figures in Sg_3^6 are equal to 5. From the decomposition (1.170) for the group G_2 we can also find the geometric characteristics of fundamental figures in the space Sg_3^6, whose fundamental group is the split group G_2.

Theorem 7.34. *The α_1-figures in Sg_3^6 are the special rectilinear generators on the absolute hyperquadric of this space which are the intersections of this absolute hyperquadric with planes determined by associative subalgebras of $\mathbb{O}'$ isomorphic to the algebras $\mathbb{H}^0$ and $\mathbb{H}'^0$ of semiquaternions and split semiquaternions (see* **1.2.11***).*

The α_2-figures in Sg_3^6 are points of the absolute hyperquadric, together with the special planar generators of this absolute hyperquadric determined by associative subalgebras of $\mathbb{O}'$ isomorphic to the algebra $\mathbb{H}^{00}$ of $\frac{1}{4}$-quaternions (see also **1.2.11***) and passing through these points.*

PROOF. From the decomposition (1.170) for the tangent Lie algebra of the split group G_2 we find that the matrices in the corresponding parabolic subalgebra for α_1 coincide with the matrices of the motions preserving the rectilinear generator E_2E_6 on the absolute hyperquadric of Sg_3^6. For α_2 the matrices in the corresponding parabolic subalgebra coincide with the matrices of the motions preserving the planar generator $E_2E_3E_6$ on the absolute hyperquadric of Sg_3^6 and the point E_6 in this plane. This plane corresponds to a subalgebra in $\mathbb{O}'$ isomorphic to $\mathbb{H}^{00}$. This subalgebra consists of the split octonions satisfying the condition $xa = 0$, where a is a split octonion with zero modulus. The point in this plane represented by this split octonion is called the *center* of the special planar generator. Special rectilinear generators on the absolute hyperquadric of Sg_3^6 are lines lying in special planar generators on this hyperquadric and passing through the center of this planar generator. ■

Note that through each point on the absolute hyperquadric of Sg_3^6 (whose dimension is 5) there passes only one special plane generator on this hyperquadric. This point is the center of a pencil of special rectilinear generators on this hyperquadric. Therefore the dimension of the manifold of these lines is equal to $5 + 1 - 1 = 5$.

The semisimple parts of stabilizers of both these fundamental figures are the groups of collineations in P^1, which are isomorphic to the stabilizer of a pencil of lines in P^2.

We can analogously prove that the α_1-figures in Sg^6 are imaginary special rectilinear generators on the absolute hyperquadric of this space, and that the α_2-figures in Sg^6 are points on the imaginary absolute hyperquadric of this space together with special planar generators on this imaginary hyperquadric.

7.6.3. Parabolic Figures of the Group $\mathbf{F}_4$. Since the rank of the simple group F_4 is 4, this group has four simple roots, $\alpha_1, \alpha_2, \alpha_3$, and α_4. The expressions of these roots as linear combinations of the basis vectors E_1, E_2, E_3, E_4 are shown by formula (1.109, F_4). All positive roots of this group are linear combinations of these roots, which can be written as

$$\begin{array}{llllll} 10 => 00 & 01 => 00 & 00 => 10 & 00 => 01 & 11 => 00 & 01 => 10 \\ 00 => 11 & 01 => 20 & 11 => 10 & 01 => 11 & 11 => 20 & 11 => 11 \\ 01 => 21 & 12 => 20 & 01 => 22 & 11 => 21 & 11 => 22 & 12 => 21 \\ 12 => 22 & 12 => 31 & 12 => 32 & 12 => 42 & 13 => 42 & 23 => 42 \end{array} \qquad (7.29)$$

[FdV, p. 531].

If we remove the simple roots $\alpha_1, \alpha_2, \alpha_3$, and α_4 from the Dynkin diagram of this group, we obtain the semisimple groups $C_3, A_1 \otimes A_2, A_2 \otimes A_1$, and B_3, respectively. From (7.29) we can find that the dimensions $\dim[\alpha_1], \dim[\alpha_2], \dim[\alpha_3]$, and $\dim[\alpha_4]$ of the manifolds of α_1-figures, α_2-figures, α_3-figures, and α_4-figures in Ms and $\mathbb{O}'\overline{S}^2$ are equal to 15, 20, 20, and 15, respectively, and that the dimensions $\dim[\alpha_1, \alpha_2], \dim[\alpha_1, \alpha_3], \dim[\alpha_1, \alpha_4], \dim[\alpha_2, \alpha_4]$, and $\dim[\alpha_3, \alpha_4]$ are equal to 21, 22, 20, 22, and 21, respectively.

Theorem 7.35 [Ro9]. *The α_1-figures of the real metasymplectic geometry Ms, which coincides with the geometry of the split octonionic plane $\mathbb{O}'\overline{S}^2$, are the symplecta in Ms, the α_2-figures in Ms are the planes, the α_3-figures in Ms are the lines, the α_4-figures in Ms are the points, which coincide with the points on the absolute Hermitian conic of $\mathbb{O}'\overline{S}^2$.*

The fundamental figures in the planes $\mathbb{O}\overline{S}^2$ and $\mathbb{O}\overline{H}^2$ are analogous, but in $\mathbb{O}\overline{S}^2$ all these figures are imaginary, and in $\mathbb{O}\overline{H}^2$ the α_4-figures, the points on the absolute conic, are real and all remaining α_i-figures are imaginary.

PROOF. The assertion of this theorem on points on the absolute Hermitian conic can be proved analogously to the theorem on points on the absolute Hermitian hyperquadrics of spaces over associative algebras.

The assertion on points, lines, planes, and symplecta in Ms follow from the equality of the dimensions $\dim[\alpha_i]$ of the manifolds of α_i-figures in this geometry with the dimensions found in Theorem I of Freudenthal (see **7.4.2**), and from the equality of the dimensions $\dim_j[\alpha_i]$ calculated by (1.171) with the corresponding dimensions calculated by Freudenthal in [Fr2]. In particular, the dimensions $\dim_j[\alpha_4]$ for $j = 3, 2, 1$ are equal to 1, 2, and 5, respectively, that is they are equal to the dimensions of P^1, P^2, and Sy^5. ∎

Note that the stabilizers of all these fundamental figures contain the groups of symplectic transformations in Sy^5, of collineations in P^2 and P^1, and of motions in S^6, respectively.

We can analogously prove that the α_4-figures in $\mathbb{O}\overline{S}^2$ are the points on the imaginary absolute Hermitian conic of this plane, while the α_3-figures, α_2-figures, and α_1-figures in this plane are the lines $\mathbb{C}P^1$, planes $\mathbb{C}P^2$, and 5-spaces $\mathbb{C}Sp^5$. The Satake diagram (Fig. 1.9) of the noncompact group FII shows that the α_4-figures in $\mathbb{O}\overline{H}^2$ are the real points on the real absolute Hermitian conic of this plane, and the α_3-figures, α_2-figures, and α_1-figures are imaginary; they are the same figures as in $\mathbb{O}\overline{S}^2$.

Note that the α_4-figures in $\mathbb{O}\overline{H}^2$ are the points on the absolute Hermitian conic of this plane, and that the stabilizers of these figures coincide with the stabilizers of the Hermitian horocycles in this plane tangent to the absolute Hermitian conic at these points. The Borel group of this parabolic group coincides with the analog of the *Heisenberg group* for $\mathbb{O}\overline{H}^2$.

7.6.4. Parabolic Figures of the Groups $\mathbf{E_6}$. Since the rank of the simple group E_6 is 6, this group has six simple roots, $\alpha_1, \alpha_2, \alpha_3, \alpha_4, \alpha_5$, and α_6. The expressions of these roots as linear combinations of the basis vectors E_0, E_1, E_2, E_3, E_4,

E_5, E_6 are shown in (1.109, E_6). All positive roots of this group are linear combinations of these roots, which can be written as

$$\begin{array}{cccccc}
0 & 0 & 0 & 1 & 0 & 0 \\
10000 & 01000 & 00100 & 00000 & 00010 & 00001 \\
0 & 0 & 1 & 0 & 0 & \\
11000 & 01100 & 00100 & 00110 & 00011 & \\
0 & 1 & 0 & 1 & 0 & \\
11100 & 01100 & 01110 & 00110 & 00111 & \\
1 & 0 & 1 & 0 & 1 & \\
11100 & 11110 & 01110 & 01111 & 00111 & \\
1 & 0 & 1 & 1 & & \\
11110 & 11111 & 01111 & 01210 & & \\
1 & 1 & 1 & & & \\
11111 & 11210 & 01211 & & & \\
1 & 1 & 1 & & & \\
11211 & 12210 & 01221 & & & \\
1 & 1 & 1 & & & \\
12211 & 11221 & 12221 & & & \\
1 & 2 & & & & \\
12321 & 12321 & & & &
\end{array} \tag{7.30}$$

[FdV, p. 528].

If we remove the simple roots $\alpha_1, \alpha_2, \alpha_3, \alpha_4, \alpha_5$, and α_6 from the Dynkin diagram of this group, we obtain the semisimple groups $D_5, A_1 \otimes A_4$, $A_2 \otimes A_1 \otimes A_2$, $A_4 \otimes A_1$, D_5, and A_5, respectively. From (7.30) we find that the dimensions $\dim[\alpha_1], \dim[\alpha_2], \dim[\alpha_3], \dim[\alpha_4], \dim[\alpha_5]$, and $\dim[\alpha_6]$ of the manifolds of α_1-figures, α_2-figures, α_3-figures, α_4-figures, α_5-figures, and α_6-figures in $\mathbb{C}'\overline{M}s$ and $\mathbb{O}'P^2$ are equal to 16, 25, 29, 25, 16, and 21, respectively, while the dimensions $\dim[\alpha_1, \alpha_2]$, $\dim[\alpha_1, \alpha_3]$, $\dim[\alpha_1, \alpha_4]$, $\dim[\alpha_1, \alpha_5]$, $\dim[\alpha_1, \alpha_6]$, and $\dim[\alpha_2, \alpha_4]$ are equal to 26, 31, 29, 24, 26, and 31, respectively.

Theorem 7.36 [Ro9]. *The α_1-figures in the split octonionic plane $\mathbb{O}'P^2$, which coincides with the plane $(\mathbb{C}' \otimes \mathbb{O}')\widetilde{\overline{S}}^2$ and with the split complex metasymplectic geometry $\mathbb{C}'\overline{M}s$, are the points in $\mathbb{O}'P^2$. The α_2-figures in $\mathbb{C}'\overline{M}s$ are the real lines. The α_3-figures in $\mathbb{C}'\overline{M}s$ are the split complex planes. The α_4-figures in $\mathbb{C}'\overline{M}s$ are the real 4-planes. The α_5-figures in $\mathbb{O}'P^2$ are the lines of this plane. The α_6-figures in $\mathbb{C}'\overline{M}s$ are the split complex symplecta.*

The (α_1, α_5)-figures in $\mathbb{C}'\overline{M}s$ are the points that coincide with the points on the absolute Hermitian conic of $(\mathbb{C}' \otimes \mathbb{O}')\widetilde{\overline{S}}^2$. The (α_2, α_4)-figures in $\mathbb{C}'\overline{M}s$ are the split complex lines.

The fundamental figures in $(\mathbb{C} \otimes \mathbb{O})\widetilde{\overline{S}}^2$ and $(\mathbb{C} \otimes \mathbb{O})\widetilde{\overline{H}}^2$ are analogous to the fundamental figures in $(\mathbb{C}' \otimes \mathbb{O}')\widetilde{\overline{S}}^2$; however in $(\mathbb{C} \otimes \mathbb{O})\widetilde{\overline{S}}^2$ all these figures are imaginary, and in $(\mathbb{C} \otimes \mathbb{O})\widetilde{\overline{H}}^2$ the (α_1, α_5)-figures, i.e. the points on the absolute conic, are real, the α_1-figures and α_5-figures are conjugate imaginary, the α_6-figures are real, and all remaining α_i-figures are imaginary.

The fundamental figures in $\mathbb{O}P^2$*, which coincides with* $(\mathbb{C}' \otimes \mathbb{O})\overline{\overline{S}}^2$*, are analogous to the fundamental figures in* $\mathbb{O}' P^2$*; however only the* α_1*-figures and the* α_5*-figures, i.e. the points and the lines in* $\mathbb{O}P^2$*, are real, and all remaining* α_i*-figures in this plane are imaginary.*

The (α_1, α_5)*-figures, the* (α_2, α_4)*-figures, the* α_3*-figures, and the* α_6*-figures in the complex metasymplectic geometry* $\mathbb{C}\overline{M}s$*, which coincides with the geometry of* $(\mathbb{C}\otimes\mathbb{O}')\overline{\overline{S}}^2$*, are real, and the* α_1*-figures and* α_5*-figures, the* α_2*-figures and* α_4*-figures are, respectively, conjugate imaginary.*

PROOF. The assertion of this theorem on points on the absolute Hermitian conic can be proved analogously to the theorem on points of the absolute Hermitian hyperquadrics of split complex spaces.

The assertions on points, lines, planes, and symplecta in $\mathbb{C}'\overline{M}s$ follow from the equality the dimensions $\dim[\alpha_i]$ and $\dim[\alpha_i, \alpha_{6-i}]$ of the manifolds of (α_1, α_5)-figures, (α_2, α_4)-figures, α_3-figures, and α_6-figures in this geometry with the dimensions found in Theorem I of Freudenthal (see **7.4.2**), and from the equality of the dimensions $\dim_j[\alpha_i]$ calculated by (1.171) with the analogous dimensions calculated by Freudenthal in [Fr2].

In particular, the dimensions $\dim_j[\alpha_1, \alpha_5]$ for $j = 3$ and 6 are equal 4 and 9, respectively, that is, they are equal to the dimensions of $\mathbb{C}'P^2$ and of the absolute Hermitian hyperquadric of $\mathbb{C}'\overline{S}^5$.

We can analogously prove that the (α_1, α_5)-figures in $(\mathbb{C} \otimes \mathbb{O})\overline{\overline{S}}^2$ are the points on the imaginary absolute Hermitian conic of this plane, and the (α_2, α_4)-figures, α_3-figures, and α_6-figures in this plane are the lines $(\mathbb{C} \otimes \mathbb{C})P^1$, planes $(\mathbb{C} \otimes \mathbb{C})P^2$, and 5-spaces $(\mathbb{C} \otimes \mathbb{C})\overline{S}p^5$.

The Satake diagrams of the noncompact groups $EII, EIII$, and EIV (Fig. 1.9) show that the (α_1, α_5)-figures in $(\mathbb{C} \otimes \mathbb{O})\ \overline{\overline{H}}\ {}^2$ are the real points on the real absolute Hermitian conic of this plane, the α_1-figures and α_5-figures of this plane are conjugate imaginary.

These Satake diagrams show also that the α_1-figures and α_5-figures, i.e. the points and lines in $\mathbb{O}P^2$, are real, and all remaining α_i-figures of this plane are imaginary.

These diagrams show also that the (α_1, α_5)-figures, the (α_2, α_4)-figures, the α_3-figures, and the α_6-figures in $\mathbb{C}\overline{M}s$, that is, the points, lines, planes, and symplecta in $\mathbb{C}\overline{M}s$, are real, and all remaining α_i-figures of this geometry are imaginary. ■

7.6.5. Parabolic Figures of the Groups $\mathbf{E_7}$. Since the rank of the simple group E_7 is 7, this group has seven simple roots $\alpha_1, \alpha_2, \alpha_3, \alpha_4, \alpha_6$ and α_7. Their expressions as linear combinations of the basis vectors $E_0, E_1, E_2, E_3, E_4, E_5, E_6, E_7$ are shown in (1.109, E_7). All positive roots of this group are linear combinations of these roots, which can be written as

0	0	0	0	1	0	0	
100000	010000	001000	000100	000000	000010	000001	
0	0	0	1	0	0		
110000	011000	001100	000100	000110	000011		
0	0	1	0	1	0		
111000	011100	001100	001110	000110	000111		
0	1	0	1	0	1		
111100	011100	011110	001110	001111	000111		
1	0	1	0	1	1		
111100	111110	011110	011111	001111	001210		
1	0	1	1	1			
111110	111111	011210	011111	001210			
1	1	1	1	1			
111210	111111	012210	011211	001221			(7.31)
1	1	1	1				
112210	111211	012211	011221				
1	1	1	1				
122210	112211	111221	012221				
1	1	1					
122211	112221	012321					
1	1	2					
122221	112321	012321					
1	2	1					
122321	112321	123321					
2	2	2					
122321	123321	123421					
2	2						
123431	123432						

[FdV, p. 529].

If we remove the simple roots $\alpha_1, \alpha_2, \alpha_3, \alpha_4, \alpha_5, \alpha_6$, and α_7 from the Dynkin diagram of this group, we obtain the semisimple groups $D_6, A_1 \otimes A_5$, $A_2 \otimes A_1 \otimes A_3$, $A_4 \otimes A_2$, $D_5 \otimes A_1, E_6$, and A_6, respectively. From (7.31) we see that the dimensions $\dim[\alpha_1], \dim[\alpha_2], \dim[\alpha_3], \dim[\alpha_4], \dim[\alpha_5], \dim[\alpha_6]$, and $\dim[\alpha_7]$ of the manifolds of α_1-figures, α_2-figures, α_3-figures, α_4-figures, α_5-figures, α_6-figures and α_7-figures in $\mathbb{H}'\overline{M}s, \mathbb{O}'\overline{S}p^5$, and $(\mathbb{H}' \otimes \mathbb{O}')\overline{\overline{S}}^2$, are equal to 33, 47, 53, 50, 42, 27, and 42, respectively, and that the dimensions $\dim_j[\alpha_5]$, that is, the dimensions of the α_j-figures themselves, are equal to 17, 8, 4, 2, 10, and 11 for $j = 1, 2, 3, 4, 6, 7$, respectively.

Theorem 7.37 [Ro9]. *The α_1-figures in the split quaternionic metasymplectic geometry $\mathbb{H}'\overline{M}s$, which coincides with the geometry in the plane $(\mathbb{H}' \otimes \mathbb{O}')\overline{\overline{S}}^2$ and in the split octonionic symplectic space $\mathbb{O}'\overline{S}y^5$, are the symplecta in $\mathbb{H}'\overline{M}s$ and the null points in $\mathbb{O}'\overline{S}y^5$.*

The α_2-figures in $\mathbb{H}'\overline{M}s$ are the split quaternionic planes.

The α_3-figures in $\mathbb{H}'\overline{M}s$ are the split quaternionic lines.

The α_4-figures in $\mathbb{H}'\overline{M}s$ are the real planes.

The α_5-figures in $\mathbb{O}'\overline{S}y^5$ are the null lines, the points in $\mathbb{H}'\overline{M}s$, and the points on the absolute Hermitian conic in $(\mathbb{H}'\otimes\mathbb{O}')\widetilde{\overline{S}}^2$.

The α_6-figures in $\mathbb{O}'\overline{S}y^5$ are the null planes.

The α_7-figures in $\mathbb{H}'\overline{M}s$ are the manifolds of lines in the real 6-spaces.

The fundamental figures in the planes $(\mathbb{H}\otimes\mathbb{O})\widetilde{\overline{\overline{S}}}^2, (\mathbb{H}\otimes\mathbb{O})\widetilde{\overline{\overline{H}}}^2$, and $(\mathbb{H}\otimes\mathbb{O}')\widetilde{\overline{\overline{S}}}^2$ are analogous to the fundamental figures in $(\mathbb{H}'\otimes\mathbb{O}')\widetilde{\overline{S}}^2$.

In $(\mathbb{H}\otimes\mathbb{O})\widetilde{\overline{\overline{S}}}^2$ all these figures are imaginary.

In $(\mathbb{H}\otimes\mathbb{O})\widetilde{\overline{\overline{H}}}^2$ and $(\mathbb{H}\otimes\mathbb{O}')\widetilde{\overline{\overline{S}}}^2$, whose fundamental groups are isomorphic, the α_5-figures, which are the points on the absolute conic of $(\mathbb{H}\otimes\mathbb{O})\widetilde{\overline{\overline{H}}}^2$ and the points in $\mathbb{H}\overline{M}s$, are real, as are the α_1-figures, the α_2-figures, and the α_3-figures, which are the symplecta, planes and lines in $\mathbb{H}\overline{M}s$, respectively; the α_4-figures, α_6-figures, and α_7-figures in these planes are imaginary.

In the plane $(\mathbb{H}'\otimes\mathbb{O})\widetilde{\overline{S}}^2$, which coincides with the space $\mathbb{O}\overline{S}y^5$, the α_1-figures, α_5-figures, and α_6-figures, which are the null points, null lines, and null planes in $\mathbb{O}\overline{S}y^5$, are real, and all remaining α_i-figures are imaginary.

PROOF. The assertion of this theorem on points on the absolute Hermitian conic is proved analogously to the theorem on points on absolute Hermitian hyperquadrics of split quaternionic spaces. The assertion on points, lines, planes, and symplecta in $\mathbb{H}'\overline{M}s$ follow from the equality the dimensions $\dim[\alpha_i]$ of the manifolds of α_3-figures, α_2-figures, and α_1-figures of this geometry with the dimensions found in Theorem I of Freudenthal (see **7.4.2**), from the equality of the dimensions 4,8, and 17 of the α_3-figures, α_2-figures, and α_1-figures in $(\mathbb{H}'\otimes\mathbb{O}')\widetilde{\overline{S}}^2$ with the dimensions of the split quaternionic lines, planes, and the manifold of null points on the absolute linear complex in $\mathbb{H}'\overline{S}y^5$, and from the isomorphism of the stabilizers of these figures. The assertions on the remaining α_i-figures are proved similarly.

We can analogously prove that the α_5-figures in $(\mathbb{H}\otimes\mathbb{O})\widetilde{\overline{\overline{S}}}^2$ are the points on the imaginary absolute Hermitian conic of this plane, and that the α_3-figures, α_2-figures, and α_1-figures in this plane are the lines $(\mathbb{C}\otimes\mathbb{H})P^1$, planes $(\mathbb{C}\otimes\mathbb{H})P^2$, and 5-spaces $(\mathbb{C}\otimes\mathbb{H})\overline{S}p^5$. The Satake diagrams of the noncompact groups EVI and $EVII$ (Fig. 1.9) show that the α_5-figures in $(\mathbb{H}\otimes\mathbb{O})\widetilde{\overline{\overline{H}}}^2$ and $(\mathbb{H}'\otimes\mathbb{O})\widetilde{\overline{S}}^2$, which are the points on the real absolute Hermitian conic of $(\mathbb{H}\otimes\mathbb{O})\widetilde{\overline{\overline{H}}}^2$ and the points in $\mathbb{H}\overline{M}s$, as well as the α_3-figures, α_2-figures, and α_1-figures, which are the lines, planes, and symplecta in $\mathbb{H}\overline{M}s$, are real, and that all remaining α_i-figures in these planes are imaginary. The Satake diagrams show also that only the α_1-figures, α_5-figures, and α_6-figures of $(\mathbb{H}'\otimes\mathbb{O})\widetilde{\overline{S}}^2$, which are the null points, null lines, and null planes of $\mathbb{O}\overline{S}y^5$, are real, and that all remaining α_i-figures of these geometries are imaginary. ■

7.6.6. Parabolic Figures of the Groups E_8. Since the rank of the simple group E_8 is 8, this group has eight simple roots, $\alpha_1, \alpha_2, \alpha_3, \alpha_4, \alpha_5, \alpha_6, \alpha_7$, and α_8. The expressions as linear combinations of the basis vectors E_0,E_1,E_2,E_3,E_4,E_5,E_6, E_7,E_8 are shown in (1.109, E_8). All positive roots of this group are linear combinations of these roots, which can be written as:

0	0	0	0	0	1	0	0
1000000	0100000	0010000	0001000	0001000	0000100	0000010	0000001
0	0	0	0	1	0	0	
1100000	0110000	0011000	0001100	0000100	0000110	0000011	
0	0	0	1	0	1	0	
1110000	0111000	0011100	0001100	0001110	0000110	0000111	
0	0	1	0	1	1	0	
1111000	0111100	0011100	0011110	0001110	0000111	0001111	
0	1	0	1	0	1	1	
1111100	0111100	0111110	0011110	0011111	0001210	0001111	
1	0	1	0	1	1	1	
1111100	1111110	0111110	0111111	0011210	0011111	0001211	
1	0	1	1	1	1	1	
1111110	1111111	0111210	0111111	0012210	0011211	0001221	
1	1	1	1	1	1		
1111210	1111111	0112210	0111211	0012211	0011221		
1	1	1	1	1	1		
1112210	1111211	0122210	0112211	0111221	0012221		
1	1	1	1	1	1		
1122210	1112211	1111221	0122211	0112221	0012321		
1	1	1	1	1	2		
1222210	1122211	1112221	0122221	0112321	0012321		
1	1	1	1	2			
1222211	1122221	1112321	0122321	0112321			
1	1	2	1	2			
1222221	1122321	1112321	0123321	0122321			
1	1	2	2				
1222321	1123321	1122321	0123321				
1	2	2	2				
1223321	1222321	1123321	0123421				
1	2	2	2				
1233321	1223321	1123421	0123431				
2	2	2	2				
1233321	1223421	1123431	0123432				
2	2	2	2				
1233421	1223431	1123432	1234421				
2	2	2	2				
1233431	1223432	1234431	1233432				
2	2	3	2				
1234531	1234432	1234531	1234532				
3	2	3	3				
1234532	1234542	1234542	1234642				
3	3	3	3				
1235642	1245642	1345642	2345642				

(7.32)

[FdV, p. 529–530].

If we remove the simple roots $\alpha_1, \alpha_2, \alpha_3, \alpha_4$, $\alpha_5, \alpha_6, \alpha_7$, and α_8 from the Dynkin diagram of this group, we obtain the semisimple groups $E_7, A_1 \otimes E_6$, $A_2 \otimes D_5 A_3 \otimes A_4$, $A_4 \otimes A_1 \otimes A_2$, $A_6 \otimes A_1$, D_7, and A_7, respectively. From (7.32) we see that the dimensions dim$[\alpha_1]$, dim$[\alpha_2]$, dim$[\alpha_3]$, dim$[\alpha_4]$,dim$[\alpha_5]$, dim$[\alpha_6]$, dim$[\alpha_7]$, and dim$[\alpha_8]$ of the manifolds of α_1-figures, α_2-figures, α_3-figures, α_4-figures, α_5-figures, α_6-figures, α_7-figures, and α_8-figures in $\mathbb{O}'\overline{M}s$ and in $(\mathbb{O}' \otimes \mathbb{O}')\widetilde{\overline{S}}^2$ are equal to 57, 83, 97, 104, 106, 98, 78, and 92, respectively, and that the dimensions $\dim_j[\alpha_7]$, that is, the dimensions of the α_j-figures themselves, are equal to 33, 16, 8, 4, 2,1, and 7 for $j = 1, 2, 3, 4, 5, 6, 8$, respectively.

Theorem 7.38 [Ro9]. *The α_1-figures in the split octonionic metasymplectic geometry $\mathbb{O}'\overline{M}s$, which coincides with the geometry in the plane $(\mathbb{O}' \otimes \mathbb{O}')\widetilde{\overline{S}}^2$, are the symplecta in $\mathbb{O}'\overline{M}s$.*

The α_2-figures in $\mathbb{O}'\overline{M}s$ are the split octonionic planes.

The α_3-figures in $\mathbb{O}'\overline{M}s$ are the split octonionic lines.

The α_4-figures in $\mathbb{O}'\overline{M}s$ are the real 4-planes.

The α_5-figures in $\mathbb{O}'\overline{M}s$ are the real planes.

The α_6-figures in $\mathbb{O}'\overline{M}s$ are the real lines.

The α_7-figures in $\mathbb{O}'\overline{M}s$ are the points and the points of the absolute Hermitian conic in $(\mathbb{O}' \otimes \mathbb{O}')\widetilde{\overline{S}}^2$.

The α_8-figures in $\mathbb{O}'\overline{M}s$ are the real 7-planes.

The fundamental figures in $(\mathbb{O} \otimes \mathbb{O})\widetilde{\overline{S}}^2$ are analogous to the fundamental figures in $(\mathbb{O}' \otimes \mathbb{O}')\widetilde{\overline{S}}^2$ and $\mathbb{O}'\overline{M}s$; however in $(\mathbb{O} \otimes \mathbb{O})\widetilde{\overline{S}}^2$ all these figures are imaginary.

Since the fundamental groups in the planes $(\mathbb{O} \otimes \mathbb{O})\widetilde{\overline{H}}^2$ and $(\mathbb{O}' \otimes \mathbb{O}')\widetilde{\overline{S}}^2$ are isomorphic, all α_i-figures in $(\mathbb{O} \otimes \mathbb{O})\widetilde{\overline{H}}^2$ are real.

The α_7-figures in the geometry $\mathbb{O}\overline{M}s$, which coincides with the geometry in $(\mathbb{O}' \otimes \mathbb{O})\widetilde{\overline{S}}^2$, which are the points on the absolute Hermitian conic of $(\mathbb{O}' \otimes \mathbb{O})\widetilde{\overline{S}}^2$ and the points in $\mathbb{O}\overline{M}s$, are real, as are the α_3-figures, α_2-figures, and α_1-figures of these geometries, i.e. the lines, planes, and symplecta in $\mathbb{O}\overline{M}s$.

All remaining α_i-figures of these geometries are imaginary.

PROOF. The assertion of this theorem on points on the absolute Hermitian conic can be proved analogously to the theorem on points on the absolute Hermitian conic of the split octonionic plane.

The assertion on points, lines, planes, and symplecta in $\mathbb{O}'\overline{M}s$ follow from the equality of the dimensions $\dim[\alpha_1], \dim[\alpha_2], \dim[\alpha_3]$, and $\dim[\alpha_7]$ of this geometry with the dimensions found in Theorem I of Freudenthal (see **7.4.2**), from the equality of the dimensions 8, 16, and 33 of the α_3-figures, α_2-figures, and α_1-figures in $(\mathbb{O}' \otimes \mathbb{O}')\widetilde{\overline{S}}^2$ with the dimensions of the split octonionic lines and planes, and manifold of null points in $\mathbb{O}\overline{S}y^5$, and from the isomorphism of the stabilizers of these figures.

The assertions on the remaining α_i-figures are proved analogously.

We can analogously prove that the α_1-figures in $(\mathbb{O} \otimes \mathbb{O})\tilde{\overline{S}}^2$ are the points in the imaginary absolute Hermitian conic of this plane, and that the α_2-figures, α_3-figures, and α_7-figures in this plane are the lines $(\mathbb{C} \otimes \mathbb{O})P^1$, planes $(\mathbb{C} \otimes \mathbb{O})P^2$, and 5-spaces $(\mathbb{C} \otimes \mathbb{O})\overline{S}p^5$. The Satake diagrams of the noncompact group EIX (Fig. 1.9) show that the α_7-figures in $(\mathbb{O}' \otimes \mathbb{O})\ \tilde{\tilde{\overline{S}}}^{\,2}$, which are the points in the real absolute Hermitian conic of this plane and the points of $\mathbb{O}\overline{M}s$, are real, as are the α_3-figures, α_2-figures, and α_1-figures, i.e. the lines, planes, and symplecta in $\mathbb{O}\overline{M}s$, and that all remaining α_i-figures of these planes are imaginary. ■

Note that if we denote the dimensions of the algebras $\mathbb{R}, \mathbb{C}, \mathbb{C}', \mathbb{H}, \mathbb{H}', \mathbb{O}$, and $\mathbb{O}'$ by p, the real dimensions of the lines, planes, and symplecta in the metasymplectic geometries Ms, $\mathbb{C}\overline{M}s$ and $\mathbb{C}'\overline{M}s, \mathbb{H}\overline{M}s$ and $\mathbb{H}'\overline{M}s$, and $\mathbb{O}\overline{M}s$ and $\mathbb{O}'\overline{M}s$ are equal to $p, 2p$, and $4p + 1$, respectively.

7.6.7. Construction of Manifolds of Fundamental Figures. The construction of the tangent spaces to the manifolds of fundamental figures of the simple Lie groups G is given by (1.170): these spaces are direct sums

$$\mathbb{G}_{-h} \oplus \mathbb{G}_{-h+1} \oplus \cdots \oplus \mathbb{G}_{-1}$$

of linear subspaces in the tangent Lie algebra $\mathbb{G}$ of the group G.

This decomposition corresponds to a fibration of the manifold of fundamental figures.

We illustrate this fibration of the manifold of fundamental figures and the corresponding decomposition of its tangent space by the example of the absolute Hermitian conics of the Hermitian hyperbolic planes $\mathbb{O}\overline{H}^2, (\mathbb{C}\otimes\mathbb{O})\tilde{\overline{H}}^2, (\mathbb{H}\otimes\mathbb{O})\tilde{\overline{H}}^2$, and $(\mathbb{O}\otimes\mathbb{O})\tilde{\overline{H}}^2$.

Theorem 7.39 (V. V. MALYUTIN [RMal]). *The absolute Hermitian conics of the Hermitian hyperbolic planes $\mathbb{O}\overline{H}^2, (\mathbb{C}\otimes\mathbb{O})\tilde{\overline{H}}^2, (\mathbb{H}\otimes\mathbb{O})\tilde{\overline{H}}^2$, and $(\mathbb{O}\otimes\mathbb{O})\tilde{\overline{H}}^2$ are the fiber spaces whose bases are the lines $\mathbb{O}P^1, (\mathbb{C}\otimes\mathbb{O})P^1, (\mathbb{H}\otimes\mathbb{O})P^1$, and $(\mathbb{O}\otimes\mathbb{O})P^1$, and whose fibers are the absolutes of the lines $\mathbb{O}\overline{H}^1$, $(\mathbb{C}\otimes\mathbb{O})\tilde{\overline{H}}^1$, $(\mathbb{H}\otimes\mathbb{O})\tilde{\overline{H}}^1$, and $(\mathbb{O}\otimes\mathbb{O})\tilde{\overline{H}}^1$.*

In the corresponding decompositions $\mathbb{G}_{-2} \oplus \mathbb{G}_{-1}$, the subspaces $\mathbb{G}_{-1}$ are the algebras $\mathbb{O}, \mathbb{C}\otimes\mathbb{O}$, $\mathbb{H}\otimes\mathbb{O}$, and $\mathbb{O}\otimes\mathbb{O}$ and the subspaces $\mathbb{G}_{-2}$ are the Jordan algebras $\mathbb{J}\bar{j}(\mathbb{O}), \mathbb{J}\tilde{\bar{j}}(\mathbb{C}\otimes\mathbb{O}), \mathbb{J}\tilde{\bar{j}}(\mathbb{H}\otimes\mathbb{O})$, and $\mathbb{J}\tilde{\bar{j}}(\mathbb{O}\otimes\mathbb{O})$, respectively.

<u>PROOF</u>. Stereographically project from a point on the absolute Hermitian conic of the plane $\mathbb{O}\overline{H}^2, (\mathbb{C}\otimes\mathbb{O})\tilde{\overline{H}}^2, (\mathbb{H}\otimes\mathbb{O})\tilde{\overline{H}}^2$, or $(\mathbb{O}\otimes\mathbb{O})\tilde{\overline{H}}^2$ onto a line B on this plane, which can be regarded as the projective line over the corresponding algebra. The projecting lines, which are the Hermitian hyperbolic lines over the same algebras since they intersect the absolute Hermitian conics of the planes, meet the line B at one point and cut out from the absolute conics their absolutes.

From (7.29), (7.30), (7.31), and (7.32) we find that the dimensions of the subspaces $\mathbb{G}_{-1}$ and $\mathbb{G}_{-2}$ are 8 and 7, 16 and 8, 32 and 10, and 64 and 14, respectively (the sums of these dimensions, 15, 24, 42, and 78, are equal to the dimensions of the absolute Hermitian conics themselves).

The subspaces $\mathbb{G}_{-1}$, that is, the tangent spaces to the bases of these fibrations, can be regarded as the affine lines over the above mentioned algebras or as these algebras themselves.

The subspaces $\mathbb{G}_{-2}$, that is, the tangent spaces to the fibers of these fibrations, can be regarded as the manifolds of elements in these algebras that satisfy the condition $a = -\overline{a}$ in the case of $\mathbb{O}$ and the condition $a = -\tilde{\overline{a}}$ for the tensor products $\mathbb{C}\otimes\mathbb{O}, \mathbb{H}\otimes\mathbb{O}$, and $\mathbb{O}\otimes\mathbb{O}$. These manifolds can be regarded as the above mentioned Jordan algebras. ■

Since the absolute of the line $\mathbb{O}\overline{H}^1$ is homeomorphic to the hypersphere Σ^7, the first assertion of this theorem is equivalent to the assertion on fibration of Hermitian circles in $\mathbb{O}\overline{H}^2$ in **7.2.1**.

Note that the subspaces $\mathbb{G}_{-1}$ are tangent to the polar lines of the points on the absolute Hermitian conics with respect to these conics.

There are analogous decompositions for the Hermitian elliptic planes $\mathbb{O}'\overline{S}^2, (\mathbb{C}'\otimes \mathbb{O}')\tilde{\overline{S}}^2, (\mathbb{H}'\otimes\mathbb{O}')\tilde{\overline{S}}^2$, and $(\mathbb{O}'\otimes\mathbb{O}')\tilde{\overline{S}}^2$, and for each Hermitian elliptic plane whose group of motions is an exceptional simple Lie group.

By applying formula (1.171) to the individual subspaces $\mathbb{G}_{-1}$ and $\mathbb{G}_{-2}$ of the tangent spaces to the absolute conics of the planes $\mathbb{O}'\overline{S}^2, (\mathbb{C}'\otimes\mathbb{O}')\tilde{\overline{S}}^2, (\mathbb{H}'\otimes\mathbb{O}')\tilde{\overline{S}}^2$, and $(\mathbb{O}'\otimes\mathbb{O}')\tilde{\overline{S}}^2$ we can prove.

Theorem 7.40 (T.A.STEPASHKO [RoSt]). *The lines and planes in the metasymplectic geometries $Ms, \mathbb{C}\overline{M}s, \mathbb{H}\overline{M}s$, and $\mathbb{O}\overline{M}s$, regarded as submanifolds of the absolute Hermitian conics of $\mathbb{O}'\overline{S}^2, (\mathbb{C}'\otimes\mathbb{O}')\tilde{\overline{S}}^2, (\mathbb{H}'\otimes\mathbb{O}')\tilde{\overline{S}}^2$, and $(\mathbb{O}'\otimes\mathbb{O}')\tilde{\overline{S}}^2$, are located in the polar lines of their points with respect to the absolute conics.*

For the symplecta in $Ms, \mathbb{C}\overline{M}s, \mathbb{H}\overline{M}s$, and $\mathbb{O}\overline{M}s$, regarded as $(4p+1)$-dimensional submanifolds of the absolute Hermitian conics of these planes, the intersections of these submanifolds with the polar lines of the points on these conics are $4p$-dimensional, and the intersections of these manifolds with fibers in the fibrations of these conics are 1-dimensional.

We mention also

Theorem II of Freudenthal [Fr2]. *In the decompositions $\mathbb{G}_{-2}\oplus\mathbb{G}_{-1}$ of the tangent spaces to the manifolds of symplecta in the metasymplectic geometries $Ms, \mathbb{C}\overline{M}s, \mathbb{H}\overline{M}s$, and $\mathbb{O}\overline{M}s$, the subspaces $\mathbb{G}_{-1}$ have the structure of the Freudenthal algebras $\mathbb{F}r(\mathbb{R}_3), \mathbb{F}r(\mathbb{C}_3), \mathbb{F}r(\mathbb{H}_3)$, and $\mathbb{F}r(\mathbb{O}_3)$, respectively, and the subspaces $\mathbb{G}_{-2}$ coincide with the field $\mathbb{R}$.*

Note that in **1.5.4** we have seen that the dimensions of the Freudenthal algebras $\mathbb{F}r(\mathbb{R}_3), \mathbb{F}r(\mathbb{C}_3), \mathbb{F}r(\mathbb{H}_3)$, and $\mathbb{F}r(\mathbb{O}_3)$, are equal to 14, 20, 32, and 58, respectively, and in Theorem I of Freudenthal (see **7.4.2**) and in **7.6.3**, **7.6.4**, **7.6.5**, and **7.6.6** we have seen that the dimensions of the manifolds of symplecta in the geometries $Ms, \mathbb{C}\overline{M}s, \mathbb{H}\overline{M}s$, and $\mathbb{O}\overline{M}s$ are 15, 21, 33, and 59, respectively.

An analogous theorem is valid also for the manifolds of symplecta in the metasymplectic geometries $\mathbb{C}'\overline{M}s, \mathbb{H}'\overline{M}s$, and $\mathbb{O}'\overline{M}s$.

7.6.8. Fundamental Linear Representations of Exceptional Lie Groups. We have defined fundamental representations of simple Lie groups in **1.6.2**, and have given in Table 1.1 the dimensions of the spaces of these representations.

For the fundamental representations φ_1 and φ_2 of the group G_2 the dimensions of the spaces of representations are equal to 14 and 7. The first of these linear representations is the representation by the adjoint group, the second one is the representation by rotations in the space R^7 or R^7_3.

The fundamental representations φ_1 of the groups F_4, E_7 and E_8 and the fundamental representation φ_6 of the group E_6 are representations by the adjoint groups, the dimensions of the spaces of these representations are equal to 52, 78, 133 and 248, respectively, that is, to the dimensions of these groups.

Since the fundamental representation φ_i corresponds to the fundamental root α_i Theorems 7.32, 7.33, 7.34, and 7.34 imply

Theorem 7.41. *The fundamental representations of the groups F_4, E_6, E_7, and E_8 by their adjoint groups correspond to the symplecta of the metasymplectic geometries.*

The linear representations of the exceptional simple Lie groups with minimal dimensions of the space of the representation are: for the group F_4 the representation φ_4 corresponding to the points in the absolute Hermitian conic of the Hermitian plane (this dimension is equal to 26), for the group E_6 – the representations φ_1 and φ_5 corresponding to the points and lines in the projective plane (this dimension is equal to 27), for the group E_7 – the representation φ_6 corresponding to null planes of the symplectic 5-space (this dimension is equal to 56), for the group E_8 it is the above mentioned representation φ_1.

The role of vectors of the representations φ_1 and φ_5 for E_6 is played by the Hermitian symmetric (3×3)*-matrices (7.2), representing the points and lines in the projective plane (see* **7.2.2***).*

The role of such vectors of the representation φ_4 for F_4 is played by the same (3×3)*-matrices but, in addition satisfying the condition* $x^{00} + x^{11} + x^{22} = 0$ *for elliptic planes and the condition* $-x^{00} + x^{11} + x^{22} = 0$ *for hyperbolic plane (these conditions are equivalent to the equations of the absolute Hermitian conics of these planes).*

The role of vectors of the representations φ_6 for E_7 is played by the Hermitian symmetric (3×3)*-matrices that represent the null planes in the symplectic 5-space, they are analogous to the* $(n \times n)$*-matrices considered in* **6.4.2**.

Note that the representations φ_4 for $F_4, \varphi_1 \otimes \varphi_5$ for E_6, φ_5 for E_7, and φ_7 for E_8 are linear representations by transformations of coordinates of points of the corresponding metasymplectic geometries and of points of the absolute Hermitian conics of the corresponding Hermitian planes.

§7.7. Finite Geometries

7.7.1. Exceptional Simple Finite Groups of Lie Type. In **1.4.15** we have defined simple finite groups of Lie type. Among these groups there are exceptional groups $\mathbb{F}_q G_2, \mathbb{F}_q F_4, \mathbb{F}_q E_6, \mathbb{E}_q E_7$, and $\mathbb{F}_q E_8$. There is also the group $\mathbb{F}_q E_6^{(2)}$, whose Satake diagram coincides with the diagram of the noncompact *semisplit* group EII (see Fig. 1.9).

7.7.2. Finite Algebras over Galois Fields. Since finite division rings must be commutative, algebras over the Galois field $\mathbb{F}_q$ with bases of the algebras $\mathbb{H}$ and $\mathbb{O}$, cannot be division rings, that is, fields. We denote by $\mathbb{A}_q$ an algebra over $\mathbb{F}_q$ with basis of algebra $\mathbb{A}$. Therefore, we denote by $\mathbb{H}_q$ and $\mathbb{O}_q$, respectively, the algebras of quaternions and octonions over $\mathbb{F}_q$. O. Lyubinskaya [Lyu] has proved that the algebra $\mathbb{H}_q$ is isomorphic to the algebra $(\mathbb{F}_q)_2$ of (2×2)-matrices over $\mathbb{F}_q$. In **2.9.3** we have seen that algebras $\mathbb{C}_q$ can be the field $\mathbb{F}_{q^2}$ and the ring $\mathbb{F}_q \oplus \mathbb{F}_q$, there we have also considered the algebra $\mathbb{C}_q^0$.

7.7.3. Finite Geometries with Fundamental Exceptional Simple Finite Groups of Lie Type. Analogously to the construction of the plane $\mathbb{O}P^2$ and Hermitian elliptic planes over $\mathbb{O}, \mathbb{C} \otimes \mathbb{O}, \mathbb{H} \otimes \mathbb{O}$, and $\mathbb{O} \otimes \mathbb{O}$, we can define the planes $\mathbb{O}_q P^2$ and Hermitian elliptic planes over the algebras $\mathbb{O}_q, \mathbb{C}_q \otimes \mathbb{O}_q, \mathbb{H}_q \otimes \mathbb{O}_q$, and $\mathbb{O}_q \otimes \mathbb{O}_q$. Since the algebras $\mathbb{C}_q$ have two forms, $\mathbb{F}_{q^2}$ and $\mathbb{F}_q \oplus \mathbb{F}_q$, the algebras $\mathbb{C}_q \oplus \mathbb{O}_q$, also have two forms, $\mathbb{O}_{q^2}$ and $\mathbb{O}_q \oplus \mathbb{O}_q$.

Analogously to Theorems 7.4, 7.6, 7.12, and 7.16 we can prove that the group of collineations in $\mathbb{O}_q P^2$ is isomorphic to the group $\mathbb{F}_q E_6$, and that the groups of motions in $\mathbb{O}_q \overline{S}^2, \mathbb{O}_{q^2} \overline{\widetilde{S}}^2, (\mathbb{H}_q \oplus \mathbb{O}_q)\overline{\widetilde{S}}^2$, and $(\mathbb{O}_q \oplus \mathbb{O}_q)\overline{\widetilde{S}}^2$ are isomorphic to the groups $\mathbb{F}_q F_4$, $\mathbb{F}_q E_6^{(2)}$, $\mathbb{F}_q E_7$, and $\mathbb{F}_q E_8$. Note also that the plane $\mathbb{O}_q P^2$ admits an interpretation as the plane $(\mathbb{O}_q \oplus \mathbb{O}_q)\overline{\widetilde{S}}^2$ with group of motions isomorphic to the group $\mathbb{F}_q E_6$.

§7.8. Applications to Physics

7.8.1. Geometry of Exceptional Lie Group and Supergravity Theory. The geometry of the exceptional simple Lie groups has applications to *supergravity theory*, that is, to the attempts to create a new unified field theory including not only gravity and electromagnetic field theories but also strong and weak interactions theories.

M. Günaydin and his collaborators [GST] consider in their supergravity theory twelve supergravity fields, connected with the twelve symmetric Riemannian spaces whose groups of motions are noncompact groups at the three last lines in the *Freudenthal magic square* (7.19) and whose stabilizers of points are the compact groups at the first line in this square, the direct products of compact groups at the second line in this square by the group of complex numbers with unit modulus, and the direct products of compact groups at the third line in this square by the group of quaternions with unit modulus.

Geometric models of the first four of these spaces are given by the manifolds of imaginary conics and Hermitian conics in the projective planes $P^2, \mathbb{C}P^2, \mathbb{H}P^2$, and $\mathbb{O}P^2$, or, if these planes are interpreted as the planes $\mathbb{C}'\overline{S}^2, (\mathbb{C}'\otimes\mathbb{C})\widetilde{\overline{S}}^2, (\mathbb{C}'\otimes\mathbb{H})\widetilde{\overline{S}}^2$, and $(\mathbb{C}'\otimes\mathbb{O})\widetilde{\overline{S}}^2$, by the manifold of normal 2-chains in $\mathbb{C}'\overline{S}^2$, by the manifold of normal complex 2-chains in $(\mathbb{C}'\otimes\mathbb{C})\widetilde{\overline{S}}^2$, by the manifold of normal quaternionic 2-chains in $(\mathbb{C}'\otimes\mathbb{H})\widetilde{\overline{S}}^2$, and by the manifold of normal octonionic 2-chains in $(\mathbb{C}'\otimes\mathbb{O})\widetilde{\overline{S}}^2$.

Geometric models of the second four of these spaces are given by the manifolds of elliptic symplectic congruences of the symplectic spaces $Sy^5, \mathbb{C}\overline{S}y^5, \mathbb{H}\overline{S}y^5$, and $\mathbb{O}\overline{S}y^5$, or, if these spaces are interpreted as the planes $\mathbb{H}'\overline{S}^2$, $(\mathbb{H}'\otimes\mathbb{C})\widetilde{\overline{S}}^2$, $(\mathbb{H}'\otimes\mathbb{H})\widetilde{\overline{S}}^2$, and $(\mathbb{H}'\otimes\mathbb{O})\widetilde{\overline{S}}^2$, by the manifold of normal complex 2-chains in $\mathbb{H}'\overline{S}^2$, by the manifold of normal bicomplex 2-chains in $(\mathbb{H}'\otimes\mathbb{C})\widetilde{\overline{S}}^2$, by the manifold of normal biquaternionic 2-chains in $(\mathbb{H}'\otimes\mathbb{H})\widetilde{\overline{S}}^2$, and by the manifold of normal bioctonionic 2-chains in $(\mathbb{H}'\otimes\mathbb{O})\widetilde{\overline{S}}^2$.

Geometric models of the third four of these spaces are given by the manifolds of the symmetry figures with compact stabilizers in the metasymplectic geometries $Ms, \mathbb{C}\overline{M}s, \mathbb{H}\overline{M}s$, and $\mathbb{O}\overline{M}s$, or, if these geometries are interpreted as the geometries of the planes $\mathbb{O}'\overline{S}^2, (\mathbb{O}'\otimes\mathbb{C})\widetilde{\overline{S}}^2, (\mathbb{O}'\otimes\mathbb{H})\widetilde{\overline{S}}^2$, and $(\mathbb{O}'\otimes\mathbb{O})\widetilde{\overline{S}}^2$, by the manifold of normal quaternionic 2-chains in $\mathbb{O}'\overline{S}^2$, by the manifold of normal biquaternionic 2-chains in $(\mathbb{O}'\otimes\mathbb{C})\widetilde{\overline{S}}^2$, by the manifold of normal quaterquaternionic 2-chains in $(\mathbb{O}'\otimes\mathbb{H})\widetilde{\overline{S}}^2$, and by the manifold of normal quateroctonionic 2-chains in $(\mathbb{O}'\otimes\mathbb{O})\widetilde{\overline{S}}^2$.

The 1st, 3rd, and 4th of these symmetric spaces are noncompact Riemannian symmetric spaces AI, AII, and EIV. The 5th, 6th, 7th, and 8th spaces are noncompact Riemannian symmetric spaces $CI, AIII, DIII$, and $EVII$. The 9th, 10th, 11th, and 12th spaces are noncompact Riemannian symmetric spaces FI, EII, EVI, and EIX.

References

[Ab1] Abbasov N. T., *Bicomplex Elliptic Spaces,* **2** (1962), Uch. Zap. Azerb. Univ., ser. fiz.-mat. nauk, 3-9 (Russian).

[Ab2] Abbasov N. T., *Biquaternionic Elliptic Spaces* **2** (1963), Uch. Zap. Azerb. Univ., ser. fiz.-mat. nauk, 3-9. (Russian).

[Ad] Adamushko N. N., *Geometry of simple and quasisimple groups G_2* **253** (1969), Uch. Zap. Mosk. Obl. Ped. Inst., 23-42 (Russian).

[Af1] Afanas'yev V. V., *Construction of Extended Finite Projective Planes* **180** (1979), Konstruktivnaya i algebraicheskaya geometriya, Yaroslavl', 3-10 (Russian).

[Af2] Afanas'yev V. V., *Extension of Extended Finite Projective Planes* **190** (1980), Konstruktivnaya i algebraicheskaya geometriya, Yaroslavl', 10-21 (Russian).

[Af3] Afanas'yev V. V., *Three Times Extended Finite Projective Planes* **194** (1981), Konstruktivnaya i algebraicheskaya geometriya, Yaroslavl', 16-21 (Russian).

[AG] Akivis M. A., Goldberg V. V., *Conformal Differential Geometry and Its Generalizations* to appear in New York - Toronto:, Wiley-Interscience Publ.

[Al1] Albert A.A., *Modern Higher Algebra,* Univ. Press, Chicago, 1947.

[Al2] Albert A. A., *Quadratic Forms Permitting Composition,* Ann. of Math. **43** (1942), 161-177.

[Alx] Alexeyeva N. K., *The Plane of Dual Variable and the Line Geometry of a Pseudo-Euclidean Space* **20** (1979), Geom. Sb., Tomsk, 97-100 (Russian).

[ASh] Andreyeva L. P., Shestyreva L. V., *Limit Symplectic Geometries, Spaces.* **8** (1964), Proyektivnye metriki. Uch. Zap. Kolom. ped. instituta, 23-44 (Russian).

[Ari] Aristotle, *Metaphysics,* Grinell Iowa: The Peripatetic Press, 1979.

[ArG] Arnold V. I., Givental' A. B., *Symplectic Geometry,* in book: Arnold V. I., Novikov S. P. (Eds), *Dynamical Systems IV* Encyclopaedia of Mathematical Sciences **4** (1990), New York - Berlin a.o.: Springer-Verlag, 1-136.

[Ba] Barbilian D., *Zur Axiomatik der projektiven Ringgeometrie* **50** (1940), Jahresber. Deut. Math. Verein., 179-229, **51** (1941), 34–76.

[Bel] Beltrami E., *Saggio di interpetrazione della geometria non euclidea* **1** (1902), Opere matematiche, Milano: Hoepli, 374-402.

[Bec] Berceanu S., *On the Geometry of Complex Grassmann Manifold, Its Noncompact Dual and Coherent States* (1995), Bucharest: Institute of Atomic Physics.

[Ber] Berger M., *Les espaces symétriques noncompacts* **74** (1957), Ann. Sci École Normale Sup., (3), no. 2 85-117.

[Brt] Bertini E., *Einführung in die projektive Geometrie mehrdiminsionaler Räume,* Wien: Seidel, 1924.

[Bet] Betti E., *Sopra gli spazi di un numero qualunque di dimensioni* **2** (1914), Opere matematiche, Milano: Hoepli, 273-290.

[Bi] Bilo J., *Onderzoekingen betreffende de Meetkundige Grondslagen van de Projectieve Quaternionenmeetkunde,* Brussel: Vanderlinden, 1949.

[Bl] Blaschke W., *Euklidische Kinematik und nichteuklidische Geometrie* **2** (1983), Gesammelte Werke, Essen: Thales-Verlag, 53-66.

[BlT] Blaschke W., Terheggen H., *Trigonometria Hermitiana* **2** (1983), Blaschke W., Gesammelte Werke, Essen: Thales-Verlag, 277-285.

[Blu] Blunck A., *Chain Spaces via Clifford Algebras,* Darmstadt: Technische Hochschule, 1995 to appear in Monatshefte für Mathematik..

[Boi] Bolyai J., *Appendix Exhibiting the Absolutely True Science of Space Independent of Axiom XI of Euclid (not to be decided a priori) with the Geometrical Quadrature of a Circle in the Case of Its Falsity* (1959), Smith D. E. *A Source Book in Mathematics,* New York: Dover, 375-388.

[Bor] Borel A., *Le plan projective des octaves et les sphères comme les espaces homogènes* **1** (1983), Oeuvres/Collected Papers, Berlin - New York: Springer-Verlag, 39-41.

[Bot] Bortolotti E., *Sulle forme differenziali quadratiche spezializzate* **12** (1930), Rend. Accad. Lincei, (6), 541-547.

[Bou1] Bourbaki N., *Elements of Mathematics. Book 1, Theory of Sets*, Paris: Hermann - Reading, MA: Addison - Wesley, 1968.

[Bou2] Bourbaki N., *Elements of Mathematics. Book 2, Algebra*, Ch. 1–7, Berlin, New York a.o.: Springer-Verlag, 1985–1990. *Éléments de Mathématiques, Livre 2, Algèbre*, Ch. 8–10., Paris - New-York a.o.: Masson, 1980–1981.

[Bou3] Bourbaki N., *Elements of Mathematics. Book 3, General Topology*, Paris: Hermann - Reading, MA: Addison-Wesley, 1960–1966.

[Bou4] Bourbaki N., *Éléments de Mathématiques. Livre 4, Fonctions d'une variable réelle*, Paris: Hermann, 1960–1961.

[Bou5] Bourbaki N., *Elements of Mathematics, Book 5, Topological Vector Spaces*, Ch. 1–5, Berlin - New York a.o.: Springer-Verlag, 1987.

[Bou6] Bourbaki N., *Éléments de Mathématiques, Livre 6, Intégration*, Paris: Hermann, 1956–1969.

[Bou7] Bourbaki N., *Éléments de Mathématiques, L'algèbre commutative*, Paris - New York a.o: Masson, 1961–1983.

[Bou8] Bourbaki N., *Elements of Mathematics, Lie Groups and Lie Algebras*, Ch. 1–3, Berlin - New York a.o.: Springer-Verlag, 1989. *Éléments de Mathématiques, Groupes et Algèbres de Lie*, Ch. 4–8, Paris: Hermann, 1968–1985, Ch. 9, Paris: New York a.o.: Masson, 1972.

[Bou9] Bourbaki N., *Éléments de Mathématiques, Théories spectrales*, Paris: Hermann, 1967.

[BrWe] Brauer R., Weyl H., *Spinors in n Dimensions* **3** (1968), Weyl H., Gesammelte Abhandlungen, Berlin - New York: Springer-Verlag, 493-516.

[Bre] Brehm U., *The Shape Invariant of Triangles and Trigonometry in Two-Point Homogeneous Spaces* **33** (1990), Geom. Dedic., no. 1, 59-76.

[Bur] Burtseva T. A., *Real 2-Surfaces in the Octonionic Hermitian Plane* **19** (1989), Trudy Geom. Sem. Kazan. Univ., 40-52 (Russian).

[Car1] Cartan É., *Sur la structure des groupes de transformations finis et continus* Thèse, **1** (1984), Oeuvres complètes, Paris: CNRS, 37-286.

[Car2] Cartan É., *On Manifolds with an Affine Connection and the Theory of General Relativity*, Naples: Bibliopolis, 1986.

[Car3] Cartan É., *Leçons sur la géométrie projective complexe*, Paris: Gauthier-Villars 2nd ed., 1950.

[Car4] Cartan É., *The Theory of Spinors*, 2nd ed., New York: Dover, 1980.

[Car5] Cartan É., *Les groupes bilinéaires et systèmes de nombres complexes* **2** (1984), Oeuvres complètes, Paris: CNRS, 7-105.

[Car6] Cartan É., *Nombres complexes* **2** (1984), Oeuvres complètes, Paris: CNRS, 857-925.

[Car7] Cartan É., *Les groupes projectifs qui ne laissent invariante aucune multiplicité plane* **1** (1984), Oeuvres complètes, Paris: CNRS, 355-398.

[Car8] Cartan É., *Les groupes réels et simples finis et continus* **1** (1984), Oeuvres complètes, Paris: CNRS, 399-491.

[Car9] Cartan É., *Les groupes projectifs continus réels qui ne laissent invariante aucune multiplicité plane* **1**, Oeuvres complètes, Paris: CNRS, 493-530.

[Car10] Cartan É., *La théorie des groupes continus et la géométrie* **3** (1984), Oeuvres complètes, Paris: CNRS, 1727-1861.

[Car11] Cartan É., *Le principe de dualité et la théorie des groupes simples et semisimples* **1** (1984), Oeuvres complètes, Paris: CNRS, 555-568.

[Car12] Cartan É., *Sur une classe remarquable d'espaces de Riemann* **1** (1984), Oeuvres complètes, Paris: CNRS, 587-650.

[Car13] Cartan É., *Sur certaines formes riemanniennes remarquables des géométries à groupe fondamental simple* **1** (1984), Oeuvres complètes, Paris: CNRS, 867-989.

[Car14] Cartan É., *La géométrie des groupes de transformations* **1** (1984), Oeuvres complètes, Paris, CNRS, 673-791.

[Car15] Cartan É., *Groupes simples clos et ouverts et géométrie riemannienne* **1** (1984), Oeuvres complètes, Paris: CNRS, 1011-1043.

[Car16] Cartan É., *La théorie de groupes finis et continus et l'Analysis situs* **1** (1984), Oeuvres complètes, Paris: CNRS, 1165-1225.

[Car17] Cartan É., *Sur une dégénérescence de la géométrie euclidienne* Akivis M.A., Rosenfeld B. A., *Élie Cartan (1869–1951)* (1993), Providence, RI: AMS., 273-274.

[Car18] Cartan É., *Les surfaces isotropes d'une quadrique de l'espace à 7 dimensions* In memoriam N. I. Lobatschevskii **3** (1995), Kazan: Univ.,, no. 1, 10-119.

[Cay1] Cayley A., *On Jacobi's Elliptic Functions and on Quaternions* **1** (1889), Collected Papers, Cambridge: Univ. Press, 127-130.

[Cay2] Cayley A., *Memoir on the Theory of Matrices* **1** (1889), Collected Papers, Cambridge: Univ. Press, 475-496.

[Cay3] Cayley A., *Sixth Memoir upon Quantics* **2** (1890), Collected Papers, Cambridge: Univ. Press, 561-592.

[Cha] Chakhlenkova T. G., *Geometry of m-Euclidean Spaces* (1958), Izv.VUZov, Matem., no. 1(2), 174-184 (Russian).

[Chkh] Chakhtauri I. A., *Projective and Elliptic Spaces of Integer and Fractional Dimension over Matrix Algebras* **63** (1971), Soobshcheniya Akad. Nauk Gruz. SSR, no.1, 541-544 (Russian).

[Chv] Chevalley C., *Sur certains groupes simples* **7** (1955), Tôhoku Math. J., 14-66.

[Cli1] Clifford W. K., *Synthetic Proof of Miquel's Theorem* (1968), Mathematical Papers, New York: Chelsea, 38-55.

[Cli2] Clifford W. K., *Preliminary Sketch on Biquaternions* (1968), Mathematical Papers, New York: Chelsea, 181-200.

[Cli3] Clifford W. K., *Applications of Grassmann's Extensive Algebra* (1968), Mathematical Papers, New York: Chelsea, 266-276.

[Coo1] Coolidge J. L., *The Geometry of the Complex Domain*, Oxford: Clarendon Press, 1924.

[Coo2] Coolidge J. L., *Hermitian Metrics* **22** (1921), Ann. of Math., 11-28.

[Cox1] Coxeter H. S. M., *Introduction to Geometry*, New York - London: J. Wiley & Sons, 1961.

[Cox2] Coxeter H. S. M., *Regular Polytopes,* - 2nd ed., New York - London: Colliers - Macmillan, 1963.

[Cox3] Coxeter H. S. M., *Discrete Groups Generated by Reflections* **35** (1934), Ann. of Math., 588-621.

[Cox4] Coxeter H. S. M., *Discrete Groups Generated by Reflections,* Weyl H. *The Structure and Representations of Continuous Groups* (1935), Mimeogr. notes ed. by R. Brauer, Princeton: Inst. for Adv. Study, 186-212.

[Cox5] Coxeter H.S.M., *The Product of the Generators of Finite Groups Generated by Reflections* **18** (1951), Duke Math. J., 765-782.

[Cox6] Coxeter H. S. M., *Regular Honeycombs in Elliptic Space* **4** (1954), Proc. London Math. Soc. (3), 471-501.

[Cox7] Coxeter H. S. M., *Regular Honeycombs in Hyperbolic Space* **3** (1954), Proc. Internat. Congress of Math., Amsterdam, 155-169.

[Cox8] Coxeter H. S. M., *Integral Cayley Numbers* **13** (1946), Duke Math. J., no. 4, 561-578.

[Cox9] Coxeter H. S. M., *Regular Complex Polytopes*, Cambridge: University Press, 1991.

[CuRe] Curtis C. W., Reiner I., *Representation Theory of Finite Groups and Associative Algebras*, New York: Interscience Publ., 1962.

[Def] Defrise P., *Espaces-temps à temps absolu* **13** (1961), Bull. Soc. Math. Belg., no. 1/2, 6-37.

[DeV1] De Vries J., *Über gewisse ebene Configurationen* **12** (1889), Acta Math., 63-81.

[DeV2] De Vries J., *Räumliche Configurationen* **3** (1895), Verh. Konink. Akad. Wet., Amsterdam, 101-107.

[Dic] Dickson L. E., *Algebren und ihre Zahlentheorie* **4** (1975), Mathematical Papers, New York: Chelsea, 7-320.

[Dob] Dobromyslov V. A., *On the Geometry of k-Quasiaffine Space* (1988), Tkani i Kvazigruppy, Kalinin, 147-165 (Russian).

[Dy1] Dynkin E., *The Structure of Semisimple Lie Algebras*, vol. 17, Amer. Math. Soc. Translations (1), 1950.

[Dy2] Dynkin E., *Certain Propertiesof the System of Weights of Semisimple Lie Algebras* **71** (1950), Dokl. AN SSSR, 221-224 (Russian).

[Ehr] Ehresmann Ch., *Sur la topologie de certains espaces homogènes* **24** (1981), Oeuvres complètes et commentées, I-1, Cahiers Topologie, Géométrie Différentielle Categoriques, 5-53.

[Ei1] Einstein A., *On the Electrodynamics of Moving Bodies,* Lorentz H. A., Einstein A., Minkowski H., Weyl H., *The Principle of Relativity* (1923), New York: Dover, 37-65.

[Ei2] Einstein A., *The Foundations of the General Theory of Relativity,* Lorentz H. A., Einstein A., Minkowski H., Weyl H. *The Principle of Relativity* (1923), New York: Dover, 109-164.

[Euc] [Euclid], *The Thirteen Books of Euclid's Elements*, vol. 1–3, transl. with introd. and comm. by T. L. Heath., Cambridge: Univ. Press, 1926.

[Fed1] Fedenko A.S., *Spaces with Symmetries*, Minsk: Univ. Press, 1977 (Russian).

[Fed2] Fedenko A.S., *Symmetric spaces with simple noncompact fundamental groups* **108** (1956), Dokl. AN SSSR, 1026-1028 (Russian).

[Fr1] Freudenthal H., *Oktaven, Ausnahmegruppen und Oktavengeometrie* **19** (1985), Geom. Dedic., 1-63.

[Fr2] Freudenthal H., *Beziehungen der E_7 und E_8 zur Oktavenebene,* Proc. Kon. Ned. Akad. Wet. (A), **57** (1954), 218–230, 363–368, **58** (1955), 151–157, 277–285, **62** (1959), 165–201, 447–474 **66** (1963), 457-487.

[FdV] Freudenthal H., De Vries H., *Linear Lie Groups*, New York - London: Academic Press, 1969.

[Fro1] Frobenius G. F., *Über lineare Substitutionen und bilineare Formen* **1** (1958), Gesammelte Abhandlungen, Berlin - Heidelberg - New York: Springer-Verlag, 343-405.

[Fro2] Frobenius G. F., *Über die Darstellung der endlichen Gruppen durch lineare Substitutionen* **3** (1958), Gesammelte Abhandlungen, Berlin - Heidelberg - New York: Springer-Verlag, 82-103.

[Frö] Fröhlicher A., *Zur Differentialgeometrie der komplexen Strukturen* **129** (1955), Math. Ann., 50-95.

[Fu1] Fubini G. G., *Il parallelismo di Clifford negli spazî elittici* **1** (1957), Opere scelte, Roma: Cremonese, 36-94.

[Fu2] Fubini G. G., *Sulle metriche definite da una forma Hermitiana* **2** (1959), Opere scelte, Roma: Cremonese, 10-20.

[Gal] Galois E., *On Groups and Equations and Abelian Integrals,* Smith D. E. *A Source Book in Mathematics* (1959), New York: Dover, 278-285.

[Gau] Gauss C. F., *Grundlagen der Geometrie. Nachlass und Briefwechsel* **8** (1900), Werke, Göttingen: Teubner, 160-268.

[Gin] Gindikin S. G., *The Complex Universe of Roger Penrose* **5** (1983), Math. Intelligencer, no. 1, 27-35.

[Gol] Goldman W. M., *Complex Hyperbolic Geometry*, Preprint: Maryland Univ. Press, 1985 To appear in Oxford: Univ. Press..

[Gor] Gorbunova M. I., *Geometry of m-Horospheres in Non-Euclidean Spaces* **8** (1964), Proyektivnye metriki. Uch. Zap. Kolom. ped. instituta, 63-74 (Russian).

[Gr] Grassmann H., *On the Ausdehnungslehre* - Smith D. E., *A Source Book in Mathematics* (1959), New York: Dover, 684-696.

[GST] Günaydin M., Sierra G., Townsend P. K., *Exceptional Supergravity Theories and the Magic Square* **133B** (1983), Physics Letters, no. 12, 72-76.

[Ham] Hamilton W. R., *Elements of Quaternions*, New York: Chelsea,, 1969.

[Han] Hangan Th., *Formules de trigonométrie sur la variété de Grassmann* **50** (1992), Rend. Sem. Mat. Univ. Politec. Torino, no. 4, Diff. Geometry, 367-380.

[Hel] Helgason S., *Differential Geometry, Lie Groups, and Symmetric Spaces*, New York - San Francisco - London: Academic Press, 1978.

[He1] Hesse O., *Über die Wendepuncte der Curven dritter Ordnung* **28** (1844), J. reine angew. Math., 97-107.

[He2] Hesse O., *Über ein Übertragungsprinzip* **66** (1866), J. reine angew. Math., 15-21.

[Hil] Hilbert D., *The Foundation of Geometry,* LaSalle, Ill.: Open Court, 1971.

[Hir] Hirsch G., *La géométrie projective des espaces fibrés* (1949), Colloques Intern. du Centre Nah. de Recherches Sc., Bruxelles XII, Paris, 35-42.

[Hje] Hjelmslev J., *Einleitung in die allgemeine Kongruezenlehre Mitt. Danske Vid. Selsk., mat.-fys. medd.*, **8** (1929), **10** (1942), **22** (1945), **25** (1949).

[Hsi] Hsiang Wu-Yi, *On the Laws of Trigonometries of Two-Point Homogeneous Spaces* **7** (1989), Ann. Global Analysis and Geometry, no. 1, 29-45.

[Hua] Hua Loo-Kehg, *Geometries of Matrices* (1983), Selected papers. New York: Springer Verlag, 285-415.

[Hur] Hurwitz A., *Über die Komposition der quadratischen Formen mit beliebig vielen Variabeln* (1898), Göttinger Nachrichten, 309-316.

[Iw] Iwasawa K., *On Some Types of Topological Groups* **50** (1949), Ann. Math., (2), 507-558.

[Jac] Jacobi C.G.J., *Gesammelte Werke* **3** (1884), Berlin: Reimer, 191–268.

[Ja] Jacobson N., *Structure and Representations of Jordan Algebras* **71** (1949), Amer. Math. Soc. Colloquium Publications, 149-170.

[JaJa] Jacobson F. D., Jacobson N.,, *Classifications and Representations of Semi-simple Jordan Algebras* **65** (1949), Transactions of Amer. Math. Soc., no. 2, 141-169.

[Jan] Jankiewicz Cz., *On Degenerated Riemannian Geometries* **2** (1954), Bull. Pol. Ac. Sci., sect. 3, 305-308 (Russian).

[Jav1] Javadov M. A., *Conformal Transformations in Euclidean and Pseudo-Euclidean Spaces of Arbitrary Dimension as Linear Fractional Transformations* **84** (1952), Dokl. AN SSSR, no. 4, 653-656 (Russian).

[Jav2] Javadov M. A., *Projective Spaces over Algebras* (1957), Uch. Zap. Azerb. Univ., no. 2, 3-18 (Russian).

[Jo] Jordan. P., *Ueber eine Klasse nichtassoziativen hyperkomplexer Algebren* (1932), Göttinger Nachrichten, 569-575.

[Käh] Kähler E., *Über eine bemerkungswerte Hermitische Metrik* **9** (1933), Abh. Math. Sem. Hamburg, 171-186.

[Kal] Kalish G. K., *On Special Jordan Algebras* **61** (1947), Transactions of Amer. Math. Soc., 482-494.

[Kan1] Kantor I.L., *Transitively Differential Groups and Invariant Conections on Homogeneous Spaces* **13** (1966), Trudy Sem. Vekt. i Tenz. Anal. Mosk. Univ., 310–398 (Russian).

[Kan2] Kantor I. L., *Some Generalizations of Jordan Algebras* **16** (1972), Trudy Sem. Vekt. i Tenz. Anal. Mosk. Univ., 407-499 (Russian).

[Kan3] Kantor I. L., *Models of Exceptional Lie Algebras* **208** (1973), Dokl. AN SSSR, 1276-1278 (Russian).

[Kap] Kapralova S. B., *Dual Galois Spaces* **12** (1980), Trudy Geom. Sem. Kazan. Univ., 38-44 (Russian).

[Kat] Katok S., *Fuchsian Groups*, Chicago-London: Chicago Univ. Press,, 1992.

[KOV] Karpelevich F. I., Onishchik A. L., Vinberg E. B., *On the Work of E. B. Dynkin in the Theory of Lie Groups,* Gindikin S. G., Vinberg E. B. (Eds), Lie Groups and Lie Algebras: E. B. Dynkin's Seminar, Providence: AMS Press, 1995.

[Kha] Kharitonova N. I., *Projective and Non-Euclidean Spaces over Galois Square* **2** (1976), Aktual'nye problemy geometrii i yeyo prilozheniy, Cheboksary, 44-49 (Russian).

[Ki1] Killing W., *Die nicht-euklidische Raumformen in analytischer Behandlung*, Leipzig: Teubner, 1885.

[Ki2] Killing W., *Einführung in die Grundlagen der Geometrie* **1–2**, Padenborn: Schöhning, 1893-1897.

[Ki3] Killing W., *Die Zusammensetzung der stetigen endlichen Transformations-gruppen*, Math. Ann. **31** (1888), 252–290, **33** (1989), 1–48, **34** (1889), 57–122, **36** (1890), 161–189.

[Kl1] Klein F., *Über die sogenannte Nicht-Euklidische Geometrie* **1** (1921), Gesammelte Math. Abhandlungen, Berlin: Springer-Verlag,, 254-305.

[Kl2] Klein F., *Vergleichende Betrachtungen über neuere geometrische Forschungen (Das Erlanger Programm)* **1** (1921), Gesammelte Math. Abhandlungen, Berlin: Springer-Verlag, 460-497.

[Kl3] Klein F., *Zur Nicht-Euklidische Geometrie* **1** (1921), Gesammelte Math. Abhandlungen, Berlin: Springer-Verlag, 353-383.

[Klm] Klimanova T. M., *Unitary Semielliptic Spaces* (1963), Izv. AN Azerb. SSR, no. 3, 21-29 (Russian).

[Kli] Klingenberg W., *Projektive und affine Ebenen mit Nachbarelementen* **60** (1954), Math. Z., 384-409.

[Kne] Kneser H., *Der Simplexinhalt in der nichteuklidischen Geometrie* **1** (1936), Deutsche Mathematik, 337-340.

[KoN] Kobayashi S., Nomizu K., *Foundations of Differential Geometry*, vol. 2, New York - London - Sidney : Interscience Publ., 1969.

[Koe] Koecher M., *Imbedding of Jordan algebras into Lie algebras*, Amer. J. of Math. **39** (1967), no. 3, 787–816, **40** (1968), no. 2, 476–510.

[Kol] Kolmogorov A. N., *Zur Begründung der projektiven Geometrie* **33** (1932), Ann. of Math., 175-176.

[Kor] Korányi A., *Geometric Aspects of Analysis on the Heisenberg Group* (1983), Topics in Modern Harmonic Analysis, Roma: Istituto Naz. di Alta Matem., 209-258.

[Kos] Kostant B., *The Principal Three-dimentional Subgroup and the Betti Numbers of a Complex Simple Lie Group* **81** (1959), Amer. J. of Math., 973-1032.

[Kot1] Kotel'nikov A. P., *Screw Calculus and its Applications to Geometry and Mechanics*, Kazan: Univ., 1895 (Russian).

[Kot2] Kotel'nikov A. P., *Projective Theory of Vectors*, Kazan: Univ., 1899 (Russian).

[Kot3] Kotel'nikov A.P., *The Principle of Relativity and Lobachevskian Geometry* **2** (1927), In memoriam N.I.Lobatschevskii, Kazan: Glavnauka, 34-66 (Russian).

[Kow] Kowalski O., *Generalized Symmetric Spaces*, Berlin - Heidelberg - New Yotk: Springer-Verlag, 1980.

[Lan] Lanner F., *On Complexes with Transitive Groups of Automorphisms* **11** (1950), Comm. Sem. Math. Univ. Lund, 1-71.

[Led] Ledger A. J., *Espaces de Riemann symétriques généralisés* **A 264** (1964), C. R. Acad. sci. Paris, 947-948.

[LeO] Ledger A. J., Obata M., *Affine and Riemannian s-manifolds* **2** (1968), J. Diff. Geometry, no. 4, 451-459.

[Lev] Levy H., *Forma canonica dei ds^2 per i quali si annulano i simboli di Riemann a cinque indici* **3** (1925), Rend. Accad. Lincei, 65-69.

[Leu] Leuzinger E., *On the Trigonometry of Symmetric Spaces* **67** (1992), Comment. Math. Helvetici, 252-286.

[Lib] Liebmann H., *Nichteuklidische Geometrie*, 2nd ed., Leipzig: Teubner, 1912.

[LiE] Lie S., Engel F., *Theorie der Transformationsgruppen*, vol. 1–3, Leipzig: Teubner, 1888-1893.

[Lin] Linnik Yu. V., *Quaternions and Cayley Numbers; Some Applications of the Arithmetics of Quaternions* **4** (1949), Uspekhi mat. nauk., no. 5 (33), 49-98 (Russian).

[Lio] Liouville J., *Note VI. Extension au cas des trois dimensions de la question du tracé géographique* Monge G., *Application de l'analyse à la géométrie* (1850), cinqième édition, revue, corrigée et annotée par M. Liouville, Paris: Bachelier, 608-616.

[Lip] Lipschitz R., *Untersuchungen über die Summen von Quadraten*, Bonn: Univ., 1886.

[Lob] Lobachevsky N. I., *Pangeometry* Smith D. E. *A Source Book in Mathematics* (1959), New York: Dover, 360-374.

[Los] Loos O., *Symmetric Spaces*, vol 1-2, New York - Amsterdam: Benjamin, 1969.

[Lün] Lüneburg H., *Über die Gruppen von Matthieu* **10** (1968), J. of Algebra, 194-210.

[Lyu] Lyubinskaya O. M., *Projective Soace over Quaternionic Extension of a Finite Field* (1979), Izbrannye voprosy matem. fiziki i teorii funktsiy, Moscow, 54-68 (Russian).

[Mas] Maslov V. P., *Théorie des perturbations et méthodes asymptotiques*, Paris: Dunod, 1972.

[Mat1] Mathieu É., *Mémoire sur l'étude de fonctions de plusieurs quantités* **6** (1861), J. Math. pures et appl., 241-323.

[Mat2] Mathieu É., *Sur la fonction cinq fois transitive de 24 quantités* **18** (1873), J. Math. pures et appl., 25-46.

[Miq1] Miquel A., *Théorèmes de géométrie* **3** (1838), J. Math. pures et appl., 485-498.

[Miq2] Miquel A., *Théorèmes sur intersections des circles et des sphères* **3** (1838), J. Math. pures et appl., 517–522.

[Moi] Moisil G., *Sur les géodesiques des espaces de Riemann singuliers* **42** (1940), Bull. Math. de Soc. Roum. Sci., 33-52.

[Mol] Molien Th., *Über Systeme höherer complexen Zahlen* **41** (1892), Math. Ann., 41-156, **42** (1893), 308-312.

[Mos] Mostow G. D., *Rigidity Theorems*, Seminar in Geometric Rigidity 1992-93 Lecture Notes, Penn State Univ, no. 3, 1993.

[Mou] Moufang R., *Alternativkörper und der Satz vom vollständigen Vierseit* (D_9) **9** (1933), Abh. Math. Sem. Univ. Hamburg, 207-222.

[Mü] Müntz Ch., *Aufbau der gesamten Geometrie auf Grunde der projektiven Axiome allein* **64** (1912), Sitzungsber. Sächs. Ges. Wiss., Math.-Phys. Kl., 223-266.

[Nom] Nomizu K., *Invariant Affine Connexions on Homogeneous Spaces* **76** (1954), Amer. J. Math, no. 1, 33-65.

[Nor1] Norden A. P., *On interpretation of a Space with Degenerate Metric* **50** (1945), Dokl. AN SSSR, 57-60 (Russian).

[Nor2] Norden A. P., *Generalization of the Geometry of the Two-dimensional Line Space* **18 (60)** (1946), Mat. Sb., 139-152.

[Pen] Penrose R., *Twistors Theory. Its Aims and Achievements* (1975), Quantum Gravity. Oxford Symposium, Oxford: Clarendon Press, 268-404.

[Pet] Petsko N. D., *Biquaternionic Elliptic Spaces and their Application to Real Spaces* **8** (1964), Proyektivnye metriki. Uch. Zap. Kolom. ped. instituta, 144-164 (Russian).

[Pim1] Pimenov R. I., *On the Definition of a Semi-Riemannian Space* **14** (1965), Vestnik Leningr. Univ., 137-140 (Russian).

[Pim2] Pimenov R. I., *Application of the Semi-Riemannian Geometry to the Unified Field Theory* **157** (1954), Dokl. AN SSSR, no. 4, 795-798 (Russian).

[Pim3] Pimenov R. I., *Principles of the Theory of the Temporal Universum*, Syvtyvkar: Komi Nauchny Centr AN SSSR, 1991 (Russian).

[Plü] Plücker J., *Neue Geometrie des Raumes gegründet auf die Betrachtung der geraden Linie als Raumelement*, Leipzig: Teubner, 1868-1869.

[Poi1] Poincaré H., *Théorie des groupes fuchsiens* **2** (1916), Oeuvres, Paris: Gauthier-Villars, 108-168.

[Poi2] Poincaré H., *Mémoire sur les groupes kleinéens* **2** (1916), Oeuvres, Paris: Gauthier-Villars, 258-299.

[Poi3] Poincaré H., *Analysis situs* **6** (1953), Oeuvres, Paris: Gauthier-Villars, 435-498.

[Poi4] Poincaré H., *Cinquième complement à l'Analysis situs* **6** (1953), Oeuvres, Paris: Gauthier-Villars, 435-498.

[Poi5] Poincaré H., *Sur les hypothèses fondamentales de la géométrie* **11** (1956), Oeuvres, Paris: Gauthier-Villars, 79-91.

[Pon] Pontryagin L. S., *Über stetige algebraische Körper* **1** (1986), Selected Works, New York - London a. o.: Gordon & Breach, 85-102.

[Pop] Popović M., *Cyclographical Methods of the Geometry of the Hyperbolic Space*, Kazan: Univ., 1994 (Russian).

[Prv] Prvanović M., *Holomorphically Projective Transformations in a Locally Product Spaces* **1** (1971), Math. Balkanica, 105-213.

[Ra] Rashevsky P. K., *On Geometry of Homogeneous Spaces* **80** (1951), Dokl. AN SSSR, 169-171 (Russian).

[Re] Reye Th., *Das Problem der Configurationen* **1** (1882), Acta Math., 93-96.

[Ri1] Riemann B., *On Riemann's Surfaces and Analysis Situs*. Smith D. E., *A Source Book in Mathematics* (1959), New York: Dover, 404-410.

[Ri2] Riemann B., *On the Hypotheses which Lie at the Foundations of Geometry*. Smith D. E., *A Source Book in Mathematics* (1959), New York: Dover, 411-435.

[Ro1] Rosenfeld B. A., *Non-Euclidean Geometries*, Moscow: GITTL, 1955 (Russian).

[Ro2] Rosenfeld B. A., *Multidimensional Spaces*, Moscow: Nauka, 1966 (Russian).

[Ro3] Rosenfeld B. A., *Non-Euclidean Spaces*, Moscow: Nauka, 1969 (Russian).

[Ro4] Rosenfeld B. A., *A History of Non-Euclidean Geometry*, New York - Heidelberg - Berlin: Springer-Verlag, 1988.

[Ro5] Rosenfeld B. A., *Symmetric Spaces and Their Geometric Applications*. Appendix to: Kartan É. *Geometriya grupp Li i simmetricheskie prostranstva. Sbornik rabot* (Russian translations by B.A. Rosenfeld of papers [Car12, 14-16]) (1949), Moscow: IIL, 331-368 Russian.

[Ro6] Rosenfeld B. A., *The Compact Simple Group E_6 as the Group of Motions in the Complex Octonionic Non-Euclidean Plane* **10** (1954), Dokl. Akad. Nauk Azerb. SSR, no. 12, 829-833 (Russian).

[Ro7] Rosenfeld B. A., *Geometric Interpretation of Simple Lie Groups of the Class E* **106** (1956), Dokl. AN SSSR, 600–603 (Russian).

[Ro8] Rosenfeld B. A., *On the Theory of Symmetric Spaces of Rank 1* **41(83)** (1957), Mat. Sb., 373-380 (Russian).

[Ro9] Rosenfeld B. A., *Spaces with Exceptional Fundamental Groups* **54 (68)** (1993), Publications de l'Institut Mathématique, Beograd, 97-119.

[Ro10] Rosenfeld B.A., *Which is Geometry of the Space-Time of Classical Mechanics?* **3** (1995), In memoriam N. I. Lobatschevskii, Kazan: Univ., no. 2 62-66.

[Ro11] Rosenfeld B.A., *Sur la loi de sinus sur la variété de Grassmann*, vol. (1955) 53 no. 1, Rend. Sem. Mat. Univ. Politec. Torino, pp. 69-72.

[RoB1] Rosenfeld B. A., Burtseva T. A., *Lipschitzians, Octonionic Geometry and Lipschitz-affine Spaces* **34** (1990), Sov. Math. (Iz. VUZ), no. 6, 97-100.

[RoB2] Rosenfeld B. A., Burtseva T. A., *Application of Lipschitzians to the Geometry of Exceptional Simple Groups of the Class E* **3** (1995), In memoriam N. I. Lobatschevskii, Kazan: Univ., no.2, 67-78.

[RBMY] Rosenfeld B. A., Burtseva T. A., Dushina N. V., Kostrikina L. P., Malyutin V. V., Yukhtina T. I., *Curvature Tensors of Hermitian Elliptic Spaces* **3** (1989), Zbornik radova filozofskog fakulteta u Nišu, 83-98 (Russian).

[RDKM] Rosenfeld B. A., Denisova N. S., Klimentova N. P., Mirzo-zade D. I., *Geometry of Certain Quasisimple and Biquasisimple Lie Groups of the Class A* **17** (1976), Geom. Sb., Tomsk, 8-19 (Russian).

[RKP] Rosenfeld B. A., Klimanova T. M., Petsko N. D., *Projective Theory of Vectors* (1962), Matem. Izv. VUZov, no. 2 (27), 120–141, no. 3 (28), 122-130 (Russian).

[RoK] Rosenfeld B. A., Karpova L. M., *Symmetric Quasi-Riemannian Spaces* (1964), Matem. Izv. VUZov, no. 1 (38), 108-116 (Russian).

[RoKK] Rosenfeld B. A., Kashirina I. N., Kharitonova N. I., *Finite Geometries with Simple, Semisimple, and Quasisimple Fundamental Groups* (1982), Geometric Vein. The Coxeter Festschrift, New York - Heidelberg - Berlin: Springer-Verlag, 477-484.

[RKMV] Rosenfeld B. A., Kolokol'tseva I. I., Malyutin V. V., Vyplavina R. P., *Linear Fractional Transformations in Jordan Algebras* **18** (1974), Sov. Math. (Iz. VUZ), no. 5, 142-153.

[RKSY] Rosenfeld B. A., Kostrikina L. P., Stepanova G. V., Yukhtina T. I., *Focally Affine Spaces* **34** (1990), Sov. Math. (Iz. VUZ), no. 5, 70-78.

[RMal] Rosenfeld B. A., Malyutin V. V., *Hyperbolic Planes over Tensor Products* **18** (1977), Geom. Sb., Tomsk, 93-97 (Russian).

[RoMa] Rosenfeld B. A., Masagutova A. F., *Focally Euclidean and Focally Riemannian Spaces* **35** (1991), Sov. Math. (Iz. VUZ), no. 7, 35-38.

[RPRY] Rosenfeld B. A., Polovtseva M. A., Ryazanova L. A., Yukhtina T. I., *Segreans and Quasisegreans and Their Applications to Geometry of Families of Lines and Planes* **32** (1988), Sov. Math. (Iz. VUZ), no. 5, 60-69.

[RoSt] Rosenfeld B. A., Stepashko T. A., *Metasymplectic Geometries as Geometries on Absolutes of Hermitian Planes* **27** (1983), Sov. Math. Dokl., no. 1, 132-135.

[RoU] Rosenfeld B. A., Uzdenov O. M., *Application of the Static of Lobachevskian Space to the Theory of Electromagnetic Field* **33** (1989), Sov. Math. (Iz. VUZ), 48-53.

[RoVZ] Rosenfeld B. A., Vyplavina R. P., Zernova T. A., *Elliptic Straight Lines and Euclidean Planes over Tensor Products of Fields* **19** (1975), Sov. Math. (Iz. VUZ), no. 12, 35-42.

[RoZ] Rosenfeld B. A., Zamakhovsky M. P., *Homogeneous k-Symmeteric Spaces of Interior Type with Simple Real Fundamental Groups and Their Connection with Parabolic Spaces*, to appear in Publications de l'Institut Mathématique, Beograd.

[RoZT] Rosenfeld B. A., Zamakhovsky M. P., Timoshenko T. A., *Parabolic Spaces* **47** (1990), J. Sov. Math., no. 3, 1014-1034.

[Ru1] Rumyantseva L. V., *Quaternionic Symplectic Geometry* **12** (1963), Trudy Sem. Vekt. i Tenz. Anal. Mosk. Univ., 287-314 (Russian).

[Ru2] Rumyantseva L. V., *Quadriquaternionic Elliptic Spaces* (1963), Uch. Zap. Azerb. Univ., ser. fiz.-mat. nauk, no. 3, 35-38 (Russian).

[Rys] Ryser H. J., *Combinatorial Mathematics*, New York: Wiley, 1963.

[Sat] Satake I., *On Representation and Compactification of Symmetric Riemannian Spaces* **71** (1960), Ann. of Math.(2), 77-110.

[Scf] Scheffers G., *Verallgemeinerung der Grundlagen der gewöhnlichen komplexen Funktionen* **45** (1893), Sitzungsber. Sächs. Gesellsch. Wiss. Math. - phys. Kl., 328-342.

[Sck] Schellekens G. J., *On a Hexagonic Structure* **A 65** (1962), Proc. Kon. Ned. Akad. Wet., 201-234.

[Scl] Schläfli L., *Theorie der vielfachen Kontinuität* **1** (1950), Gesammelte Math. Abhandlungen, Basel: Birkhäuser, 169-387.

[Sco] Schouten J.A., *Lëber die verschiedenen Arten der Lbërtragung, die einer Differetialgeometrie zugrunde gelegt wrden Können* **13** (1922), Math Z., 56–81.

[SeB] Segre B., *Le geometrie di Galois* **48** (1959), Annali di Mat., 1-96.

[SeC] Segre C., *Varietà che rappresentano le coppie di punti di 2 piani o spazi* **1** (1957), Opere, Roma: Cremonese, 173-184.

[Sem] Semyonova I. N., *Limit Projective Spaces* **8** (1964), Proyektivnye metriki. Uch. Zap. Kolom. ped. instituta, 165-174 (Russian).

[ShN] Shestyreva L. V., Nikitina L. S., *Real Interpretation of Symplectic 3-Spaces over Algebras of Quaternions and Their Analogs Based on the Triality Principle in 7-Spaces* **3** (1971), Trudy ob"yedineniya matem. kafedr ped. institutov Central. zony RSFSR, Tambov, 223-240 (Russian).

[ShA1] Shirokov A. P., *Spaces over Associative Unital Algebras* **123** (1963), Geometriya obobshchennykh prostranstv. Uch. Zap. Kazan. Univ., no. 1, 222-247 (Russian).

[ShA2] Shirokov A. P., *On Symmetric Spaces Determined by Commutative Algebras of Fourth Order* **126** (1966), Uch. Zap. Kazan. Univ., no. 1, 60-69 (Russian).

[ShP1] Shirokov P. A., *Constant Fields of Vectors and Tensors of Second Order in Riemannian Manifolds* (1966), Selected Works in Geometry, Kazan: Univ., 256-280 (Russian).

[ShP2] Shirokov P. A., *On a Certain Type of Symmetric Spaces* (1966), Selected Works in Geometry, Kazan: Univ., 408-418 (Russian).

[Sie] Siegel C. L., *Symplectic Geometry*, New York: Academic Press, 1964.

[SiM] Siegel C. L., Moser J. K., *Lectures on Celestial Mechanics*, New York - Berlin e.a.: Springer-Verlag, 1971.

[Sk1] Skopets Z. A., *Cyclographical Map of the Lobachevskian Space* (1952), 125 let neevklidovoy geometrii Lobachevskogo. Moscow: GITTL, 129-150 (Russian).

[Sk2] Skopets Z. A., *Principles of the Descriptive Geometry of the Lobachevskian Space* (1955), Metody nachertatel'noy geometrii i yeyo prilozheniya, Moscow: GITTL, 312-368 (Russian).

[Smf] Sommerfeld A., *Über die Zusammensetzung der Geschwindigkeiten in der Relativtheorie* **10** (1910), Phys. Z., no. 22, 826-829.

[Smv] Sommerville D. M. Y., *Classification of Geometries with Projective Metrics* **28** (1910–1911), Proc. Edinburgh Math. Soc., 25-41.

[Ste] Steiner J., *Combinatorische Aufgabe* **2** (1882), Gesammelte Werke, Berlin: Reimer, 435-438.

[Stp] Stepanov N. A., *Fundamental Facts of the Theory of Φ-Spaces* (1967), Izv. VUZov, Matem., no. 3, 70-79 (Russian).

[Str1] Strubecker, K., *Differentialgeometrie des isotropen Raumes*, I Sitzungsber. Akad. Wiss. Wien (IIa) **150** (1941), 1–53; II - V Math. Z. **47** (1942), 743–777, **48** (1942), 369–427, **50** (1944), 1–92, **52** (1949), 525–573.

[Str2] Strubecker K., *Geometrie in einer isotropen Ebene*, I - III Math. und naturwiss. Unterricht **15** (1962), no. 7, 297–306, (1963), no. 8, 343–351, no. 9, 385–394.

[Stu1] Study E., *Geometrie der Dynamen*, Leipzig: Teubner, 1903.

[Stu2] Study E., *Kürzeste Wege im komplexen Gebiete* **60** (1905), Math. Ann., 321-377.

[Stu3] Study E., *Ein Seitenstück der Theorie der linearen Transformationen einer komplexen Veränderlichen* **18** (1923), Math. Z., 55-86, **21** (1924), 45-71, 174-194.

[Th] Thas J. A., *The m-dimensional Projective Space $S_m(M_n(GF(q)))$ over the Total Matrix Algebra $M_n(GF(q))$ of the $n \times n$-matrices with Elements in the Galois Field $GF(q)$* **4** (1971), Rendiconti di Matematica (6), no. 3, 459-532.

[Ti1] Tits J., *Sur certaines classes d'espaces homogènes de groupes de Lie* **29** (1955), Mém. Acad. Roy. Belg., Cl. Sci., no. 3, 1-268.

[Ti2] Tits J., *Les groupes de Lie exceptionnels et leur interprétation géométrique* **8** (1956), Bull. Soc. Math. Belg., no. 1, 48-81.

[Ti3] Tits J., *Sur la géométrie de R-espaces* **36** (1957), J. Math. pures et appl., no. 5, 17-38.

[Tl] Tlupova R. G., *Stationary Groups of Points in G-Non-Euclidean and in G-Quasi-non-Euclidean Spaces* (1981), Geometriya pogruzhennykh mnogoobraziy, Moscow: MGPI, 89–94 (Russian).

[Tou] Toupin R. A., *World Invariant Kinematics* **1** (1958), Archives for Rat. Mechanics and Analysis, no. 3, 181-211.

[VeB] Veblen O., Bussey W., *Finite Projective Geometries* **7** (1906), Transactions of Amer. Math. Soc., 241-259.

[Ved1] Vedernikov V.I., *Symmetric Spaces and Conjugate Connections* **125** (1965), Uch. Zap. Kazan. Univ., no. 1, 7-59 (Russian).

[Ved2] Vedernikov V.I., *On a Special Class of Homogeneous Spaces* (1972), Izv. VUZov, Matem, no. 12, 117-122 (Russian).

[Ver] Veronese G., *La superficie omaloide normale a due dimensioni e del quatro ordine dello spazio a cinque dimensioni e le sue proiezioni nel piano e nel spazio ordinario* **19** (1884), Mem. Accad. dei Lincei, 344-371.

[Vin] Vinberg E. B., *Constructions of Exceptional Simple Lie Algebras* **13** (1966), Trudy Sem. Vekt. i Tenz. Anal. Mosk. Univ., 7-9 (Russian).

[Vol] Volman A.Ya., *A Surface of Revolution of Constant Curvature in the Lobachevskian Space* **1** (1949), Uch. zap. Kishinev. gos. Univ., no.1, 65-93 (Russian).

[Wan] Wan Zhe-xian, *Geometry of Classical Groups over Finite Fields*, Lund: Chartwell-Bratt, 1993.

[WeS] Weber C., Seifert H., *Die beiden Dodekaederräume* **37** (1932), Math. Z., 237-253.

[Wei] Weierstrass C., *Zur Theorie der aus n Haupteinheiten gebildeten complexen Grössen* **2** (1895), Math. Werke, Berlin, Bauer & Müller, 311-332.

[Wes] Weiss E., *Oktaven, Engelscher Komplex und Trialitätsprinzip* **44** (1938), Math. Z., 580-611.

[Wey1] Weyl H., *Space, Time, Matter*, New York: Dover, 1950.

[Wey2] Weyl H., *The Classical Groups. Their Invariants and Representations*, Princeton, NJ: Univ. Press, 1939.

[Wey3] Weyl H., *Symmetry*, Princeton, NJ: Univ. Press, 1969.

[Wey4] Weyl H., *Reine Infinitesimaltheorie* **2** (1968), Gesammelte Abhandlungen, Springer-Verlag: Berlin - New York, 1-28.

[Wey5] Weyl H., *Theorie der Darstellung der kontinuierlichen halbeinfachen Gruppen durch lineare Transformationen* **2** (1968), Gesammelte Abhandlungen, Springer-Verlag: Berlin - New York:, 548-641.

[Wit1] Witt E., *Die 5-fach transitiven Gruppen von Matthieu* **12** (1938), Abh. Math. Sem. Univ. Hamburg, 256-264.

[Wit2] Witt E., *Über Steinersche Systeme* **12** (1938), Abh. Math. Sem. Univ. Hamburg, 265-275.

[Wo1] Wolf J. A., *Spaces of Constant Curvature*, 5th ed., Wilmington, Delaware: Perish or Publish, 1984.

[Wo2] Wolf J. A., *The Action of a Real Semisimple Group on a Complex Flag Manifold* **75** (1969), Bull. Amer. Math. Soc., no. 6, 1121-1237.

[WoG] Wolf J. A., Gray A., *Homogeneous Spaces Defined by Lie Group Automorphisms* **2** (1968), J. Diff. Geom., no. 1-2, 77-159.

[YRY] Yaglom I. M., Rosenfeld B.A., Yasinskaya E. U., *Projective Metrics* **19** (1964), Russian Math. Surveys, no. 5, 49-107.

[Zab] Zablotskikh N. M., *Octonionic Geometries with Classical Fundamental Groups* **253** (1969), Uch. Zap. MOPI, 62-76 (Russian).

[Zh1] Zhelezina I. I., *Line Quasi-non-Euclidean Geometry as Complex Euclidean Geometry* (1965), Voprosy diff. i neevkl. geometrii, Uch. zap. MGPI, 233-240 (Russian).

[Zh2] Zhelezina I. I., *Line Galilean Geometry as Dual Euclidean Geometry* (1965), Voprosy diff. i neevkl. geometrii, Uch. zap. MGPI, 241-247 (Russian).

[Zo] Zorn M., *Theorie der alternativen Ringe* **8** (1931), Abh. Math. Sem. Univ. Hamburg, 123-147.

INDEX OF PERSONS

INDEX OF SUBJECTS

Other *Mathematics and Its Applications* titles of interest:

V.I. Istratescu: *Fixed Point Theory. An Introduction.* 1981, 488 pp.
out of print, ISBN 90-277-1224-7

A. Wawrynczyk: *Group Representations and Special Functions.* 1984, 704 pp.
ISBN 90-277-2294-3 (pb), ISBN 90-277-1269-7 (hb)

R.A. Askey, T.H. Koornwinder and W. Schempp (eds.): *Special Functions: Group Theoretical Aspects and Applications.* 1984, 352 pp. ISBN 90-277-1822-9

A.V. Arkhangelskii and V.I. Ponomarev: *Fundamentals of General Topology. Problems and Exercises.* 1984, 432 pp. ISBN 90-277-1355-3

J.D. Louck and N. Metropolis: *Symbolic Dynamics of Trapezoidal Maps.* 1986, 320 pp. ISBN 90-277-2197-1

A. Bejancu: *Geometry of CR-Submanifolds.* 1986, 184 pp.
ISBN 90-277-2194-7

R.P. Holzapfel: *Geometry and Arithmetic Around Euler Partial Differential Equations.* 1986, 184 pp. ISBN 90-277-1827-X

P. Libermann and Ch.M. Marle: *Sympletic Geometry and Analytical Mechanics.* 1987, 544 pp. ISBN 90-277-2438-5 (hb), ISBN 90-277-2439-3 (pb)

D. Krupka and A. Svec (eds.): *Differential Geometry and its Applications.* 1987, 400 pp. ISBN 90-277-2487-3

Shang-Ching Chou: *Mechanical Geometry Theorem Proving.* 1987, 376 pp.
ISBN 90-277-2650-7

G. Preuss: *Theory of Topological Structures. An Approach to Categorical Topology.* 1987, 318 pp. ISBN 90-277-2627-2

V.V. Goldberg: *Theory of Multicodimensional (n+1)-Webs.* 1988, 488 pp.
ISBN 90-277-2756-2

C.T.J. Dodson: *Categories, Bundles and Spacetime Topology.* 1988, 264 pp.
ISBN 90-277-2771-6

A.T. Fomenko: *Integrability and Nonintegrability in Geometry and Mechanics.* 1988, 360 pp. ISBN 90-277-2818-6

L.A. Cordero, C.T.J. Dodson and M. de Leon: *Differential Geometry of Frame Bundles.* 1988, 244 pp. *out of print*, ISBN 0-7923-0012-2

E. Kratzel: *Lattice Points.* 1989, 322 pp. ISBN 90-277-2733-3

E.M. Chirka: *Complex Analytic Sets.* 1989, 396 pp. ISBN 0-7923-0234-6

Kichoon Yang: *Complete and Compact Minimal Surfaces.* 1989, 192 pp.
ISBN 0-7923-0399-7

A.D. Alexandrov and Yu.G. Reshetnyak: *General Theory of Irregular Curves.* 1989, 300 pp. ISBN 90-277-2811-9

Other *Mathematics and Its Applications* titles of interest:

B.A. Plamenevskii: *Algebras of Pseudodifferential Operators*. 1989, 304 pp. ISBN 0-7923-0231-1

Ya.I. Belopolskaya and Yu.L. Dalecky: *Stochastic Equations and Differential Geometry*. 1990, 288 pp. ISBN 90-277-2807-0

V. Goldshtein and Yu. Reshetnyak: *Quasiconformal Mappings and Sobolev Spaces*. 1990, 392 pp. ISBN 0-7923-0543-4

A.T. Fomenko: *Variational Principles in Topology. Multidimensional Minimal Surface Theory*. 1990, 388 pp. ISBN 0-7923-0230-3

S.P. Novikov and A.T. Fomenko: *Basic Elements of Differential Geometry and Topology*. 1990, 500 pp. ISBN 0-7923-1009-8

B.N. Apanasov: *The Geometry of Discrete Groups in Space and Uniformization Problems*. 1991, 500 pp. ISBN 0-7923-0216-8

C. Bartocci, U. Bruzzo and D. Hernandez-Ruiperez: *The Geometry of Supermanifolds*. 1991, 242 pp. ISBN 0-7923-1440-9

N.J. Vilenkin and A.U. Klimyk: *Representation of Lie Groups and Special Functions. Volume 1: Simplest Lie Groups, Special Functions, and Integral Transforms*. 1991, 608 pp. ISBN 0-7923-1466-2

A.V. Arkhangelskii: *Topological Function Spaces*. 1992, 206 pp. ISBN 0-7923-1531-6

Kichoon Yang: *Exterior Differential Systems and Equivalence Problems*. 1992, 196 pp. ISBN 0-7923-1593-6

M.A. Akivis and A.M. Shelekhov: *Geometry and Algebra of Multidimensional Three-Webs*. 1992, 358 pp. ISBN 0-7923-1684-3

A. Tempelman: *Ergodic Theorems for Group Actions*. 1992, 400 pp. ISBN 0-7923-1717-3

N.Ja. Vilenkin and A.U. Klimyk: *Representation of Lie Groups and Special Functions, Volume 3. Classical and Quantum Groups and Special Functions*. 1992, 630 pp. ISBN 0-7923-1493-X

N.Ja. Vilenkin and A.U. Klimyk: *Representation of Lie Groups and Special Functions, Volume 2. Class I Representations, Special Functions, and Integral Transforms*. 1993, 612 pp. ISBN 0-7923-1492-1

I.A. Faradzev, A.A. Ivanov, M.M. Klin and A.J. Woldar: *Investigations in Algebraic Theory of Combinatorial Objects*. 1993, 516 pp. ISBN 0-7923-1927-3

M. Puta: *Hamiltonian Mechanical Systems and Geometric Quantization*. 1993, 286 pp. ISBN 0-7923-2306-8

V.V. Trofimov: *Introduction to Geometry of Manifolds with Symmetry*. 1994, 326 pp. ISBN 0-7923-2561-3

Other *Mathematics and Its Applications* titles of interest:

J.-F. Pommaret: *Partial Differential Equations and Group Theory. New Perspectives for Applications*. 1994, 473 pp. ISBN 0-7923-2966-X

Kichoon Yang: *Complete Minimal Surfaces of Finite Total Curvature*. 1994, 157 pp. ISBN 0-7923-3012-9

N.N. Tarkhanov: *Complexes of Differential Operators*. 1995, 414 pp. ISBN 0-7923-3706-9

L. Tamássy and J. Szenthe (eds.): *New Developments in Differential Geometry*. 1996, 444 pp. ISBN 0-7923-3822-7

W.C. Holland (ed.): *Ordered Groups and Infinite Permutation Groups*. 1996, 255 pp. ISBN 0-7923-3853-7

K.L. Duggal and A. Bejancu: *Lightlike Submanifolds of Semi-Riemannian Manifolds and Applications*. 1996, 308 pp. ISBN 0-7923-3957-6

D.N. Kupeli: *Singular Semi-Riemannian Geometry*. 1996, 187 pp. ISBN 0-7923-3996-7

L.N. Shevrin and A.J. Ovsyannikov: *Semigroups and Their Subsemigroup Lattices*. 1996, 390 pp. ISBN 0-7923-4221-6

C.T.J. Dodson and P.E. Parker: *A User's Guide to Algebraic Topology*. 1997, 418 pp. ISBN 0-7923-4292-5

B. Rosenfeld: *Geometry of Lie Groups*. 1997, 412 pp. ISBN 0-7923-4390-5

MIX
Papier aus verantwortungsvollen Quellen
Paper from responsible sources
FSC® C105338

If you have any concerns about our products,
you can contact us on
ProductSafety@springernature.com

In case Publisher is established outside the EU,
the EU authorized representative is:
Springer Nature Customer Service Center GmbH
Europaplatz 3, 69115 Heidelberg, Germany

Printed by Libri Plureos GmbH
in Hamburg, Germany